Harmonic Analysis on Symmetric Spaces and Applications II

Audrey Terras

Harmonic Analysis on Symmetric Spaces and Applications II

With 26 Illustrations

Springer-Verlag
Berlin Heidelberg New York
London Paris Tokyo

Audrey Terras
Department of Mathematics
University of California at San Diego
La Jolla, CA 92093
USA

Mathematics Subject Classification (1980): 42-01, 43-01, 44-01, 10Dxx, 51M10

Library of Congress Cataloging-in-Publication Data
(Revised for vol. 2)
Terras, Audrey.
　Harmonic analysis on symmetric spaces and applications I.
　Includes bibliographies and indexes.
　1. Harmonic analysis.　2. Symmetric spaces.
I. Title.
QA403.T47　1985　　515′.2433　　84-23568

© 1988 by Springer-Verlag New York Inc.
All rights reserved. This work may not be translated or copied in whole or in part without the written permission of the publisher (Springer-Verlag, 175 Fifth Avenue, New York, NY 10010, USA), except for brief excerpts in connection with reviews or scholarly analysis. Use in connection with any form of information storage and retrieval, electronic adaptation, computer software, or by similar or dissimilar methodology now known or hereafter developed is forbidden.
The use of general descriptive names, trade names, trademarks, etc. in this publication, even if the former are not especially identified, is not to be taken as a sign that such names, as understood by the Trade Marks and Merchandise Marks Act, may accordingly be used freely by anyone.

Typeset by Asco Trade Typesetting Ltd., Hong Kong.
Printed and bound by R.R. Donnelley & Sons, Harrisonburg, Virginia.
Printed in the United States of America.

9 8 7 6 5 4 3 2 1

ISBN 0-387-96663-3 Springer-Verlag New York Berlin Heidelberg
ISBN 3-540-96663-3 Springer-Verlag Berlin Heidelberg New York

To the Vulcans and other logical creatures

There's always some error, even with experts. There's got to be, since there are so many variables. Put it this way—the geometry of space is too complicated to handle and hyperspace compounds all these complications with a complexity of its own that we can't even pretend to understand.

>From I. Asimov, *Foundation's Edge*, Doubleday & Co., Inc., N.Y., 1982, p. 162.

Preface

Well, finally, here it is—the long-promised "Revenge of the Higher Rank Symmetric Spaces and Their Fundamental Domains." When I began work on it in 1977, I would probably have stopped immediately if someone had told me that ten years would pass before I would declare it "finished." Yes, I am declaring it finished—though certainly not perfected. There is a large amount of work going on at the moment as the piles of preprints reach the ceiling. Nevertheless, it is summer and the ocean calls. So I am not going to spend another ten years revising and polishing. But, gentle reader, do send me your corrections and even your preprints. Thanks to your work, there is an Appendix at the end of this volume with corrections to Volume I.

I said it all in the Preface to Volume I. So I will try not to repeat myself here. Yes, the "recent trends" mentioned in that Preface are still just as recent. And there are newer, perhaps even more pernicious tendencies here in the U.S. This is the age of the billion dollar "defense" funding of research, the "initiatives" to put more power and money in the hands of fewer and fewer, the boondoggles to spend huge sums on supercomputers and to bring space war movies into the university. Yes, and compartmentalized research is still in the ascendency. But I do not feel much happier looking to the international mathematical community that just declared most female, minority, and third world mathematicians unfit to speak at the international congress in Berkeley last summer. Oh well, many fields were not represented either. But, for me, the best conference is one run democratically and covering a wide spectrum of viewpoints, a conference in which anyone who wishes can speak on their research. Infinite diversity in infinite combinations!

Well, so much for purple prose. Clearly I am hoping for some forgiving readers. I also need readers who are willing to work out lots of exercises on a large variety of topics. Yes, once more there are lots of exercises. But, isn't it boring to read other people's proofs? In Chapter 4 this can mean some rather complicated calculations on matrix space. In Section 1 of Chapter 5 this will require some familiarity with beginning differential geometry—tangent spaces, differentials, and the like. Some parts of Section 2 of Chapter 5 will demand a little knowledge of beginning algebraic number theory. I have added a few things that were not present in the last edition of these notes—a rather strange analogue of the central limit theorem for rotation-invariant random variables on the symmetric space for $GL(3, \mathbb{R})$, pictures of projections of Grenier's 5-dimensional fundamental domain for $GL(3, \mathbb{Z})$. I hope someday that a movie can be made of the tessellation of the 5-dimensional symmetric space that is produced by letting $GL(3, \mathbb{Z})$ act on that fundamental domain.

Perhaps I should repeat one thing from the Preface to Volume I—the warning that I am very bad at proofreading. And the formulas in Volume II are much worse than in Volume I. So please do remember this when a formula looks weird.

Thanks again to all who helped me write this. You know who you are, I hope. Live long and prosper!

Encinitas, California   AUDREY TERRAS
August 1987

Contents of Volume II

Preface   vii

CHAPTER IV

The Space $\mathscr{P}_n$ of Positive $n \times n$ Matrices   1

4.1.	Geometry and Analysis on $\mathscr{P}_n$	1
4.1.0.	Introduction	1
4.1.1.	Elementary Results	8
4.1.2.	Geodesics and Arc Length	11
4.1.3.	Measure and Integration on $\mathscr{P}_n$	18
4.1.4.	Differential Operators on $\mathscr{P}_n$	26
4.1.5.	A List of the Main Formulas Derived in Section 1	33
4.1.6.	An Application to Multivariate Statistics	36
4.2.	Special Functions on $\mathscr{P}_n$	38
4.2.1.	Power and Gamma Functions	38
4.2.2.	K-Bessel Functions	49
4.2.3.	Spherical Functions	64
4.2.4.	The Wishart Distribution	83
4.3.	Harmonic Analysis on $\mathscr{P}_n$ in Polar Coordinates	86
4.3.1.	Properties of the Helgason-Fourier Transform on $\mathscr{P}_n$	87
4.3.2.	Beginning of the Discussion of Part (1) of Theorem 1—Steps 1 and 2	94
4.3.3.	End of the Discussion of Part (1) of Theorem 1—Steps 3 and 4	99
4.3.4.	Applications	106
4.3.5.	Other Directions in the Labyrinth	110
4.4.	Fundamental Domains for $\mathscr{P}_n/GL(n, \mathbb{Z})$	112
4.4.1.	Introduction	113
4.4.2.	Minkowski's Fundamental Domain	125

4.4.3.	Grenier's Fundamental Domain	146
4.4.4.	Integration over Fundamental Domains	164
4.5.	Automorphic Forms for $GL(n,\mathbb{Z})$ and Harmonic Analysis on $\mathscr{P}_n/GL(n,\mathbb{Z})$	181
4.5.1.	Analytic Continuation of Eisenstein Series by the Method of Inserting Larger Parabolic Subgroups	181
4.5.2.	Hecke Operators and Analytic Continuation of L-Functions Associated to Modular Forms by the Method of Theta Functions	198
4.5.3.	Fourier Expansions of Eisenstein Series	215
4.5.4.	Remarks on Harmonic Analysis on the Fundamental Domain	245

CHAPTER V

The General Noncompact Symmetric Space   258

5.1.	Geometry and Analysis on G/K	258
5.2.	Geometry and Analysis on $\Gamma\backslash G/K$	314

APPENDIX

Corrections to Volume I	351
Bibliography for Volume II	354
Index	375

Contents of Volume I

Preface

CHAPTER I

Flat Space. Fourier Analysis on $\mathbb{R}^m$

1.1. Distributions or Generalized Functions
1.2. Fourier Integrals
1.3. Fourier Series and the Poisson Summation Formula
1.4. Mellin Transforms, Epstein and Dedekind Zeta Functions

CHAPTER II

A Compact Symmetric Space—The Sphere

2.1. Spherical Harmonics
2.2. $O(3)$ and $\mathbb{R}^3$. The Radon Transform

CHAPTER III

The Poincaré Upper Half-Plane

3.1. Hyperbolic Geometry
3.2. Harmonic Analysis on H
3.3. Fundamental Domains for Discrete Subgroups Γ of $G = SL(2, \mathbb{R})$
3.4. Automorphic Forms—Classical
3.5. Automorphic Forms—Not So Classical—Maass Waveforms
3.6. Automorphic Forms and Dirichlet Series. Hecke Theory and Generalizations
3.7. Harmonic Analysis on the Fundamental Domain. The Roelcke-Selberg Spectral Resolution of the Laplacian, and the Selberg Trace Formula

Bibliography
Index

CHAPTER IV

The Space $\mathscr{P}_n$ of Positive $n \times n$ Matrices

4.1. Geometry and Analysis on $\mathscr{P}_n$

The story so far:

> In the beginning the Universe was created. This has made a lot of people very angry and been widely regarded as a bad move. Many races believe that it was created by some sort of god, though the Jatravartid people of Viltvodle VI believe that the entire Universe was in fact sneezed out of the nose of a being called the Great Green Arkleseizure.
>
> From *The Restaurant at the End of the Universe*, by Douglas Adams, Harmony Books, NY, 1980. Reprinted by permission of The Crown Publishing Group.

4.1.0. Introduction

In this chapter, our universe is a higher rank symmetric space $\mathscr{P}_n$, *the space of positive $n \times n$ real matrices*:

$$\mathscr{P}_n = \{Y = (y_{ij})_{1 \le i,j \le n} | y_{ij} \text{ real}, {}^tY = Y, Y \text{ positive definite}\}. \quad (1.1)$$

Here and throughout this book, "Y *is positive*" means that the quadratic form

$$Y[x] = {}^t x Y x = \sum_{i,j=1}^{n} x_i y_{ij} x_j > 0, \quad \text{if } x \in \mathbb{R}^n, x \ne 0. \quad (1.2)$$

We shall always write vectors x in $\mathbb{R}^n$ as column vectors and so the transpose is ${}^t x = (x_1, \ldots, x_n)$.

The universe which we now enter is one of dimension at least 6 and this means that we will have trouble drawing meaningful pictures, not to mention keeping our calculations on small pieces of paper. The reader is advised to get some big sheets of paper to do some of the exercises (or maybe an account on a supercomputer).

Exercise 1. Some books define positive matrices as non-zero matrices all of whose entries are nonnegative integers (see Berman and Plemmons [1]* and Pullman [1]). Show that this concept is totally different from (1.2) in the sense that neither implies the other when $n > 1$.

We can lower the dimension of our space by one (and make the Lie group involved simple) by looking at the symmetric space of *special positive matrices*:

$$\mathscr{SP}_n = \{Y \in \mathscr{P}_n | |Y| = \det Y = 1\}. \tag{1.3}$$

Recall that we can identify $\mathscr{SP}_2$ with the Poincaré upper half plane H (see Exercise 9 on p. 125 of Vol. I).

Our goal in this chapter is to extend to $\mathscr{P}_n$ as many of the results of Chapter 3, Vol. I, as possible. For example, we will study analogues of our favorite special functions—gamma, K-Bessel, and spherical. The last two functions will be eigenfunctions of the Laplacian on $\mathscr{P}_n$. They will display a bit more complicated structure than the functions we saw in Vol. I, but they and related functions have had many applications. For example, James [1-2] and others have used zonal polynomials and hypergeometric functions of matrix argument to good avail in multivariate statistics (see also Muirhead [1]).

There are, in fact, many applications of analysis on $\mathscr{P}_n$ in *multivariate statistics*, which is concerned with data on several aspects of the same individual or entity; e.g., reaction times of one subject to several stimuli. Multivariate statistics originated in the early part of the century with Fisher and Pearson. In Section 4.1 we will consider a very simple application of one of our coordinate systems for $\mathscr{P}_n$ in the study of partial correlations, with an example from agriculture. Limit theorems for products of random matrices are also of interest; e.g. in demography (see Cohen [1]). We will be able to say something about this subject as well, by making use of harmonic analysis on $\mathscr{P}_n$.

We will see that $\mathscr{P}_n$ is a *symmetric space*; i.e., a Riemannian manifold with a geodesic-reversing isometry at each point. Moreover it is a homogeneous space of the Lie group $GL(n, \mathbb{R})$—the *general linear group* of all $n \times n$ non-singular real matrices. By *homogeneous space*, we essentially refer to the identification of $\mathscr{P}_n$ with the quotient $K\backslash G$ in Exercise 5 below. Because $\mathscr{P}_n$ is a symmetric space, harmonic analysis on $\mathscr{P}_n$ will be rather similar to that on the spaces considered in Volume I—$\mathbb{R}^n$, S^2, and H. For example, the ring of G-invariant differential operators on $\mathscr{P}_n$ is easily shown to be commutative from the existence of a geodesic-reversing isometry. There are 4 main types of

---

* This reference number style will be used throughout the book. In this instance, reference is made to the first Berman and Plemmons title in the Bibliography section at the end of the book.

symmetric spaces: compact Lie groups, quotients of compact Lie groups, quotients of non-compact semisimple Lie groups, and Euclidean spaces. E. Cartan classified these types further (see Helgason [1, 2, 3]). We are mainly considering a special example of the third type here (if we restrict to the determinant 1 surface (1.3)). In the next chapter we will look at the more general theory.

Work of Harish-Chandra and Helgason allows us to extend the results of Section 3.2 of Volume I to $\mathscr{P}_n$, by obtaining an analogue of the Fourier or Mellin transform on $\mathscr{P}_n$ (see Section 4.3 or Helgason [1]). Of course this result is more complicated than those discussed in Volume I and the available tables of this transform are indeed very short. Still we will be able to generalize some of our earlier applications.

We will also study the fundamental domain in $\mathscr{P}_n$ for $GL(n, \mathbb{Z})$, the discrete group of $n \times n$ matrices with integer entries and determinant $+1$ or -1. But our attempt to generalize Section 3.7 of Volume I and so obtain a theory of harmonic analysis on $\mathscr{P}_n/GL(n, \mathbb{Z})$ will not be anywhere near as satisfactory as it was for $n = 2$ in Chapter 3, although work of Langlands [1], Selberg [1], and others allows us to complete many of the foundational results.

One of the reasons for singling out the symmetric space $\mathscr{P}_n$ is that arguments for $\mathscr{P}_n$ can often be generalized to arbitrary symmetric spaces. One such example is the characterization of geodesics (see Chapter 5). This sort of approach is used by Mostow [1], for example.

There are also many applications in *physics*. Some are analogues of applications considered in Chapter 3. One can solve the heat and wave equations on $\mathscr{P}_n$ and consider central limit theorems. There are applications in quantum mechanics for systems of coherent states coming from higher rank symmetric spaces such as $\mathscr{P}_n$ (see Hurt [1–2], Monastyrsky and Perelmonov [1]). And it is possible to make a statistical study of eigenvalues of random symmetric matrices with implications for quantum mechanics (see Mehta [1]).

Many of our applications will be in *number theory*. For example, in Section 4.4, we will begin with work of Minkowski [1, Vol. II, pp. 53–100] on fundamental domains for $\mathscr{P}_n/GL(n, \mathbb{Z})$—work that is basic for algebraic number theory: e.g. in the discussions of the finiteness of the class number and Dirichlet's unit theorem (see Vol. I, Section 1.4). We will find that integrals over $\mathscr{P}_n/GL(n, \mathbb{Z})$ lead to some higher dimensional integral tests for the convergence of matrix series. We will also find that some of the integrals that we need for number-theoretic applications were computed independently by number theorists and statisticians in the early part of this century. For example, one of Wishart's formulas will be very useful in applying our matrix integral test. Wishart was a statistician and his distribution is of central importance in multivariate statistics (see Section 4.2, Farrell [1, 2], and Muirhead [1]). It may also come as a surprise to number theorists and physicists to learn that some of the integrals needed by physicists in their study of eigenvalues of random Hermitian matrices (see Mehta [1]) were computed 23 years earlier by the number theorist Selberg [2]. Many number theorists are, in fact, interested in such eigenvalue problems, thanks to possible connections with the Riemann hypothesis (see Montgomery [1] and Hejhal [3]).

There are many sorts of fundamental domains for $GL(n, \mathbb{Z})$ besides that of Minkowski. We will also consider the domain obtained by Grenier [1–2] using a generalization of the highest point method discussed for $n = 2$ in Vol. I, Section 3.3. This method is similar to that used by Siegel for the Siegel modular group $Sp(n, \mathbb{Z})$ (see Section 5.2, Maass [2, p. 168], or Siegel [1]). The domain obtained by Grenier has an advantage over Minkowski's in that it has an exact box shape at infinity. This allows one to use it to compute the integrals arising in the parabolic terms of the Selberg trace formula, for example (see Vol. I, pp. 216 and 289). One would hope to be able to generalize many of the other results in Volume I, Chapter 3, which depended on knowing the explicit shape of the fundamental domain. There are many places to look for applications of the study of the fundamental domain for $GL(n, \mathbb{Z})$:

1. algorithms for class numbers and units in number fields,
2. higher dimensional continued fraction algorithms,
3. geometric interpretations of $\zeta(3)$,
4. the study for $Y \in \mathscr{P}_n$ of the numbers $a_m(Y) = \#\{a \in \mathbb{Z}^n | Y[a] = m\}$, where m is a given integer. In 1883, Smith and Minkowski won a prize for proving Eisenstein's formula on the subject. This work was greatly generalized by C.L. Siegel [1] and later reinterpreted adelically by Tamagawa and Weil [2, Vol. III, pp. 71–157].
5. densest lattice packings of spheres in Euclidean space, Hilbert's 18th problem. This has applications in coding (see for example, Sloane [1]).
6. study of matrix analogues of the Riemann zeta function and corresponding automorphic forms generalizing the theta function. For example, we will investigate the following zeta function:

$$Z(s) = \sum_{\substack{A \in \mathbb{Z}^{n \times n} \text{ rank } n/GL(n, \mathbb{Z})}} |A|^{-s}, \quad \text{for } s \in \mathbb{C} \text{ with } \operatorname{Re} s > n \quad (1.4)$$

where $|A|$ = determinant of A. Here the sum is over $n \times n$ integral matrices A of rank n running through a complete set of representatives for the equivalence relation

$$A \sim B \text{ iff } A = BU \quad \text{for some } U \in GL(n, \mathbb{Z}).$$

This zeta function might be called the Dedekind zeta function attached to the simple algebra $\mathbb{Q}^{n \times n}$. It turns out to have an analytic continuation and functional equation and one of Riemann's original proofs can (with some difficulty) be made to show this. But actually $Z(s)$ is a product of Riemann zeta functions and this leads to a formula for the volume of the fundamental domain for $GL(n, \mathbb{Z})$ involving a product of values of Riemann zeta functions at the integers from 2 to n.

Perhaps the main number theoretical application of harmonic analysis on $\mathscr{P}_n/GL(n, \mathbb{Z})$ is the study of L-functions corresponding to modular forms for

4.1. Geometry and Analysis on $\mathscr{P}_n$

$GL(n, \mathbb{Z})$ and analogues of Hecke's correspondence. Results along these lines in the classical manner, using matrix Mellin transforms, can be found in Section 4.5 (see also Maass [2,4], Imai [1], and Bump [1]). This sort of result goes back to Koecher [1], who studied matrix Mellin transforms (over $\mathscr{P}_n/GL(n, \mathbb{Z})$) of Siegel modular forms, but not enough complex variables were present to invert Koecher's transform. The L-function (1.4) above is an example of one of the functions considered in Koecher's theory. Imai [1] and Weissauer [1] show how to obtain a converse theorem for such a transform. See Chapter 5, Section 2. Solomon [1] and Bushnell and Reiner [1] obtain generalizations of such zeta functions and applications to the theory of algebras and combinatorics.

We will not discuss the adelic-representation theoretic view of Hecke theory on higher rank groups such as $GL(n)$—a theory which has been developed by Jacquet, Langlands, Piatetski-Shapiro, and others (see the Corvallis conference volume edited by Borel and Casselman [1]). Jacquet, Piatetski-Shapiro and Shalika [1] show that "roughly speaking, all infinite Euler products of degree 3 having a suitable analytic behavior are attached to automorphic representations of $GL(3)$." For $GL(n)$, $n \geq 4$, they must "twist" the Euler product by all automorphic cuspidal representations of $GL(j)$, $1 \leq j \leq n - 2$. The $GL(3)$ result is applied to attach automorphic representations to cubic number fields, in a similar way to that in which Maass [3] found non-holomorphic cusp forms for congruence subgroups of $SL(2, \mathbb{Z})$ (of level greater than one) corresponding to Hecke L-functions of real quadratic fields. The Langlands philosophy indicates that non-abelian Galois groups of number fields should have Artin L-functions coming from modular forms for matrix groups.

One might also hope for higher rank analogues of the classical applications of harmonic analysis in number theory; e.g., in the proof of the prime number theorem (see, for example, Grosswald [1]). There is not much in print in this direction. However, already Herz [1] had been motivated by the desire to understand the asymptotics of

$$\#\{T \in \mathbb{Z}^{k \times m} | A - {}^tTT \in \mathscr{P}_m\}, \quad \text{as } A \text{ approaches infinity.}$$

Herz hoped to use his theory of special functions on $\mathscr{P}_m$ to carry out this research, but he did not apparently manage to do this. We will be able to use the Poisson sum formula for $GL(n, \mathbb{Z})$ to study the related question of the asymptotics of

$$\#\{\gamma \in GL(n, \mathbb{Z}) | \text{Tr}({}^t\gamma\gamma) \leq x\}, \quad \text{as } x \to \infty.$$

Bartels [1] has obtained results on this question for very general discrete groups Γ acting on symmetric spaces. One goal of the theory developed in Sections 4.4–4.5 is the search for higher rank analogues of the work of Sarnak on units in real quadratic fields (see Vol. I, p. 292). Wallace [1–6] gives results in this direction. One would also hope for higher degree analogues of results of Hejhal [4] on the distribution of solutions of quadratic congruences, results that were first proved by Hooley [1]. And there should be analogues of the work of Elstrodt, Grunewald, and Mennicke [1–3]. Bump [1], Bump and Goldfeld [1], Bump and Hoffstein [1], and Bump, Friedberg and Goldfeld [1] have generalized many classical results to $GL(3)$.

Warning to the Reader. This chapter and the next will be much more sketchy and demanding than the preceding chapters. The theory is still being developed and it is not possible to say the final word. There are many research problems here.

I have chosen a point of view that I think best approximates that of Volume I, as well as that of Maass [1–4] and Siegel [1, Vol. III, pp. 97–137]. Many experts might disagree with me here. And I'm sure that some would tell the reader to do everything adelically or in the language of group representations or both. We urge the reader to look at other references for a broader perspective on the subject.

We have sought to provide a large number of examples rather than general theorems. We ask the reader to consider special cases, look at examples using pencil and paper or even a computer. We found that it is helpful to use a computer to understand fundamental domains by drawing pictures of them (see Section 4.4.3).

We shall assume that the reader has read about Haar or G-invariant measures on Lie groups G as well as G-invariant measures on quotient spaces G/H for closed subgroups H of G. Some references for this are Vol. I, Dieudonné [1], Helgason [1–4], Lang [1], and Weil [1]. For example, Haar measure on the multiplicative abelian group

$$A = \{a \in \mathscr{P}_n \mid a \text{ is diagonal}\}$$

is given by

$$da = \prod_{j=1}^{n} a_j^{-1} da_j \quad \text{if} \quad a = \begin{pmatrix} a_1 & & & 0 \\ & a_2 & & \\ & & \ddots & \\ 0 & & & a_n \end{pmatrix}. \tag{1.5}$$

Here the group operation on A is matrix multiplication and da_j is Lebesgue measure on the real line.

Exercise 2 (A Nilpotent Lie Group). Define N to be the group of all upper triangular $n \times n$ real matrices with ones on the diagonal. Show that N forms a group under matrix multiplication and that this group is not abelian unless $n = 2$. Show that Haar measure on N is given by:

$$dn = \prod_{1 \leq i < j \leq n} dx_{ij} \quad \text{if} \quad n = \begin{pmatrix} 1 & x_{12} & x_{13} & \cdots & x_{1n} \\ 0 & 1 & x_{23} & \cdots & x_{2n} \\ \vdots & \vdots & \ddots & & \vdots \\ 0 & 0 & \cdots & 1 & x_{n-1,n} \\ 0 & 0 & \cdots & & 1 \end{pmatrix}.$$

Here dx_{ij} is the Lebesgue measure on the real line. Show that right and left Haar measures on N are the same.

Exercise 3. Show that the Haar measure on $G = GL(n, \mathbb{R})$ is given by
$$dg = |g|^{-n} \prod_{i,j} dg_{ij}.$$
Show that the right and left Haar measures are the same in this case.

From the theory of integrals on quotients G/H where H is a closed subgroup of the Lie group G, we will need an understanding of the formula:
$$\int_G f(g)\, dg = \int_{G/H} \int_H f(gh)\, dh\, d\bar{g}. \tag{1.6}$$
Here dg and dh are Haar measures on G, H, respectively. And this formula really defines the G-invariant measure $d\bar{g}$ on the quotient space G/H. Such an integral is determined up to a positive constant. Formula (1.6) holds provided that, for example, both G and H are unimodular (i.e., left and right Haar measures are the same), and even more generally (see Helgason [4, p. 91]). If there is a fundamental domain S for H in G; i.e. a measurable subset S of G which can be identified with G/H, then the integral over G/H is the same as the integral over S and (1.6) is clear since G is a disjoint union of cosets sH over $s \in S$. It is formula (1.6) that leads to our integral test for the convergence of matrix series in Section 4.4.4. Helgason [4, pp. 139ff.] uses this result to find some interesting formulas involving generalizations of the Radon transform. Weil [2, Vol. I, pp. 339–357] employed (1.6) to prove an integral formula of Siegel, which leads to the existence of dense lattice packings of spheres in Euclidean spaces of high dimensions—a result that we will discuss in Section 4.4.4.

It would also be useful for the reader to make herself or himself comfortable with calculus on manifolds (see Choquet-Bruhat et al. [1], Dieudonné [1], or Lang [1], for example). We will need to be able to change variables in Laplace operators, in particular. Courant and Hilbert [1, p. 224] give an easy method to do this using the calculus of variations. Suppose that the old variables are $x \in \mathbb{R}^n$ and the new ones are $y \in \mathbb{R}^n$. Let the Jacobian matrix of the change of variables be
$$J = (\partial x_i / \partial y_j).$$
Then we have the following formulas for the arc length, volume, and Laplacian in the new coordinates:

$$ds^2 = \sum_{j=1}^n dx_j^2 = \sum_{i,j=1}^n g_{ij}\, dy_i\, dy_j, \quad \text{where } G = {}^tJJ,$$

$$dv = \prod_{j=1}^n dx_j = g^{1/2} \prod_{j=1}^n dy_j, \quad \text{if } g = |G| = \det G, \tag{1.7}$$

$$\Delta = g^{-1/2} \sum_{i=1}^n \frac{\partial}{\partial y_i} g^{1/2} \sum_{j=1}^n g^{ij} \frac{\partial}{\partial y_j}, \quad \text{for } G^{-1} = (g^{ij}).$$

Exercise 4. Prove (1.7) using the calculus of variations, as in Courant and Hilbert [1, p. 224].

Hint. See formulas (2.2)–(2.6) in Vol. I. The calculus of variations says that if u minimizes the integral

$$F(u) = \int \sum_{i=1}^{n} u_{x_i}^2 \, dx,$$

subject to the constraint

$$K(u) = \int u^2 \, dx = \text{constant},$$

then u must satisfy the Euler-Lagrange equation:

$$\Delta u = \sum_{i=1}^{n} u_{x_i x_i} = \lambda u;$$

i.e., u is an eigenfunction for the Euclidean Laplacian.

In this section we shall collect the basic facts about $\mathscr{P}_n$. An outline of the section is:

1.1. Elementary Results,
1.2. Geodesics and Arc Length,
1.3. Measure and Integration on $\mathscr{P}_n$,
1.4. Differential Operators on $\mathscr{P}_n$,
1.5. A List of the Main Formulas Derived in Section 1,
1.6. An Application to Multivariate Statistics.

Most of the section consists of exercises in advanced (or not so advanced) calculus on matrix space. We have tried to give extensive hints. The summary §1.5 includes tables of formulas which put these exercises into a nutshell. Many readers may want to skip to the summary, as these exercises can be time consuming. But we do this at our peril, of course.

4.1.1. Elementary Results

Now we begin the detailed consideration of the geometry and analysis of $\mathscr{P}_n$, the space of positive $n \times n$ real matrices (1.1). Other references for this section are Helgason [1–4], Hua [1], Maass [2], and Selberg [1].

Now $\mathscr{P}_n$ is the symmetric space of the general linear group $G = GL(n, \mathbb{R})$ of non-singular $n \times n$ real matrices. The *action* of $g \in G$ on $Y \in \mathscr{P}_n$ is given by:

$$Y \mapsto Y[g] = {}^t g Y g, \quad \text{where } {}^t g = \text{transpose of } g. \tag{1.8}$$

Let $K = O(n)$ be the *orthogonal group* of matrices $k \in G = GL(n, \mathbb{R})$ such that ${}^t k k = I$. Consider the map

4.1. Geometry and Analysis on $\mathscr{P}_n$

$$K\backslash G \to \mathscr{P}_n$$
$$Kg \mapsto I[g].$$
(1.9)

This map provides an identification of the two spaces as *homogeneous spaces* of G. By a homogeneous space X of G, we mean that X is a differentiable manifold with transitive differentiable G-action

$$G \times X \to X$$
$$(g, x) \mapsto gx$$

such that

1. $ex = x$ for e = identity of G and all $x \in X$

and

2. $(gh)x = g(hx)$ for all $g, h \in G$ and $x \in X$.

Exercise 5. Prove that the map (1.9) identifies the two spaces as homogeneous spaces of $GL(n, \mathbb{R})$. Make sure that you show that the mapping preserves the group actions.

Hint. You need to use the general result which says that having fixed a point p in a homogeneous space X of the group G, then, defining the isotropy group $G_p = \{g \in G | g \cdot p = p\}$, we see that the map

$$G/G_p \to X$$
$$gG_p \mapsto g \cdot p$$

is a diffeomorphism onto. There is a proof in Broecker and tom Dieck [1, p. 35].

In order to see that the map (1.9) is onto, you can use the spectral theorem for positive matrices (Exercise 7).

Throughout this book, we will use the notation

$$|Y| = \text{determinant}(Y).$$
(1.10)

Exercise 6 (Criteria for the Positivity of a Real Symmetric Matrix). Show that if $Y = {}^tY$ is a real $n \times n$ matrix, then Y is positive definite (as in (1.2)) if and only if one of the following equivalent conditions holds:

(a) All of the eigenvalues of Y are positive.
(b) If

$$Y = \begin{pmatrix} Y_k & * \\ * & * \end{pmatrix}, \quad \text{with } Y \in \mathbb{R}^{k \times k},$$

then $|Y_k| > 0$, for all $k = 1, 2, \ldots, n$.

Hint. It is helpful to use the partial Iwasawa decomposition discussed in Exercise 11.

Exercise 7 (Spectral Theorem for a Symmetric Matrix). Suppose $Y = {}^tY$ is an $n \times n$ real matrix. Then there is an orthogonal matrix k in $O(n)$ such that $Y = {}^tkDk$, where

$$D = \begin{pmatrix} d_1 & & 0 \\ & \ddots & \\ 0 & & d_n \end{pmatrix}$$

and d_j is real and equal to the jth eigenvalue of Y. We can characterize the d_j by a sequence of maximum problems:

$$d_j = \max\{{}^txYx \mid x \in \mathbb{R}^n, \|x\| = 1, {}^txx_i = 0, i = 1,\ldots,j-1\} = {}^tx_jYx_j,$$

with $Yx_j = d_j x_j$. Here we maximize over x in $\mathbb{R}^n$ of length 1 and orthogonal to all the preceding $j-1$ eigenvectors of Y. Then $k = (x_1 x_2 \cdots x_n)$.

Hint. See Courant and Hilbert [1, Chapter I].

There are many familiar applications of Exercise 7; e.g., to the discussion of normal modes of vibrating systems (see Courant and Hilbert [1]). Exercise 7 also leads to a numerical method for the solution of partial differential equations known as the Rayleigh-Ritz method (see Arfken [1, pp. 800–803]), a method which has developed into the modern finite element method (see Strang and Fix [1]).

Exercise 8 (Mini-Max Principle for Eigenvalues of $Y = {}^tY$).

(a) Let Y and d_j be as in Exercise 7. Show that

$$d_j = \min_{w_1,\ldots,w_{j-1} \in \mathbb{R}^n} \left(\max_{\substack{x \in \mathbb{R}^n, \|x\|=1 \\ {}^tx \cdot w_i = 0, \text{ all } i=1,\ldots,j-1}} {}^txYx \right).$$

(b) As an application of part (a) show that if

$$Y = \begin{pmatrix} Y_1 & * \\ * & * \end{pmatrix}$$

where Y is as in Exercise 7 and Y_1 is a $k \times k$ symmetric matrix, then (jth eigenvalue of Y_1) $\leq$ (jth eigenvalue of Y) $\leq$ (($j-1$)st eigenvalue of Y_1).

The mini-max principle can be generalized greatly and leads to many qualitative results about solutions of boundary value problems of mathematical physics. For example, it can be used to compare fundamental tones for vibrating strings. The string of larger density will have the lower fundamental tones (see Courant and Hilbert [1, Ch. VI]).

4.1. Geometry and Analysis on $\mathscr{P}_n$

Exercise 9. Given two $n \times n$ real symmetric matrices P and Y with $P \in \mathscr{P}_n$, show that there is a matrix g in $GL(n, \mathbb{R})$ such that both $Y[g]$ and $P[g]$ are diagonal.

Hint. Use Exercise 7.

The space of symmetric real $n \times n$ matrices is easily seen to be identifiable with $\mathbb{R}^m$, where $m = n(n + 1)/2$. And it follows from Exercise 9 that $\mathscr{P}_n$ is an *open cone* in $\mathbb{R}^m$, $m = n(n + 1)/2$; i.e., $X, Y \in \mathscr{P}_n$ and $t > 0$ implies that $X + Y$ and tX are also in $\mathscr{P}_n$.

4.1.2. Geodesics and Arc Length

In order to view the space $\mathscr{P}_n$ as an analogue of the Poincaré upper half plane which was studied in Chapter 3 of Volume I, we need a notion of arc length. This will turn the space into a *Riemannian manifold*. Now $\mathscr{P}_n$ is an open set in $n(n + 1)/2$ dimensional Euclidean space as can be seen from Exercise 6 in the last section. Thus it is indeed a differentiable manifold of the easiest sort to consider. The choice of coordinates is not a problem. To define the Riemannian structure, we need to define an arc length element (thinking as physicists). Differential geometers would consider a bilinear form on the tangent space or a continuous 2-covariant tensor field, or whatever. We want to think as students of advanced calculus, however. But duty impels us to note that Lie groups people would say that it all comes from the Killing form on the Lie algebra. We will consider that approach in Chapter 5.

Well, anyway, suppose we just define the *arc length* element ds by the formula:

$$ds^2 = \text{Tr}((Y^{-1} dY)^2), \quad \text{where } Y = (y_{ij})_{1 \leq i,j \leq n}$$
$$\text{and } dY = (dy_{ij})_{1 \leq i,j \leq n}, \text{ Tr} = \text{Trace}. \tag{1.11}$$

It is easy to show that ds is invariant under the group action of $g \in GL(n, \mathbb{R})$ on $Y \in \mathscr{P}_n$ given by (1.8). For $W = Y[g]$ is a linear function of Y. Recall from advanced calculus that a linear map is its own differential (see for example Lang [1, Chapter 5]). Therefore $dW = dY[g]$ and it follows that upon plugging into the definition of ds, we obtain:

$$ds^2 = \text{Tr}((Y^{-1} dY)^2) = \text{Tr}((gW^{-1} {}^t g \, {}^t g^{-1} \, dW \, g^{-1})^2) = \text{Tr}((W^{-1} dW)^2).$$

Thus we have shown that our arc length on $\mathscr{P}_n$ behaves like that on the Poincaré upper half plane as far as group invariance properties go. Moreover, the arc length element is positive definite, since it is positive definite at $Y = I$ and the action of $GL(n, \mathbb{R})$ on $\mathscr{P}_n$ is transitive (see Exercise 5).

Recall that in the case of the Poincaré upper half plane H, the geodesics through the point gi, for g in $SL(2, \mathbb{R})$, have the form

$$g\begin{pmatrix} \exp(at) & 0 \\ 0 & \exp(-at) \end{pmatrix} i, \qquad \text{for } t \in \mathbb{R},$$

where the action of the 2×2 matrix on i is by fractional linear transformation (see Volume I, page 122). Here a is any real number. Thus it will not come as a surprise to you that in the case of $\mathcal{P}_n$ the geodesics through the point $I[g]$, for $g \in GL(n, \mathbb{R})$ are given by:

$$\begin{pmatrix} \exp(a_1 t) & & 0 \\ & \ddots & \\ 0 & & \exp(a_n t) \end{pmatrix} [g], \qquad \text{for } t \in \mathbb{R},$$

where the numbers a_i are arbitrary real numbers. It may not surprise you either if we reveal that the method of proof of this fact (Theorem 1 below) is just a rather straightforward generalization of the proof that works in the Euclidean case and in the upper half plane case. The only problem will be the computation of a rather nasty Jacobian. Mercifully Hans Maass [2] computes the Jacobian for us.

It is perhaps more convenient to view a geodesic through the identity in $\mathcal{P}_n$ as the *matrix exponential*:

$$\exp(tX) = \sum_{n \geq 0} \frac{(tX)^n}{n!}, \qquad \text{for } t \in \mathbb{R}, X \in \mathbb{R}^{n \times n}, {}^t X = X. \qquad (1.12)$$

The following exercise reviews the basic properties of exp. There is an analogue for any Riemannian manifold (see Helgason [2, 3]). And in Chapter 5 we will consider the analogue for Lie groups G.

Exercise 10 (The Matrix Exponential).

(a) Show that the series below converges absolutely for all $A \in \mathbb{R}^{n \times n}$.

$$\exp(A) = \sum_{n \geq 0} \frac{A^n}{n!}.$$

(b) Show that if $AB = BA$, for A, B in $\mathbb{R}^{n \times n}$, then

$$\exp(A + B) = \exp(A) \exp(B).$$

Show that this equality may fail if $AB \neq BA$.

(c) Show that if U is in $GL(n, \mathbb{R})$, then $\exp(U^{-1} A U) = U^{-1} \exp(A) U$.

(d) Using the notation (1.10), show that

$$|\exp(A)| = \exp(\text{Tr}(A)).$$

Hint. See Lang [2, pp. 295–296].

There are applications of the matrix exponential to the solution of ordinary differential equations (see Apostol [1, Vol. II, pp. 213–214]).

4.1. Geometry and Analysis on $\mathscr{P}_n$

In order to prove the characterization of geodesics in $\mathscr{P}_n$, we need to have an analogue of rectangular coordinates in the Poincaré upper half plane. This analogue is what I call *partial Iwasawa coordinates* defined as follows for $Y \in \mathscr{P}_n$, with the notation (1.8):

$$Y = \begin{pmatrix} V & 0 \\ 0 & W \end{pmatrix} \begin{bmatrix} I_p & 0 \\ X & I_q \end{bmatrix}, \quad \text{for } V \in \mathscr{P}_p, W \in \mathscr{P}_q, X \in \mathbb{R}^{q \times p}, n = p + q. \quad (1.13)$$

When $n = 2$, Exercise 9 of Section 3.1 in Volume I shows that these coordinates (or those from Exercise 11(b) below) correspond to rectangular coordinates in the Poincaré upper half plane. There are as many versions of these partial Iwasawa coordinates as there are partitions of n into 2 parts. We feel impelled to say that somehow all these different coordinate systems measure ways of approaching the boundary of the symmetric space. Thus if we manage to try to do the analysis of behavior of functions like Eisenstein series for $GL(n)$, we will have to make serious use of these coordinates.

Partial Iwasawa coordinates generalize the well known technique from high school algebra known as "completing the square," since if $a \in \mathbb{R}^p$ and $b \in \mathbb{R}^q$,

$$Y \begin{bmatrix} a \\ b \end{bmatrix} = V[a] + W[Xa + b],$$

which is indeed a sum of squares if $n = 2$ and $p = q = 1$.

At the end of this section, we will see an application for partial Iwasawa coordinates in multivariate statistics—with results going back to Pearson in 1896 and Yule in 1897 (see also T.W. Anderson [1, p. 27]).

There is also another type of partial Iwasawa coordinates, as the next exercise shows.

Exercise 11.

(a) Show that the equation (1.13) can always be solved uniquely for V, W, X, once Y in $\mathscr{P}_n$ is given, along with $p, q \in \mathbb{Z}^+$, such that $n = p + q$.
(b) Obtain a second set of partial Iwasawa coordinates, using the equality:

$$\begin{pmatrix} F & 0 \\ 0 & G \end{pmatrix} \begin{bmatrix} I_p & H \\ 0 & I_q \end{bmatrix}, \quad \text{with } F \in \mathscr{P}_p, G \in \mathscr{P}_q, H \in \mathbb{R}^{p \times q}, n = p + q.$$

We have seen that there are many sorts of partial Iwasawa coordinates. In fact, you can continue the idea to decompose Y into an analogous product, with any number of block matrices along the diagonal. This gives coordinate systems associated to any partition of n. Such coordinate systems will be needed in our study of Eisenstein series in Section 4.3. But let's restrain ourselves since we do not need these more general decompositions for our goal of characterizing the geodesics in $\mathscr{P}_n$. However, later in this section, we will need the full Iwasawa decomposition corresponding to the partition of n given by

$$n = 1 + \cdots + 1$$

—a decomposition which is studied in the following exercise.

Exercise 12 (The Full Iwasawa Decomposition). Show that repetition of the partial Iwasawa decomposition in Exercise 11(b) leads to the full Iwasawa decomposition: $Y = a[n]$, with a positive diagonal and n upper triangular with 1's on the diagonal.

The coordinate system of Exercise 12 corresponds to the full Iwasawa decomposition on the group $G = GL(n, \mathbb{R})$ (see Helgason [1–3] and Iwasawa [1] as well as Chapter 5). Siegel [2, p. 29] calls the full Iwasawa decomposition the "Jacobi transformation." Weil [3, p. 7] calls it "Babylonian reduction." Numerical analysts call it the "LDU factorization" and obtain it by Gaussian elimination (see Strang [1]). The Gram-Schmidt orthogonalization process can also be used to derive the full Iwasawa decomposition.

Exercise 13. Suppose that Y_k is the upper left hand $k \times k$ corner of the matrix Y. For $1 \leq i_1 \leq \cdots \leq i_h \leq n$ and $1 \leq j_1 \leq \cdots \leq j_h \leq n$, set

$$Y(i_1,\ldots,i_h|j_1,\ldots,j_h) = \det(Y_{i_a j_b})_{1 \leq a, b \leq h}.$$

Then show that the matrices a and n from Exercise 12 are given by:

$$a = \begin{pmatrix} a_1 & & 0 \\ & \ddots & \\ 0 & & a_n \end{pmatrix} \quad n = \begin{pmatrix} 1 & & x_{ij} \\ & \ddots & \\ 0 & & 1 \end{pmatrix}, \quad \text{with}$$

$a_i = |Y_i|/|Y_{i-1}|$ and $x_{ij} = Y(1,\ldots,i-1,i|1,\ldots,i-1,j)/|Y_i|$.

Hint. (See Minkowski [1, Vol. II, p. 55] and Hancock [1, Vol. II, pp. 536–539].) Note that if we choose ${}^t u = (u_1,\ldots,u_i,0,0,\ldots,0,1,0,\ldots,0)$ with the one in the jth place, then we can minimize $Y[u]$ over such vectors u. If $Y = a[n]$, then $Y[u] = a_1(u_1 + x_{12}u_2 + \cdots + x_{1n}u_n)^2 + a_2(u_2 + x_{23}u_3 + \cdots + x_{2n}u_n)^2 + \cdots + a_n u_n^2$, and thus to minimize $Y[u]$, we need

$$u_i + x_{ij} = 0,$$

$$2Yu = \operatorname{grad} Y[u] = \left(\frac{\partial Y[u]}{\partial u_1}, \ldots, \frac{\partial Y[u]}{\partial u_i}\right) = 0.$$

Cramer's rule completes the proof of the formula for x_{ij}. The formula for a_i is easily proved.

Now, in order to obtain the geodesics on $\mathscr{P}_n$, we must express the *arc length element in partial Iwasawa coordinates* (1.13):

$$ds_Y^2 = ds_V^2 + ds_W^2 + 2\operatorname{Tr}(V^{-1}\, {}^t dX\, W dX), \tag{1.14}$$

4.1. Geometry and Analysis on $\mathcal{P}_n$

where ds_V is the element of arc length in $\mathcal{P}_p$, ds_W is the analogue for $W \in \mathcal{P}_q$, and dX is the matrix of differentials dx_{ij} if $X = (x_{ij}) \in \mathbb{R}^{p \times q}$.

To prove (1.14) write:
$$Y = \begin{pmatrix} V & 0 \\ 0 & W \end{pmatrix} \begin{bmatrix} I & 0 \\ X & I \end{bmatrix} = \begin{pmatrix} V + W[X] & {}^tXW \\ WX & W \end{pmatrix}.$$

Then
$$Y^{-1} = \begin{pmatrix} V^{-1} & -V^{-1}\,{}^tX \\ -XV^{-1} & V^{-1}[{}^tX] + W^{-1} \end{pmatrix}$$

and
$$dY = \begin{pmatrix} dV & 0 \\ 0 & dW \end{pmatrix} \begin{bmatrix} I & 0 \\ X & I \end{bmatrix} + \begin{pmatrix} 0 & {}^tdX \\ 0 & 0 \end{pmatrix} \begin{pmatrix} V & 0 \\ 0 & W \end{pmatrix} \begin{pmatrix} I & 0 \\ X & I \end{pmatrix}$$
$$+ \begin{pmatrix} I & {}^tX \\ 0 & I \end{pmatrix} \begin{pmatrix} V & 0 \\ 0 & W \end{pmatrix} \begin{pmatrix} 0 & 0 \\ dX & 0 \end{pmatrix}.$$

This allows one to compute
$$dY\,Y^{-1} = \begin{pmatrix} L_0 & L_1 \\ L_2 & L_3 \end{pmatrix}.$$

We might wish for MACSYMA (the symbol manipulation language) and a nice friendly computer to help us out here.

Exercise 14 (Proof of Formula (1.14)).

(a) Prove that
$$L_0 = dV\,V^{-1} + {}^tXW\,dX\,V^{-1},$$
$$L_1 = -dV\,V^{-1}\,{}^tX - {}^tXW\,dX\,V^{-1}\,{}^tX + {}^tX\,dW\,W^{-1} + {}^tdX,$$
$$L_2 = W\,dX\,V^{-1},$$
$$L_3 = -W\,dX\,V^{-1}\,{}^tX + dW\,W^{-1}.$$

(b) Use part (a) to show that
$$ds^2 = \text{Tr}(L_0^2 + L_1 L_2 + L_2 L_1 + L_3^2)$$
$$= \text{Tr}((V^{-1}\,dV)^2 + (W^{-1}\,dW)^2 + 2V^{-1}\,{}^tdX\,W\,dX).$$

(c) Find ds^2 for the other partial Iwasawa decomposition given in Exercise 11(b).

Answer. $ds_Y^2 = \text{Tr}((F^{-1}\,dF)^2 + (G^{-1}\,dG)^2 + 2\,dH\,G^{-1}\,{}^tdH\,F)$, for
$$Y = \begin{pmatrix} F & 0 \\ 0 & G \end{pmatrix} \begin{bmatrix} I & H \\ 0 & I \end{bmatrix}.$$

Exercise 15.

(a) Show that if $V = {}^t\!AA$, $W = {}^t\!BB$, $H = B\,dX\,A^{-1}$, for $A \in GL(p, \mathbb{R})$, $B \in GL(q, \mathbb{R})$, then

$$\operatorname{Tr}(V^{-1}\,{}^t\!dX\,W\,dX) = \operatorname{Tr}(H\,{}^t\!H) \geq 0.$$

(b) Conclude from Exercise 14 and part (a) that when seeking a curve $T(t) \in \mathscr{P}_n$, such that $T(0) = I$, $T(1) = D$, a diagonal matrix, the arc length will only be decreased by taking the X-coordinate in the partial Iwasawa decomposition of $T(t)$ in (1.13) to be zero.

(c) Use induction to conclude from part (b) that we only decrease the arc length by taking $T(t)$ to be diagonal for all $t \in [0, 1]$.

Theorem 1 (Geodesics in $\mathscr{P}_n$). *A geodesic segment $T(t)$ through I and Y in $\mathscr{P}_n$ has the form:*

$$T(t) = \exp(t\,A[U]), \qquad 0 \leq t \leq 1,$$

where Y has the spectral decomposition:

$$Y = (\exp A)[U] = \exp(A[U]), \qquad \text{for } U \in O(n)$$

and

$$A = \begin{pmatrix} a_1 & & & 0 \\ & a_2 & & \\ & & \ddots & \\ 0 & & & a_n \end{pmatrix}, \qquad a_j \in \mathbb{R}, \quad j = 1, 2, \ldots, n.$$

The length of the geodesic segment is:

$$\left(\sum_{j=1}^n a_j^2 \right)^{1/2}.$$

PROOF. Use Exercise 7 to write $Y = D[U]$ with U in the orthogonal group $O(n)$ and

$$D = \begin{pmatrix} d_1 & & & 0 \\ & d_2 & & \\ & & \ddots & \\ 0 & & & d_n \end{pmatrix}, \qquad d_j > 0, \quad j = 1, 2, \ldots, n.$$

Then $D = \exp A$ with $d_j = \exp(a_j)$, $j = 1, \ldots, n$. From Exercise 10 we know that $Y = (\exp A)[U] = \exp(A[U])$.

And Exercise 15 tells us that a geodesic $T(t)$ in $\mathscr{P}_n$ such that $T(0) = I$ and $T(1) = Y[U]$ has the form:

4.1. Geometry and Analysis on $\mathscr{P}_n$

$$T(t) = \begin{pmatrix} d_1(t) & & 0 \\ & d_2(t) & \\ & & \ddots \\ 0 & & d_n(t) \end{pmatrix} [U], \qquad d_j(0) = 0, \qquad d_j(1) = d_j,$$
$$j = 1, \ldots, n.$$

Let $d_j(t) = \exp(a_j(t))$. This is possible, since all the d_j's are positive. Using the definition (1.11) of arc length in $\mathscr{P}_n$, it is easily seen that the arc length of the curve $T(t)$ is:

$$\int_0^1 \left(\sum_{j=1}^n a_j'^2 \right)^{1/2} dt.$$

The curve $a(t) \in \mathbb{R}^n$ such that $a(0) = 0$ and $a(1) = (a_1, \ldots, a_n)$ which minimizes this distance is the straight line:

$$a_j(t) = t a_j, \qquad j = 1, \ldots, n.$$

This completes the proof of Theorem 1. □

Exercise 16. Use Exercise 9 and Theorem 1 to find the geodesics through two arbitrary points in $\mathscr{P}_n$.

In Chapter 5 we will generalize Theorem 1 to an arbitrary noncompact symmetric space. For example, define the *symplectic group* $Sp(n, \mathbb{R})$ of all $(2n) \times (2n)$ real matrices g such that if

$$J = \begin{pmatrix} 0 & I_n \\ -I_n & 0 \end{pmatrix}, \qquad {}^t g J g = J. \tag{1.15}$$

And we can consider the *symmetric space* attached to $Sp(n, \mathbb{R})$ to be the subset of $\mathscr{P}_{2n}$ consisting of all positive $(2n) \times (2n)$ symplectic matrices. In fact, this subset of will be seen to be a *totally geodesic submanifold* of $\mathscr{P}_{2n}$; i.e., the geodesics in the larger space which join two points in the smaller space must actually lie entirely in the smaller space.

Exercise 17 (The Determinant One Surface).

(a) As in Exercise 5, show that the determinant one surface $\mathscr{SP}_n$, defined in (1.3), can be identified with G/K, for $G = SL(n, \mathbb{R})$, $K = SO(n)$. Here $SL(n, \mathbb{R})$ denotes the *special linear group* of all determinant one matrices in $GL(n, \mathbb{R})$ and $SO(n)$ denotes the special orthogonal group which is defined analogously as the determinant one matrices in $O(n)$.
(b) Show that $\mathscr{SP}_n$ is a totally geodesic submanifold of $\mathscr{P}_n$.

Hint. You can relate the geometric structures on $\mathscr{P}_n$ and $\mathscr{SP}_n$ using formula (1.7) and the substitution

$$Y = tW, \qquad \text{for } t > 0 \text{ and } W \in \mathscr{SP}_n.$$

The group $SL(n, \mathbb{R})$ is a *simple* Lie group and thus certainly semisimple—a standard hypothesis for many theorems in the theory of group representations and harmonic analysis (see the next chapter). But $GL(n, \mathbb{R})$ has a center and is thus not semisimple. $GL(n, \mathbb{R})$ is only *reductive*. But this should not make us abandon $GL(n)$ for its "simple" relative. For many theorems about simple and semisimple Lie groups are proved by first settling the case of $GL(n, \mathbb{R})$ and then using the Adjoint representation. For examples of this phenomenon, see the next chapter and Helgason [2, pp. 234–237]. Moreover it is often true that calculations are easier on $GL(n, \mathbb{R})$ than $SL(n, \mathbb{R})$. Thus we will usually consider the general rather than special linear group.

Exercise 18 (A Geodesic-Reversing Isometry). Show that the map $\sigma_I(Y) = Y^{-1}$ fixes the point I, preserves arc length, and reverses geodesics through the point I.

Hint. You could use the fact that $W = Y^{-1}$ implies that $Y \, dW + dY \, W = 0$.

Note that one can easily find a geodesic-reversing isometry of $\mathscr{P}_n$ at each point Y in $\mathscr{P}_n$ by translating the result of Exercise 18 using the action of $GL(n, \mathbb{R})$. It is the existence of these *geodesic-reversing isometries* that turns the Riemannian manifold $\mathscr{P}_n$ into a *symmetric space*. Such spaces were classified by E. Cartan in the 1920's (see Cartan [1, 2], Helgason [1–3], and Chapter 5). The geodesic-reversing isometry is used, for example, in the proof that the G-invariant differential operators on $\mathscr{P}_n$ form a commutative algebra (see part 2 of Lemma 2 in Section 4.1.4). Selberg [1, p. 51] notes that for this commutativity, it would suffice to assume that the Riemannian manifold S is only a "weakly symmetric space." That is, S is assumed to have a group G of isometries such that G is locally compact and transitive on S and having a fixed isometry μ (which may not lie in G) such that $\mu G \mu^{-1} = G$, and $\mu^2 \in G$. And for S to be weakly symmetric it is assumed that for every pair $x, y \in S$, there is an element $m \in G$ such that $mx = \mu y$ and $my = \mu x$.

4.1.3. Measure and Integration on $\mathscr{P}_n$

Now that we have completed our study of geodesics on $\mathscr{P}_n$, it is time to consider the $GL(n, \mathbb{R})$-invariant volume element and integration on $\mathscr{P}_n$. For $Y = (y_{ij}) \in \mathscr{P}_n$, let dy_{ij} = Lebesgue measure on $\mathbb{R}$. Then we define the $GL(n, \mathbb{R})$-invariant volume element $d\mu$ on $\mathscr{P}_n$ by:

$$d\mu = d\mu_n(Y) = |Y|^{-(n+1)/2} \prod_{1 \le i \le j \le n} dy_{ij}. \tag{1.16}$$

One can show that $d\mu_n$ in (1.16) is the volume element associated by (1.7) to the Riemannian metric tensor defined by (1.11).

To prove that the *volume element given in formula (1.16) is invariant under*

4.1. Geometry and Analysis on $\mathcal{P}_n$

$GL(n, \mathbb{R})$, suppose that $g \in GL(n, \mathbb{R})$ and $W = Y[g]$. The *Jacobian* of this mapping is:

$$J(g) = \|dW/dY\| = \|g\|^{n+1}, \quad \text{for } \|g\| = \text{absolute value of } \det(g). \quad (1.17)$$

To prove this, note that $J(g)$ is multiplicative; i.e.,

$$J(gh) = J(g)J(h), \quad \text{for all } g, h \in G.$$

And $J(g)$ is clearly a polynomial in the entries of g. So we can assume that g lies in the set of matrices of the form $U^{-1}DU$, with D diagonal and $U \in GL(n, \mathbb{R})$, as such matrices are dense in $GL(n, \mathbb{R})$. Suppose that D has jth diagonal entry d_j. Then $Y[D] = (d_i y_{ij} d_j)$, which implies that

$$J(U^{-1}DU) = J(D) = \prod_{1 \leq i \leq j \leq n} |d_i d_j| = \|D\|^{n+1} = \|U^{-1}DU\|^{n+1}.$$

This completes the proof of formula (1.17). One quickly deduces from (1.17) the fact that the volume element in (1.16) is invariant under the action of $GL(n, \mathbb{R})$.

Exercise 19 (The Jacobian for Partial Iwasawa Coordinates).

(a) Show that if

$$Y = \begin{pmatrix} V & 0 \\ 0 & W \end{pmatrix} \begin{bmatrix} I_p & 0 \\ X & I_q \end{bmatrix}, \quad V \in \mathcal{P}_p, \quad W \in \mathcal{P}_q, \quad X \in \mathbb{R}^{q \times p}, \quad n = p + q,$$

then

$$\left| \frac{\partial Y}{\partial(V, W, X)} \right| = |W|^p.$$

(b) If instead, we write

$$Y = \begin{pmatrix} F & 0 \\ 0 & G \end{pmatrix} \begin{bmatrix} I_p & H \\ 0 & I_q \end{bmatrix}, \quad F \in \mathcal{P}_p, \quad G \in \mathcal{P}_q, \quad H \in \mathbb{R}^{p \times q}, \quad n = p + q,$$

then

$$\left| \frac{\partial Y}{\partial(F, G, H)} \right| = |F|^q.$$

Hint. (a) Set

$$Y = \begin{pmatrix} A & B \\ {}^t B & C \end{pmatrix}, \quad \text{with } {}^t B = WX,$$

and note that $|\partial(WX)/\partial X| = |W|^p$—you get one $|W|$ for each column of X.

We can use Exercise 19 to obtain *the relation between invariant volumes in partial Iwasawa coordinates* (1.13):

$$d\mu_n(Y) = |V|^{-q/2} |W|^{p/2} d\mu_p(V) d\mu_q(W) dX, \quad (1.18)$$

where $d\mu_n$ is the invariant volume on $\mathcal{P}_n$ defined by (1.16) and dX is Lebesgue measure on $\mathbb{R}^{q \times p}$.

Exercise 20 (The Jacobian for the Full Iwasawa Decomposition of $GL(n, \mathbb{R})$).
(a) Define for $G = GL(n, \mathbb{R})$, $K = O(n)$,

$$A = \{a \in G \mid a \text{ is positive and diagonal}\},$$

$$N = \{n \in G \mid n \text{ is upper triangular with ones on the diagonal}\}.$$

Write for $x \in G$, the ANK-Iwasawa decomposition as $x = a(x)n(x)k(x)$, with $a(x) \in A$, $n(x) \in N$, and $k(x) \in K$. Show that the integral formula for the ANK-Iwasawa decomposition is:

$$\int_G f(x)\, dx = \int_A \int_N \int_K f(ank)\, dk\, dn\, da,$$

where all the measures are left ($=$ right) Haar measures.

(b) For $x \in G$, the KAN-Iwasawa decomposition of x can be written $x = K(x)A(x)N(x)$, with $K(x) \in K$, $A(x) \in A$, and $N(x) \in N$. Show that the relation between the two Iwasawa decompositions is:

$$A(x^{-1}) = a(x)^{-1}, \quad K(x^{-1}) = k(x)^{-1}, \quad \text{and} \quad N(x^{-1}) = a(x)n(x)^{-1}a(x)^{-1}.$$

(c) Show that if we set $n^a = ana^{-1}$ for $n \in N$, $a \in A$, then $n^a \in N$ and the Jacobian is

$$\alpha(a) \stackrel{(\text{defn})}{=} |dn^a/dn| = \prod_{1 \le i < j \le n} a_i/a_j = \prod_{i=1}^n a_i^{n-2i+1}.$$

Here, if

$$n = \begin{pmatrix} 1 & & x_{ij} \\ & \ddots & \\ 0 & & 1 \end{pmatrix}, \quad \text{then} \quad dn = \prod dx_{ij} = \text{the left } (=\text{right}) \text{ Haar measure on } N, \text{ by Exercise 2}.$$

(d) Prove the integral formula for the KAN-Iwasawa decomposition:

$$\int_G f(x)\, dx = \int_K \int_A \int_N f(kan)\alpha(a)\, dn\, da\, dk, \quad \text{with } \alpha(a) \text{ as in (c)}.$$

Hints. (See Lang [3, pp. 37–40].) Note that one can normalize the left Haar measures on G and K along with the G-invariant measure on G/K to obtain as in (1.6):

$$\int_G f(x)\, dx = \int_{gK = \bar{g} \in G/K} \int_{k \in K} f(gk)\, dk\, d\bar{g}.$$

Now G/K can be identified with AN, and we need only show that

4.1. Geometry and Analysis on $\mathcal{P}_n$

$\int_A \int_N f(an)\, dn\, da$ gives a left AN-invariant integral on AN. To see this, note that if $a_1 \in A$ and $n_1 \in N$, then we have the equality below for $n_2 = a^{-1} n_1 a$:

$$\int_A \int_N f(a_1 n_1 an)\, dn\, da = \int_A \int_N f(a_1 a n_2 n)\, dn\, da.$$

Then use the left invariance of da and dn, to complete the proof of (a).
It is also clear that if

$$n = \begin{pmatrix} 1 & & x_{ij} \\ & \ddots & \\ 0 & & 1 \end{pmatrix} \quad \text{and} \quad a = \begin{pmatrix} a_1 & & 0 \\ & \ddots & \\ 0 & & a_n \end{pmatrix},$$

then

$$ana^{-1} = \begin{pmatrix} 1 & & y_{ij} \\ & \ddots & \\ 0 & & 1 \end{pmatrix}, \quad \text{for } y_{ij} = a_i x_{ij} a_j^{-1}.$$

Now the left Haar measure on N is just $dn = \prod dx_{ij}$ (see Exercise 2), where dx_{ij} is just the usual Lebesgue measure on $\mathbb{R}$. The formula for the Jacobian in part (c) follows easily from these considerations. And it implies that

$$\int_N f(an)\, dn = \alpha^{-1}(a) \int_N f(na)\, dn.$$

Part (d) follows from this and part (a), along with the fact that all the groups A, N, K, G are unimodular (i.e., the right and left Haar measures are the same). See the discussion before Proposition 1 in Section 5.1 below. Thus

$$\int_G f(x)\, dx = \int_G f(x^{-1})\, dx = \int_N \int_A \int_K f(k^{-1} a^{-1} n^{-1}) \alpha^{-1}(a)\, dk\, da\, dn.$$

Exercise 21 (Relation Between Measures on $\mathcal{P}_n$ and AN).

(a) Show that, if we write $da = \prod da_j / a_j$ and $dn = \prod dx_{ij}$, for

$$a = \begin{pmatrix} a_1 & & & 0 \\ & a_2 & & \\ & & \ddots & \\ 0 & & & a_n \end{pmatrix} \quad \text{and} \quad n = \begin{pmatrix} 1 & x_{12} & \cdots & x_{1n} \\ 0 & 1 & \cdots & x_{2n} \\ \vdots & \vdots & \ddots & \vdots \\ 0 & 0 & \cdots & 1 \end{pmatrix},$$

then

$$\int_{\mathcal{P}_n} f(Y)\, d\mu_n(Y) = c \int_A \int_N f(I[(an)^{-1}])\, dn\, da = c \int_A \int_N f(I[an]) \alpha(a)\, dn\, da.$$

Use induction on Exercise 19(b) to see that $c = 2^n$.

(b) Another approach to this problem goes as follows. Make the change of variables from Y in $\mathcal{P}_n$ to an upper triangular matrix T via

$$Y = {}^t TT, \quad T = \begin{pmatrix} t_1 & & t_{ij} \\ & \ddots & \\ 0 & & t_n \end{pmatrix}, \quad t_i > 0, \quad t_{ij} \in \mathbb{R}.$$

Show that the Jacobian of this change of variables is:

$$\left| \frac{\partial Y}{\partial T} \right| = 2^n \prod_{j=1}^{n} t_j^{n-j+1}.$$

Later (see §4.3), we will need to consider integrals over the *boundary B of the symmetric space* $\mathscr{P}_n$, which is defined to be the compact space $B = K/M$, where M is the subgroup of K consisting of diagonal matrices with entries ± 1. Using the Iwasawa decomposition, we can identify the boundary B with G/MAN. This is done as follows:

$$\begin{aligned} G/MAN &\to K/M \\ gMAN &\mapsto K(g)M. \end{aligned} \tag{1.19}$$

Here $K(g)$ is the K-part of g in the Iwasawa decomposition $G = KAN$, as in Exercise 20. This mapping is well-defined, as is easily checked, since $xN = Nx$, for $x \in MA$. Now an *element* $g \in G$ *acts on the boundary element* kM via:

$$g(kM) = K(gk)M. \tag{1.20}$$

More information on boundaries of symmetric spaces and compactifications can be found in Chapter 5 (see also Furstenberg [2], Gérardin [1], Helgason [1, 5], Koranyi [1], Moore [1]). In the case of $SL(2, \mathbb{R})$, the symmetric space is the Poincaré upper half plane H, and the boundary can be identified with the circle, which can clearly be thought of as the compactified boundary of H.

Exercise 22 (The Jacobian of the Action of G on the Boundary B). Show that using the notation of Exercise 20 we have

$$\int_B h(b)\, db = \int_B h(g(b)) \alpha^{-1}(A(gk))\, db, \quad \text{if } b = kM \in B.$$

Hint. (See Helgason [1, pp. 50–51].) We want to use Exercise 20, and apply the integral formula for the KAN Iwasawa decomposition to a function on the boundary of the form

$$\int_A \int_N f(kan) \alpha(a)\, dn\, da.$$

The whole theory of harmonic functions on the unit disc can be extended to symmetric spaces such as $\mathscr{P}_n$ (see Chapter 5, Theorems 5–7 or Helgason [1, 5] or [4, pp. 36 and 78 for the history]). Godement [1] defines a function

4.1. Geometry and Analysis on $\mathscr{P}_n$

u which is infinitely differentiable on a symmetric space G/K to be *harmonic* if $Lu = 0$ for all G-invariant differential operators L on G/K such that L annihilates constants. Furstenberg [2] shows that, when u is bounded, it suffices for the Laplacian of u to be zero in order for u to be harmonic. The *Poisson kernel* on $G/K \times B$ can be defined by:

$$P(gK, b) = d(g^{-1}(b))/db.$$

And Furstenberg [2] proved the *Poisson integral representation* for a bounded harmonic function u on G/K:

$$u(x) = \int_{b \in B} P(x, b) \mu(b)\, db,$$

where μ denotes some bounded measurable function on the boundary B of G/K. Helgason vastly generalized this result in 1970 (see his book [5, pp. 279–280]).

Exercise 23 (Relation Between Invariant Measures on $\mathscr{P}_n$ and the Determinant One Surface $\mathscr{SP}_n$). Show that we can define an $SL(n, \mathbb{R})$-invariant measure dW on the symmetric space $\mathscr{SP}_n$ defined in (1.3) by setting, for $Y \in \mathscr{P}_n$,

$$Y = t^{1/n} W, \qquad \text{with } t > 0 \text{ and } W \in \mathscr{SP}_n,$$
$$d\mu_n(Y) = |Y|^{-(n+1)/2} \prod dy_{ij} = t^{-1}\, dt\, dW. \qquad (1.21)$$

Another useful coordinate system on $\mathscr{P}_n$ is *polar coordinates*:

$$Y = a[k], \qquad \text{for } a \in A \text{ and } k \in K, \qquad (1.22)$$

with A and K as in Exercise 20. The existence of this decomposition follows from the spectral theorem (Exercise 7). On the group level, formula (1.22) becomes $G = KAK$. Physicists often call this the *Euler angle decomposition* (see Wigner [1]) Numerical analysts call it the *singular value decomposition* (see Strang [1, p. 139]). Polar coordinates have been very useful in multivariate statistics (see James [1] and Muirhead [1]). They are also the coordinates used by Harish-Chandra and Helgason to do harmonic analysis on $\mathscr{P}_n$ (see §4.3).

Exercise 24 (The Invariant Volume Element in Polar Coordinates). Show that in polar coordinates $Y = a[k]$, for $a \in A, k \in K$, the invariant volume (1.16) is:

$$d\mu_n(Y) = c_n \prod_{j=1}^{n} a_j^{-(n-1)/2} \prod_{1 \le i < j \le n} |a_i - a_j|\, da\, dk,$$

where

$$a = \begin{pmatrix} a_1 & & 0 \\ & \ddots & \\ 0 & & a_n \end{pmatrix}, \qquad da = \prod_{j=1}^{n} da_j/a_j, \qquad \text{and}$$

$$dk = \text{Haar measure on } K = O(n),$$

normalized so that $\int_K dK = 1$. The positive constant c_n will be determined in Proposition 3 of Section 4.2.1.

Hint. Note that if $Y = a[k]$, for $k \in K$ and $a \in A$, then
$$dY = da[k] + {}^tdk\, ak + {}^tka\, dk,$$
and
$$ {}^tdk\, k + {}^tk\, dk = 0.$$
Thus
$$dY = \{da - (dk\, k^{-1})a + a(dk\, k^{-1})\}[k].$$

Now $X = dk\, k^{-1}$ is a skew-symmetric matrix. And $(aX - Xa)_{ij} = (a_i - a_j)x_{ij}$ if $X = (x_{ij})$.

Polar coordinates give a $(2^n n!)$-fold covering of $\mathscr{P}_n$, since the entries of a are the eigenvalues of $Y = a[k]$. Thus they are unique up to the action of the *Weyl group* W of permutations of the a_j, $j = 1, 2, \ldots, n$, as well as the action of the group M of orthogonal diagonal matrices. The latter are matrices which are diagonal with $+1$ or -1 as the entries.

Exercise 25 (The Arc Length in Polar Coordinates). Show that in polar coordinates (1.22), $Y = a[k]$, for $a \in A$, $k \in K$, the arc length (1.11) is:
$$ds_Y^2 = \sum_{j=1}^n a_j^{-2} da_j^2 + \text{Tr}((a^{-1} dk\, k^{-1} a - dk\, k^{-1})^2).$$
Set $dk\, k^{-1} = X =$ a skew-symmetric matrix. Then show that you can rewrite the formula as:
$$ds_Y^2 = \sum_{j=1}^n a_j^{-2} da_j^2 + 2 \sum_{1 \leq i < j \leq n} \frac{(a_i - a_j)^2}{a_i a_j} x_{ij}^2, \quad \text{for } X = (x_{ij}) = dk\, k^{-1}.$$
You might want to write out the case $n = 2$ first. Then
$$k = \begin{pmatrix} \cos\theta & \sin\theta \\ -\sin\theta & \cos\theta \end{pmatrix}, \quad dk\, k^{-1} = \begin{pmatrix} 0 & d\theta \\ -d\theta & 0 \end{pmatrix}.$$

Hint. See the hint for Exercise 24 or see Muirhead [1, p. 241]. Thus
$$Y^{-1} dY = {}^tka^{-1}(da - (dk\, k^{-1})a + a(dk\, k^{-1}))k$$
and if $X = dk\, k^{-1}$, we have
$$\text{Tr}((Y^{-1} dY)^2) = \text{Tr}((a^{-1} da - a^{-1} Xa + X)^2)$$
$$= \text{Tr}((a^{-1} da)^2 - 2\,\text{Tr}(Xa^{-1} Xa) + 2\,\text{Tr}(X^2)).$$

The formula of Exercise 24 was first proved in 1939 by three statisticians working independently (see Fisher [1], Hsu [1], and Roy [1]).

4.1. Geometry and Analysis on $\mathscr{P}_n$

It is a familiar fact from calculus that integral operators are easier to deal with than differential operators. Integrals tend to make sequences converge while derivatives tend to make sequences diverge. Thus people use integral operators to study spectral theory for differential operators—the theory of Green's functions or resolvent kernels. See Courant and Hilbert [1], Lang [1], for example. In fact, some mathematicians, perhaps motivated by the words of Hermann Weyl, in some of his early papers, have decided to throw out the differential operators altogether (cf. our remarks on p. 105 of Vol I). In particular, there is a lot of work on spherical functions that does not discuss differential equations at all. For p-adic groups, this may be a good idea, but I don't think it is so clever to forget about differential operators entirely—even if they may be ugly and/or hairy. Anyway, in this section we want to discuss the integral operators that can be used to replace the differential operators. That brings us to our next topic—convolution operators, as in Section 3.7 of Volume I. Of course, these convolution operators will be of central importance for analysis on $\mathscr{P}_n$ and on fundamental domains thereof, just as in Chapter 3, Vol. I.

In order to define convolution, we must use Exercise 5 to identify $\mathscr{P}_n$ with the homogenous space $K\backslash G$ where $G = GL(n, \mathbb{R})$ and $K = O(n)$, and then we must think of functions on $\mathscr{P}_n$ as functions on G by writing:

$$f(x) = f(I[x]), \qquad \text{for } x \in G. \tag{1.23}$$

Suppose that f and g are in $L^1(\mathscr{P}_n, d\mu_n)$ and define the *convolution* (splat) of f and g by:

$$(f * g)(a) = C_g f(a) = \int_{G=GL(n, \mathbb{R})} f(b)g(ab^{-1})\, db. \tag{1.24}$$

Here db is the right or left (they are equal) Haar measure on G as in Exercise 3. The difference between the definitions given in formula (1.24) above and in formula (3.103) in Vol. I is due to the fact that we are thinking that $\mathscr{P}_n$ has a right G-action while the Poincaré upper half plane H has a left G-action.

Lemma 1 (Properties of Convolution Operators). *Throughout this list of properties we assume that g is a right $K = O(n)$-invariant function on $G = GL(n, \mathbb{R})$ which is infinitely differentiable with compact support (to be cautious) i.e., $g \in C_c^\infty(G/K)$. We will ultimately need to generalize this, however. And g will be convolved with integrable right K-invariant functions f.*

(1) *The operator C_g defined by (1.24) commutes with the action of $c \in GL(n, \mathbb{R})$ on functions $f(a)$, $a \in G$, defined by*

$$f^c(a) = f(ac).$$

Thus we say that C_g is a G-invariant integral operator.

(2) *If $g(a) = g(a^{-1})$, for all $a \in G$ and if g is real-valued, then the convolution operator C_g is a self-adjoint operator with respect to the usual inner product on $L^2(\mathscr{P}_n, d\mu_n)$. Here $d\mu_n$ is the G-invariant volume element defined by (1.16).*

(3) $C_{g*h}f = C_g C_h f$.
(4) The operators C_g commute for functions g which are K-bi-invariant (or radial in the sense of polar coordinates (1.22)); i.e., $g(kak') = g(a)$ for all $a \in G$, $k, k' \in K$. Such functions g must be symmetric functions of the diagonal elements of $a \in A$. Thus considering g as a function of Y in $\mathscr{P}_n$, we see that $g(Y)$ must be a function of $\text{Tr}(Y^j)$, for $j = 1, 2, \ldots, n$.
(5) $C_g: L^2(\mathscr{P}_n, d\mu_n) \to C^\infty(\mathscr{P}_n)$.

Exercise 26. Prove Lemma 1 by imitating the proof of Lemma 2 of Section 3.7 in Volume I. You may have to modify some of the arguments slightly since the action of G on $\mathscr{P}_n$ is a right rather than a left action.

Selberg [1] and Maass [2] consider these convolution integral operators from a slightly different point of view. They define an *invariant integral operator* L to be given by:

$$Lf(Y) = \int_{W \in \mathscr{P}_n} k(Y, W) f(W) \, d\mu_n(W), \quad \text{for } Y, W \in \mathscr{P}_n, k \in C_c^\infty(\mathscr{P}_n \times \mathscr{P}_n)$$

(where we are now considering only integrable functions f on $\mathscr{P}_n$) provided that $L(f^a) = (Lf)^a$ for all $a \in G$. Here for $f: \mathscr{P}_n \to \mathbb{C}$ we define $f^a(Y) = f(Y[a])$, for all $Y \in \mathscr{P}_n$. Clearly L is an invariant operator if and only if the kernel k satisfies:

$$k(Y[a], W[a]) = k(Y, W), \quad \text{for all } a \in G, \quad Y, W \in \mathscr{P}_n. \quad (1.25)$$

Kernels $k: \mathscr{P}_n \times \mathscr{P}_n \to \mathbb{C}$ which satisfy (1.25) are called *point-pair invariants* by Selberg and Maass. Note that for such k

$$k(Y, W) = k(I[a], I[b]) = k(I[ab^{-1}], I) = g(ab^{-1}), \quad \text{when } a, b \in G. \quad (1.26)$$

Thus the invariant operator L is really a convolution operator. Moreover, the function g in (1.26) must be K bi-invariant, since $Y = I[k_1 a]$ and $W = I[k_2 b]$ for $k_1, k_2 \in K$ implies that

$$g(ab^{-1}) = g(k_1(ab^{-1})k_2^{-1}).$$

4.1.4. Differential Operators on $\mathscr{P}_n$

Given a Riemannian manifold, one always has a Laplacian Δ defined using (1.7) once one knows the arc length element. The goal of the chapter is the resolution of functions on $\mathscr{P}_n$ or $\mathscr{P}_n/GL(n, \mathbb{Z})$ in eigenfunctions of this Laplacian. This is the analogue of the main result of Chapter 3 in Volume I. However, life is more complicated in $\mathscr{P}_n$. There are G-invariant differential operators which are not polynomials in the Laplacian. Here a differential operator L on $\mathscr{P}_n$ is said to be *invariant* with respect to $G = GL(n, \mathbb{R})$ if L commutes with the action of G; i.e., if for $a \in G$ and $f \in C^\infty(\mathscr{P}_n)$, defining

$$f^a(Y) = f(Y[a]), \quad \text{when } Y \in \mathscr{P}_n,$$

4.1. Geometry and Analysis on $\mathscr{P}_n$

then we have $(Lf)^a = (Lf^a)$. And we define

$$D(\mathscr{P}_n) = \text{the algebra of } G\text{-invariant differential operators on } \mathscr{P}_n. \quad (1.27)$$

Proceeding as in Maass [2], we can find examples of invariant differential operators on $\mathscr{P}_n$ using the total differential of $f \in C^\infty(\mathscr{P}_n)$ defined by:

$$df = \text{Tr}\left(dY \frac{\partial}{\partial Y} f\right) = \sum_{1 \le i \le j \le n} \frac{\partial f}{\partial y_{ij}} dy_{ij}, \quad \text{writing} \quad (1.28)$$

$$dY = (dy_{ij})_{1 \le i,j \le n} \quad \text{and} \quad \frac{\partial}{\partial Y} = \left(\frac{1}{2}(1 + \delta_{ij}) \frac{\partial}{\partial y_{ij}}\right)_{1 \le i,j \le n}.$$

Here $\delta_{ij} = 1$ if $i = j$ and 0 otherwise. It follows that if $a \in G = GL(n, \mathbb{R})$ and $W = Y[a]$, then

$$df = \text{Tr}\left(dW \frac{\partial}{\partial W}\right) f = \text{Tr}\left({}^t a \, dY \, a \frac{\partial}{\partial W}\right) f = \text{Tr}\left(dY \, a \frac{\partial}{\partial W} {}^t a\right) f$$

$$= \text{Tr}\left(dY \frac{\partial}{\partial Y}\right) f.$$

The transformation formula for the matrix differential operator in (1.28) is thus:

$$\frac{\partial}{\partial W} = a^{-1} \frac{\partial}{\partial Y} {}^t a^{-1}, \quad \text{if } W = Y[a], \quad \text{for } Y \in \mathscr{P}_n \text{ and } a \in G. \quad (1.29)$$

From this formula it is easily proved that the following differential operators are G-invariant:

$$\text{Tr}\left(\left(Y \frac{\partial}{\partial Y}\right)^j\right), \quad j = 1, 2, 3, \ldots. \quad (1.30)$$

Exercise 27. Prove this last statement. Show also that the case $j = 2$ is the Laplacian on $\mathscr{P}_n$. You may want to postpone this last verification until we have discussed what happens to these differential operators when we express them in partial Iwasawa coordinates (see (1.32)).

Lemma 2.

(1) *A differential operator L in $D(\mathscr{P}_n)$ is uniquely determined by its action on K bi-invariant or radial functions $f(Y) = f(Y[k])$ for all $Y \in \mathscr{P}_n$ and $k \in K$.*
(2) *$D(\mathscr{P}_n)$ is a commutative algebra.*

PROOF.

(1) Given $g \in C^\infty(\mathscr{P}_n)$, we can construct a radial function $g^\#$ as follows:

$$g^\#(Y) = \int_{k \in K} g(Y[k]) \, dk, \quad \text{where } dk = \text{Haar measure on } K, \text{ and}$$

$$\int_{k \in K} dk = 1.$$

Then, for $L \in D(\mathcal{P}_n)$, we have $L(g^\#) = (Lg)^\#$, so that $L(g^\#)(I) = (Lg)(I)$. Suppose that $L, M \in D(\mathcal{P}_n)$ are identical on radial or K bi-invariant functions. Then for any $g \in C^\infty(\mathcal{P}_n)$ and any a in G, we have

$$Lg(a) = (Lg^a)(I) = (Mg^a)(I) = Mg(a).$$

Thus $L = M$.

(2) In this proof, which follows Selberg [1], let us use the point-pair invariant kernel notation $f(ab^{-1}) = k(a, b)$. Then, if $L \in D(\mathcal{P}_n)$, write $L_1 k$ when L acts on the first argument of k and $L_2 k$ when L acts on the second argument of k. Note that:

$$L_1 k(a, b) = L(f^{b^{-1}})(a) = (Lf)^{b^{-1}}(a).$$

Now we want to make use of the geodesic-reversing isometry of $\mathcal{P}_n$ at I which is given by $\sigma(Y) = Y^{-1}$ (see Exercise 18). So we set

$$f^\sigma(a) = f(a^{-1}) \quad \text{and} \quad k^\sigma(a, b) = f^\sigma(ab^{-1}) = k(b, a).$$

Define, for $L \in D(\mathcal{P}_n)$, the differential operator L^σ by:

$$L^\sigma f = [(L(f^{\sigma^{-1}})]^\sigma.$$

In Section 4.2, we will show that

$$L^\sigma = \overline{L^*} = \text{the complex conjugate adjoint operator.}$$

Next we need to prove the following fact.

Claim.

$$L_1 k(a, b) = L_2^\sigma k(a, b).$$

PROOF. In order to prove this claim, we need to use the fact that $\mathcal{P}_n$ is indeed a weakly symmetric space in the sense of Selberg; i.e., we need to know that for each $X, Y \in \mathcal{P}_n$, there is a matrix $g \in G$ such that

$$X[g] = Y^{-1} \quad \text{and} \quad Y[g] = X^{-1}.$$

To see this, use the fact that there is a matrix $h \in G$ such that

$$X[h] = I \quad \text{and} \quad Y[h] = D \text{ positive diagonal.}$$

If we replace g by $hg\,{}^th$, the equations we seek to solve become:

$$X[hg\,{}^th] = Y^{-1} \quad \text{and} \quad Y[hg\,{}^th] = X^{-1}.$$

This means that we need to find g such that

$$I[g] = (Y[h])^{-1} = D^{-1},$$

and

$$D[g] = (X[h])^{-1} = I.$$

4.1. Geometry and Analysis on $\mathscr{P}_n$

The solution is thus $g = D^{-1/2}$. Since D is a positive diagonal matrix, we can indeed take its square root.

Now to prove the claim, note the following sequence of equalities:

$$L_1 k(X, Y) = k'(X, Y) = k'(Y^{-1}[g], X^{-1}[g]) = k'(Y^{-1}, X^{-1})$$
$$= k'(Y, \sigma X)|_{Y \to \sigma Y} = (L_Y) k(Y, \sigma X)|_{Y \to \sigma Y}$$
$$= (L_Y)^\sigma k(Y^{-1}, X^{-1}) = (L_Y)^\sigma k(X, Y).$$

This completes the proof of the claim.

Thanks to the claim, we see that if L and M are both in $D(\mathscr{P}_n)$, we can write

$$L_1 M_1 k = L_1 M_2^\sigma k = M_2^\sigma L_1 k = M_1 L_1 k,$$

since differential operators acting on different arguments certainly commute. The proof of (2) is thus accomplished. $\square$

Theorem 2 (Structure of the Ring of G-Invariant Differential Operators on $\mathscr{P}_n$). *The differential operators*

$$\operatorname{Tr}\left(\left(Y \frac{\partial}{\partial Y}\right)^j\right), \quad j = 1, 2, \ldots, n$$

form an algebraically independent basis for the ring $D(\mathscr{P}_n)$ of $GL(n, \mathbb{R})$-invariant differential operators on $\mathscr{P}_n$. Thus $D(\mathscr{P}_n)$ can be identified with $\mathbb{C}[X_1, \ldots, X_n] = $ the ring of polynomials in n indeterminates. In particular, $D(\mathscr{P}_n)$ is a commutative ring.

PARTIAL PROOF. This is a result of Selberg [1, pp. 49–51, 57]. We follow the discussion given by Maass [2, pp. 64–67]. An invariant differential operator L on $\mathscr{P}_n$ has the form $L = L(Y, \partial/\partial Y)$. We want to show that L is a polynomial in $\operatorname{Tr}((Y \partial/\partial Y)^j), j = 1, 2, \ldots, n$. The proof involves induction on the degree of $L(Y, X)$ considered as a polynomial in the entries of the matrix X. The invariance of L implies that

$$L(Y[a], X[{}^t a^{-1}]) = L(Y, X), \quad \text{for all } a \in G.$$

So we may assume that $Y = I$ and X is diagonal with diagonal entries x_j, $j = 1, \ldots, n$. If $k \in K$, then $L(I, X) = L(I, X[k])$. Thus L must be a symmetric polynomial in $x_1, \ldots, x_n$, since we can take k to be any permutation matrix.

Then the fundamental theorem on symmetric polynomials says that L is a polynomial in $x_1^j + \cdots + x_n^j, j = 1, 2, \ldots, n$. Going back to the old variables, we find that there is a polynomial p with complex coefficients such that $L(Y, X) = p(\operatorname{Tr}(YX), \operatorname{Tr}((YX)^2), \ldots, \operatorname{Tr}((YX)^n))$. Replacing X by $\partial/\partial Y$, we have

$$L(Y, \partial/\partial Y) = p(L_1, \ldots, L_n) + M, \quad \text{where } L_j = \operatorname{Tr}((Y \partial/\partial Y)^j)$$

and $M \in D(\mathscr{P}_n)$ has lower degree than L, for the homogeneous terms of highest degree in two differential operators must commute. The induction hypothesis completes the proof that the $L_j, j = 1, 2, \ldots, n$, do indeed generate $D(\mathscr{P}_n)$.

See Maass [2, pp. 64–67] for the proof that the operators $L_j, j = 1, 2, \ldots, n$, are algebraically independent. □

Theorem 2 is rather surprising since we have found a large number of differential operators which behave like the Laplacian for the Riemannian structure on $\mathscr{P}_n$ as in (1.7). One of them is the Laplacian, of course. But there are also others, including a differential operator of degree one. This result can also be generalized to arbitrary symmetric spaces (see Chapter 5 and Helgason [2, p. 432]). The generalization requires results of Harish-Chandra on the algebra of invariant differential operators for a semi-simple Lie group as well as Chevalley's generalization of the fundamental theorem on symmetric polynomials. Chevalley's result can be found in Carter [1, Ch. 9].

The next exercise is useful is showing that the spectrum of the invariant differential operators in $D(\mathscr{P}_n)$ is the same as that of the convolution integral operators in formula (1.24).

Exercise 28. Suppose that $g: \mathscr{P}_n \to \mathbb{C}$ is infinitely differentiable with compact support. Identify it with a function on $G = GL(n, \mathbb{R})$ as in Lemma 1. Suppose that $L \in D(\mathscr{P}_n)$. If $g(a) = g(a^{-1})$ for all $a \in G$, show that

$$LC_g = C_g L \quad \text{and} \quad LC_g = C_{Lg},$$

where C_g denotes the convolution integral operator defined by (1.24).

Hint. Recall Lemma 2 of Section 3.7 in Volume I.

Now consider what happens to the Laplace operator in the various coordinate systems which have been introduced. We begin with partial Iwasawa coordinates:

$$Y = \begin{pmatrix} V & 0 \\ 0 & w \end{pmatrix} \begin{bmatrix} I & x \\ 0 & 1 \end{bmatrix}, \quad V \in \mathscr{P}_{n-1}, \quad w > 0, \quad x \in \mathbb{R}^{n-1}. \quad (1.31)$$

We know from Exercise 14 that

$$ds_Y^2 = \operatorname{Tr}((V^{-1}\,dV)^2) + (w^{-1}\,dw)^2 + 2w^{-1}V[dx])$$
$$= ds_V^2 + (w^{-1}\,dw)^2 + 2w^{-1}V[dx].$$

So the Riemannian metric tensor G_Y for $\mathscr{P}_n$ is:

$$G_Y = (g_{ij}) = \begin{pmatrix} G_V & 0 & 0 \\ 0 & 2w^{-1}V & 0 \\ 0 & 0 & w^{-2} \end{pmatrix},$$

where

$$|G_Y| = 2^{n-1}|V|^{1-n}w^{-n-1}$$

4.1. Geometry and Analysis on $\mathscr{P}_n$

and G_V is the Riemannian metric tensor for $V \in \mathscr{P}_{n-1}$, $G_V^{-1} = (g_V^{ijkl})$. Thus, by formula (1.7), we find that if $V^{-1} = (v^{ij})$, then the *Laplacian in partial Iwasawa coordinates* (1.31) is:

$$\Delta_Y = w^{(n+1)/2}\frac{\partial}{\partial w}w^2 w^{(-n-1)/2}\frac{\partial}{\partial w} + \frac{1}{2}w\sum_{i,j=1}^{n-1} v^{ij}\frac{\partial^2}{\partial x_i \partial x_j} + L_V,$$

where

$$L_V = |V|^{(n-1)/2}\sum_{\substack{i,j,k,l=1 \\ i \le j \\ k \le l}}^{n-1} \frac{\partial}{\partial v_{ij}}|V|^{(1-n)/2} g_V^{ijkl}\frac{\partial}{\partial v_{kl}}. \tag{1.32}$$

Note that L_V is not the Laplacian Δ_V on $\mathscr{P}_{n-1}$, since:

$$\Delta_V = |V|^{n/2}\sum_{\substack{i,j,k,l=1 \\ i \le j \\ k \le l}}^{n-1} \frac{\partial}{\partial v_{ij}}|V|^{-n/2} g_V^{ijkl}\frac{\partial}{\partial v_{kl}}.$$

However, we can rewrite (1.32) as:

$$\Delta_Y = \left(w\frac{\partial}{\partial w}\right)^2 + \frac{1-n}{2}w\frac{\partial}{\partial w} + \frac{1}{2}w\sum_{i,j=1}^{n-1} v^{ij}\frac{\partial^2}{\partial x_i \partial x_j} \tag{1.33}$$
$$+ \Delta_V + \frac{1}{2}\operatorname{Tr}\left(V\frac{\partial}{\partial V}\right).$$

Here $V^{-1} = (v^{ij})_{1 \le i,j \le n-1}$.

Exercise 29. Deduce formula (1.33) from formula (1.32).

Now let's consider what happens to the differential operator $\operatorname{Tr}((Y\partial/\partial Y)^2)$ in the partial Iwasawa decomposition (1.31). Let

$$Y = \begin{pmatrix} F & h \\ {}^t h & g \end{pmatrix} \quad \text{and} \quad \frac{\partial}{\partial Y} = \begin{pmatrix} \partial/\partial F & \frac{1}{2}\partial/\partial h \\ \frac{1}{2}{}^t(\partial/\partial h) & \partial/\partial g \end{pmatrix}.$$

Then $F = V$, $g = w + V[x]$, $h = Vx$, and $dF = dV$, $dg = dw + dV[x] + {}^t dx\, Vx + {}^t x V\, dx$, $dh = dV \cdot x + V\, dx$. Substitute this into the total differential (1.28)

$$\operatorname{Tr}\left(dY\frac{\partial}{\partial Y}\right) = \operatorname{Tr}\left(dF\frac{\partial}{\partial F}\right) + dg\frac{\partial}{\partial g} + {}^t dh\frac{\partial}{\partial h},$$

and compare the result with

$$\operatorname{Tr}\left(dY\frac{\partial}{\partial Y}\right) = \operatorname{Tr}\left(dV\frac{\partial}{\partial V}\right) + dw\frac{\partial}{\partial w} + {}^t dx\frac{\partial}{\partial x}.$$

This leads to the following formulas:

$$\frac{\partial}{\partial V} = \frac{\partial}{\partial F} + x\frac{\partial}{\partial g}{}^t x + \frac{1}{2}\left(x\frac{{}^t\partial}{\partial h} + {}^t\left(x\frac{{}^t\partial}{\partial h}\right)\right),$$

$$\frac{\partial}{\partial w} = \frac{\partial}{\partial g}, \quad \frac{\partial}{\partial x} = 2Vx\frac{\partial}{\partial g} + V\frac{\partial}{\partial h}.$$

It follows that

$$\frac{\partial}{\partial g} = \frac{\partial}{\partial w}, \quad \frac{\partial}{\partial h} = V^{-1}\frac{\partial}{\partial x} - 2x\frac{\partial}{\partial w},$$

$$\frac{\partial}{\partial F} = \frac{\partial}{\partial V} + x\frac{\partial}{\partial w}{}^t x - \frac{1}{2}\left(x\frac{{}^t\partial}{\partial x}V^{-1} + V^{-1}\,{}^t\!\left(x\frac{{}^t\partial}{\partial x}\right)\right).$$

The preceding calculation is a little tricky since $\partial/\partial F$ must be symmetric. So you must put in

$$\frac{1}{2}\left(x\frac{{}^t\partial}{\partial x}V^{-1} + V^{-1}\,{}^t\!\left(x\frac{{}^t\partial}{\partial x}\right)\right) \text{ and not just } x\frac{{}^t\partial}{\partial x}V^{-1}.$$

In the term ${}^t(x\,{}^t\partial/\partial x)$, the order is to differentiate first, then multiply.

Our calculation implies that

$$Y\frac{\partial}{\partial Y} = \begin{pmatrix} V\dfrac{\partial}{\partial V} - \dfrac{1}{2}{}^t\!\left(x\dfrac{{}^t\partial}{\partial x}\right) & \dfrac{1}{2}\dfrac{\partial}{\partial x} \\ {}^t\!xV\dfrac{\partial}{\partial V} - w\dfrac{\partial}{\partial w}{}^t x - \dfrac{1}{2}{}^t x\,{}^t\!\left(x\dfrac{{}^t\partial}{\partial x}\right) + \dfrac{1}{2}w\dfrac{{}^t\partial}{\partial x}V^{-1} & \dfrac{1}{2}{}^t x\dfrac{\partial}{\partial x} + w\dfrac{\partial}{\partial w} \end{pmatrix}.$$

If we square this matrix operator and take the trace, we find via mathematical induction that $\operatorname{Tr}((Y\,\partial/\partial Y)^2)$ has the same partial Iwasawa decomposition as that of Δ_Y which was given in (1.33). Therefore the two operators are indeed the same.

Next consider the Laplacian in polar coordinates (1.22). We saw in Exercises 24 and 25 that if $X = -{}^t X = dk\,k^{-1},\ k \in K$, then

$$ds_Y^2 = \sum_{j=1}^n (da_j/a_j)^2 + 2\sum_{1 \le i < j \le n} \frac{(a_i - a_j)^2}{a_i a_j} x_{ij}^2.$$

So the Riemannian metric tensor in (1.7) becomes:

$$G = \begin{pmatrix} a_1^{-2} & & & & & & \\ & \ddots & & & & & \\ & & a_n^{-2} & & & & 0 \\ & & & 2(a_1-a_2)^2/(a_1 a_2) & & & \\ & & & & \ddots & & \\ & & & & & 2(a_1-a_n)^2/(a_1 a_n) & \\ & 0 & & & & & \ddots \\ & & & & & & & 2(a_{n-1}-a_n)^2/(a_{n-1}a_n) \end{pmatrix}$$

4.1. Geometry and Analysis on $\mathscr{P}_n$

with $|G| = 2^{n(n-1)/2} \prod_{j=1}^n a_j^{-(n+1)} \prod_{1 \le i < j \le n}(a_i - a_j)^2$. By formula (1.7), if $a_1 > a_2 > \cdots > a_n$, then the Laplacian in polar coordinates is:

$$\Delta = \prod_{j=1}^n a_j^{(n+1)/2} \prod_{1 \le i < j \le n}(a_i - a_j^{-1}) \sum_{k=1}^n \frac{\partial}{\partial a_k} a_k^2 \prod_{j=1}^n a_j^{-(n+1)/2} \prod_{1 \le i < j \le n}(a_i - a_j)\frac{\partial}{\partial a_k}$$

$$+ \sum_{1 \le i < j \le n} \frac{a_i a_j}{2(a_i - a_j)^2} \frac{\partial^2}{\partial x_{ij}^2}.$$

Clearly we can rewrite this as:

$$\Delta = \sum_{k=1}^n \left\{ a_k^2 \frac{\partial^2}{\partial a_k^2} + \left(\sum_{\substack{j=1 \\ j \ne k}}^n a_k^2 \frac{1}{a_k - a_j} - \frac{n-3}{2} a_k\right) \frac{\partial}{\partial a_k} \right\}$$

$$+ \frac{1}{2} \sum_{1 \le i < j \le n} \frac{a_i a_j}{(a_i - a_j)^2} \frac{\partial^2}{\partial x_{ij}^2}.$$

Therefore (cf. Muirhead [1, p. 242])

$$\Delta = \sum_{k=1}^n a_k^2 \frac{\partial^2}{\partial a_k^2} - \frac{n-3}{2} \sum_{k=1}^n a_k \frac{\partial}{\partial a_k} + \frac{1}{2} \sum_{1 \le i < j \le n} \frac{a_i a_j}{(a_i - a_j)^2} \frac{\partial^2}{\partial x_{ij}^2}$$
$$+ \sum_{k=1}^n \left(\sum_{\substack{j=1 \\ j \ne k}}^n a_k^2 \frac{1}{a_k - a_j}\right) \frac{\partial}{\partial a_k}. \tag{1.34}$$

4.1.5. A List of the Main Formulas Derived in Section 1

Now we can finally summarize our results. This will be convenient for future reference.

First *the results of changing to partial Iwasawa coordinates* are:

$$Y = \begin{pmatrix} V & 0 \\ 0 & w \end{pmatrix} \begin{bmatrix} I & x \\ 0 & 1 \end{bmatrix}, \quad \text{for } V \in \mathscr{P}_{n-1}, w > 0, \text{ and } x \in \mathbb{R}^{n-1},$$

$$ds_Y^2 = \text{Tr}((Y^{-1} dY)^2) = ds_V^2 + (w^{-1} dw)^2 + 2w^{-1} V[dx],$$

$$d\mu_n(Y) = |Y|^{-(n+1)/2} \prod_{1 \le i \le j \le n} dy_{ij} = w^{(1-n)/2} |V|^{1/2} d\mu_{n-1}(V) \frac{dw}{w} dx, \tag{1.35}$$

$$\Delta_Y = \text{Tr}((Y \partial/\partial Y)^2)$$

$$= \Delta_V + \frac{1}{2} \text{Tr}(V \partial/\partial V) + \left(w \frac{\partial}{\partial w}\right)^2 - \frac{n-1}{2} w \frac{\partial}{\partial w} + \frac{1}{2} w \frac{{}^t\partial}{\partial x} V^{-1} \frac{\partial}{\partial x}.$$

For discussions of these formulas, see Exercises 14, 19, and 29.

Then *the results of changing to polar coordinates* are:

$$Y = a[k], \quad a \in A, k \in K,$$

$$ds_Y^2 = \sum_{j=1}^n (a_j^{-1} da_j)^2 + 2 \sum_{1 \le i < j \le n} \frac{(a_i - a_j)^2}{a_i a_j} x_{ij}^2, \quad \text{for } X = dk\, k^{-1} = (x_{ij}),$$

$$d\mu_n(Y) = c_n \prod_{j=1}^n a_j^{-(n+1)/2} \prod_{1 \le i < j \le n} |a_i - a_j|\, da\, dk,$$

$$da = \prod_{j=1}^n da_j/a_j, \quad \int_K dk = 1, \quad dk = \text{Haar measure},$$

$$\Delta = \sum_{i=1}^n a_i^2 \frac{\partial^2}{\partial a_i^2} - \frac{n-3}{2} \sum_{i=1}^n a_i \frac{\partial}{\partial a_i} + \frac{1}{2} \sum_{1 \le i < j \le n} \frac{a_i a_j}{(a_i - a_j)^2} \frac{\partial^2}{\partial x_{ij}^2}$$

$$+ \sum_{k=1}^n \left(\sum_{\substack{j=1 \\ j \ne k}}^n a_k^2 \frac{1}{a_k - a_j} \right) \frac{\partial}{\partial a_k}.$$

(1.36)

Here $K = O(n)$ and A is the multiplicative group of positive $n \times n$ diagonal matrices. For discussions of these results, see Exercises 24 and 25 as well as the discussion before formula (1.34). The positive constant c_n will be determined in Section 4.2.1, where we will show that it is given by (1.43) below.

Next we list a few integral formulas. For $x \in G = GL(n, \mathbb{R})$, let dx denote a Haar measure (as in Exercise 3). For $k \in K = O(n)$, let dk be a Haar measure chosen so that the volume of K is one.

Let N be the nilpotent group of matrices n of the form:

$$n = \begin{pmatrix} 1 & x_{12} & \cdots & x_{1n} \\ & 1 & \cdots & x_{2n} \\ & & \ddots & \vdots \\ 0 & & & 1 \end{pmatrix},$$

with Haar measure $dn = \prod dx_{ij}$, dx_{ij} = Lebesgue measure on $\mathbb{R}$ (see Exercise 2). Let A be the abelian multiplicative group of matrices a of the form

$$a = \begin{pmatrix} a_1 & & & 0 \\ & a_2 & & \\ & & \ddots & \\ 0 & & & a_n \end{pmatrix},$$

with Haar measure $da = \prod da_i/a_i$. Let T_n be the multiplicative group of matrices of the form

$$t = \begin{pmatrix} t_{11} & \cdots & t_{1n} \\ & \ddots & \vdots \\ 0 & & t_{nn} \end{pmatrix},$$

with measure $dt = \prod dt_{ij}$, dt_{ij} = Lebesgue measure on $\mathbb{R}$. Let dW denote an $SL(n, \mathbb{R})$-invariant measure on $\mathcal{SP}_n$, the determinant one surface in $\mathcal{P}_n$ defined in (1.3). And set $B = K/M$, the boundary of $\mathcal{P}_n$, with K-invariant volume db. For $b = kM$ in B and $g \in G$, define $A(g(b))$ to be the A-part of gk in its

4.1. Geometry and Analysis on $\mathscr{P}_n$

KAN-Iwasawa decomposition as in Exercise 20. Define

$$\alpha(a) = \prod_{i=1}^{n} a_i^{n-2i+1}, \quad \text{for } a \in A, \text{ as above;}$$

$$\beta(t) = \prod_{i=1}^{n} t_{ii}^{-i}, \quad \text{for } t \in T, \text{ as above;} \tag{1.37}$$

$$\gamma(a) = \prod_{i=1}^{n} a_i^{-(n-1)/2} \prod_{1 \leq i < j \leq n} |a_i - a_j|, \quad \text{for } a \in A, \text{ as above.}$$

Then we have the following integral formulas (using the notation in (1.37)):

$$\int_G f(x)\,dx = \int_A \int_N \int_K f(ank)\,dk\,dn\,da = \int_K \int_A \int_N f(kan)\alpha(a)\,dn\,da\,dk; \tag{1.38}$$

$$\int_{\mathscr{P}_n} f(Y)\,d\mu_n(Y) = 2^n \int_A \int_N f(I[(an)^{-1}])\,dn\,da$$

$$= 2^n \int_A \int_N f(I[an])\alpha(a)\,dn\,da \tag{1.39}$$

$$= 2^n \int_{T_n} f(I[t])\beta(t)\,dt;$$

$$\int_{\mathscr{P}_n} f(Y)\,d\mu_n(Y) = \int_{t>0} \int_{W \in \mathscr{SP}_n} f(t^{1/n}W)\,dW\,t^{-1}\,dt; \tag{1.40}$$

$$\int_B h(b)\,db = \int_B h(g(b))\alpha^{-1}(A(g(b)))\,db; \tag{1.41}$$

$$\int_{\mathscr{P}_n} f(Y)\,d\mu_n(Y) = c_n \int_K \int_A f(a[k])\gamma(a)\,da\,dk. \tag{1.42}$$

Formulas (1.38)–(1.42) are proved in Exercises 20–24. The positive constant c_n in formula (1.42) will be determined in Section 4.2.1, where we will show that:

$$c_n^{-1} = \pi^{-(n^2+n)/4} \prod_{j=1}^{n} j\Gamma(j/2). \tag{1.43}$$

Exercise 30. Show that right and left Haar measures are different for the multiplicative group $T = AN$ of upper triangular matrices with positive diagonal entries. The group T is called a "solvable" Lie group.

Exercise 31. Is the Laplacian Δ a negative operator on the square integrable functions $f \in L^2(\mathscr{P}_n)$ such that $\Delta f \in L^2(\mathscr{P}_n)$?

Exercise 32 (Grenier [1]). When Y lies in the determinant one surface $\mathscr{SP}_{n+1}$, write:

$$Y = \begin{pmatrix} v & 0 \\ 0 & v^{-1/n}W \end{pmatrix} \begin{bmatrix} 1 & {}^tx \\ 0 & I_n \end{bmatrix}, \quad v > 0, \quad x \in \mathbb{R}^n, \quad W \in \mathscr{SP}_n.$$

Show that if ds^2 is the arc length $\text{Tr}((Y^{-1}\,dY)^2)$ on the determinant one surface, $d\mu_{n+1}$ is the G-invariant volume, and Δ_{n+1} is the corresponding Laplacian, we have the following expressions relating these quantities for rank $n+1$ and those for the rank n case:

$$ds_Y^2 = \frac{n+1}{n} v^{-2}\,dv^2 + 2v^{(n+1)/n} W^{-1}[dx] + ds_W^2,$$

$$d\mu_{n+1}(Y) = v^{(n-1)/2}\,dv\,dx\,d\mu_n(W),$$

$$\Delta_{n+1} = \frac{n}{n+1}\left\{v^2 \frac{\partial}{\partial v^2} + \frac{n+3}{2} v \frac{\partial}{\partial v}\right\} + \frac{1}{2} v^{-(n+1)/n} W\left[\frac{\partial}{\partial x}\right] + \Delta_n.$$

4.1.6. An Application to Multivariate Statistics

References for this application are Anderson [1], Morrison [1], and Muirhead [2].

A random variable X in $\mathbb{R}^n$ is *normal with mean* $\mu \in \mathbb{R}^n$ *and covariance* Σ in $\mathscr{P}_n$, which is written $N(\mu, \Sigma)$ if it has the probability density:

$$(2\pi)^{-n/2} |\Sigma|^{-1/2} \exp\{-\tfrac{1}{2}\Sigma^{-1}[x-\mu]\}$$

(see Anderson [1, p. 17]). It follows that if X is a normal random variable in $\mathbb{R}^n$ distributed according to $N(\mu, \Sigma)$ and if $A \in \mathbb{R}^{m \times n}$, with $m \leq n$, then $Y = AX$ is normal and distributed according to $N(A\mu, \Sigma[{}^tA])$.

Notions of partial and multiple correlation are quite important in the analysis of data. We can use the Iwasawa decomposition to aid in this analysis. Such results go back to Pearson (1896) and Yule (1897) (see Anderson [1, pp. 27–28]). If we partition the random variable ${}^tX = ({}^tX_1, {}^tX_2)$, mean ${}^t\mu = ({}^t\mu_1, {}^t\mu_2)$, and covariance

$$\Sigma = \begin{pmatrix} \Sigma_{11} & \Sigma_{12} \\ {}^t\Sigma_{12} & \Sigma_{22} \end{pmatrix}, \quad \text{with } X_1 \in \mathbb{R}^p,\ X_2 \in \mathbb{R}^q,\ n=p+q,\ \Sigma_{11} \in \mathscr{P}_p,$$

then (using the result at the end of the last paragraph), we can see that X_i is normal and distributed according to $N(\mu_i, \Sigma_{ii})$, for $i = 1, 2$. The *conditional distribution* of X_1 holding $X_2 = x_2$ constant is *normally distributed* according to $N(\mu_1 + H(x_2 - \mu_2), V)$, where H and V are defined by the Iwasawa decomposition:

$$\Sigma = \begin{pmatrix} V & 0 \\ 0 & W \end{pmatrix} \begin{bmatrix} I & 0 \\ {}^tH & I \end{bmatrix}, \quad \text{for } V \in \mathscr{P}_p,\ W \in \mathscr{P}_q,\ n = p + q.$$

To see this, note that the conditional distribution of X_1 holding $X_2 = x_2$ constant is:

$$g(x_1|x_2) = \frac{(2\pi)^{-n/2} |\Sigma|^{-1/2} \exp\{-\tfrac{1}{2}\Sigma^{-1}[x-\mu]\}}{(2\pi)^{-q/2} |W|^{-1/2} \exp\{-\tfrac{1}{2}W^{-1}[x_2-\mu_2]\}}.$$

4.1. Geometry and Analysis on $\mathscr{P}_n$

But we know that
$$\Sigma^{-1}[x] = V^{-1}[x_1 - Hx_2] + W^{-1}[x_2].$$
Therefore
$$g(x_1|x_2) = (2\pi)^{-p/2}|V|^{-1/2}\exp\{-\tfrac{1}{2}V^{-1}[x_1 - \mu_1 - H(x_2 - \mu_2)]\}.$$
Note that $V = \Sigma_{11} - \Sigma_{22}^{-1}['\Sigma_{12}]$ and $H = \Sigma_{12}\Sigma_{22}^{-1}$. The matrix H is called the *matrix of regression coefficients* of X_1 on x_2. The entries of V are called the *partial covariances*. The *partial correlation* between the ith entry of X_1 and the jth entry of X_1, holding $X_2 = x_2$ fixed is:
$$\rho_{ij} = \frac{v_{ij}}{\sqrt{v_{ii}v_{jj}}}.$$

A fundamental problem in statistics is the estimation of the mean and the covariance, after making N sample observations of the random variable X. Suppose X is distributed according to $N(\mu, \Sigma)$ and we have N observations $x_1, \ldots, x_N$, $N > n$. The *likelihood function* is
$$L = (2\pi)^{-nN/2}|\Sigma|^{-N/2}\exp\left\{-\frac{1}{2}\sum_{i=1}^{N}\Sigma^{-1}[x_i - \mu]\right\}.$$

Maximizing L over Σ, μ gives *the maximum likelihood estimates* for Σ, μ and which are:
$$\hat{\mu} = \bar{x} = \frac{1}{N}\sum_{k=1}^{N}x_k, \qquad \hat{\Sigma} = \frac{1}{N}\sum_{k=1}^{N}I['(x_i - \bar{x})]$$

(see Anderson [1, pp. 44–48]). Correlation coefficients can be estimated from $\hat{\Sigma}$.

Now we want to consider an example discussed in 1907 by Hooker (see Anderson [1, p. 82]). Suppose that X_1 represents hay yield in hundredweights per acre, X_2 represents spring rainfall in inches, X_3 represents accumulated spring temperature over $42°F$, for a certain English region, measured over a period of 20 years. One looks at the data and uses maximum likelihood estimates for the mean, covariance, and correlation coefficients. The result of these calculations is:

$$\hat{\mu} = \bar{x} = \begin{pmatrix} 28.02 \\ 4.91 \\ 594.00 \end{pmatrix}, \quad \begin{pmatrix} \hat{\sigma}_1 \\ \hat{\sigma}_2 \\ \hat{\sigma}_3 \end{pmatrix} = \begin{pmatrix} 4.42 \\ 1.10 \\ 85.00 \end{pmatrix}, \quad \hat{\sigma}_i^2 = \hat{\sigma}_{ii}, \quad \hat{\rho}_{ij} = \hat{\sigma}_{ij}/(\hat{\sigma}_i\hat{\sigma}_j),$$

$$\begin{pmatrix} 1 & \hat{\rho}_{12} & \hat{\rho}_{13} \\ \hat{\rho}_{21} & 1 & \hat{\rho}_{23} \\ \hat{\rho}_{31} & \hat{\rho}_{32} & 1 \end{pmatrix} = \begin{pmatrix} 1.00 & 0.80 & -0.40 \\ 0.80 & 1.00 & -0.56 \\ -0.40 & -0.56 & 1.00 \end{pmatrix}.$$

We can then ask: Is high temperature correlated with low yield or is high temperature correlated with low rainfall and thus with low yield? To answer this question, one estimates the partial correlation between X_1 and X_3 while holding $X_2 = $ constant. The Iwasawa decomposition method discussed above then leads to the result that this correlation is:

$$\frac{\hat{\sigma}_1 \hat{\sigma}_3 (\hat{\rho}_{13} - \hat{\rho}_{12}\hat{\rho}_{23})}{\sqrt{\hat{\sigma}_1^2(1 - \hat{\rho}_{12}^2)\hat{\sigma}_3^2(1 - \hat{\rho}_{23}^2)}} = 0.0967.$$

Thus, if the effect of the rainfall is removed, yield and temperature are positively correlated. So both high temperature and high rainfall increase yield, but usually high rainfall occurs with low temperature.

4.2. Special Functions on $\mathscr{P}_n$

Attempts have been made to generalize hypergeometric functions to the case of several variables, based on the construction of a many-dimensional analogue to the hypergeometric series [P. Appell and J. Kampé de Fériet, Fonctions Hypergéométriques et Hypersphériques, Polynomes d'Hermite, Gauthier-Villars, Paris, 1926]. However, this approach leads to functions which, in the opinion of the author, do not sufficiently fully reflect the multi-dimensionality of the domain. The present article is concerned with another approach to the theory of special functions for several variables. Special functions of a single variable can be expressed, as we know, in terms of elementary functions, viz., the power and exponential functions, by use of simple integral representations. It is precisely these integral representations that are taken as the pattern for the definition of the many-dimensional analogues of special functions.

From Gindikin [1, p. 1].

4.2.1. Power and Gamma Functions

This section concerns the matrix argument analogues of functions we encountered in Volume I—gamma, K-Bessel, and spherical functions. The approach is similar to that of Gindikin [1] quoted above—an approach that emphasizes integral representations for the functions. The main references for this section are Bengtson [1], Bump [1], Gindikin [1], Helgason [1-7], James [1-3], Maass [2], Muirhead [1], and Selberg [1]. Of course, our chief concern will always be eigenfunctions of the invariant differential operators $D(\mathscr{P}_n)$ studied in Theorem 2 of the preceding section.

The most basic special function on $\mathscr{P}_n$ is a generalization of the power function y^s, $y \in \mathscr{P}_1 = \mathbb{R}^+$, $s \in \mathbb{C}$, appearing in the Mellin transform of Vol. I, Section 1.4. The *power function* $p_s(y)$, for $Y \in \mathscr{P}_n$, and $s = (s_1, \ldots, s_n) \in \mathbb{C}^n$, is defined by:

$$p_s(Y) = \prod_{j=1}^n |Y_j|^{s_j}, \tag{2.1}$$

where $Y_j \in \mathscr{P}_j$ is the $j \times j$ upper left hand corner in Y, $j = 1, 2, \ldots, n$. Note that

4.2. Special Functions on $\mathscr{P}_n$

when $n = 2$ and $Y \in \mathscr{SP}_2$, the power function can be identified with the function on the upper half plane defined by $p_s(x + iy) = y^s$ for $y > 0$ (see Exercise 9 of Section 3.1, Volume I).

The power functions (2.1) were introduced by Selberg [1, pp. 57–58]. Tamagawa [1, p. 369] calls them right spherical functions. In the language of Harish-Chandra and Helgason [1, p. 52], the power functions are:

$$\exp[\lambda(H(gk))], \quad \text{for } g \in G, k \in K,$$

where $H(x) = \log A(x)$, if $x = K(x)A(x)N(x)$ is the Iwasawa decomposition of $x \in G$ (see Exercise 20 of Section 4.1.3). So H maps the group A of positive diagonal matrices into arbitrary diagonal matrices; i.e., into $\mathbb{R}^n$. Now λ is a linear functional on $\mathbb{R}^n$, which can be identified with an n-tuple of complex numbers. Thus the composition of $\exp$, λ, and the H-function does indeed become a power function.

It is also possible to view the power function $p_s(Y)$ as a homomorphism of the *group T_n of upper triangular matrices with positive diagonal entries*:

$$t = \begin{pmatrix} t_{11} & t_{12} & \cdots & t_{1n} \\ 0 & t_{22} & \cdots & t_{2n} \\ \vdots & & \ddots & \vdots \\ 0 & 0 & \cdots & t_{nn} \end{pmatrix}, \quad t_{jj} > 0. \tag{2.2}$$

If $r = (r_1, \ldots, r_n) \in \mathbb{C}^n$ and $t \in T_n$, define the *homomorphism* $\tau_r : T_n \to \mathbb{C} - 0$ by:

$$\tau_r(t) = \prod_{j=1}^n t_{jj}^{r_j}. \tag{2.3}$$

Clearly $\tau_r(t_1 t_2) = \tau_r(t_1)\tau_r(t_2)$. So you can think of τ_r as a homomorphism or, more significantly, as a degree one *representation* of T_n. The following proposition relates (2.1) and (2.3).

Proposition 1 (Properties of the Power Function).

(1) Relation of p_s and τ_r.
   Suppose that $Y = I[t]$ for $t \in T_n$ as in (2.2). Then $p_s(Y) = \tau_r(t)$, if $r_j = 2(s_j + \cdots + s_n)$ and upon setting $r_{n+1} = 0$, we have $s_j = (r_j - r_{j+1})/2$.
(2) Action of T_n on p_s.
   If $Y \in \mathscr{P}_n$ and $t \in T_n$, then $p_s(Y[t]) = p_s(Y)p_s(I[t])$.
(3) Power Functions Are Eigenfunctions of Invariant Differential Operators.
   If $L \in D(\mathscr{P}_n)$, then $Lp_s = \lambda_L(s)p_s$; i.e., p_s is an eigenfunction of L with eigenvalue $\lambda_L(s) = Lp_s(I)$.
(4) A Symmetry.
   Set $s = (s_1, \ldots, s_n)$, $s^* = (s_{n-1}, \ldots, s_2, s_1, -(s_1 + \cdots + s_n))$, and

$$\omega = \begin{pmatrix} 0 & & 1 \\ & \reflectbox{$\ddots$} & \\ 1 & & 0 \end{pmatrix}.$$

Then $p_s(Y^{-1}[\omega]) = p_{s*}(Y)$, for all $Y \in \mathscr{P}_n$. Also $\omega^2 = I$ and $s^{**} = s$.

PROOF.

(1) Note that $|Y_j| = t_1^2 \cdots t_j^2$. The result follows easily.
(2) This follows from part (1) and the fact that τ_r is a homomorphism.
(3) Set $W = Y[t]$ for $Y \in \mathscr{P}_n$, $t \in T_n$. Then if $L \in D(\mathscr{P}_n)$, write L_W for L acting on the W-variable. We have the following equalities, if we make use of the G-invariance of L as well as part (2):

$$L_W p_s(W) = L_Y p_s(Y[t]) = p_s(I[t]) L_Y p_s(Y).$$

Then set $Y = I$ to obtain $W = I[t]$ and $Lp_s = \lambda_L(s) p_s$, with the eigenvalue stated in the proposition.
(4) See Exercise 1. □

Note. The formula for the eigenvalue of the invariant differential operator L acting on the power function p_s is not a very useful one. In the case of the Poincaré upper half plane we had $\Delta y^s = s(s-1)y^s$ and it was clear that we could find powers $s \in \mathbb{C}$ to match any eigenvalue $\lambda = s(s-1)$, using high school algebra. To generalize this, we need a better formula for the eigenvalues $\lambda_L(s)$ in part (3). We will do better soon, with a fair amount of work, which was done for us by Maass [2].

Exercise 1. Show that if $t \in T_n$ and ω is as defined in part (4) of the preceding proposition, then

$$t[\omega] = \begin{pmatrix} t_{nn} & & & & 0 \\ & t_{n-1\,n-1} & & & \\ & & \ddots & & \\ & & & t_{22} & \\ & * & & & t_{11} \end{pmatrix}.$$

Use this result to prove part (4) of the preceding proposition.

The power function is the appropriate kernel for the $\mathscr{P}_n$ analogue of the Mellin transform which we will call the Helgason-Fourier transform. We will see in Section 4.3 that the Helgason-Fourier transform does indeed have many of the properties of the usual Fourier and Mellin transforms. For example, the $\mathscr{P}_n$ Fourier transform does have an inversion formula (if one also includes a variable from $K = O(n)$ or the boundary K/M). It is also possible to consider an analogue of the Laplace transform on $\mathscr{P}_n$, a transform with a more elementary inversion formula, whose proof requires only ordinary Euclidean Fourier transforms and Cauchy's theorem in one variable as is seen in the next exercise.

Exercise 2 (The Laplace Transform). Define the *Laplace transform* of $f: \mathscr{P}_n \to \mathbb{C}$ at the symmetric matrix $Z \in \mathbb{C}^{n \times n}$ by:

4.2. Special Functions on $\mathscr{P}_n$

$$\mathscr{L}f(Z) = \int_{Y \in \mathscr{P}_n} f(Y) \exp[-\text{Tr}(YZ)] \, dY, \quad \text{where } dY = \prod_{1 \le i \le j \le n} dy_{ij}.$$

For sufficiently nice functions f, the integral above converges in a right half plane, $\text{Re } Z > X_0$, meaning that $\text{Re } Z - X_0 \in \mathscr{P}_n$. Show that the inversion formula for this transform is:

$$(2\pi i)^{-n(n+1)/2} \int_{\text{Re } Z = X_0} \mathscr{L}f(Z) \exp[\text{Tr}(YZ)] \, dZ = \begin{cases} f(Y), & \text{for } Y \in \mathscr{P}_n, \\ 0, & \text{otherwise.} \end{cases}$$

Here $dZ = \prod dz_{ij}$. And the integral is over symmetric matrices Z with fixed real part (in the domain of absolute convergence).

Hints. This result is discussed by Bochner [1, pp. 686–702], Bochner and Martin [90–92, 113–132], Herz [1, pp. 479–480], and Muirhead [1, p. 252]. You can use inversion of the Euclidean Fourier transform on the space of symmetric $n \times n$ real matrices and imitate the proof that worked for $n = 1$ (see Exercise 18 of Vol. I, Section 1.2).

The most basic example of a Helgason-Fourier transform (or of a Laplace transform) on $\mathscr{P}_n$ is the *gamma function* for $\mathscr{P}_n$ defined by:

$$\Gamma_n(s) = \int_{Y \in \mathscr{P}_n} p_s(Y) \exp[-\text{Tr}(Y)] \, d\mu_n(Y), \tag{2.4}$$

for $s \in \mathbb{C}^n$ with $\text{Re } s_j$ sufficiently large; that is,

$$\text{Re}(s_j + \cdots + s_n) > (j-1)/2, \quad j = 1, \ldots, n.$$

In fact, we can use the Iwasawa decomposition to write $\Gamma_n(s)$ as a product of ordinary gamma functions $\Gamma_1 = \Gamma$:

$$\Gamma_n(s) = \pi^{n(n-1)/4} \prod_{j=1}^n \Gamma\left(s_j + \cdots + s_n - \frac{j-1}{2}\right). \tag{2.5}$$

Exercise 3. Prove formula (2.5) by making the change of variables $Y = I[t]$ for $t \in T_n$ defined by (2.2), using formulas (1.37) and (1.39) from Section 4.1.5 to get

$$\Gamma_n(s) = 2^n \int_{T_n} \exp\{-\text{Tr}(I[t])\} \tau_r(t) \prod_{j=1}^n t_{jj}^{-j} \prod_{1 \le i \le j \le n} dt_{ij}.$$

Exercise 4. Show that

$$\int_{Y \in \mathscr{P}_n} p_s(Y) \exp\{-\text{Tr}(YX^{-1})\} \, d\mu_n(Y) = p_s(X) \Gamma_n(s).$$

Exercise 4 will be useful in the study of the algebra $D(\mathscr{P}_n)$ of invariant differential operators on $\mathscr{P}_n$—a study which was begun in §4.1.4. It will also be necessary when we consider analogues of Hecke's correspondence between modular forms and Dirichlet series in later sections of this chapter and the next.

A special case of the product formula (2.5) for Γ_n was found in 1928 by the statistician Wishart [1]. A more general result is due to Ingham [1]. Later Siegel needed this special case of (2.5) in his work on quadratic forms (see [1, Vol. I, pp. 326–405]). Such gamma functions for $\mathcal{P}_n$ and more general domains of positivity are considered by Gindikin [1]. More general integrals of this type appear in quantum statistical mechanics (see Mehta [1, p. 40]) where a conjecture is given for the value of

$$\int_{\mathbb{R}^n} \exp\{-k\,{}^t xx\} \prod_{i<j} |x_i - x_j|^s \, dx. \tag{2.6}$$

Selberg [2] had already proved the conjecture 23 years earlier. Macdonald and Dyson have generalized the conjecture to arbitrary groups. Regev [1] has used a result of Bechner to prove Macdonald's conjecture for the main types of simple Lie groups. See Macdonald [1]. There is a special case of Selberg's formula in Exercise 5.

Next we compute the constant in the integral formula for polar coordinates (see formulas (1.42) and (1.43) from Section 4.1.5).

Proposition 2 (Volume of $O(n)$). *Let $dk\, k^{-1} = (dh_{ij}(k))_{1 \le i < j \le n}$ and set $dh(k) = \prod_{1 \le i < j \le n} dh_{ij}(k)$. Then*

$$\mathrm{Vol}(O(n)) = \int_{k \in K = O(n)} dh(k) = 2^n \pi^{n^2/2} \Gamma_n(0, \ldots, 0, n/2)^{-1}.$$

PROOF. (From Muirhead [1, pp. 63–71].) First note that

$$\int_{X \in \mathbb{R}^{n \times n}} \exp[-\mathrm{Tr}(X^t X)] \, dX = \pi^{n^2/2}.$$

Now change variables via $X = kt$, for $k \in K = O(n)$ and $t \in T_n$ (the upper triangular matrices with positive diagonal entries). This is possible by the Gram-Schmidt orthogonalization process. Then

$$k^{-1} \, dX = dt + (k^{-1} \, dk)t,$$

and one finds the Jacobian of this change of variables to be:

$$\prod_{i=1}^{n} t_{ii}^{n-i}.$$

If we use the formula for changing variables from $Y \in \mathcal{P}_n$ to $t \in T_n$ via $Y = I[t]$ in Exercise 3, then we obtain

$$\pi^{n^2/2} = \mathrm{Vol}(K) \int_{T_n} \exp[-\mathrm{Tr}(I[t])] \prod_{i=1}^{n} t_{ii}^{n-i} \, dt = \mathrm{Vol}(K) 2^{-n} \Gamma_n\!\left(0, \ldots, 0, \frac{n}{2}\right).$$

This completes the proof. □

4.2. Special Functions on $\mathscr{P}_n$

Note that the measure $dh(k)$ in Proposition 2 is a Haar measure for K. However, in the integral formula for polar coordinates ((1.42) and (1.43) of Section 4.1.5) we normalized the Haar measure on K to obtain:

$$\int_{k \in K} dk = 1.$$

Now we can compute the constant in this integral formula.

Proposition 3 (The Constant in the Integral Formula for Polar Coordinates). *Let dk denote Haar measure on $K = O(n)$, normalized so that $\int_{k \in K} dk = 1$, and let*

$$\gamma(a) = \prod_{i=1}^{n} a_i^{-(n-1)/2} \prod_{1 \le i < j \le n} |a_i - a_j|, \quad \text{if } a = \begin{pmatrix} a_1 & & 0 \\ & \ddots & \\ 0 & & a_n \end{pmatrix} \in A,$$

with $a_i > 0$ and $da = \prod_{i=1}^{n} da_i/a_i$. Then the integral formula for polar coordinates (1.42) and (1.43) of Section 4.1.5 is:

$$\int_{Y \in \mathscr{P}_n} f(Y) \, d\mu_n(Y) = c_n \int_{a \in A} \int_{k \in K} f(a[k]) \gamma(a) \, da \, dk,$$

with constant c_n given by

$$c_n^{-1} = \pi^{-n^2/2} n! \Gamma_n\left(0, \ldots, 0, \frac{n}{2}\right) = \pi^{-(n^2+n)/4} \prod_{j=1}^{n} j\Gamma(j/2).$$

PROOF. It suffices to prove the result for K bi-invariant functions f; i.e., $f(a) = f(a[k])$ for $a \in A$ and $k \in K$. Then, from Exercise 24 of §4.1.3, we have

$$\int_{Y \in \mathscr{P}_n} f(Y) \, d\mu_n(Y) = \text{Vol}(K) 2^{-n} (n!)^{-1} \int_{a \in A} f(a) \gamma(a) \, da,$$

where the volume of K is computed as in Proposition 2. So the formula for this volume which is given in Proposition 2 completes the proof of Proposition 3. □

Exercise 5 (Evaluation of a Special Case of a Selberg Integral). Set

$$D(a) = \prod_{1 \le i < j \le n} |a_i - a_j|, \quad \text{for } a \in A = \text{positive diagonal matrices}.$$

Then a limiting case of an integral evaluated by Selberg [2] is:

$$S(p, z) = \int_{a \in A} |a|^p D(a)^{2z} \exp[-\text{Tr}(a)] \, da$$

$$= \prod_{k=1}^{n} \Gamma(1 + kz) \Gamma(p + (k-1)z)/\Gamma(1 + z).$$

Check that the formula (2.5) for $\Gamma_n(s)$ and Proposition 3 give Selberg's result in the case $z = \frac{1}{2}$. Selberg [2] gives evaluations of integrals which appear in Mehta [1].

Now we want to return to our study of $D(\mathscr{P}_n)$, the G-invariant differential operators on $\mathscr{P}_n$ (see Section 4.1.4). First we need a few definitions. We define the *adjoint* L^* of a differential operator L in $D(\mathscr{P}_n)$ by the following formula, assuming that f, g are such that the integrals converge:

$$\int_{Y \in \mathscr{P}_n} (Lf)(Y)\overline{g(Y)}\, d\mu_n(Y) = \int_{Y \in \mathscr{P}_n} f(Y)\overline{(L^*g)(Y)}\, d\mu_n(Y). \quad (2.7)$$

The geodesic-reversing isometry σ of $\mathscr{P}_n$ at the identity is $\sigma(Y) = Y^{-1}$ and we can define L^σ for $L \in D(\mathscr{P}_n)$ by:

$$L^\sigma f = L(f \circ \sigma^{-1}) \circ \sigma, \qquad \text{where "$\circ$" denotes composition of functions.} \quad (2.8)$$

In Theorem 1 we will show that $L^\sigma = \bar{L}^*$. In order to do this, we will need a result about the eigenvalues of the invariant differential operators acting on power functions. To put this result in its best form, we need the *proper normalization of variables in the power function* (Selberg [1, p. 57]):

$$\varphi_r(t) = \prod_{i=1}^n t_{ii}^{2r_i + i - (n+1)/2}, \qquad \text{for } t \in T_n, r \in \mathbb{C}^n. \quad (2.9)$$

Exercise 6. Find $s = s(r)$ such that $p_s(I[t]) = \varphi_r(t)$.

Selberg [1, p. 58] states the following theorem.

Theorem 1 (Normalized Power Functions and Invariant Differential Operators).

(1) If $L_j = \mathrm{Tr}((Y \partial/\partial Y)^j)$, for $j = 1, 2, \ldots, n$, as in Theorem 2 of Section 4.1.4, and φ_r is the normalized power function from formula (2.9), then

$$L_i \varphi_r = \lambda_i(r) \varphi_r$$

and $\lambda_i(r)$ is a symmetric polynomial in r_j of degree i and having the form:

$$\lambda_i(r_1, \ldots, r_n) = r_1^i + \cdots + r_n^i + \text{terms of lower degree}.$$

(2) The effect of $L \in D(\mathscr{P}_n)$ on power functions $p_s(Y)$ determines L uniquely.

(3) For $L \in D(\mathscr{P}_n)$, using the notation (2.7) and (2.8), we have

$$L^\sigma = \bar{L}^*,$$

with "$\bar{}$" denoting complex conjugation.

PROOF (Maass [2, pp. 70–76]).
(1) We shall use induction on n. Recall that we showed in Section 4.1.4 that if $Y \in \mathscr{P}_n$ has partial Iwasawa decomposition

4.2. Special Functions on $\mathscr{P}_n$

$$Y = \begin{pmatrix} V & 0 \\ 0 & w \end{pmatrix} \begin{bmatrix} I & x \\ 0 & 1 \end{bmatrix}, \quad \text{for } V \in \mathscr{P}_{n-1}, w > 0, x \in \mathbb{R}^{n-1}, \quad (2.10)$$

then

$$Y\frac{\partial}{\partial Y} = \begin{pmatrix} V\dfrac{\partial}{\partial V} - \dfrac{1}{2}{}^t\!\left(x\dfrac{{}^t\partial}{\partial x}\right) & \dfrac{1}{2}\dfrac{\partial}{\partial x} \\ {}^t\!xV\dfrac{\partial}{\partial V} - w\dfrac{\partial}{\partial w}{}^t\!x - \dfrac{1}{2}{}^t\!x\left(x\dfrac{{}^t\partial}{\partial x}\right) + \dfrac{1}{2}w\dfrac{{}^t\partial}{\partial x}V^{-1} & \dfrac{1}{2}{}^t\!x\dfrac{\partial}{\partial x} + w\dfrac{\partial}{\partial w} \end{pmatrix}. \quad (2.11)$$

If L_1 and L_2 are matrix differential operators, we will write $L_1 \sim L_2$ if they agree on functions on $\mathscr{P}_n$ which are independent of the x-variable in (2.10).

It can be proved inductively that if

$$A_h = \left(V\frac{\partial}{\partial V} + \frac{1}{2}I\right)^h - \frac{1}{2}\sum_{j=0}^{h-1}\left(V\frac{\partial}{\partial V} + \frac{1}{2}I\right)^j\left(w\frac{\partial}{\partial w}\right)^{h-1-j}, \quad (2.12)$$

then

$$\left(Y\frac{\partial}{\partial Y}\right)^h \sim \begin{pmatrix} A_h & 0 \\ {}^t\!x\,A_h - \left(w\dfrac{\partial}{\partial w}\right)^h {}^t\!x & \left(w\dfrac{\partial}{\partial w}\right)^h \end{pmatrix}. \quad (2.13)$$

Exercise 7. Prove formula (2.13), using (2.12) to define A_h and induction.

Hint. Note that if c is a real variable which does not depend on x, but may depend on the other variables, then

$$\frac{\partial}{\partial x}(c\,{}^t\!x) = cI,$$

where I denotes the $(n-1) \times (n-1)$ identity matrix.

It follows from (2.13) that

$$\operatorname{Tr}\left(\left(Y\frac{\partial}{\partial Y}\right)^h\right) \sim \operatorname{Tr}\left(\left(V\frac{\partial}{\partial V} + \frac{1}{2}I\right)^h\right) + \left(w\frac{\partial}{\partial w}\right)^h$$
$$-\frac{1}{2}\sum_{j=0}^{h-1}\operatorname{Tr}\left(\left(V\frac{\partial}{\partial V} + \frac{1}{2}I\right)^j\right)\left(w\frac{\partial}{\partial w}\right)^{h-1-j}. \quad (2.14)$$

This formula is peculiarly unsymmetric.

Next, set up the notation, $Y = I[t]$, for $t \in T_n$,

$$V = \begin{pmatrix} t_{11}^2 & & 0 \\ & \ddots & \\ 0 & & t_{n-1\,n-1}^2 \end{pmatrix}\begin{bmatrix} 1 & & * \\ & \ddots & \\ 0 & & 1 \end{bmatrix}, \quad w = t_{nn}^2,$$

$$\varphi_r(Y) = \prod_{j=1}^{n} t_{jj}^{2r_j+j-(n+1)/2}, \qquad \varphi_a(V) = \prod_{j=1}^{n-1} t_{jj}^{2a_j+j-n/2},$$

$$r = (a, b), \, a \in \mathbb{C}^{n-1}, \, b \in \mathbb{C}.$$

Then

$$\varphi_r(Y) = |V|^{-1/4} w^{b+(n-1)/4} \varphi_a(V). \tag{2.15}$$

Now we need the following exercise.

Exercise 8. Show that

$$\operatorname{Tr}\left(\left(V \frac{\partial}{\partial V}\right)^h\right) |V|^m = |V|^m \operatorname{Tr}(m^h I),$$

where I denotes the $(n-1) \times (n-1)$ identity matrix.

Hint. Use expansion of $|V|$ by minors to find $\partial |V|/\partial v_{ij}$ is the i,j cofactor of V, that is,

$$N_{ij} = (-1)^{i+j} |M_{ij}|,$$

where M_{ij} is the matrix obtained from V by crossing out the ith row and the jth column of V. Then

$$|V| = \sum_{i=1}^{n} v_{ij} |N_{ij}|$$

and $V^{-1} = |V|^{-1} \, {}^t(N_{ij})$. Note that the $\frac{1}{2}$'s in the off-diagonal entries of $\partial/\partial V$ are necessary to make this exercise work for symmetric matrices. Putting together formulas (2.14), (2.15) and Exercise 8, we obtain (2.17), with

$$\operatorname{Tr}\left(\left(Y \frac{\partial}{\partial Y}\right)^h\right) \varphi_r(Y)$$

$$= |V|^{-1/4} w^{(n-1)/4} \left\{ \operatorname{Tr}\left(\left(V \frac{\partial}{\partial V} + \frac{1}{4} I\right)^h\right) + \left(w \frac{\partial}{\partial w} + \frac{n-1}{4}\right)^h \right. \tag{2.16}$$

$$\left. - \frac{1}{2} \sum_{j=0}^{h-1} \operatorname{Tr}\left(\left(V \frac{\partial}{\partial V} + \frac{1}{4} I\right)^j\right) \left(w \frac{\partial}{\partial w} + \frac{n-1}{4}\right)^{h-1-j} \right\} \varphi_a(V) w^b.$$

Now the eigenvalue of interest is:

$$\lambda_h(r) = \left[\operatorname{Tr}\left(\left(Y \frac{\partial}{\partial Y}\right)^h\right) \varphi_r(Y) \right]_{Y=I}. \tag{2.17}$$

Exercise 9. Check that when $n = 2$ the eigenvalue defined by formula (2.17), with $Y \in \mathscr{P}_2$, is a symmetric polynomial in r_1 and r_2 having highest degree term

$$r_1^h + r_2^h, \qquad h = 1, 2, 3, \ldots.$$

4.2. Special Functions on $\mathcal{P}_n$

The complete polynomial is:

$$\left(r_1 + \frac{1}{4}\right)^h + \left(r_2 + \frac{1}{4}\right)^h - \frac{1}{2}\sum_{j=0}^{h-1}\left(r_1 + \frac{1}{4}\right)^j\left(r_2 + \frac{1}{4}\right)^{h-1-j}.$$

Now suppose $n \geq 3$ and proceed by induction, assuming that

$$\mathrm{Tr}\left(\left(V\frac{\partial}{\partial V} + \frac{1}{4}I\right)^h\right)\varphi_a(V) = \xi_h(a)\varphi_a(V), \qquad \text{where } a_1 = r_1, \ldots, a_{n-1} = r_{n-1}$$

and $\xi_h(a)$ is a symmetric polynomial in a with highest degree term

$$r_1^h + \cdots + r_{n-1}^h.$$

Clearly

$$\left(w\frac{\partial}{\partial w} + \frac{n-1}{4}\right)^h w^b = v_h(b)w^b, \qquad b = r_n, \qquad v_h(b) = \left(b + \frac{n-1}{4}\right)^h.$$

Then, by formula (2.16), $\lambda_h(r)$ is invariant under permutations of $r_1, \ldots, r_{n-1}$ and $\lambda_h(r)$ has highest degree term $r_1^h + \cdots + r_{n-1}^h + r_n^h$, for

$$\lambda_h(r) = \xi_h(a) + v_h(b) - \frac{1}{2}\sum_{j=0}^{h-1}\xi_j(a)v_{h-1-j}(b).$$

In order to see that λ_h is symmetric in the last variable too, we have to note that you could also do this expansion with

$$Y = \begin{pmatrix} v & 0 \\ 0 & W \end{pmatrix}\begin{bmatrix} 1 & {}^t x \\ 0 & I \end{bmatrix}, \qquad v > 0, \qquad W \in \mathcal{P}_{n-1}, \qquad x \in \mathbb{R}^{n-1}.$$

Then you would find $\lambda_h(r)$ to be symmetric in $r_2, \ldots, r_n$, by the induction assumption. In fact, you can make the same argument with an arbitrary partial Iwasawa decomposition as in Exercise 10 below. This completes the proof of (1), since we have checked the case $n = 2$, by Exercise 9.

Exercise 10. Run through the preceding proof for a general Iwasawa decomposition

$$Y = \begin{pmatrix} V & 0 \\ 0 & W \end{pmatrix}\begin{bmatrix} I & X \\ 0 & I \end{bmatrix}, \qquad \text{for } V \in \mathcal{P}_r,\ W \in \mathcal{P}_{n-r},\ X \in \mathbb{R}^{r\times(n-r)}.$$

Hint. (See Maass [2, pp. 70–76].)

(2) Suppose $L = p(L_1, \ldots, L_n)$, where p is a polynomial in n indeterminates and $L_j = \mathrm{Tr}((Y\,\partial/\partial Y)^j)$, for $j = 1, 2, \ldots, n$. Then $L\varphi_r = 0$ implies that $p(\lambda_1(r), \ldots, \lambda_n(r))$ vanishes for all $r \in \mathbb{C}^n$. But then p must vanish identically, since the eigenvalues $\lambda_j(r)$ form a basis for the symmetric polynomials in $r_1, \ldots, r_n$ by part (1). Thus the mapping from $\mathbb{C}^n$ to $\mathbb{C}^n$ which takes r to $(\lambda_1(r), \ldots, \lambda_n(r))$ is onto.

Exercise 11. Suppose that $\lambda_1(r), \ldots, \lambda_n(r)$ form a basis for all the symmetric polynomials in $r_1, \ldots, r_n$.

(a) Prove that the mapping from $\mathbb{C}^n$ to $\mathbb{C}^n$ taking r to $(\lambda_1(r), \ldots, \lambda_n(r))$ is onto.
(b) Prove that $\lambda_j(r) = \lambda_j(r')$ for all $j = 1, \ldots, n$, implies that $r' = (r_{\sigma(1)}, \ldots, r_{\sigma(n)})$ for some permutation σ of n elements.

Hint. Let $u_j(r) =$ the jth elementary symmetric polynomial. Then

$$\sum_{j=0}^{n} u_j(s) x^j = \prod_{j=1}^{n} (x - s_j), \qquad \text{for } s \in \mathbb{C}^n.$$

(3) First, note that $k(X, Y) = \exp(-\operatorname{Tr}(YX^{-1}))$ is a point-pair invariant or convolution operator as in §4.1.3. If $L \in D(\mathscr{P}_n)$, write $L_X k(X, Y)$ when L acts on the X-variable and $L_Y k(X, Y)$ when L acts on the Y-variable. Just as in the proof of part 2 of Lemma 2 in §4.1.4, we use the fact that

$$L_X^\sigma k(X, Y) = L_Y k(X, Y).$$

This fact implies the second in the following sequence of equalities which stem from Exercise 4:

$$(L^\sigma p_s)(X)\Gamma_n(s) = \int_{Y \in \mathscr{P}_n} \{L_X^\sigma \exp(-\operatorname{Tr}(YX^{-1}))\} p_s(Y) \, d\mu_n(Y)$$

$$= \int_{Y \in \mathscr{P}_n} \{L_Y \exp(-\operatorname{Tr}(YX^{-1}))\} p_s(Y) \, d\mu_n(Y)$$

$$= \int_{Y \in \mathscr{P}_n} \exp(-\operatorname{Tr}(YX^{-1}))(\bar{L}^* p_s(Y)) \, d\mu_n(Y)$$

$$= (\bar{L}^* p_s)(X)\Gamma_n(s).$$

Here we use the fact that the adjoint operator is also G-invariant. This completes the proof of Theorem 1. □

Our study of the gamma function for $\mathscr{P}_n$ is now at an end. It will find applications in statistics at the end of the section. And these Γ-functions will also appear in functional equations of L-functions and Eisenstein series for $GL(n, \mathbb{Z})$ in Section 4.3.

One can also study *matrix incomplete gamma functions*. We saw an example of these incomplete gamma functions in Exercise 4, Section 3.6, Volume I when we obtained the analytic continuation of L-functions corresponding to Maass wave forms. These incomplete gamma functions appear in the analytic continuation of Dedekind zeta functions of number fields as well as in the analytic continuation of Eisenstein series for $GL(n, \mathbb{Z})$ (see Section 1.4 of Volume I and Section 4.5 which follows). More information on incomplete gamma functions can be found in Terras [1]. And *matrix beta functions* will arise in Section 4.3 as part of the computation of the Plancherel or spectral measure for Mellin inversion on $\mathscr{P}_n$. See also Gindikin [1].

4.2. Special Functions on $\mathscr{P}_n$

Exercise 12.

(a) Consider the power function given by formula (2.9) and the operator $L_j = \text{Tr}((Y\partial/\partial Y)^j)$, for $Y \in \mathscr{P}_n$. If $L_j \varphi_r(Y) = \lambda_j^n(r)\varphi_r(Y)$, show that the eigenvalue of L_1 is $\lambda_1^n(r) = r_1 + \cdots + r_n$ and the eigenvalue of $\Delta_Y = L_2$ is $\lambda_2^n(r) = r_1^2 + \cdots + r_n^2 + (n - n^3)/48$.

(b) Then show that if instead we consider the Laplacian on the determinant one surface $\mathscr{SP}_3$ as in Exercise 32 of Section 4.1.5, we find that

$$\Delta p_s(Y) = \left\{ \frac{2}{3}(s_1^2 + s_1 s_2 + s_2^2) + s_1 + s_2 \right\} p_s(Y).$$

Hint. (a) You can use formulas (2.15) and (2.16) or you can use formula (1.33) of Section 4.1.4 to see that $\Delta_Y \sim (w\partial/\partial w)^2 + ((1-n)/2)w\partial/\partial w + \Delta_V + \frac{1}{2}\text{Tr}(V\partial/\partial V)$. Now formula (2.15) implies that:

$$\lambda_2^n(r) = [\Delta_Y |V|^{-1/4} \varphi_a(V) w^{b+(n-1)/4}]_{Y=I}$$

$$= \left(b + \frac{n-1}{4} \right)^2 + \frac{1-n}{2}\left(b + \frac{n-1}{4} \right) + \lambda_2^{n-1}\left(r_1 - \frac{1}{4}, \ldots, r_{n-1} - \frac{1}{4} \right)$$

$$+ \frac{1}{2}\lambda_1^{n-1}\left(r_1 - \frac{1}{4}, \ldots, r_{n-1} - \frac{1}{4} \right).$$

4.2.2. K-Bessel Functions

K-Bessel functions for $\mathscr{P}_n$ have been discussed by various authors with vastly different points of view. We will attempt to say a little more about some of the other developments at the end of this subsection. The closest references to our treatment are: Herz [1], Bengtson [1], Bump [1], Imai and Terras [1], Maass [2, Ch. 18], Terras [2–5]. Bessel functions analogous to the classical J-Bessel function are to be found in Bochner [1], Godement's article in Séminaire Cartan [1, exposé 9], Gelbart [1], and Gross, Holman and Kunze [1]. The classical Whittaker functions are confluent hypergeometric functions generalizing K-Bessel functions (see Lebedev [1]). Whittaker functions and Whittaker models for representations of real, complex, p-adic and adelic groups are discussed by Bump [1], Hashizume [1], Jacquet [1], Piatetski-Shapiro in Borel and Casselman [1, Vol. I, pp. 209–212], Schiffman [1], Shalika [1], and Shintani [1]. Related references are Goodman [1], Goodman and Wallach [1], and Kostant [1]. Hypergeometric functions of matrix argument are also considered by Gindikin [1], Gross and Richards [1], Herz [1], James [1–3], Maass [2, Chapter 18], Muirhead [1], and Shimura [1].

Many of the preceding references are motivated by the number-theoretical problem of obtaining Fourier expansions of automorphic forms and this will be our main application (see Section 4.5 and the references mentioned there).

Others seek to solve statistical problems such as that of finding the non-central Wishart distribution (see Herz [1], Muirhead [1], and the discussion at the end of this subsection). Still others seek uniqueness results about representations.

The K-Bessel functions which we study are not the most general of those mentioned above, but the suffice for our purposes and to give an introduction to the subject. Consideration of Kirillov's theory of the representations of the nilpotent group N (see Kirillov [1], Proskurin [1], and Moore [1]) serves to clarify the concepts. It is also useful to view Bessel and Whittaker functions in the light of the theory of the operators intertwining pairs of representations (see Dieudonné [1, Vol. VI], Hashizume [1], Kirillov [2], Mackey [1, pp. 363ff.], and Vilenkin [1, Ch. VIII]). However, we will not delve into group representations in this volume.

Define the abelian group $N(m, n - m)$, for $1 \leq m < n$, and define the character χ_A of $N(m, n - m)$ for fixed $A \in \mathbb{R}^{m \times (n-m)}$ by:

$$N(m, n - m) = \left\{ U = \begin{pmatrix} I & X \\ 0 & I \end{pmatrix} \middle| X \in \mathbb{R}^{m \times (n-m)} \right\}, \tag{2.18}$$

$$\chi_A \begin{pmatrix} I & X \\ 0 & I \end{pmatrix} = \exp(2\pi i \operatorname{Tr}({}^t\!AX)), \quad \text{for } X \in \mathbb{R}^{m \times (n-m)}.$$

Using the notation (2.18), we will say that $f: \mathscr{P}_n \to \mathbb{C}$ is a *K-Bessel function* if, for some fixed $A \in \mathbb{R}^{m \times (n-m)}$, we have:

(a) f transforms by $N(m, n - m)$ according to the character χ_A;
i.e., $f(Y[U]) = \chi_A(U)f(Y)$ for all $Y \in \mathscr{P}_n$, $U \in N(m, n - m)$;

(b) f is an eigenfunction for all the G-invariant (2.19)
differential operators $L \in D(\mathscr{P}_n)$;

(c) f grows at most like a power function at the boundary.

This definition is analogous to (3.12) in Volume I. However, if one simply thinks of the behavior of $K_s(y)$, as y approaches infinity, one might think that the growth condition (c) is somewhat weak. But recall that in the case of $SL(2, \mathbb{R})$ we found that, for f as in (2.19),

$$f\left(\begin{pmatrix} 1/y & 0 \\ 0 & y \end{pmatrix} \begin{bmatrix} 1 & x \\ 0 & 1 \end{bmatrix} \right) = cy^{1/2} K_{s-1/2}(2\pi |R| y),$$

if $R \neq 0$ (see Exercise 1 of Vol. I, Section 3.2). Here $K_s(y)$ denotes the ordinary K-Bessel function. As y approaches infinity, the function $K_s(y)$ approaches zero exponentially. But (c) in (2.19) is still O.K., since for $\operatorname{Re} s > 0$, as y approaches 0, $K_s(y)$ blows up like y^{-s} (see Exercise 2 of Section 3.2 in Vol. I). Moreover, if $R = 0$, then we obtain (for the case of $SL(2, \mathbb{R})$):

$$f\left(\begin{pmatrix} 1/y & 0 \\ 0 & y \end{pmatrix} \begin{bmatrix} 1 & x \\ 0 & 1 \end{bmatrix} \right) = cy^s + dy^{1-s}.$$

4.2. Special Functions on $\mathscr{P}_n$

Just as in the case $n = 2$ (see Vol. I, Section 3.5) these K-Bessel functions appear in Fourier expansions of automorphic forms for $GL(n, \mathbb{Z})$, that is, expansions with respect to the abelian groups $N(m, n - m)$ in (2.18) above (cf. Imai and Terras [1], and Terras [2–5]). Such expansions are analogous to those used by Siegel in his study of Siegel modular forms (see Siegel [1, Vol. III, pp. 97–137] and Section 4.5).

Property (a) of (2.19) says that we are studying a special function corresponding to a representation of G induced from the character of $N(m, n - m)$ given by $\chi_A(U)$, for $U \in N(m, n - m)$, using the notation of (2.18).

Kirillov [1] shows that (up to equivalence) one obtains the infinite dimensional irreducible unitary representations of the nilpotent group N of upper triangular matrices with one on the diagonal by inducing the representations corresponding to a character of $N(m, n - m)$ with $m = [n/2]$. The finite dimensional (actually 1-dimensional) irreducible unitary representations of N come from a different construction which we shall discuss at the end of this subsection in connection with Whittaker functions. It is only the infinite dimensional representations that contribute to the Plancherel formula for N.

In a sense, the K-Bessel functions considered here are analogous to the Eisenstein series for maximal parabolic subgroups of $GL(n)$ which will be discussed in Section 4.5. The Whittaker functions to be considered at the end of this section are similar to the Eisenstein series for minimal parabolic subgroups of $GL(n)$ which will also be studied in Section 4.5.

It is easy to give examples of functions satisfying the conditions in (2.19). In what follows we will find it natural to define two sorts of K-Bessel functions. To distinguish them, we use the capital "K" for the function in (2.21) below and the small "k" for the function in (2.20) below. Part (2) of Theorem 2 shows that the two functions are really essentially the same. Imitating formula (3.14) in Section 3.2 of Vol. I, we define *the first type of matrix k-Bessel function* to be:

$$k_{m,n-m}(s|Y, A) = \int_{X \in \mathbb{R}^{m \times (n-m)}} p_{-s}\left(Y^{-1} \begin{bmatrix} I & 0 \\ {}^tX & I \end{bmatrix}\right) \exp\{2\pi i \operatorname{Tr}({}^tAX)\} \, dX, \quad (2.20)$$

for $s \in \mathbb{C}^n$ with coordinates restricted to suitable half planes, $Y \in \mathscr{P}_n$, $A \in \mathbb{R}^{m \times (n-m)}$, $1 \leq m < n$. Here $p_s(Y)$ denotes the power function (2.1). Formula (2.20) is useful for demonstrating that $k_{m,n-m}$ satisfies (2.19a, b), since it is clearly an eigenfunction for any differential operator in $D(\mathscr{P}_n)$ and has the correct invariance property under transformation by elements of $N(m, n - m)$.

The *second type of K-Bessel function* is defined by:

$$K_m(s|V, W) = \int_{Y \in \mathscr{P}_m} p_s(Y) \exp\{-\operatorname{Tr}(VY + WY^{-1})\} \, d\mu_m(Y), \quad (2.21)$$

for $V, W \in \mathscr{P}_m, s \in \mathbb{C}^m$, or W singular with Re s_j suitably restricted. The function (2.21) generalizes the formula in part (a) of Exercise 1 in Section 3.2 of Vol. I. This second type of K-Bessel function is useful in the study of convergence properties and analytic continuation in the s-variable. Herz [1, p. 506] considers the special case of (2.21) with $s_j = 0$ for $j \neq n$.

At this point, it is not clear how K_m is related to $k_{m,n-m}$. It will turn out that Bengtson's formula in Theorem 2 relates the two functions and thus gives a generalization of the result in Exercise 1 of Section 3.2, Vol. I. We review this result in the next example.

Example 1 (The One Variable Case). When $m = 1$, formula (2.21) is the ordinary K-Bessel function K_s defined in Exercise 1(a) of Section 3.2 of Volume I, since for $a, b > 0$, $s \in \mathbb{C}$:

$$K_1(s|a,b) = \int_0^\infty y^{s-1} \exp\{-(ay + b/y)\}\, dy = 2(b/a)^{s/2} K_s(2\sqrt{ab}).$$

When $n = 2$ and $m = 1$, $a \in \mathbb{R}$, by part (a) of the same exercise, we have:

$$k_{1,1}\left(s, 0 \middle| \begin{pmatrix} 1/y & 0 \\ 0 & y \end{pmatrix}, a\right)$$

$$= \int_{x \in \mathbb{R}} p_{-s}\left(\begin{pmatrix} y & 0 \\ 0 & 1/y \end{pmatrix} \begin{bmatrix} 1 & 0 \\ x & 1 \end{bmatrix}\right) \exp(2\pi i a x)\, dx$$

$$= y^s \int_{x \in \mathbb{R}} (y^2 + x^2)^{-s} \exp(2\pi i a x)\, dx$$

$$= \begin{cases} 2\pi^{1/2} \Gamma(s)^{-1} |\pi a|^{s-1/2} y^{1/2} K_{s-1/2}(2\pi |a| y), & \text{if } a \neq 0, \\ \Gamma(\tfrac{1}{2}) \Gamma(s - \tfrac{1}{2}) \Gamma(s)^{-1} y^{1-s}, & \text{if } a = 0. \end{cases}$$

In the next example, we see that our matrix argument k-Bessel functions can sometimes be factored into products of ordinary K-Bessel functions and Γ-functions. However, we must caution the reader that this does not seem to be a general phenomenon. Thus these Bessel functions differ greatly from the gamma functions considered in the last section.

Example 2 (A Factorization in a Special Case). Using the first remarks in the proof of part (5) in Theorem 2, we find that:

$$k_{2,1}(s_1, s_2, 0|I, (a, 0)) = \iint (1 + x_1^2)^{-s_1}(1 + x_1^2 + x_2^2)^{-s_2} \exp(2\pi i a x_1)\, dx_1\, dx_2$$

$$= \int (1 + x_1^2)^{-s_1 - s_2 + 1/2} \exp(2\pi i a x_1)\, dx_1 \int (1 + y^2)^{-s_2}\, dy$$

$$= k_{1,1}(s_1 + s_2 - \tfrac{1}{2}, 0|I, a) k_{1,1}(s_2, 0|I, 0)$$

$$= k_{1,1}(s_1 + s_2 - \tfrac{1}{2}, 0|I, a) B(\tfrac{1}{2}, s_2 - \tfrac{1}{2}),$$

where we have used the substitution $x_2 = (1 + x_1^2)^{1/2} y$ and $B(p, q) = \Gamma(p)\Gamma(q)/\Gamma(p+q)$. The method of Example 2 extends to $k_{m,1}$ by part (5) of Theorem 2:

4.2. Special Functions on $\mathscr{P}_n$

$$k_{m,1}(s_1, s_2, 0 | I_{m+1}, (a_1, 0)) = k_{m-1,1}(s_2, 0 | I_m, 0) k_{1,1}(-b, 0 | I, a_1),$$

where $s_1 \in \mathbb{C}$, $s_2 \in \mathbb{C}^{m-1}$, $a_1 \in \mathbb{R}$, $b = (m-1)/2 - \sum_{j=1}^m s_j$. However when $a_2 \neq 0$, there does not appear to be such a factorization. Thus the k- and K-Bessel functions for $\mathscr{P}_n$ do not, in general, factor into products of ordinary K-Bessel functions.

Exercise 13. Prove that the first matrix k-Bessel function $k_{m,n-m}(s|Y, A)$ in (2.20) is an eigenfunction for all the differential operators in $D(\mathscr{P}_n)$ when considered as a function of $Y \in \mathscr{P}_n$. And show that it has the invariance property (2.19a) (again when considered as a function of Y).

The following exercise generalizes the first asymptotic result on the ordinary K-Bessel function in Exercise 2, Section 3.2, Vol. I.

Exercise 14.

(a) Show that $K_m(s|I, 0) = \Gamma_m(s)$, where the Γ-function is defined in (2.4).
(b) Show that $K_m(s|A, B) \sim p_s(A^{-1}) \Gamma_m(s)$, as $B \to 0$, for fixed $A \in \mathscr{P}_m$.

Exercise 15. Show that if $y > 0$ and $a \in \mathbb{R}$, then

$$k_{1,1}\left(s, 0 \left| \begin{pmatrix} 1/y & 0 \\ 0 & y \end{pmatrix}, a \right.\right) = y^{1-s} k_{1,1}(s, 0 | I_2, ay), \quad \text{for } s \in \mathbb{C}.$$

Exercise 16. Suppose that $t \in T_m$ defined in (2.2). Show that

$$K_m(s | V[{}^t t], W[t^{-1}]) p_s(I[t]) = K_m(s | V, W).$$

Exercise 16 shows that we can reduce one of the positive matrix arguments in K_m to the identity. However, it is convenient for our purposes to separate the arguments V and W. An illustration of this convenience can be found, for example, in Exercise 14, where we see that we can treat the case that one of the arguments is singular.

The following theorem gives the main properties (known to the author) of these matrix argument K-Bessel functions. It is mainly due to Tom Bengtson [1].

Theorem 2 (Properties of Matrix K-Bessel Functions).

(1) Convergence and Decay at Infinity.
Suppose that λ is the smallest element in the set of eigenvalues of V and W in $\mathscr{P}_m$. Then

$$K_m(s | V, W) = O(\lambda^{-m(m+1)/4} \exp(-2m\lambda)), \quad \text{as } \lambda \to \infty,$$

for fixed s. In particular, the integral (2.21) converges for all $s \in \mathbb{C}^n$ if V, $W \in \mathscr{P}_m$. And $K_m(s | V, W) \to 0$ exponentially as the eigenvalues of V and W all go to infinity.

(2) Bengtson's Formula Relating the Two Bessel Functions.
Let $s \in \mathbb{C}^m$, $s\# = -s + (0,\ldots,0,(n-m)/2)$ and $s^* = (s_{m-1},\ldots,s_1, -(s_1 + \cdots + s_m))$. Then, assuming that the coordinates of s are restricted to suitable half planes:

$$\Gamma_m(-s^*)k_{m,n-m}\left(s,0\left|\begin{pmatrix} V & 0 \\ 0 & W \end{pmatrix}, A\right.\right) = \pi^{m(n-m)/2}|W|^{m/2}K_m(s\#|W[\pi\,^t A], V^{-1}).$$

(3) K-Bessel Functions with a Singular Argument Reduce to Lower Rank K-Bessel Functions.

Let $P_s\begin{pmatrix} V & 0 \\ 0 & W \end{pmatrix} = p_{\sigma_1}(V)p_{\sigma_2}(W)|W|^{m/2}$, for $\sigma_1 \in \mathbb{C}^m$, $\sigma_2 \in \mathbb{C}^{n-m}$, $V \in \mathscr{P}_m$, $W \in \mathscr{P}_{n-m}$. Then

$$K_n\left(s\left|\begin{pmatrix} A & 0 \\ 0 & B \end{pmatrix}\begin{bmatrix} I & 0 \\ {}^t C & I \end{bmatrix}, \begin{pmatrix} 0 & 0 \\ 0 & D \end{pmatrix}\right.\right)$$
$$= \pi^{m(n-m)/2}|B|^{-m/2}p_{\sigma_1}(A^{-1})\Gamma_m(\sigma_1)K_{n-m}(\sigma_2|B,D).$$

We need to assume that $\operatorname{Re}\sigma_1$ is sufficiently large for the convergence of Γ_m.

(4) The Argument in $\mathscr{P}_n$ of the Matrix k-Bessel Function Can Be Reduced to I.
Let $V = g\,^t g$ for $g \in T_m$; i.e., g is upper triangular with positive diagonal. If $a \in \mathbb{R}^n$, $V \in \mathscr{P}_m$, $w > 0$, then

$$k_{m,1}\left(s,0\left|\begin{pmatrix} V & 0 \\ 0 & w \end{pmatrix}, a\right.\right) = p_{-s}(V^{-1})|V|^{-1/2}w^{m/2}k_{m,1}(s,0|I_{m+1}, w^{1/2}g^{-1}a).$$

Here $s \in \mathbb{C}^m$.

(5) An Inductive Formula for k-Bessel Functions.
For $s_1 \in \mathbb{C}$, $s_2 \in \mathbb{C}^{m-1}$, $a_1 \in \mathbb{R}$, $a_2 \in \mathbb{R}^{m-1}$, we have the following formula, if $a = (a_1, a_2)$ and $s = (s_1, s_2)$ are suitably restricted for convergence and $b = (m-1)/2 - \sum_{j=1}^m s_j$:

$$k_{m,1}(s,0|I_{m+1}, a)$$
$$= \int_{u \in \mathbb{R}} (1+u^2)^b k_{m-1,1}(s_2,0|I_m, a_2\sqrt{1+u^2})\exp(2\pi i a_1 u)\,du.$$

PROOF (Bengtson [1]).
(1) Since λ is the smallest element of the set of eigenvalues of V and W, $V[x] \geq \lambda\,^t xx$, for $x \in \mathbb{R}^m$ and $\operatorname{Tr}(V[X]) \geq \lambda\operatorname{Tr}(I[X])$ if $X \in \mathbb{R}^{m \times m}$. By the integral formula for the Iwasawa decomposition (see formulas (1.37) and (1.39) of §4.1.5), we have, upon setting $Y = I[t]$, $t \in T_n$:

$$K_m(s|V,W) \leq 2^m \int_{t \in T_m} \exp\{-\lambda\operatorname{Tr}(I[t] + I[t^{-1}])\} \prod_{i=1}^m t_{ii}^{\operatorname{Re} r_i - i} \prod_{1 \leq i \leq j \leq m} dt_{ij}.$$

The variables $r \in \mathbb{C}^m$ are related to $s \in \mathbb{C}^m$ by the formula given in part (1) of

4.2. Special Functions on $\mathscr{P}_n$

Proposition 1. Write

$$t^{-1} = \begin{pmatrix} t_{11}^{-1} & & t^{ij} \\ & \ddots & \\ 0 & & t_{mm}^{-1} \end{pmatrix}.$$

Then

$$K_m(s|V, W) \leq 2^m \prod_{j=1}^m \int_{t_{jj}>0} \exp\{-\lambda(t_{jj}^2 + t_{jj}^{-2})\} t_{jj}^{\operatorname{Re} r_j - j} dt_{jj}$$

$$\times \prod_{1 \leq i < j \leq m} \int_{t_{ij} \in \mathbb{R}} \exp\{-\lambda(t_{ij}^2 + (t^{ij})^2)\} dt_{ij}$$

$$\leq (\pi/\lambda)^{m(m-1)/4} \prod_{j=1}^m K_{(\operatorname{Re} r_j - j + 1)/2}(2\lambda), \qquad s \in \mathbb{R}.$$

For the final estimate, we need to know that $K_s(y) \leq (\pi/(2y))^{1/2} e^{-y}$, for $y > 0$, $s \in \mathbb{R}$ (see Lebedev [1]).

(2) Let Ξ denote the left hand side of the equality that we are trying to prove. Then

$$\Xi = \Gamma_m(-s^*) k_{m,n-m}\left(s, 0 \middle| \begin{pmatrix} V & 0 \\ 0 & W \end{pmatrix}, A\right)$$

$$= \int_{Y \in \mathscr{P}_m} p_{-s^*}(Y) \exp\{-\operatorname{Tr}(Y)\} d\mu_m(Y) \int_{X \in \mathbb{R}^{m \times (n-m)}} p_{-s}(V^{-1} + W^{-1}[{}^t X])$$

$$\times \exp\{2\pi i \operatorname{Tr}({}^t A X)\} dX.$$

Now we want to use Exercise 4 of Section 4.2.1. In order to do this, we utilize another property of power functions from part (4) of Proposition 1 in Section 4.2.1 and obtain:

$$p_{-s}(V^{-1} + W^{-1}[{}^t X]) = p_{-s^*}(((V^{-1} + W^{-1}[{}^t X])[\omega])^{-1}),$$

$$\omega = \begin{pmatrix} 0 & & 1 \\ & \ddots & \\ 1 & & 0 \end{pmatrix}.$$

Then, by Exercise 4, we have the following equalities, letting $Z = Y[\omega]$:

$$\Xi = \int_{Y \in \mathscr{P}_m} \int_{X \in \mathbb{R}^{m \times (n-m)}} p_{-s^*}(Y) \exp\{-\operatorname{Tr}((V^{-1} + W^{-1}[{}^t X])[\omega] Y)\}$$

$$\times \exp\{2\pi i \operatorname{Tr}({}^t A X)\} dX \, d\mu_m(Y)$$

$$= \int_{Z \in \mathscr{P}_m} \int_{X \in \mathbb{R}^{m \times (n-m)}} p_{-s}(Z^{-1}) \exp\{-\operatorname{Tr}((V^{-1} + W^{-1}[{}^t X])Z$$

$$- 2\pi i {}^t A X)\} dX \, d\mu_m(Z).$$

Now complete the square in the exponent. Let $Z = Y^2$ with $Y \in \mathcal{P}_m$, $W = Q^2$, $Q \in \mathcal{P}_{n-m}$, and change variables via $U = YXQ^{-1}$ to obtain:

$$\Xi = |W|^{m/2} \int_{Z \in \mathcal{P}_m} |Z|^{-(n-m)/2} \int_{U \in \mathbb{R}^{m \times (n-m)}} p_{-s}(Z^{-1})$$

$$\times \exp\{-\mathrm{Tr}(V^{-1}Z + {}^tUU - 2\pi i\, {}^tAY^{-1}UQ)\}\, dU\, d\mu_m(Z).$$

Let $C = i\pi Y^{-1}AQ$ and observe that

$$\mathrm{Tr}({}^tUU - 2\pi i\,{}^t(Y^{-1}AQ)U) = \mathrm{Tr}(I[U-C] + W[\pi {}^tA]Z^{-1}).$$

Thus

$$\Xi = |W|^{m/2} \int_{Z \in \mathcal{P}_m} |Z|^{-(n-m)/2} p_{-s}(Z^{-1}) \exp\{-\mathrm{Tr}(V^{-1}Z + W[\pi {}^tA]Z^{-1})\}\, d\mu_n(Z)$$

$$\times \int_{U \in \mathbb{R}^{m \times (n-m)}} \exp\{-\mathrm{Tr}(I[U-C])\}\, dU$$

$$= |W|^{m/2} \pi^{m(n-m)/2} K_m(s \# | W[\pi {}^tA], V^{-1}).$$

(3) Then set Ξ equal to the left hand side of the equality we are trying to prove. By definition then

$$\Xi = \int_{\mathcal{P}_n} p_s(Y) \exp\left\{-\mathrm{Tr}\left(\begin{pmatrix} A & 0 \\ 0 & B \end{pmatrix}\begin{bmatrix} I & 0 \\ {}^tC & I \end{bmatrix} Y + \begin{pmatrix} 0 & 0 \\ 0 & D \end{pmatrix} Y^{-1}\right)\right\} d\mu_n(Y).$$

Let Y be expressed according to the appropriate partial Iwasawa decomposition:

$$Y = \begin{pmatrix} V & 0 \\ 0 & W \end{pmatrix}\begin{bmatrix} I & X \\ 0 & I \end{bmatrix}.$$

By Exercise 19 of Section 4.1.3, Ξ is

$$\int_{X \in \mathbb{R}^{m \times (n-m)}} \int_{V \in \mathcal{P}_m} \int_{W \in \mathcal{P}_{n-m}} p_{\sigma_1}(V) p_{\sigma_2}(W) |W|^{m/2}$$

$$\times \exp\{-\mathrm{Tr}(VA + V[X+C]B + WB + W^{-1}D)\}$$

$$\times |V|^{(n-2m-1)/2} |W|^{-(n+1)/2}\, dw_{ij}\, dv_{ij}\, dx_{ij}$$

$$= \pi^{m(n-m)/2} |B|^{-m/2} p_{\sigma_1}(A^{-1}) \Gamma_m(\sigma_1) K_{n-m}(\sigma_2 | B, D).$$

(4) The left hand side of the equality that we wish to prove is:

$$\int_{x \in \mathbb{R}^m} p_{-s}(V^{-1} + xw^{-1}\,{}^tx) \exp(2\pi i\,{}^tax)\, dx$$

which equals:

$$w^{m/2} |V|^{-1/2} p_{-s}(V^{-1}) \int_{u \in \mathbb{R}^m} p_{-s}(I + I[{}^tu]) \exp(2\pi i\,{}^ta w^{1/2}\,{}^tg^{-1}u)\, du,$$

upon setting $u = {}^tgxw^{-1/2}$.

4.2. Special Functions on $\mathscr{P}_n$

(5) First note that the upper left $j \times j$ corner of the matrix $(I + x\,'x)$, $x = a$ column vector in $\mathbb{R}^m$, is

$$(I_j + w_j\,'w_j), \qquad \text{where } 'w_j = (x_1 \cdots x_j).$$

And the matrix $w_j\,'w_j$ is a $j \times j$ matrix of rank one. The unique non-zero eigenvalue of $w_j\,'w_j$ is $\|w_j\|^2 = x_1^2 + \cdots + x_j^2$. We can therefore find $k \in O(j)$ such that

$$(I_j + w_j\,'w_j)[k] = I_j + \begin{pmatrix} \|w_j\|^2 & 0 \\ 0 & 0 \end{pmatrix}.$$

Thus $|I_j + w_j\,'w_j| = 1 + \|w_j\|^2$.

It follows from these considerations that

$$k_{m,1}(s|I_{m+1}, a) = \int_{x \in \mathbb{R}^m} p_{-s}(I + x\,'x) \exp(2\pi i\,'ax)\,dx$$

$$= \int_{x \in \mathbb{R}^m} (1 + x_1^2)^{-s_1}(1 + x_1^2 + x_2^2)^{-s_2} \cdots (1 + x_1^2 + \cdots + x_m^2)^{-s_m}$$

$$\times \exp(2\pi i\,'ax)\,dx.$$

Now make the change of variables $x_j^2 = (1 + x_1^2)u_j^2$, $j = 2, \ldots, m$, to complete the proof. $\square$

Exercise 17.

(a) Can you generalize property (4) of Theorem 2 to $k_{m,n-m}$?
(b) (Bengtson [1]). Show that $K_2(s|\,'qq, I_2)$ converges for $s \in \mathbb{C}^2$, $q \in \mathbb{R}^2$, when

$$\text{Re}\,s_2 \text{ and } \text{Re}(s_1 + s_2) < -\tfrac{1}{2}, \qquad \text{if } q = 0;$$

$$\text{Re}\,s_2 < 0, \qquad \text{if } q_2 = 0, q_1 \neq 0;$$

$$\text{Re}(s_1 + s_2) < 0, \qquad \text{if } q_2 \neq 0.$$

Note that this is the function $k_{2,1}$ essentially.
(c) Obtain a functional equation for $k_{2,1}$.

Hint. (c) Use property (5) of Theorem 2 and the functional equation of the ordinary K-Bessel function.

Remaining Questions.
1. *Concerning the K-Bessel Functions.*

(a) *Are these K-Bessel functions products of ordinary K-Bessel functions as was the case for the gamma function of matrix argument?* The answer must be "No, except under very special circumstances, as in Example 2 above."

(b) *Does (2.19) lead to a unique function?* Here the answer appears to be "Yes" and "No." For many functions satisfying (2.19) can be constructed out of the same basic function; e.g.,

$$f(Y) = k_{m,n-m}\left(s\left|Y\begin{bmatrix} {}^tA^{-1} & 0 \\ 0 & {}^tB \end{bmatrix}, C\right.\right),$$

with $A \in GL(m, \mathbb{R})$, $B \in GL(n-m, \mathbb{R})$, $C \in \mathbb{R}^{m \times (n-m)}$, such that $R = ACB$. For it is easily seen that

$$f\left(Y\begin{bmatrix} I_m & X \\ 0 & I_{n-m} \end{bmatrix}\right) = \exp\{2\pi i \operatorname{Tr}({}^t(ACB)X)\} f(Y).$$

(c) *Are there relations between $k_{m,n-m}$ and $k_{n-m,m}$? What functional equations do the $k_{m,n-m}$ satisfy?* See Exercise 17(c). The theory of Eisenstein series for maximal parabolic subgroups of $GL(n)$, which will be discussed in Section 4.5, leads us to expect that there is essentially only one functional equation (e.g., that of Exercise 17). Note that $m > n - m$ implies that $k_{m,n-m}(s, 0 | Y, A)$, $s \in \mathbb{C}^m$, which is the function related to $K_m(s^\# | W[\pi\, {}^tA], V^{-1})$ by part (2) of Theorem 2, has more s-variables than the same function with m and $n - m$ interchanged. This means that you cannot use parts (2) and (3) of Theorem 2 to write $k_{m,n-m}$ as a product of lower rank functions.

(d) *Can one generalize the Kontorovich-Lebedev inversion formula* (3.15) *of Section 3.2 in Vol. I to $\mathscr{P}_n$ and then obtain harmonic analysis on $\mathscr{P}_n$ in partial Iwasawa coordinates, thus generalizing Theorem 1 in Vol. I, Section 3.2?* This leads one to ask again: "What functional equations do matrix K-Bessel functions satisfy?" One is also led to attempt to generalize the Laplace transform relations between K-Bessel functions and spherical functions (Exercise 13 of Vol. I, Section 3.2) to a matrix version involving the spherical functions for $\mathscr{P}_n$ to be considered in the next subsection. Gelbart [1], Gross and Kunze [1], and Herz [1] generalize the Hankel transform to a transform involving matrix J-Bessel functions (which are operator valued in the 1st 2 references) and show that such a transform can be used to generalize Theorem 2, Vol. I, §2.2 and decompose the Fourier transform on matrix space $\mathbb{R}^{k \times m}$ in polar coordinates for that space. Define the compact Stiefel manifold $V_{k,m} = \{X \in \mathbb{R}^{k \times m} | {}^tXX = I\} \cong O(k)/O(k - m)$. Then polar coordinates for $X \in \mathbb{R}^{k \times m}$ are $R \in \mathscr{P}_m$ and $V \in V_{k,m}$ with $X = VR^{1/2}$. Here $k \geq m$. Of course, harmonic analysis on the Stiefel manifold involves representations of the orthogonal group. Thus one expects to see matrix-valued J-Bessel functions. In any case, this work on J-Bessel functions and inversion formulas for Hankel transforms certainly leads one to expect a similar theory for K-Bessel functions.

2. *More General Hypergeometric Functions for $\mathscr{P}_n$.*

(a) *Can one relate the K-Bessel functions with the J-Bessel functions considered by Gelbart* [1], *Gross, Holman and Kunze* [1], *Herz* [1], *and Muirhead* [1, Chapter 10]? Gelbart [1] and Gross, Holman and Kunze [1] consider matrix-valued *J-Bessel functions* defined for an irreducible unitary representation λ of a compact Lie group U acting on a real finite dimensional inner product space X by orthogonal linear transformations via:

$$J_\lambda(w, z) = \int_U \exp\{i(w|uz)\} \lambda(u) \, du. \tag{2.22}$$

4.2. Special Functions on $\mathscr{P}_n$

for $w, z \in X^{\mathbb{C}} = X \otimes_{\mathbb{R}} \mathbb{C}$, the complexification of X. Here $(w|z)$ is the complex bilinear form on $X^{\mathbb{C}}$ that uniquely extends the inner product on X. James (see Muirhead [1, p. 262]) defines a function $_0F_1$ which is the case $\lambda(u) \equiv 1$ identically in (2.22), with $(w|z) = \text{Tr}(wz)$.

And Herz [1] considers an analogue of the J-Bessel function given by

$$A_\delta(M) = (2\pi i)^{-n} \int_{\text{Re } Z = X_0} \exp\{\text{Tr}(Z - MZ^{-1})\} |Z|^{-\delta - p} \, dZ, \qquad (2.23)$$

for $n = m(m+1)/2$, $p = (m+1)/2$, $\delta \in \mathbb{C}$ with $\text{Re } \delta > p - 1$, Z in the Siegel upper half plane H_m with fixed real part $X_0 \in \mathscr{P}_m$. Herz finds that this function is needed to express the non-central Wishart distribution in multivariate statistics (see also Muirhead [1, Ch. 10]). In addition, such functions arise in summation formulas considered by Bochner [1] in his study of matrix analogues of the circle problem. But there do not appear to be good estimates for the error terms in these formulas. And such integrals appear in Fourier coefficients of Eisenstein series for $Sp(n, \mathbb{Z})$. See Godement's article in Séminaire Cartan [1, Exposé 9].

How are (2.22) and (2.23) related? The answer is to be found in Herz [1, p. 493] and Muirhead [1, p. 262]. See also Gelbart [1] and the references indicated there. Gelbart applies his results to the construction of holomorphic discrete series representations of $Sp(n, \mathbb{R})$. Gross, Holman, and Kunze [1] apply their results similarly to $U(n, n)$.

When considering the central limit theorem for $\mathscr{P}_n$ and in the next section, we will develop an asymptotic relation between spherical functions for $\mathscr{P}_n$ and J-Bessel type functions (which are spherical functions for the Euclidean group of the tangent space to $\mathscr{P}_n$ at the point I). Such a relation comes from Lemma 4.3 in Helgason [8]. Actually, to be precise, we will relate spherical functions for $\mathscr{P}_n$ with James and Muirhead's $_0F_0$ function of 2 arguments for $\mathscr{P}_n$. Presumably this function is related to the $_0F_1$ for $\mathscr{P}_{n-1}$. We will use the same methods as Helgason [8] specialized to our case.

(b) *How do the K-Bessel functions relate to matrix argument confluent hypergeometric functions* considered by Gindikin [1], Herz [1], James [1–3], and Muirhead [1, pp. 264, 472]? One can define a *matrix argument confluent hypergeometric function of the first kind* $\Phi_n = {_1F_1}$ by:

$$\Phi_n(a, c; X) = \frac{\Gamma_n(c)}{\Gamma_n(a) \Gamma_n(c-a)} \int_{0 < Y < I} \exp[\text{Tr}(XY)] |Y|^a \qquad (2.24)$$
$$\times |I - Y|^{c - a - (n+1)/2} \, d\mu_n(Y)$$

for $a, b \in \mathbb{C}$, with a, b suitably restricted and X a symmetric $n \times n$ matrix. The domain of integration is the subset of $Y \in \mathscr{P}_n$ such that $I - Y \in \mathscr{P}_n$. Muirhead [1, p. 447] shows that Φ_n gives the moment of the generalized variance of the noncentral Wishart distribution. This is due to Herz [1] and Constantine.

A *matrix confluent hypergeometric function of the 2nd kind* can be defined for $a, c \in \mathbb{C}$ and $X \in \mathscr{P}_n$ by:

$$\Psi_n(a,c;X) = \Gamma_n(a)^{-1} \int_{Y \in \mathcal{P}_n} \exp[-\mathrm{Tr}(XY)] |Y|^a |I + Y|^{c-a-(n+1)/2} d\mu_n(Y). \tag{2.25}$$

Muirhead [1, p. 474] uses this function to express certain statistical quantities coming from the T_0^2-statistic, which was proposed by Lawley in 1938 and Hotelling in 1947 in connection with a military problem—the air testing of bombsights.

If we do ask for a relation between our K-Bessel function (2.21) and the confluent hypergeometric function (2.25), we find that it is only clear for the case $n = 1$, when the functions are the classical ones considered by Lebedev [1].

One can show, for example, that the classical K-Bessel function is a special case of Ψ_1:

$$K_s(z) = \sqrt{\pi} (2z)^s e^{-z} \Psi_1(s + 1/2, 2s + 1; 2z).$$

This fact is proved in Lebedev [1, pp. 118 and 274]. The main results needed to prove it are:

(i) $K_s(z) = \dfrac{1}{2} \displaystyle\int_0^\infty u^{-s-1} \exp\left[-\dfrac{1}{2}z(u + 1/u)\right] du;$

(ii) $u^{-s-(1/2)} = \Gamma\left(s + \dfrac{1}{2}\right)^{-1} \displaystyle\int_0^\infty e^{-xu} x^{s-(1/2)} dx;$

(iii) $K_{1/2}(z) = \left(\dfrac{\pi}{2z}\right)^{1/2} e^{-z}.$

Thus it is worthwhile generalizing (iii) to $\mathcal{P}_n$. We already have the analogues of (i) and (ii). The analogous relation between the J-Bessel type function in (2.23) and $_1F_1$ in (2.24) is proved by Muirhead [1, p. 262].

Maass [2, Ch. 18] finds that the confluent hypergeometric functions of matrix argument occur in the Fourier coefficients of certain nonholomorphic automorphic forms for the Siegel modular group $Sp(n, \mathbb{Z})$. See also Shimura [1]. This suggests that we could relate Ψ and K by relating (non-holomorphic) Eisenstein series for $Sp(n, \mathbb{Z})$ and those for $GL(n, \mathbb{Z})$.

(c) *What is the connection between the K-Bessel functions and Whittaker functions?* Whittaker functions and Fourier expansions of automorphic forms as sums of these functions are discussed by Bump [1], Jacquet [1], Jacquet, Piatetski-Shapiro and Shalika [1], Proskurin [1], Schiffman [1], and Shalika [1]. For $r \in \mathbb{R}^{m-1}$, $Y \in \mathcal{P}_m$, and $s \in \mathbb{C}^m$, with Re s suitably restricted for convergence, the *Whittaker function* can be defined by:

$$W(s|Y,r) = \int_{n \in N} p_{-s}(Y^{-1}[^t n]) \exp\left(2\pi i \sum_{i=1}^{m-1} r_i x_{i,i+1}\right) dn, \tag{2.26}$$

where N is the nilpotent group of real $m \times m$ upper triangular matrices with

4.2. Special Functions on $\mathscr{P}_n$

ones on the diagonal,

$$n = \begin{pmatrix} 1 & & x_{ij} \\ & \ddots & \\ 0 & & 1 \end{pmatrix}, \quad \text{and } dn \text{ is found in Exercise 2, Section 4.1.0.}$$

The exponential appearing in the integral is easily seen to be a one-dimensional character of N. The integral itself can easily be shown to converge wherever the numerator in the Harish-Chandra c-function of Section 4.3 converges (i.e., when b_m given by (3.34) in Section 4.3 converges). One also sees easily that

$$W(s|Y[n], r) = \exp\left\{2\pi i \sum_{i=1}^{m-1} r_i x_{i,i+1}\right\} W(s|Y, r).$$

Thus the Whittaker function satisfies the analogues of properties (2.19) with the abelian group $N(m, n - m)$ replaced by the nilpotent group N. Kirillov [1] shows that the characters of N in the transformation formula above are (up to unitary equivalence) the only finite (actually 1) dimensional irreducible unitary representations of N. There are also infinite dimensional irreducible unitary representations, as we mentioned earlier. It is possible to view $W(s|Y, r)$ as an analogue of the Eisenstein series $E_{(n)}$ (to be defined in formula (5.5) of Section 4.5) with the largest possible number of complex variables $s \in \mathbb{C}^n$; i.e., the highest dimensional part of the spectrum of the Laplacian. Thus one can follow ideas of Jacquet and use techniques developed by Selberg for Eisenstein series in order to obtain $n!$ functional equations for the Whittaker functions (see Bump [1] and Jacquet [1]). The idea is to write the Whittaker function for $\mathscr{P}_n$ as an integral of Whittaker type functions of lower rank, such as the k-Bessel function (2.20). This is analogous to writing an Eisenstein series with n complex variables as a sum of Eisenstein series with a smaller number of complex variables (see Lemma 2 of Section 4.5, for example).

More explicitly, one can write the Whittaker function as a Fourier transform if a k-Bessel function (2.20). For example, when $n = 3$:

$$W(s|Y, r) = \int_{x_{12} \in \mathbb{R}} k_{2,1}\left(s \middle| Y \begin{bmatrix} 1 & -x_{12} & 0 \\ 0 & 1 & 0 \\ 0 & 0 & 1 \end{bmatrix}, (0, r_2)\right) \exp(2\pi i r_1 x_{12}) dx_{12}.$$

Then one can obtain properties of the Whittaker functions from those of the lower rank k-Bessel functions, and vice versa, since the k-Bessel function is also a Fourier transform of the Whittaker function. This same sort of idea relates the Fourier expansions of automorphic forms for $GL(n, \mathbb{Z})$ in formula (5.41) of Section 4.5 with those involving Whittaker functions.

As we mentioned at the beginning of this subsection, there are many papers on Whittaker functions, including those of Bump [1], Goodman and Wallach [1], Hashizume [1], Jacquet [1], Kostant [1], and Shalika [1]. For example, Hashizume considers a Whittaker model to come from intertwining operators between admissible representations of G and representations induced from a non-degenerate unitary character of N. He proves some general multiplicity results. Kostant connects the theory of Whittaker functions and the theory of Toda lattices. He obtains the complete integrability of the corresponding geometrically quantized system. Bump uses Shalika's multiplicity 1 theorem to argue that Whittaker functions give the Fourier coefficients of automorphic forms for $GL(3)$ and notes that Kostant shows that the solution space of the differential equations for $W(s|Y,r)$ coming from the operators in $D(\mathscr{P}_n)$ has dimension the order of the Weyl group, which is $n!$. But the solutions of polynomial growth form a 1-dimensional subspace (i.e., we have multiplicity one).

Piatetski-Shapiro (in Borel and Casselman [1, Vol. I, pp. 209–212]) uses the uniqueness of Whittaker models for representations to show that if π is an irreducible smooth admissible representation of the *adelized GL(n)*, then the multiplicity of π in the space of cusp forms is one or zero.

It is possible to use Theorem 2 and Propositions 2 and 3 of Section 4.2.1 to evaluate various special integrals.

Example 3. For $s \in \mathbb{C}$, define a vector $r(s) \in \mathbb{C}^{2m}$ by setting every entry of $r(s)$ equal to 0 except the mth entry which is set equal to s. Then, by part (2) of Theorem 2 in Section 4.2.1 and Exercise 14, we have:

$$k_{m,m}(r(s)|I,0) = \pi^{m^2/2} \Gamma_m\left(0,\ldots,0,\frac{m}{2} - s\right) \bigg/ \Gamma_m(0,\ldots,0,s).$$

Note that as in the proof of Proposition 2 in Section 4.2.1, we have the following equalities, where T_n is the group of upper triangular $n \times n$ matrices with positive diagonal entries:

$$k_{m,m}(r(s)|I,0) = \int_{X \in \mathbb{R}^{m \times m}} |I + X^t X|^{-s} dX$$

$$= \text{Vol}(K) \int_{t \in T_m} |I + I[t]|^{-s} \prod_{i=1}^m t_{ii}^{m-i} dt$$

$$= \text{Vol}(K) 2^{-m} \int_{Y \in \mathscr{P}_m} |I + Y|^{-s} |Y|^{m/2} d\mu_m(Y)$$

$$= \pi^{m^2/2} \Gamma_m(0,\ldots,0,m/2)^{-1} \int_{Y \in \mathscr{P}_m} |I + Y|^{-s} |Y|^{m/2} d\mu_m(Y).$$

It follows that, upon setting $A = $ the positive diagonal matrices and for $a \in A$:

$$D(a) = \prod_{1 \le i < j \le n} |a_i - a_j|, \quad da = \prod_{j=1}^n da_j/a_j,$$

$$I(s) = \int_{a \in A} |I + a|^{-s} |a|^{1/2} D(a) \, da = \frac{\Gamma_m(0,\ldots,0,\frac{m}{2} - s) \Gamma_m(0,\ldots,0,\frac{m}{2})}{\Gamma_m(0,\ldots,0,s)} c_m^{-1}.$$

4.2. Special Functions on $\mathscr{P}_n$

Here $c_m^{-1} = \pi^{-m^2/2} m! \Gamma_m(0, \ldots, 0, m/2)$. Thus

$$I(s) = \pi^{-m^2/2} m! \Gamma_m\left(0, \ldots, 0, \frac{m}{2} - s\right) \Gamma_m(0, \ldots, 0, m/2)^2 / \Gamma_m(0, \ldots, 0, s).$$

Exercise 18 (Mellin Transforms of K-Bessel Functions (A Generalization of Exercise 3 of Section 3.6 of Vol. I)). Show that if $s, r \in \mathbb{C}^n$, then if $B \in \mathscr{P}_n$,

$$\int_{A \in \mathscr{P}_n} p_s(A) K_n(r | B, A) \, d\mu_n(A) = \Gamma_n(s) \Gamma_n(s + r) p_{s+r}(B^{-1}).$$

Hint. (Bengtson [2].) Note that the left hand side is:

$$\int_{\mathscr{P}_n} \int_{\mathscr{P}_n} p_s(A) p_r(Y) \exp\{-\text{Tr}(Y + AY^{-1})\} \, d\mu_n(Y) \, d\mu_n(A).$$

Let $Y = I[t]$, $t \in T_n$ and change variables via $C = A[t^{-1}]$.

Note. Bump [1, Ch. X] computes a Mellin-type transform of his Whittaker function for $SL(3, \mathbb{R})$ and obtains a quotient of 6 gammas over one gamma. Such results are useful in the study of L-functions corresponding to automorphic forms for $GL(n)$, as we saw already in Volume I, pages 231–234.

Exercise 19 (A Functional Equation).

(a) Let ω and s^* be as defined in part (4) of Proposition 1 in §4.2.1. Show that if $A, B \in \mathscr{P}_n$, then

$$K_n(s | A, B) = K_n(s^* | B[\omega], A[\omega]).$$

(b) Show that if $k \in K = O(n)$, then for $s \in \mathbb{C}$, we have

$$K_m(0, s | A[k], B[k]) = K_m(0, s | A, B).$$

Exercise 20 (Inductive Formula for K-Bessel Functions). Prove that

$$K_m\left(s \left| \begin{pmatrix} A & 0 \\ 0 & B \end{pmatrix} \begin{bmatrix} I & 0 \\ {}^tQ & I \end{bmatrix}, I \right.\right) = \int_{X \in \mathbb{R}^{m \times (n-m)}} K_m(r_1 | A + B[{}^tX + {}^tQ], I)$$

$$\times K_{n-m}(r_2 | B, I + I[X]) \, dX.$$

Here for $s \in \mathbb{C}^n$, we have chosen $r_1 \in \mathbb{C}^m$, $r_2 \in \mathbb{C}^{n-m}$ such that:

$$p_s\begin{pmatrix} A & 0 \\ 0 & B \end{pmatrix} = p_{r_1}(A) |A|^{(m-n)/2} p_{r_2}(B) |B|^{m/2}.$$

Hint. See Terras [4].

Exercise 21 (Writing the Matrix K-Bessel Function as an Integral of the Ordinary K-Bessel Function). Show that if $r \in \mathbb{C}^{m-1}$, $s \in \mathbb{C}$,

$$K_m(r,s|A,B)$$
$$= \frac{2}{m} \int_{W \in \mathcal{SP}_m} p_r(W^{-1}) K_{ms}(2\sqrt{\text{Tr}(AW)\text{Tr}(BW^{-1})}) \left(\frac{\text{Tr}(BW^{-1})}{\text{Tr}(AW)}\right)^{ms/2} dW,$$

where the measure dW is chosen as in Exercise 23 of §4.1.3.

In Terras [4], motivated by the study of Fourier expansions of modular forms, we consider more general K-Bessel type functions with the power function p_s replaced by other sorts of eigenfunctions for $D(\mathcal{P}_n)$. We also find that we need to answer the question of whether it is possible to move the variable B over to be next to C in $K(s|W[C], V[B])$. Exercise 16 allows us to do something in this direction. But it does not seem to be exactly what we will need later. See the discussion before Exercise 27 in Section 4.5.3 below.

Thus we close this section on Bessel functions with too many questions unresolved. This will not be the last such section. See the quotation at the beginning of Section 4.5.

4.2.3. Spherical Functions

We want to find an analogue for $\mathcal{P}_n$ of the notion of spherical harmonic, the basic function for Fourier analysis on the sphere which was considered in Chapter 2 of Volume I. Of course the symmetric space under consideration is a higher rank analogue of the Poincaré upper half plane and thus it has spherical functions which generalize the Legendre or conical functions discussed in Section 3.2 of Volume I. We will see in the next section that we can use the spherical functions for $\mathcal{P}_n$ to obtain a Fourier transform on $\mathcal{P}_n$, generalizing Theorem 3 of Section 3.2, Vol. I to $\mathcal{P}_n$.

The theory of spherical functions really goes back to the study of spherical harmonics by Legendre, Laplace, and Jacobi in the late 1700's (see Chapter 2 of Vol. I). In 1916–1918 Funk [1] and Hecke [1, pp. 208–214] developed their integral formula for spherical harmonics (see Theorem 2, Ch. 2, Vol. I). In 1929 and 1934 Cartan [3] and Weyl [1, Vol. III, pp. 386–399] began the modern theory with the study of spherical functions associated with compact symmetric spaces. The compactness hypothesis was dropped in the 1950's by Gelfand [1], Godement [1], Harish-Chandra [2], and others. Selberg [1] gives the basic theory of spherical functions for the case under consideration. Other references for the general theory include: Barut and Rączka [1] (who call spherical functions "harmonic functions" on p. 302), Berezin and Gelfand [1], Dieudonné [1, Vol. V, Ch. XXI, Vol. VI, Ch. XXII], Ehrenpreis and Mautner [1], Gangolli [2], Godement [2], Helgason [1–8], Maurin [1], Mautner [1], Satake [1], Tamagawa [1], Warner [1], and Wawrzyńczyk [1].

Many of the authors listed above are motivated by the desire to understand the representation theory of Lie groups. Others are prompted by number-

4.2. Special Functions on $\mathcal{P}_n$

theoretic applications; e.g., connections with Hecke operators (see Section 3.6 of Vol. I and Section 4.5 following). Still others are inspired by the appearance of spherical functions in various statistical problems (see Farrell [1, 2], James [1–3], and Muirhead [1]). And finally, some are motivated by physical applications. The following references are indicative of some of the possibilities for applications of harmonic analysis on Lie groups in physics: Barut and Rączka [1], Mackey [1–3], Menotti, and Onofri [1].

We define a *spherical function* to be a function

$$h: \mathcal{P}_n \to \mathbb{C}$$

with the following properties:

1. $h(Y[k]) = h(Y)$, for all $Y \in \mathcal{P}_n$ and $k \in O(n)$;
2. $Lh = \lambda_L h$, $(\lambda_L \in \mathbb{C})$ for all invariant differential operators $L \in D(\mathcal{P}_n)$;
3. $h(I) = 1$.

(2.27)

That is, we are seeking a rotation-invariant eigenfunction for all the invariant differential operators, normalized to have the value 1 at the identity. This definition should be compared with (2.19) in the last section.

We should probably call the functions satisfying (2.27) "zonal spherical functions" or "spherical functions of class 1," but we will not do that here (cf. Section 2.1 of Vol. I), since we do not intend to consider the more general spherical functions transforming according to a non-trivial representation of K (but see Vol. I, pp. 89–91, 141).

As we shall see in Theorem 3, there are many ways to characterize spherical functions other than (2.27). In fact, H. Weyl has remarked that "their property as eigenfunctions of Laplace operators is merely accidental" (see Maurin [1, pp. 224–225]). However, one appears to have some difficulty in making connections with applications if one insists on throwing out the differential equations.

Example (Spherical Functions on the Poincaré Upper Half Plane). For the Poincaré upper half plane H, the spherical function is a standard special function—the Legendre or conical function discussed in Section 2 of Chapter 3. It is unique because it solves a 2nd order singular ODE whose 2nd solution has a singularity at i in H.

It is easy to write down such a spherical function by integration over $K = SO(2)$:

$$h(s|z) = \frac{1}{2\pi} \int_0^{2\pi} \text{Im}(k_{-u}(z))^s \, du, \quad \text{where } k_u = \begin{pmatrix} \cos u & \sin u \\ -\sin u & \cos u \end{pmatrix}.$$

It follows that

$$h(s|z) = P_{-s}(\cosh r) \quad \text{if} \quad z = k_u \exp(-r)i,$$

where the action of $g \in SL(2, \mathbb{R})$ on $z \in H$ is by fractional linear transformation (see page 141, Ch. 3, Vol. I). Here P_{-s} denotes the Legendre function which can be defined by the integral:

$$P_s(t) = \frac{1}{2\pi} \int_0^{2\pi} \{t + \sqrt{t^2 - 1} \cos u\}^s \, du.$$

The other solution to the 2nd order ODE satisfied by the Legendre function is called Q and it has the following asymptotic behavior as r approaches 0:

$$Q_{s-1}(\cosh r) \sim -(1/2) \log(\cosh r - 1), \qquad \text{as } r \to 0$$

(see Vol. I, pp. 269–270).

Motivated by the preceding example and the construction of k-Bessel functions satisfying (2.19) via formula (2.20); i.e., by integration over the appropriate subgroup of G, we construct a *spherical function* by integrating the power function over K; i.e.,

$$h_s(Y) = \int_{k \in K} p_s(Y[k]) \, dk, \qquad \text{for } Y \in \mathscr{P}_n, s \in \mathbb{C}^n. \tag{2.28}$$

Part (4) of Theorem 3 shows that these are the only spherical functions for $\mathscr{P}_n$.

In the following discussion (just as in (1.23) of Section 4.1.3), we will sometimes identify functions $f: \mathscr{P}_n \to \mathbb{C}$ with functions on $G = GL(n, \mathbb{R})$, by writing $Y = I[g]$, for $g \in G$. Such a function on G will be left K-invariant.

Next we will need to show that we could equivalently require the spherical functions to be eigenfunctions of convolution integral operators. In Section 3.7, Vol. I, the analogous result was proved for the Poincaré upper half plane. Recall now the definition of *convolution operators* in (1.24) of Section 4.1.3:

$$(f * g)(a) = C_g f(a) = \int_G f(b) g(ab^{-1}) \, db. \tag{2.29}$$

And recall Lemma 1 of that same section —a lemma which gave the properties of these convolution operators. The following proposition is necessary for the study of spherical functions as well as analogues of Poisson summation for $\mathscr{P}_n/GL(n, \mathbb{Z})$. In order to state it, we need to define the *Helgason-Fourier transform* of a function $f \in C_c(\mathscr{P}_n/K)$, which is:

$$\hat{f}(s) = \int_{Y \in \mathscr{P}_n} f(Y) \overline{p_s(Y)} \, d\mu, \qquad \text{for } s \in \mathbb{C}^n. \tag{2.30}$$

This transform will be scrutinized as carefully as we can manage in the next section. In the special case under consideration (i.e., when the function f is K-invariant) this transform can be identified with the spherical transform whose inversion formula was obtained by Harish-Chandra. And in the context of the present discussion it can also be called the "Selberg transform." We have named the transform for Helgason since his lectures [1] demonstrate clearly that the transform really does behave like the usual Fourier or Mellin

4.2. Special Functions on $\mathscr{P}_n$ 67

transform and can be used to solve some of the sorts of problems that Fourier transforms are traditionally used to solve in applied mathematics (e.g., those connected with the wave equation on a symmetric space). But the reader should be cautioned that this transform has a plethora of names in the literature.

Proposition 4.

(1) *The spherical function h corresponding to the eigenvalues $(\lambda_1, \ldots, \lambda_n) \in \mathbb{C}^n$, with*

$$\mathrm{Tr}((Y \partial/\partial Y)^i h) = \lambda_i h,$$

is unique. Here the invariant differential operators are from Theorem 2 of Section 4.1.4.

(2) Eigenfunctions of Invariant Differential Operators Are Eigenfunctions of Invariant Integral Operators.

Let $f \in C^\infty(\mathscr{P}_n)$ be an eigenfunction of all the G-invariant differential operators $L \in D(\mathscr{P}_n)$; i.e., $Lf = \lambda_L f$, for some $\lambda_L \in \mathbb{C}$. Define

$$s \in \mathbb{C}^n \text{ by } Lp_s = \lambda_L p_s,$$

where p_s denotes the power function (2.1). If $g \in C_c^\infty(K \backslash G / K)$; i.e., if g is K-bi-invariant and infinitely differentiable with compact support, and if we assume, in addition, that $g(x) = g(x^{-1})$, for all $x \in G$, then f is an eigenfunction of the convolution operator C_g in (2.29). More precisely,

$$C_g f = f * g = \hat{g}(\bar{s}) f,$$

with $\hat{g}$ denoting the Helgason-Fourier transform (2.30). Conversely, suppose that $f \in C(\mathscr{P}_n)$ is an eigenfunction of all the convolution operators C_g in (2.29), with $g \in C_c^\infty(\mathscr{P}_n/K)$. Then f is also an eigenfunction of all the invariant differential operators.

PROOF (Selberg [1, pp. 53-56]).

(1) Let h_i, $i = 1, 2$, be two spherical functions corresponding to the eigenvalues λ_L; i.e., $Lh_i = \lambda_L h_i$, $i = 1, 2$, for $L \in D(\mathscr{P}_n)$. Since h_i is a solution of an elliptic partial differential equation with analytic coefficients, by a theorem of Bernstein, h_i must be real analytic (see John [1, p. 142] or Garabedian [1, p. 164]). We want to show that all the terms in a Taylor expansion of $h_1 - h_2$ must be zero.

What is the Taylor expansion of a function f on a symmetric space like $\mathscr{P}_n$? It is best to view f as a function on $G = GL(n, \mathbb{R})$. The *Taylor expansion* of f is then

$$f(\exp Xg) = \sum_{n \geq 0} \frac{1}{n!} (\tilde{X}^n f)(g),$$

where $(\tilde{X}^n f)(g) = [d^n/ds^n f(\exp(sX)g)]_{s=0}$. Then $\tilde{X}^n$ is a right-invariant differential operator on G (see Helgason [1, p. 16] and Chapter V).

Suppose $L \in D(\mathcal{P}_n)$. We can think of L as a differential operator on G commuting with right translation. Form a left $K = O(n)$-invariant differential operator $L\#$ by taking the K-average of the transforms L^k of L under inner automorphism by $k \in K$. That is, let $i_k(x) = kxk^{-1}$, for $x \in G$. Then

$$L^k(f) = L(f \circ i_k) \circ i_k^{-1},$$

$$L\# = \int_{k \in K} L^k \, dk.$$

See Helgason [1, pp. 41–43] or [2, Chs. I and X], or [4] for more details on these constructions. The conclusion is that the differential operators in the Taylor series for f on the group G correspond to G-invariant differential operators on the symmetric space $K \backslash G$, with the same value at the identity.

It follows from our original hypothesis that all terms in the Taylor series for $h_1 - h_2$ must vanish at the identity. But we can translate everything by $g \in G$ to complete the proof.

(2) Here we imitate the proof of Lemma 3 of Section 3.7 in Vol. I. Define an operator M which averages functions over the compact group $K = O(n)$:

$$Mf(Y) = \int_{k \in K} f(Y[k]) \, dk.$$

Since f is assumed to be an eigenfunction of the G-invariant differential operators $L \in D(\mathcal{P}_n)$, it follows from part (1) that f is unique up to a constant. So we find that

$$Mf(Y) = f(I) h_s(Y),$$

where $h_s(Y)$ is the spherical function defined by (2.28); i.e.,

$$h_s(Y) = \int_{k \in K} p_s(Y[k]) \, dk.$$

Here we are using Exercise 11 of Section 4.2.1. It follows that

$$M(f * g)(a) = (Mf * g)(a) = f(I)(h_s * g)(a).$$

Evaluate this at $a = I$ to find that

$$\frac{(f * g)(I)}{f(I)} = \frac{(h_s * g)(I)}{h_s(I)},$$

since $h_s(I) = 1$ and $Mf(I) = f(I)$. Therefore this quotient is independent of the chosen eigenfunction f for all the $L \in D(\mathcal{P}_n)$. Thus, in particular, we can replace f by the power function p_s and get the same result. This means that

$$\frac{f * g}{f}(I) = \frac{p_s * g}{p_s}(I).$$

Next note that if $Lf = \lambda f$ and if $a \in G$, then defining f^a as in Lemma 1 of

4.2. Special Functions on $\mathcal{P}_n$

Section 1.3 or before (1.27) in Section 4.1.4, we have $Lf^a = \lambda f^a$, by the G-invariance of L. It follows that

$$\frac{f*g}{f}(a) = \frac{f^a*g}{f^a}(I) = \frac{p_s*g}{p_s}(I) = \hat{g}(\bar{s}).$$

This completes the proof of the first statement in part (2).

For the converse, look at $(f*g) = \lambda_g f$ and apply the G-invariant differential operator L to this equality to obtain:

$$\lambda_g Lf = L(f*g) = f*(Lg) = \lambda_{Lg} f.$$

This implies that

$$Lf = (\lambda_{Lg}/\lambda_g)f.$$

And it is easy to see that we can choose $\lambda_g \neq 0$ by taking g to run through a Dirac sequence at the identity. For then $f*g$ approaches $f = f*\delta$ and the eigenvalues λ_g must approach 1. □

Exercise 22.

(a) Fill in the details in the proof of part (2) of Proposition 1. For example, what happens if $f(a) = 0$?
(b) What happens in part (2) of Proposition 4 if we do not assume that $g(x) = g(x^{-1})$ for all $x \in G$?

Now we can give some other characterizations of spherical functions.

Theorem 3 (Equivalent Definitions of Spherical Functions). *Here* $G = GL(n, \mathbb{R})$, $K = O(n)$.

(1) Eigenfunctions of Convolution Operators.
   Let $\mathcal{A}$ denote the set of all $f: G \to \mathbb{C}$ which are continuous with compact support and K bi-invariant; i.e., $f(k_1 a k_2) = f(a)$ for all $k_i \in K$, $a \in G$. Then $\mathcal{A}$ is a commutative algebra under pointwise sum and convolution product. A function $h: G \to \mathbb{C}$ which is K bi-invariant with $h(I) = 1$ is a spherical function if and only if h is a common eigenfunction of all the convolution equations:

$$f*h = \lambda_L h \qquad \text{for all } f \in \mathcal{A}.$$

   Here the eigenvalue is λ_f.

(2) Homomorphisms of $\mathcal{A}$ (Gelfand).
   The spherical functions are the functions $h: G \to \mathbb{C}$ which are K bi-invariant and continuous such that the mapping

$$f \mapsto (f*h)(I)$$

   defines an algebra homomorphism of $\mathcal{A}$ onto $\mathbb{C}$.

(3) **More Integral Equations—The Analogue of the Funk-Hecke Theorem (Gelfand).**
A function $h\colon G \to \mathbb{C}$ which is continuous and K bi-invariant is spherical if and only if
$$\int_{v \in K} h(xvy)\,dv = h(x)h(y) \qquad \text{for all } x, y \in G.$$

(4) **Harish-Chandra's Integral Formula (Selberg [1, pp. 53–59]).**
A spherical function must be of the form (2.28) and if we use the r-variables from formula (2.3) in Section 4.2.1, we have for $Y \in \mathscr{P}_n$
$$h_s(Y) = \int_{k \in K} p_s(Y[k])\,dk = \int_{k \in K} \varphi_r(Y[k])\,dk, \qquad \text{if } \varphi_r(I[t]) = \prod_{j=1}^{n} t_{jj}^{2v_j}$$
where $2v_j = 2(s_j + \cdots + s_n) = 2r_j + j - (n+1)/2$, for $r \in \mathbb{C}^n$, and $t \in T_n$, the group of upper triangular $n \times n$ matrices with positive diagonal entries. Moreover $h_{s(r)} = h_{s(r')}$ if and only if $r' = (r_{\sigma(1)}, \ldots, r_{\sigma(n)})$ for some permutation σ of n elements.

(5) **Connection with Group Representations (Gelfand and Naimark).**
A spherical function h is called positive definite *if for any $f\colon G \to \mathbb{C}$ which is continuous with compact support, the following inequality holds:*
$$\int_G \int_G h(a^{-1}b)\overline{f(a)}f(b)\,da\,db \geq 0.$$
Positive definite spherical functions h can be expressed in the form
$$h(a) = (x_0 | U_a x_0), \qquad a \in G,$$
for some irreducible unitary representation U of G of class 1. Here x_0 is a K-fixed vector in the Hilbert space X on which U acts and $(x|y)$ denotes the Hilbert space inner product of x and y in X. Class 1 means that such a K-fixed x_0 must exist in X.

PROOF.
$(1) \Leftrightarrow (h \text{ is spherical})$.
This follows from Proposition 4 and the following exercise.

Exercise 23. Show that $\mathscr{A}$ is a commutative algebra.

Hint. Imitate the proof of part (4) of Lemma 2 in Section 3.7, Vol. I. This was essentially Exercise 26 of Section 4.1.3.

$(1) \Rightarrow (2)$.
Suppose that $(f * h) = \lambda_f h$ for all $f \in \mathscr{A}$. Then $(f * h)(I) = \lambda_f h(I) = \lambda_f$. Therefore
$$\lambda_{f*g} = ((f*g)*h)(I) = (f*(g*h))(I) = \lambda_f \lambda_g.$$

4.2. Special Functions on $\mathcal{P}_n$

(2) $\Rightarrow$ (3).
Suppose that h is as in (2); i.e., suppose that upon setting $\lambda_f = (f * h)(I)$, we have $\lambda_{f*g} = \lambda_f \lambda_g$, for all f, g in $\mathcal{A}$. We want to show that h satisfies the integral equation:
$$\int_K h(xvy)\, dv = h(x)h(y).$$
Now $\lambda_{f*g} = \lambda_f \lambda_g$ implies that
$$\int\int f(y^{-1})g(x^{-1})h(yx)\, dy\, dx = \int\int f(y^{-1})g(x^{-1})h(x)h(y)\, dx\, dy.$$
And the left hand side of this equality may be rewritten as:
$$\int\int f(y^{-1})g(x^{-1}) \int_K h(yvx)\, dv\, dy\, dx.$$
It follows that
$$\int_K h(yvx)\, dx = h(x)h(y) \text{ almost everywhere}$$
and continuity completes the proof.

(3) $\Rightarrow$ (h is spherical).
To see that (3) implies that h is infinitely differentiable, suppose that g is in $C_c^\infty(G)$ with the property that $\int_G g(y)h(y)\, dy \neq 0$. By (3) we have:
$$h(x)\int_G h(y)g(y)\, dy = \int_G g(y)\int_K h(xky)\, dk\, dy = \int_K g(k^{-1}x^{-1}u)\int_G h(u)\, du\, dk.$$
It follows that $h(x)$ is infinitely differentiable, since $g(x)$ is.

To see that indeed (3) implies that h is a spherical function, we must show that h is an eigenfunction for the G-invariant differential operators $L \in D(\mathcal{P}_n)$. This follows from the following considerations:
$$(L_x h(x))h(y) = L_x \int_K h(xvy)\, dv = \int_K (Lh)(xvy)\, dv.$$
Set $x = I$ to obtain $(Lh)(I)h(y) = (Lh)(y)$. This completes the proof that h is a spherical function, since the integral formula satisfied by h clearly implies that $h(I) = 1$.

This completes the proof that (1), (2), (3) are all equivalent to the definition of spherical function.

(4) $\Leftrightarrow$ (h is spherical).
Only the $\Leftarrow$ needs some discussion. We know from Theorem 2 of §4.1 that the algebra $D(\mathcal{P}_n)$ is the polynomial algebra over $\mathbb{C}$ generated by the algebraically independent operators $L_j = \text{Tr}((Y\partial/\partial Y)^j), j = 1, 2, \ldots, n$. Any spherical function h gives a homomorphism of the algebra $D(\mathcal{P}_n)$ into $\mathbb{C}$, defined by sending L in $D(\mathcal{P}_n)$ to the eigenvalue λ_L, with $Lh = \lambda_L h$.

We know from Theorem 1 of §4.2.1 that $L_j h_r = \lambda_j(r) h_r, j = 1, 2, \ldots, n$, where $\lambda_j(r)$ is a symmetric polynomial of degree j in $r_1, \ldots, r_n$, such that the highest degree homogeneous term is $r_1^j + \cdots + r_n^j$. It follows that the λ_j form a basis for the symmetric polynomials in $r_1, \ldots, r_n$. Now suppose we are given a spherical function h and thus a set of eigenvalues λ_j, with $L_j h = \lambda_j h$, $j = 1, \ldots, n$. The uniqueness of spherical functions (proved in part 1) of Proposition (2) implies that $h = h_r$.

Suppose next that $h_r = h_{r'}$ for $r, r' \in \mathbb{C}^n$. Then each polynomial $\lambda_i(r) = \lambda_i(r')$, for $i = 1, \ldots, n$. But then all the symmetric polynomials agree on r and r'. This implies that r' is obtained from r by permuting the entries r_j.

In this proof we used some facts about symmetric polynomials which are proved in Exercise 11 of Section 4.2.1.

(5) We omit the proof of part (5) of Theorem 3. Proofs can be found in Helgason [2, pp. 414–417] or Maurin [1, p. 233], for example. □

Note. Harish-Chandra's result in the preceding theorem, generalizing the unique characterization of spherical functions for the Poincaré upper half plane is very remarkable. For it is much harder to obtain uniqueness results for solutions of partial differential equations than for solutions of ordinary differential equations. One might ask whether the method of proof could be used to obtain an analogue for Bessel functions for $\mathscr{P}_n$. So far, this does not appear to be possible. For there is no obvious way to replace the operator M which averages functions over the compact group K with its analogue for the noncompact group $N(m, n - m)$ in (2.18) of Section 4.2.2. However, for Whittaker functions this might be possible. And indeed we will see in the next section that integrals over K (actually K/M) can be replaced by integrals over N (actually the opposite group $\bar{N}$ of lower triangular matrices with ones on the diagonal).

In order to imitate the discussion in Vol. I (pp. 156–162) of the central limit theorem for $SO(2)$-invariant densities on the Poincaré upper half plane H, we will need an asymptotic formula for the spherical function $h_s(Y)$ as Y approaches the identity matrix.

In Vol. I, p. 160, we claimed that the central limit theorem for rotation invariant densities on the Poincaré upper half plane followed from the following asymptotic formula:

$$h_{1/2+ip, 0}\begin{pmatrix} e^{(1/2)r} & 0 \\ 0 & e^{-(1/2)r} \end{pmatrix} = P_{-1/2+ip}(\cosh r) \sim J_0(pr), \quad \text{as } r \to 0,$$

$$\sim 1 - \tfrac{1}{4} r^2 p^2, \quad \text{as } r \to 0. \tag{2.31}$$

Here $P_\nu(z)$ is the Legendre function and $J_0(z)$ is the Bessel function. Actually for the central limit theorem we need to know 2nd order terms exactly and thus we really need the following formulas, using the standard power series for the Gauss hypergeometric function:

4.2. Special Functions on $\mathscr{P}_n$

$$P_\nu(\cosh r) = {}_2F_1(-\nu, \nu+1; 1; \tfrac{1}{2}(1-z))$$
$$= \sum_{k\geq 1} \frac{(-\nu)_k(\nu+1)_k}{k!(1)_k}\left(\frac{1-z}{2}\right)^k, \qquad |z-1| < 2.$$

It follows that, as $r \to 0$,

$$P_{-(1/2)+ip}(\cosh r) \sim 1 + \tfrac{1}{2}|\tfrac{1}{2} - ip|^2 (1 - \cosh r) \sim 1 - \tfrac{1}{4}(\tfrac{1}{4} + p^2)r^2.$$

Thus when $n = 2$ the zonal spherical function has the following asymptotic expansion:

$$h_{(1/2)+ip}\begin{pmatrix} e^r & 0 \\ 0 & e^{-r} \end{pmatrix} \sim 1 - \left(\frac{1}{4} + p^2\right)r^2, \qquad \text{as } r \to 0. \tag{2.32}$$

This does not change the fact that the central limit theorem holds for identically distributed rotation-invariant sequences of random variables on H. In fact (2.32) is even "better" than (2.31) in the sense that the coefficient of r^2 is exactly the eigenvalue of the Laplacian corresponding to the spherical function; i.e.,

$$\Delta h_{1/2+ip}\begin{pmatrix} e^r & 0 \\ 0 & e^{-r} \end{pmatrix} = -\left(\frac{1}{4} + p^2\right) h_{1/2+ip}\begin{pmatrix} e^r & 0 \\ 0 & e^{-r} \end{pmatrix}.$$

The arguments of Vol. I, pp. 156–162, go through exactly as before.

Despite the problem noted above, it is still useful to generalize (2.31) to the space $\mathscr{P}_n$. We can do this for general n using Helgason [8, Lemma 4.3] and some expansions of James [1–3]. This will not give us a central limit theorem, however. For that, we must generalize (2.32). We shall do that only in the case that $n = 3$, using the Taylor expansion of the zonal spherical function.

Let us first discuss Helgason [8, Lemma 4.3] in our case. First we will need some preliminaries from the theory of symmetric spaces and Lie groups. We saw in Exercise 12 of Section 4.1.2 that any matrix g in $GL(n, \mathbb{R})$ can be decomposed (uniquely) into *Iwasawa coordinates*:

$$g = kan, \qquad k \in K, \qquad a \in A, \qquad n \in N. \tag{2.33}$$
Write $k = K(g), a = A(g), n = N(g)$.

Corresponding to (2.33), we have an Iwasawa decomposition of the Lie algebra $\mathfrak{g}$ of G which is the tangent space to G at the identity equipped with a Lie bracket coming from identification of tangent vectors with left G-invariant first order differential operators (vector fields) on G. In our case, we have:

$$\mathfrak{g} = \mathfrak{gl}(n, \mathbb{R}) = \mathbb{R}^{n \times n}, \qquad \text{with Lie bracket } [X, Y] = XY - YX. \tag{2.34}$$

See Chapter 5 for more information on Lie algebras.

The *Lie algebra Iwasawa decomposition is:*

$$\mathfrak{g} = \mathfrak{k} \oplus \mathfrak{a} \oplus \mathfrak{n}, \quad \text{where}$$
$$\mathfrak{k} = \mathfrak{o}(n) = \{X \in \mathfrak{g} | {}^t X = -X\},$$
$$\mathfrak{a} = \{X \in \mathfrak{g} | X \text{ diagonal}\}, \quad (2.35)$$
$$\mathfrak{n} = \{X \in \mathfrak{g} | X \text{ is upper triangular, 0 on the diagonal}\}.$$

The *tangent space to* $\mathscr{P}_n$ *at* I can be identified with:

$$\mathfrak{p} = \{X \in \mathbb{R}^{n \times n} | {}^t X = X\}. \quad (2.36)$$

And we can clearly write

$$\mathfrak{p} = \mathfrak{a} \oplus \mathfrak{q}, \quad \text{with } \mathfrak{a} \text{ as in (2.35)}$$

and

$$\mathfrak{q} = \{X \in \mathfrak{p} | \text{diagonal entries of } X \text{ are } 0\}. \quad (2.37)$$

We will use the following notation for $X \in \mathfrak{p}$:

$$X = H + Y, \quad H \in \mathfrak{a}, \quad Y \in \mathfrak{q}, \quad H = a(X), \quad Y = q(X). \quad (2.38)$$

Because $\mathscr{P}_n$ is a symmetric space coming from the Lie group G, we have a *Cartan involution* $\theta: \mathfrak{g} \to \mathfrak{g}$ given by $\theta(X) = -{}^t X$. If σ_I denotes the geodesic-reversing isometry of $\mathscr{P}_n$ at the identity, then $(d\sigma)_I = \theta|_\mathfrak{p}$. We can therefore write, for $X \in \mathfrak{p}$:

$$q(X) = -(Z + \theta(Z)) + 2Z = Z + {}^t Z, \quad \text{for some } Z \in \mathfrak{n}. \quad (2.39)$$

Now, it is *not*, in general, true that for $X, Y \in \mathfrak{g}$, we have $\exp(X)\exp(Y) = \exp(X + Y)$. Instead there is an expansion called the *Campbell-Baker-Hausdorff formula*. Here we give only the first two terms:

$$\exp(tX)\exp(tY) = \exp\{t(X + Y) + \tfrac{1}{2}t^2[X, Y] + O(t^3)\}, \quad (2.40)$$

for $X, Y \in \mathfrak{g}$ and $t \in \mathbb{R}$ assumed to be small.

It follows that if $H \in \mathfrak{a}$, $k \in K$, $t \in \mathbb{R}$ (small),

$$\exp(tH[k] + O(t^2)) = \exp(-t(Z + \theta(Z)))\exp(ta(H[k]))\exp(2tZ), \quad (2.41)$$

for some $Z \in \mathfrak{n}$. Note that in (2.41),

$$\exp(-t(Z + \theta(Z))) \in K, \quad \exp(ta(H[k])) \in A, \quad \exp(2tZ) \in N.$$

Thus, the following lemma has been proved.

Lemma 1 (Helgason [8, Lemma 4.3]). *Suppose that* $H \in \mathfrak{a}$ *and* $k \in K = O(n)$, *using the notation (2.35). Then we have the following asymptotic relation as t approaches 0:*

$$A(\exp(tH[k])) \sim \exp(ta(H[k])),$$

where we use the notation set up in (2.33) and (2.38).

4.2. Special Functions on $\mathscr{P}_n$

This lemma has as an immediate consequence an asymptotic formula for the spherical function (2.28). Suppose that for $H \in \mathfrak{a}$ we write

$$H = \begin{pmatrix} h_1 & & 0 \\ & \ddots & \\ 0 & & h_n \end{pmatrix}, \qquad h_j \in \mathbb{R}.$$

Then we normalize the power function as in part (4) of Theorem 3, to obtain:

$$p_s(\exp H) = \prod_{j=1}^{n} \exp(h_j v_j), \tag{2.42}$$

with $v_j = s_j + \cdots + s_n = r_j + \frac{1}{2}j - \frac{1}{4}(n+1)$. Lemma 1 says that

$$p_s(\exp(tH[k])) \sim p_s(\exp(ta(H[k]))), \qquad \text{as } t \to 0.$$

Now $a(H[k])$ is just the diagonal part of the symmetric matrix

$$H[k] = ({}^t k_i H k_j)_{1 \le i,j \le n}, \qquad \text{for } k = (k_1 \cdots k_n) \in K. \tag{2.43}$$

Here k_j denotes the jth column of the orthogonal matrix k. So we find that

$$p_s(\exp(ta(H[k]))) = \prod_{j=1}^{n} \exp(tH[k_j] v_j) = \exp\left\{ t \sum_{j}^{n} H[k_j] v_j \right\}$$

$$= \exp\left\{ t \sum_{i,j=1}^{n} h_i v_j k_{ij}^2 \right\} = \exp\{ t \operatorname{Tr}(H[k] V) \},$$

where V is the diagonal matrix with jth diagonal entry v_j. The following result has now been proved.

Theorem 4 (An Asymptotic Formula for the Spherical Function at the Identity). *If*

$$H = \begin{pmatrix} h_1 & & 0 \\ & \ddots & \\ 0 & & h_n \end{pmatrix} \in \mathfrak{a}, \qquad s \in \mathbb{C}^n,$$

we have the following asymptotic formula for the spherical function in (2.28):

$$h_s(\exp(tH)) \sim \int_{k \in K = O(n)} \exp\{ t \operatorname{Tr}(H[k] V) \} \, dk,$$

where

$$V = \begin{pmatrix} v_1 & & 0 \\ & \ddots & \\ 0 & & v_n \end{pmatrix}, \qquad v_j = s_j + \cdots + s_n = r_j + \tfrac{1}{4}(2j - n - 1).$$

Helgason [4, pp. 423–467] considers such functions as the integral appearing in Theorem 4. And James [1–3] has extensively studied these functions with a view towards statistical applications (see also Farrell [1, 2], Muirhead [1], and Takemura [1]). We need to review some of this work. In the notation of James, the integral of interest is:

$${}_0F_0^{(n)}(X, Y) = \int_{k \in K} \exp\{\operatorname{Tr}(XkYk^{-1})\}\, dk, \quad \text{for } X, Y \in \mathfrak{p}. \quad (2.44)$$

James obtains an expansion of this integral in a series of zonal polynomials associated to partitions

$$\kappa = (k_1, \ldots, k_n) \text{ of } k; \text{ i.e., } k = k_1 + \cdots + k_n,$$

with $k_1 \geq k_2 \geq \cdots \geq k_n \geq 0$, and $k_j \in \mathbb{Z}$. Note that we are allowing the parts k_j to vanish.

If $\kappa = (k_1, \ldots, k_n)$ and $\lambda = (l_1, \ldots, l_n)$ are two partitions of k and if $k_j > l_j$ for the first index j for which the parts are unequal, then we say $\kappa > \lambda$ and the monomial

$$x_1^{k_1} \cdots x_n^{k_n}$$

is of *higher weight* than

$$x_1^{l_1} \cdots x_n^{l_n}.$$

That is, we use the lexicographic order. Now define the *zonal polynomial* $C_\kappa(Y)$, for partition κ, $Y \in \mathscr{P}_n$ with eigenvalues $a_1, \ldots, a_n$, to be a symmetric homogeneous polynomial of degree k in $a_1, \ldots, a_n$ such that

1. the term of highest weight is $d_\kappa a_1^{k_1} \cdots a_n^{k_n}$;
2. C_κ is an eigenfunction of the Laplacian and

$$\Delta^* C_\kappa = \lambda C_\kappa, \quad \text{where } \Delta^* = \text{the part of } \left\{\Delta + \frac{n-3}{2} \sum_j a_j \frac{\partial}{\partial a_j}\right\}$$

coming from the a_j variables; that is (cf. (1.36) in 4.1.5),

$$\Delta^* = \sum_j a_j^2 \frac{\partial^2}{\partial a_j^2} + \sum_{i \neq j} \left(\frac{a_i^2}{a_i - a_j}\right) \frac{\partial}{\partial a_j};$$

3. $(\operatorname{Tr} Y)^k = \sum_\kappa C_\kappa(Y),$

where the sum runs over all partitions of k into n parts some of which may vanish. Here we are using the well known property of Euler's operator

$$\sum a_i \frac{\partial}{\partial a_i}$$

(see Apostol [1, Vol. II, p. 287]).

Lemma 2. *The eigenvalue λ in the preceding definition is*

$$\lambda = \sum_i k_i(k_i - i) + k(n - 1).$$

4.2. Special Functions on $\mathscr{P}_n$

PROOF.

Exercise 24. Prove Lemma 2.

Hint. See Muirhead [18, p. 229]. □

It is possible to express the zonal polynomials C_κ in terms of the following *monomial symmetric functions* corresponding to a partition $\kappa = (k_1, \ldots, k_p)$. If $Y = a[k]$, for $a \in A$, $k \in K$, then define M_κ by:

$$M_\kappa(Y) = \sum_I a_{i_1}^{k_1} a_{i_2}^{k_2} \cdots a_{i_p}^{k_p},$$

where the sum is over all choices $I = \{i_1, \ldots, i_p\}$ of p distinct integers in $\{1, 2, \ldots, n\}$. For example,

$$M_{(1)}(Y) = a_1 + \cdots + a_n = \text{Tr}(Y),$$
$$M_{(2)}(Y) = a_1^2 + \cdots + a_n^2 = \text{Tr}(Y^2),$$
$$M_{(1,1)}(Y) = a_1 a_2 + \cdots + a_1 a_n + a_2 a_3 + \cdots + a_{n-1} a_n.$$

Clearly

$$M_{(1,1)}(Y) = \tfrac{1}{2}\{M_{(1)}^2(Y) - M_{(2)}(Y)\}.$$

Now it is easy to compute the first few zonal polynomials.

Example 1. $k = 1$.

$$C_{(1)}(Y) = M_{(1)}(Y) = \text{Tr}(Y).$$

Example 2. $k = 2$.

$$C_{(2)}(Y) = M_{(2)}(Y) + (2/3) M_{(1,1)}(Y).$$
$$C_{(1,1)}(Y) = (4/3) M_{(1,1)}(Y).$$

Exercise 25. Fill in the details for the preceding examples.

Hint. See Muirhead [18, pp. 231–232].

Finally we can state the expansion for ${}_0F_0^{(n)}(X, Y)$ obtained by James.

Proposition 5.

$${}_0F_0^{(n)}(X, Y) = \int_{k \in K} \exp\{\text{Tr}(XkYk^{-1})\} \, dk$$

$$= \sum_{k \geq 0} \frac{1}{k!} \sum_\kappa \frac{C_\kappa(X) C_\kappa(Y)}{C_\kappa(I)},$$

where the second sum is over all partitions κ of k.

PROOF (Following Muirhead [1, pp. 243–244, 258–260]).
Clearly

$$\exp\{\text{Tr}(XkYk^{-1})\} = \sum_{l \geq 0} \frac{1}{l!} \sum_{\lambda} C_\lambda(XkYk^{-1}).$$

One can complete the proof by showing that

$$\int_{k \in K} C_\kappa(XkYk^{-1}) \, dk = \frac{C_\kappa(X)C_\kappa(Y)}{C_\kappa(I)}. \tag{2.45}$$

To see this, consider the left hand side as a function of Y and call it $f(Y)$. Then $f(Y) = f(Y[k])$ for all $k \in K$. Thus $f(Y)$ depends only on the eigenvalues of Y, and moreover, it must be a homogeneous symmetric polynomial of degree k in these eigenvalues. Now use the G-invariance of Δ_Y and standard facts about the *Euler differential operator*

$$x_1 \partial/\partial x_1 + \cdots + x_n \partial/\partial x_n$$

acting on homogeneous functions (see Apostol [1, Vol. II, p. 287]) to see that

$$\Delta_Y^* f(Y) = \lambda f(Y).$$

It follows that $f(Y) = \alpha C_\kappa(Y)$ for some scalar α. Set $Y = I$ and note that $f(I) = C_\kappa(X)$ to complete the proof. □

Thus, in particular, if

$$H = \begin{pmatrix} h_1 & & 0 \\ & \ddots & \\ 0 & & h_n \end{pmatrix}, \quad V = \begin{pmatrix} v_1 & & 0 \\ & \ddots & \\ 0 & & v_n \end{pmatrix},$$

we obtain the following expansion:

$$\int_{k \in K} \exp\{\text{Tr}(HkVk^{-1})\} \, dk$$

$$= 1 + \frac{1}{n}\text{Tr}(H)\text{Tr}(V) + \frac{4}{3n(n-1)} \sum_{i<j} h_i h_j \sum_{i<j} v_i v_j$$

$$+ \frac{3}{2n(n+2)}\left(\sum_i h_i^2 + \frac{2}{3}\sum_{i<j} h_i h_j\right)\left(\sum_i v_i^2 + \frac{2}{3}\sum_{i<j} v_i v_j\right) \tag{2.46}$$

+ higher order terms.

Since Theorem 2 is only good to first order, we do not expect this expansion to hold for the spherical function $h_s(\exp H)$, though we do expect some similarity.

Next let us do the Taylor expansion of $h_s(\exp H)$ directly when $n = 3$. The differential operators appearing in the expansion will be evaluated at $H = 0$ and they can be identified with G-invariant differential operators

4.2. Special Functions on $\mathscr{P}_n$

$L \in D(\mathscr{P}_n)$ evaluated at I (see Helgason [2,4] and the proof of part (1) of Proposition 4 above). So, making use of the symmetry of the function in the h_j, we obtain the following form of the expansion, where
$$|i| = i_1 + \cdots + i_p:$$

$h_s(\exp H)$

$$= \sum_p \frac{1}{p!} \sum_{|i|=p} \left[\frac{\partial^p h_s}{\partial h_1^{i_1} \cdots \partial h_n^{i_n}} \right]_{h_j=0} h_1^{i_1} \cdots h_n^{i_n}$$

$$= 1 + (\alpha_1(r_1 + \cdots + r_n) + \beta_1)(h_1 + \cdots + h_n)$$

$$+ \{\alpha_2(r_1^2 + \cdots + r_n^2) + \beta_2(r_1 + \cdots + r_n)^2 + \gamma_2(r_1 + \cdots + r_n) + \delta_2\}(h_1^2 + \cdots + h_n^2)$$

$$+ \{\alpha_3(r_1^2 + \cdots + r_n^2) + \beta_3(r_1 + \cdots + r_n)^2 + \gamma_3(r_1 + \cdots + r_n) + \delta_3\}$$

$$\times (h_1 h_2 + \cdots + h_{n-1} h_n) + \text{higher order terms.} \qquad (2.47)$$

Here the r-variables are related to the s-variables as in Theorem 4 above.

The first order terms in (2.47) are easy to find when $n = 3$. Here let $k = (k_1 k_2 k_3) \in K = O(3)$; i.e., k_j denotes the jth column of the 3×3 rotation matrix k. Then

$$p_s(a[k]) = a[k_1]^{s_1} |a[k_1 k_2]|^{s_2} |a|^{s_3}$$
$$= a[k_1]^{s_1} \{a[k_1]a[k_2] - ({}^t k_1 a k_2)^2\}^{s_2} |a|^{s_3},$$

and

$$a[k_j] = \sum_{j=1}^{3} a_i k_{ij}^2, \qquad {}^t k_i a k_j = \sum_{l=1}^{3} k_{li} a_l k_{lj}, \qquad |a| = a_1 a_2 a_3.$$

Therefore, we have

$$\frac{\partial}{\partial h_i} p_s(a[k])$$
$$= s_1 a[k_1]^{s_1-1} k_{i1}^2 a_i |a[k_1 k_2]|^{s_2} |a|^{s_3}$$
$$+ s_2 a[k_1]^{s_1} |a[k_1 k_2]|^{s_2-1} |a|^{s_3} a_i \{k_{i1}^2 a[k_2] + k_{i2}^2 a[k_1] - 2({}^t k_1 a k_2) k_{i1} k_{i2}\}$$
$$+ s_3 a[k_1]^{s_1} |a[k_1 k_2]|^{s_2} |a|^{s_3}. \qquad (2.48)$$

If we set $h_1 = h_2 = h_3 = 0$, then $a[k] = I$ and we obtain:

$$\left[\frac{\partial}{\partial h_i} p_s(a[k]) \right]_{\substack{h_j=0 \\ j=1,2,3}} = s_1 k_{i1}^2 + s_2(k_{i1}^2 + k_{i2}^2) + s_3. \qquad (2.49)$$

It is known (see page 104 of Vol. I and Broecker and tom Dieck [1]) that since k_{ij} is the entry of an irreducible representation of $O(3)$, we have the formula:

$$\int_{k \in K} k_{ij}^2 \, dk = 1/3. \qquad (2.50)$$

You can check this using the *Euler angle decomposition* (see Vilenkin [1, pp. 106, 435–440] or Wawrzyńczyk [1, pp. 287–291]):

$$0 \le \alpha, \gamma \le 2\pi, \qquad 0 \le \beta \le \pi,$$
$$dk = (8\pi^2)^{-1} \sin \beta \, d\alpha \, d\beta \, d\gamma,$$

$$k = \begin{pmatrix} \cos\alpha & \sin\alpha & 0 \\ -\sin\alpha & \cos\alpha & 0 \\ 0 & 0 & 1 \end{pmatrix} \begin{pmatrix} 1 & 0 & 0 \\ 0 & \cos\beta & \sin\beta \\ 0 & -\sin\beta & \cos\beta \end{pmatrix} \begin{pmatrix} \cos\gamma & \sin\gamma & 0 \\ -\sin\gamma & \cos\gamma & 0 \\ 0 & 0 & 1 \end{pmatrix}$$

$$= \begin{pmatrix} \cos\alpha\cos\gamma - \cos\beta\sin\alpha\sin\gamma & \cos\alpha\sin\gamma + \cos\beta\sin\alpha\cos\gamma & \sin\alpha\sin\beta \\ -\sin\alpha\cos\gamma - \cos\alpha\cos\beta\sin\gamma & -\sin\alpha\sin\gamma + \cos\alpha\cos\beta\cos\gamma & \cos\alpha\sin\beta \\ \sin\beta\sin\gamma & -\sin\beta\cos\gamma & \cos\beta \end{pmatrix}.$$
(2.51)

Thus, for example,

$$(2\pi)^2 \frac{1}{8\pi^2} \int_0^\pi \cos^2 \beta \sin \beta \, d\beta = \frac{1}{3}.$$

One can use permutation matrices to see that all the integrals

$$\int_{k \in K} k_{ij}^2 \, dk$$

must be equal. We could also evaluate them using James' formula (2.46). It follows that

$$\int_{k \in K} \left[\frac{\partial}{\partial h_i} p_s(a[k]) \right]_{\substack{h_j = 0 \\ j = 1, 2, 3}} dk = \frac{1}{3}(s_1 + 2s_2 + 3s_3). \tag{2.52}$$

Recall that (as in (2.42))

$$s_3 = r_3 + \tfrac{1}{2} = v_3,$$
$$s_2 = r_2 - r_3 - \tfrac{1}{2} = v_2 - v_3, \tag{2.53}$$
$$s_1 = r_1 - r_2 - \tfrac{1}{2} = v_1 - v_2.$$

Therefore the right hand side of (2.52) can be rewritten as

$$\tfrac{1}{3}(r_1 + r_2 + r_3) = \tfrac{1}{3}(v_1 + v_2 + v_3),$$

which is exactly the first order term in (2.46) for $n = 3$.

Let us now compute the 2nd derivative (a sum of 10 terms):

$$\frac{\partial^2}{\partial h_j \partial h_i} p_s(a[k]) = s_1(s_1 - 1) a[k_1]^{s_1 - 2} k_{i1}^2 k_{j1}^2 a_i a_j |a[k_1 k_2]|^{s_2} |a|^{s_3}$$
$$+ \delta_{ij} s_1 a[k_1]^{s_1 - 1} k_{i1}^2 a_i |a[k_1 k_2]|^{s_2} |a|^{s_3}$$
$$+ s_1 s_2 a[k_1]^{s_1 - 1} k_{i1}^2 |a[k_1 k_2]|^{s_2 - 1} a_i a_j$$
$$\times (k_{j1}^2 a[k_2] + k_{j2}^2 a[k_1] - 2({}^t k_1 a k_2) k_{j1} k_{j2}) |a|^{s_3}$$

4.2. Special Functions on $\mathscr{P}_n$

$$+ s_1 s_2 a[k_1]^{s_1-1} k_{j1}^2 |a[k_1 k_2]|^{s_2-1} a_i a_j$$
$$\times (k_{i1}^2 a[k_2] + k_{i2}^2 a[k_1] - 2({}^t k_1 a k_2) k_{i1} k_{i2}) |a|^{s_3}$$
$$+ s_1 s_3 a[k_1]^{s_1-1} (k_{i1}^2 a_i + k_{j1}^2 a_j) |a[k_1 k_2]|^{s_2} |a|^{s_3}$$
$$+ s_2 a[k_1]^{s_1} |a[k_1 k_2]|^{s_2-1} a_i a_j \{ k_{i1}^2 k_{j2}^2 + k_{i2}^2 k_{j1}^2$$
$$- 2k_{i1} k_{i2} k_{j1} k_{j2} \} |a|^{s_3} + s_2 a[k_1]^{s_1} |a[k_1 k_2]|^{s_2-1} a_i \delta_{ij}$$
$$\times \{ k_{i1}^2 a[k_2] + k_{i2}^2 a[k_1] - 2({}^t k_1 a k_2) k_{i1} k_{i2} \} |a|^{s_3}$$
$$+ s_2(s_2 - 1) a[k_1]^{s_1} |a[k_1 k_2]|^{s_2-2} a_i a_j |a|^{s_3}$$
$$\times (k_{i1}^2 a[k_2] + k_{i2}^2 a[k_1] - 2({}^t k_1 a k_2) k_{i1} k_{i2})$$
$$\times (k_{j1}^2 a[k_2] + k_{j2}^2 a[k_1] - 2({}^t k_1 a k_2) k_{j1} k_{j2})$$
$$+ s_2 s_3 a[k_1]^{s_1} |a[k_1 k_2]|^{s_2-1} |a|^{s_3}$$
$$\times \{ a_i (k_{i1}^2 a[k_2] + k_{i2}^2 a[k_1] - 2({}^t k_1 a k_2) k_{i1} k_{i2})$$
$$+ a_j (k_{j1}^2 a[k_2] + k_{j2}^2 a[k_1] - 2({}^t k_1 a k_2) k_{j1} k_{j2}) \}$$
$$+ s_3^2 a[k_1]^{s_1} |a[k_1 k_2]|^{s_2} |a|^{s_3}.$$

If we set $h_1 = h_2 = h_3 = 0$, we obtain:

$$\left. \frac{\partial^2}{\partial h_j \partial h_i} p_s(a[k]) \right|_{\substack{h_k=0 \\ k=1,2,3}} = s_1(s_1 - 1) k_{i1}^2 k_{j1}^2 + \delta_{ij} k_{i1}^2 s_1$$
$$+ s_1 s_2 k_{i1}^2 (k_{j1}^2 + k_{j2}^2) + \delta_{ij} s_2 (k_{i1}^2 + k_{i2}^2)$$
$$+ s_1 s_2 k_{j1}^2 (k_{i1}^2 + k_{i2}^2)$$
$$+ s_2 \{ k_{i1}^2 k_{j2}^2 + k_{i2}^2 k_{j1}^2 - 2 k_{i1} k_{i2} k_{j1} k_{j2} \}$$
$$+ s_2(s_2 - 1)(k_{i1}^2 + k_{i2}^2)(k_{j1}^2 + k_{j2}^2)$$
$$+ s_1 s_3 (k_{i1}^2 + k_{j1}^2) + s_2 s_3 (k_{i1}^2 + k_{i2}^2 + k_{j1}^2 + k_{j2}^2) + s_3^2.$$

To integrate over K, we need some integral formulas:

$$\int_{k \in K} k_{ii}^4 \, dk = 1/5, \qquad \int_{k \in K} k_{i1}^2 k_{i2}^2 \, dk = 1/15. \qquad (2.54)$$

These formulas are easily proved using the Euler angle formula (2.51) or from formula (2.46).

It follows that when $i = j$ we have

$$\int_{k \in K} \left[\frac{\partial^2}{\partial h_i^2} p_s(a[k]) \right]_{\substack{h_j=0 \\ j=1,2,3}} dk = \frac{1}{5} s_1(s_1 - 1) + \frac{1}{3}(s_1 + 2s_3) + \frac{8}{15} s_1 s_2$$
$$+ \frac{8}{15} s_2(s_2 - 1) + \frac{2}{3} s_1 s_3 + \frac{4}{3} s_2 s_3 + s_3^2.$$

Using (2.53) this is the same as:

$$\frac{3}{15}(v_1^2 + v_2^2 + v_3^2) + \frac{2}{15}(v_1 v_2 + v_1 v_3 + v_2 v_3) + \frac{2}{15}(v_1 - v_3)$$

$$= \frac{3}{15}(r_1^2 + r_2^2 + r_3^2) + \frac{2}{15}(r_1 r_2 + r_1 r_3 + r_2 r_3) - \frac{1}{15}.$$

We are not interested in the terms with $i \neq j$ since they will disappear in the central limit theorem. So our expansion becomes:

$h_s(\exp H)$

$$\sim 1 + \frac{1}{3}(r_1 + r_2 + r_3)(h_1 + h_2 + h_3)$$

$$+ \frac{1}{30}\{3(r_1^2 + r_2^2 + r_3^2) + 2(r_1 r_2 + r_1 r_3 + r_2 r_3) - 1\}(h_1^2 + h_2^2 + h_3^2)$$

$$+ p(r)(h_1 h_2 + h_1 h_3 + h_2 h_3) + \text{higher order terms.} \quad (2.55)$$

Here $p(r)$ is a symmetric polynomial of degree 2 which we will not need to evaluate.

Note that the coefficient of $h_1^2 + h_2^2 + h_3^2$ is not the eigenvalue of Δ given in Exercise 12 of Section 4.2.1. This has interesting consequences for the central limit theorem. *Recently D. St. P. Richards generalized (2.55) to $\mathcal{P}_n$, for all n.*

Exercise 26. Check the evaluations of integrals over $O(3)$ given in formulas (2.50) and (2.54). Find $p(r)$ in (2.55).

This completes our discussion of spherical functions for the present. In the next section we will consider another sort of asymptotic formula for spherical functions. In the preceding formula Y approached I, but in that of Section 4.3 the variable Y will approach the boundary of the symmetric space. It will be necessary to find such an expansion in order to obtain the inversion formula for the Helgason-Fourier transform on $\mathcal{P}_n$.

It is possible to prove a Weyl character formula (cf. Chapter 2 of Vol. I) in the framework of spherical functions (see Harish-Chandra [2] and Berezin [1]). Gelfand and Naimark [1] do the special case of the symmetric space $SL(n, \mathbb{C})/SU(n)$ (cf. Chapter V). Recall that the Weyl character formula expresses the characters of the irreducible representations of compact semisimple Lie groups as ratios of exponential polynomials on a maximal abelian Lie subalgebra.

The continuous homorphisms from the algebra $\mathcal{A}$ (defined in Theorem 3) onto the complex numbers are maps $f \mapsto (f * h)(I)$, provided that h is a bounded spherical function (see Helgason [2, p. 410]). Helgason and Johnson [1] characterize the bounded spherical functions in terms of the s-variables. Flensted-Jensen discovered relations between spherical functions on real

4.2. Special Functions on $\mathcal{P}_n$

semisimple Lie groups (such as $SL(n, \mathbb{R})$) and the more elementary spherical functions on the corresponding complex groups (e.g., $SL(n, \mathbb{C})$) (see Helgason [4, pp. 489–490]). Healy [1] has made a study of relations between Fourier analysis on $SL(2, \mathbb{C})/SU(2)$ and that on the sphere. See also Chapter V.

4.2.4. The Wishart Distribution

In this subsection, we will find applications of some of the results of the preceding subsections. In particular, we will obtain a theorem of Wishart which is important both for multivariate statistics and for the study of some number theoretical results to be discussed in Section 4.4.

References for this subsection are Anderson [1], Farrell [1, 2], Herz [1], James [1–3], Morrison [1], Muirhead [1], and Press [1].

A random matrix $Y \in \mathcal{P}_n$ is said to have the (central) *Wishart distribution* $W(\Sigma, p, n)$ with scale matrix Σ and n degrees of freedom, $p \leq n$, if the joint distribution of the entries of Y has the density function:

$$f(Y) = c |Y|^{(n-p-1)/2} |\Sigma|^{-n/2} \exp(-\tfrac{1}{2} \operatorname{Tr}(\Sigma^{-1} Y)), \quad \Sigma \in \mathcal{P}_p,$$

$$\text{with } c = 2^{-np/2} \pi^{-p(p+1)/4} \left(\prod_{j=1}^{p} \Gamma\left(\frac{n+1-j}{2}\right) \right)^{-1} \quad (2.56)$$

$$= 2^{-np/2} \Gamma_p(0, \ldots, 0, n/2)^{-1}.$$

This density function was first obtained by Fisher in 1915 for the case that $p = 2$ and by Wishart [1] in 1928 for general p.

Exercise 27. Show that in order for

$$\int_{\mathcal{P}_n} f(Y) \prod_{i \leq j} dy_{ij} = 1,$$

the constant c in formula (2.56) must be as stated.

Hint. Use the formula for $\Gamma_p(0, \ldots, 0, s)$ in Section 4.2.1.

Theorem 5 (Wishart, 1928). *Let X_i be a random variable in $\mathbb{R}^p$, for $i = 1, 2, \ldots, n$, where $p \leq n$. And suppose that $X_1, \ldots, X_n$ are mutually independent and distributed according to $N(0, \Sigma)$; that is, normal with mean 0 and covariance matrix $\Sigma \in \mathcal{P}_n$ (as in Section 4.1.6). Let $X = (X_1, \ldots, X_n) \in \mathbb{R}^{p \times n}$ and $Y = X\,{}^tX$. Then, with probability one, Y is in $\mathcal{P}_p$ and is distributed according to the Wishart density $W(\Sigma, p, n)$ defined by (2.56).*

PROOF. First one must show that $Y = X\,{}^tX$ is positive definite with probability one. We leave this to the reader in Exercise 28 below.

The joint density of $X_1, \ldots, X_n$ is (cf. Section 4.1.6):

$$d(X) = (2\pi)^{-np/2} |\Sigma|^{-n/2} \exp\left(-\frac{1}{2} \sum_{j=1}^{n} {}^t X_j \Sigma^{-1} X_j\right)$$

$$= (2\pi)^{-np/2} |\Sigma|^{-n/2} \exp\left(-\frac{1}{2} \text{Tr}(\Sigma^{-1}(X\,{}^tX))\right).$$

To complete the proof, we simply need to see how to change variables from X in $\mathbb{R}^{p \times n}$ to $V = X\,{}^tX$ in the closure of $\mathscr{P}_p$. First note that

$$\int_{X \in \mathbb{R}^{p \times n}} h(X\,{}^tX) |X\,{}^tX|^{-n/2}\, dX, \quad \text{with } dX = \prod_{i,j} dx_{ij},$$

defines a $GL(p, \mathbb{R})$-invariant measure on functions $H: \mathscr{P}_p \to \mathbb{C}$. Such measures on $\mathscr{P}_p$ are unique up to a constant. Exercise 27 shows that the constant given in (2.56) is correct. This completes the proof of Theorem 5. □

The proof of Theorem 5 proves *Wishart's formula* for $p \le n$:

$$\int_{X \in \mathbb{R}^{p \times n}} h(X\,{}^tX) |X\,{}^tX|^{-n/2}\, dX = w_{p,n} \int_{Y \in \mathscr{P}_p} h(Y)\, d\mu_p(Y),$$

$$w_{p,n} = \prod_{j=n-p+1}^{n} \pi^{j/2} \Gamma(j/2)^{-1} = \pi^{np/2} \Gamma_p(0, \ldots, 0, n/2)^{-1}.$$

(2.57)

Later (see Section 4.4) this formula will be very important to us in our study of matrix series.

Exercise 28.

(a) Show that under the hypotheses of Theorem 5, $Y = X\,{}^tX$ is in $\mathscr{P}_p$ with probability one.
(b) Show that if $p > n$, $X \in \mathbb{R}^{p \times n}$, then $Y = X\,{}^tX$ is singular.

Hint. (See Muirhead [1, pp. 82–83].) Note that if the columns of X are linearly indpendent then the matrix Y is non-singular.

The Wishart distribution is a matrix analogue of the density function for the chi-square distribution (which is the case $p = 1$). Univariate statistics makes frequent use of tables of the chi-square distribution, which is, in fact, an incomplete gamma function.

If the random variables in Theorem 5 did not all have mean zero, then one would be dealing with the noncentral Wishart distribution, written in terms of the integral

$$\int_{k \in K} \exp(\text{Tr}(XkY))\, dk,$$

which is a generalization of the J-Bessel function (see Herz[1], James

4.2. Special Functions on $\mathscr{P}_n$

[1–3], Farrell [1–2], Muirhead [1, pp. 441–449]). The noncentral Wishart distribution was first studied by T.W. Anderson in 1946 in special cases. In 1955 Herz expressed the distribution in terms of the $_0F_1$ matrix argument hypergeometric function. In the early 1960's James and Constatine gave the zonal polynomial expansion for it.

The Wishart distribution is important, for example, in factor analysis, which seeks to explain correlation between a set of random variables in terms of a minimum number of factors. Such analysis is useful in many of the social sciences. There are many examples and references in Press [1, Chapter 10]. One example given by Press involves a seven-factor analysis of prices of 63 securities. One of the factors appeared to have an effect on all securities. The remaining 6 factors tended to group the stocks by industry.

It is possible to use Proposition 4 of Section 4.2.3 to evaluate some special integrals which arise in number theory and statistics (see Maass [2, Ch. 7] and Muirhead [1, Ch. 7]). We want to evaluate an integral which appears in Muirhead [1, Theorem 7.2.7, p. 248]:

$$I(B,r,s) = \int_{Y \in \mathscr{P}_n} \exp[-\mathrm{Tr}(B^{-1}Y)] |B^{-1}Y|^r h_s(Y) \, d\mu_n(Y), \qquad (2.58)$$

for $B \in \mathscr{P}_n$, $r \in \mathbb{C}$, $s \in \mathbb{C}^n$. Here $h_s(Y)$ denotes the spherical function defined by (2.28) in the preceding subsection. We can use Proposition 4 of the preceding subsection to evaluate $I(B,r,s)$ as

$$I(B,r,x) = \hat{f}(\bar{s}^*) h_s(B), \qquad \text{with } f(Y) = |Y|^r \exp[-\mathrm{Tr}(Y)], \qquad (2.59)$$

and $s^* = (s_{n-1}, \ldots, s_2, s_1, -(s_1 + \cdots + s_n))$ as in part (4), Proposition 1, Section 4.2.1.

We should perhaps discuss the proof of (2.59), since Proposition 4, Section 4.2.3 considered only functions $g \in C_c^\infty(K \backslash G / K)$ such that g is invariant under inversion. Our function $f(Y)$ does not satisfy these hypotheses, setting $Y = I[x]$, $x \in G$ to make it a function on G. We can easily do away with the hypothesis that f have compact support. But to do away with the hypothesis that f be invariant under inversion on G is impossible. But note that our function f has the property that $f(I[x]) = f(I[^tx])$ for all $x \in G$. Therefore we have the following equalities:

$$(p_{s*}f)(I) = \int_G p_s(I[x]) f(I[x^{-1}]) \, dx = \int_G p_s(I[x]) f(I[^tx^{-1}]) \, dx$$

$$= \int_{\mathscr{P}_n} p_s(Y) f(Y^{-1}) \, d\mu_n(Y) = \int_{\mathscr{P}_n} p_s(Y^{-1}) f(Y) \, d\mu_n(Y)$$

$$= \int_{\mathscr{P}_n} p_s(Y^{-1}[\omega]) f(Y) \, d\mu_n(Y) = \int_{\mathscr{P}_n} p_{s^*}(Y) f(Y) \, d\mu_n(Y),$$

where ω and s^* are as in part (4) of Proposition 1, Section 4.2.1. This last integral is $\hat{f}(\bar{s}^*)$, and our discussion of (2.59) is finished.

Now the Helgason-Fourier transform $\hat{f}$ is:

$$\hat{f}(s) = \int_{Y \in \mathscr{P}_n} \overline{p_s(Y)} |Y|^r \exp[-\text{Tr}(Y)] \, d\mu_n(Y) = \Gamma_n((0,\ldots,0,r) + \bar{s}).$$

Thus we have proved that:

$$\int_{Y \in \mathscr{P}_n} \exp[-\text{Tr}(B^{-1}Y)] |B^{-1}Y|^r h_s(Y) \, d\mu_n(Y) = \Gamma_n(s^* + (0,\ldots,0,r)) h_s(B), \tag{2.60}$$

where $B \in \mathscr{P}_n$, $r \in \mathbb{C}$, $s \in \mathbb{C}^n$. This says that a spherical function is reproduced upon taking expectations with respect to the Wishart distribution. Muirhead [1, p. 260] uses (2.60) to show that one obtains the matrix argument $_{p-1}F_q$ function as the matrix Laplace transform (as defined in Exercise 2 of Section 4.2.1) of the matrix $_pF_q$ function. Herz [1] had defined the matrix argument hypergeometric functions recursively by taking matrix Laplace transforms in this way. See James [2] for a nice summary of the facts about the matrix argument hypergeometric functions and their statistical applications. For example, James notes [2, p. 481] that the zonal polynomials C_κ that we considered in the preceding subsection can be expressed as integrals over $O(n)$ of the characters of the general linear group. James defined the matrix argument hypergeometric functions as series of these zonal polynomials.

Exercise 29. Prove that the integral $I(B, r, s)$ defined in (2.58) really is the one considered by Muirhead [loc. cit.]. To do this you must show that given a partition κ of k, one can find a special choice $s(\kappa) \in \mathbb{C}^n$ so that

$$\alpha h_s(Y) = C_\kappa(Y), \quad \text{for some constant } \alpha.$$

Here $C_\kappa(Y)$ is the zonal polynomial which we defined in the preceding subsection (see the discussion before Proposition 5).

Hint. Clearly $h_s(Y)$ is a K-invariant eigenfunction of the Laplacian. If you choose $s \in \mathbb{C}^n$ to be a vector of integers, then $h_s(Y)$ is indeed a symmetric polynomial.

4.3. Harmonic Analysis on $\mathscr{P}_n$ in Polar Coordinates

... die Schwierigkeit beginnt da, wo es sich darum handelt, aus diesem Labyrinth von Formeln einen Ausweg zu finden ....[†]

Frobenius (quoted by Siegel [1, Vol. III, p. 373]).

[†] The difficulty begins when one must find a way out of this labyrinth of formulas.

4.3.1. Properties of the Helgason-Fourier Transform on $\mathcal{P}_n$

The main goal of this section is the discussion of an inversion formula for the Helgason-Fourier transform defined in formula (2.30) of Section 4.2.3 when the function is K-invariant. The subject contains a labyrinth of formulas, similar to that occurring in any higher rank symmetric space (as in the quote of Frobenius above, which refers to the formulas for multidimensional theta functions). The discussion is intended to provide a way through the labyrinth—a route which follows that set out by Harish-Chandra and Helgason, particularly Helgason's Battelle lectures [1]. In outlining the path, we will not provide all the details of the arguments. For example, our discussion of Fourier inversion (Theorem 1) will use the analogue of the asymptotics/functional equations principle from Section 3.2 of Vol. I, pages 138–139. We will not give a rigorous justification of the principle. That would require an analysis similar to that given in discussions of the Paley-Wiener theorem. See Helgason [4, pp. 55–56, 452–453] for a careful treatment of the argument that we are omitting.

The Fourier inversion formula for $\mathcal{P}_n$ (in part (1) of Theorem 1) writes a smooth compactly supported function on $\mathcal{P}_n$ as a superposition of eigenfunctions of the generalized Laplacians in $D(\mathcal{P}_n)$. This provides the fundamentals of harmonic analysis on $\mathcal{P}_n$ and an analogue of Theorem 3 in Section 3.2 of Volume I. This result can also be viewed as an analogue of the Mellin inversion formula (see Exercise 1 of Section 1.4 of Vol. I), which is the case $n = 1$.

A reader interested in delving further into the labyrinth of formulas associated with Fourier analysis on Lie groups and symmetric spaces might consult some of the references below. We discuss some of these other points of view briefly at the end of this section.

References for this section include: Bhanu-Murthy [1], Ehrenpreis and Mautner [1], Flensted-Jensen [1], Gangolli [1–3], Gelfand and Graev [1], Harish-Chandra [2], Helgason [1–7], Herb and Wolf [1], Koornwinder [1], Rosenberg [1], Varadarajan [2, 3], Vergne [1], Wallach [1, 2], Warner [1], and Wawrzyńczyk [1].

Define the *Helgason-Fourier Transform* of a (sufficiently nice) function $f : \mathcal{P}_n \to \mathbb{C}$, for $s \in \mathbb{C}^n$, $k \in K = O(n)$, by:

$$\mathcal{H}f(s,k) = \int_{Y \in \mathcal{P}_n} f(Y) \overline{p_s(Y[k])} \, d\mu_n(Y). \tag{3.1}$$

Here $p_s(Y)$ denotes the power function in formula (2.1) of Section 4.2.1.

Note that if we set $f(Y) = \exp\{-\text{Tr}(Y)\}$ in (3.1), we obtain

$$\mathcal{H}f(s,k) = \Gamma_n(\bar{s}),$$

with the gamma function as in (2.4) of Section 4.2.1.

If we consider the Helgason-Fourier transform $\mathcal{H}f(s, k)$ as a function of its second variable $k \in K = O(n)$, the function depends only on the coset $kM = \bar{k}$

in the boundary K/M of the symmetric space, where M is the group of all diagonal matrices with entries ± 1. See the discussion of the boundary and formulas (1.19) and (1.20) as well as Exercise 22 in Section 4.1.3. The inversion formula for the Helgason-Fourier transform will involve an integral over the boundary K/M and we shall assume that the measure $d\bar{k}$ on K/M is normalized to give

$$\int_{\bar{k}\in K/M} d\bar{k} = 1. \tag{3.2}$$

Theorem 1 (Properties of the Helgason-Fourier Transform).

(1) Inversion Formula.
   Suppose that $f: \mathscr{P}_n \to \mathbb{C}$ is infinitely differentiable with compact support. If $\mathscr{H}f(s,k)$ denotes the Helgason-Fourier transform defined by formula (3.1), then

$$f(Y) = \omega_n \int_{\substack{s\in\mathbb{C}^n \\ \text{Re}\,s=-\rho}} \int_{\bar{k}\in K/M} \mathscr{H}f(s,k) p_s(Y[k]) \, d\bar{k} \, |c_n(s)|^{-2} \, ds,$$

where $\rho = (\tfrac{1}{2},\ldots,\tfrac{1}{2},\tfrac{1}{4}(1-n))$,

$$\omega_n = \prod_{j=1}^{n} \frac{\Gamma(j/2)}{j(2\pi i)\pi^{j/2}},$$

and $c_n(s)$ denotes the Harish-Chandra c-function given by:

$$c_n(s) = \prod_{1\leq i\leq j\leq n-1} \frac{B(\tfrac{1}{2}, s_i + \cdots + s_j + \tfrac{1}{2}(j-i+1))}{B(\tfrac{1}{2}, \tfrac{1}{2}(j-i+1))}.$$

Here $B(x,y) = \Gamma(x)\Gamma(y)/\Gamma(x+y)$, the beta function.

(2) Convolution Property.
   If either f or g is a K-invariant function on $\mathscr{P}_n$ satisfying the hypothesis of part (1), then defining convolution as in formula (1.24) of Section 4.1.3:

$$\mathscr{H}(f*g) = \mathscr{H}f \cdot \mathscr{H}g.$$

(3) G-invariant Differential Operators Changed to Multiplication by a Polynomial.
   If $L \in D(\mathscr{P}_n)$ and f is as in (1), then

$$\mathscr{H}(Lf)(s,k) = \overline{\lambda_{L^*}(s)} \mathscr{H}f(s,k), \qquad \text{where } Lp_s(Y) = \lambda_L(s) p_s(Y).$$

Here L^* denotes the adjoint of L (see Theorem 1 in Section 4.2.1). Note that the eigenvalue $\lambda_{L^*}(s)$ is a polynomial in s.

(4) Plancherel Theorem in the K-invariant Case.
   Let $\alpha(s) = \omega_n |c_n(s)|^{-2}$. For f as in (1) and K-invariant, we have

$$\int_{\mathscr{P}_n} |f(Y)|^2 \, d\mu_n(Y) = \int_{\text{Re}\,s=-\rho} |\hat{f}(s)|^2 \alpha(s) \, ds.$$

4.3. Harmonic Analysis on $\mathscr{P}_n$ in Polar Coordinates

Moreover, the Helgason-Fourier transform can be extended to an isometry between $L^2(\mathscr{P}_n/K, d\mu_n)$ and functions of s which are square integrable with respect to $\alpha(s)\,ds$ and invariant under permutations of the r-variables which are related to the s-variables as in (3.3) below (see also Proposition 1 of Section 4.2.1).

The discussion of the proof of Theorem 1 will extend through the next several subsections. There are four main steps. The first step (in Section 4.3.2) is a reduction to the corresponding inversion formula for the determinant one surface $\mathscr{SP}_n$. The second step (also in Section 4.3.2) is due to Helgason (see [1, pp. 60–61]). Using arguments similar to those which we gave to prove Fourier inversion on $\mathbb{R}^n$ (see the proof of Theorem 1 in Section 1.2 of Vol. I), Helgason demonstrates that it suffices to show the inversion formula at any point Y in the symmetric space and for any Dirac family of functions $g(Y)$. Thus, in particular, it suffices to show the inversion formula for $O(n)$-invariant functions. The third step in the proof of the inversion formula (found in Section 4.3.3) is due to Harish-Chandra [2] and Bhanu-Murthy [1]. It proves the inversion formula in the $O(n)$-invariant case, when the Helgason-Fourier transform is often called the spherical transform. The idea is similar to that which we used in Section 3.2 of Vol. I to prove the Kontorovich-Lebedev and Mehler-Fock inversion formulas. Thus one reduces the computation of the spectral measure or Harish-Chandra c-function to the determination of the asymptotics and functional equations of spherical functions. As we mentioned above, we will not give a rigorous justification of this principle—only a heuristic argument. We have already found that the spherical functions on $\mathscr{P}_n$ satisfy $n!$ functional equations (see Theorem 3 of Section 4.2.3). The fourth step (also in Section 4.3.3) is the determination of the asymptotic behavior of these functions as the symmetric space variable approaches the boundary. To do this we shall rewrite Harish-Chandra's integral formula in part (4) of the theorem just cited, as an integral over $\bar{N}$, which is the group of lower triangular matrices with ones on the diagonal. This is analogous to the discussion indicated in Exercise 10, Section 3.2, Vol. I for the case of the Poincaré upper half plane.

The details of the asymptotics/functional equations argument require Helgason's version of the Paley-Wiener theorem for $\mathscr{P}_n$ (see Helgason [4–7], Gangolli [3]). We will not discuss this here or give the proof of part (4) of Theorem 1 which is the Fourier inversion formula for L^2 functions on $\mathscr{P}_n$. Nor shall we discuss the Helgason-Fourier transform on Harish-Chandra's Schwartz space for $\mathscr{P}_n$. See Helgason [4, p. 489] for an exercise on the subject. Or see Gangolli [2, pp. 78–82].

Before beginning our discussion of Theorem 1, we need to make a few preliminary remarks and do some exercises.

Remarks.

(1) Using the change of variables from formula (2.9) of Section 4.2.1, we have

$$p_s(I[t]) = \varphi_r(t) = \prod_{i=1}^{n} t_{ii}^{2r_i+i-(n+1)/2},$$

for $t \in T_n$, which is the group of upper triangular $n \times n$ real matrices with positive diagonal entries. Here then we have

$$2r_i + i - (n+1)/2 = 2(s_i + \cdots + s_n). \tag{3.3}$$

It follows that if we write Harish-Chandra's c-function in terms of the r-variables, we obtain

$$c_n(s) = \prod_{1 \leq i \leq j \leq n-1} \frac{B(\tfrac{1}{2}, r_i - r_{j+1})}{B(\tfrac{1}{2}, \tfrac{1}{2}(j-i+1))}. \tag{3.4}$$

Furthermore, $\operatorname{Re} s = -\rho$ implies that $\operatorname{Re} r_i = 0$. Of course, we cannot allow the second argument of the beta function to be zero.

(2) In the case $n = 1$, Theorem 1 is just ordinary Mellin inversion (see Exercise 1 of Section 1.4 in Volume I).

(3) When $n = 2$, Theorem 1 is just ordinary Mellin inversion plus the Helgason inversion formula in Theorem 3, Section 3.2, Vol. I. To see this, suppose that $f: \mathscr{P}_2 \to \mathbb{C}$ is $O(2)$-invariant. Then the Helgason transform of f is:

$$\mathscr{H} f(s, k) = \frac{\pi}{2} \int_{a \in A} f(a) \overline{h_s(a)} \gamma(a) \, da,$$

$$= \pi \int_{a \in A^+} f(a) \overline{h_s(a)} \gamma(a) \, da,$$

where $A^+ = \{a \in A | a_1 > a_2\}$ (a positive Weyl chamber),

$$da = \frac{da_1 \, da_2}{a_1 \, a_2}, \quad \gamma(a) = |a|^{-1/2} |a_1 - a_2|, \quad h_s(a) = \int_{k \in K} p_s(a[k]) \, dk.$$

Here we have used the integral formula for polar coordinates (see formulas (1.37), (1.42), and (1.43) from Section 4.1.5). Now change variables according to

$$a = \begin{pmatrix} a_1 & 0 \\ 0 & a_2 \end{pmatrix} = v^{1/2} \begin{pmatrix} u & 0 \\ 0 & u^{-1} \end{pmatrix}, \quad \text{writing } f(a) = f(v, u).$$

Then we see that

$$a_1 = u\sqrt{v}, \quad a_2 = \sqrt{v}/u,$$

and

$$\left| \frac{\partial(a_1, a_2)}{\partial(u, v)} \right| = \left| \begin{matrix} \sqrt{v} & -\sqrt{v}/u^2 \\ u/(2\sqrt{v}) & 1/(2u\sqrt{v}) \end{matrix} \right| = 1/u.$$

If we write $f(a) = f(v, u)$, then the Helgason-Fourier transform is:

$$\mathscr{H} f(s_1, s_2) = \pi \int_{u \geq 1} \int_{v > 0} f(v, u) h_{\bar{s}}(v, u)(u - u^{-1}) \frac{du \, dv}{u \, v}.$$

4.3. Harmonic Analysis on $\mathscr{P}_n$ in Polar Coordinates

And the spherical function is:

$$h_s(v,u) = (2\pi)^{-1} v^{s_2} \int_0^{2\pi} (a_1 \cos^2\theta + a_2 \sin^2\theta)^{s_1} d\theta$$

$$= v^{s_2+s_1/2} P_{s_1}(\cosh(\log u)),$$

where P_s denotes the Legendre function from Exercise 9, Section 3.2, Vol. I. To see the last formula, write

$$Y = \begin{pmatrix} a_1 & 0 \\ 0 & a_2 \end{pmatrix} \begin{bmatrix} \cos\theta & -\sin\theta \\ \sin\theta & \cos\theta \end{bmatrix} = \begin{pmatrix} y_1 & * \\ * & * \end{pmatrix}$$

and note that

$$y_1 = a_1 \cos^2\theta + a_2 \sin^2\theta = a_1 + (a_2 - a_1)\sin^2\theta$$

$$= a_1 + (a_2 - a_1)(1 - \cos 2\theta)/2 = \tfrac{1}{2}(a_1 + a_2) - \tfrac{1}{2}(a_2 - a_1)\cos 2\theta.$$

Thus

$$y_1 = v^{1/2} \left\{ \frac{u + u^{-1}}{2} - \frac{u - u^{-1}}{2} \cos 2\theta \right\}.$$

So the inversion formula which we seek becomes, upon setting $x = \cosh \log u$ and $x_1 = \cosh \log u_1$:

$$f(v,u) = \omega_2 \int_{\operatorname{Re} s = -\rho} 2\pi \int_{v_1 > 0, x_1 > 1} f(v_1, u_1) v_1^{\overline{s_2+s_1/2}} \overline{P_{s_1}(x_1)} dx_1 \frac{dv_1}{v_1} v^{s_2+s_1/2}$$

$$\times P_{s_1}(x) |c_2(s)|^{-2} ds,$$

for the constant ω_2 of part (1) in Theorem 1. By formula (3.23) in Section 3.2 of Volume I, we see that

$$|c_2(-\tfrac{1}{2} + it)|^{-2} = \pi t \tanh \pi t.$$

Thus we obtain the result of Theorem 1 in the case $n = 2$ using ordinary Mellin inversion (Exercise 1 in Section 1.4 of Volume I) and the Mehler-Fock inversion formula (formula (3.22) in Section 3.2 of Volume I). For this, we need:

$\operatorname{Re} s_1 = -1/2$ and $\operatorname{Re} s_2 = 1/4$, with $\omega_2 = (2\pi i)^{-2}(2\pi)^{-1} = -(2\pi)^{-3}$.

(4) Note that if we consider the spectral measure $|c_n(s)|^{-2}$ from part (1) of Theorem 1, as a function of $r = it$, in formula (3.3), we are looking at:

$$\prod_{1 \le i \le j \le n-1} \left| B\left(\frac{1}{2}, r_i - r_{j+1}\right) \right|^{-2};$$

that is,

$$\prod_{1 \le i \le j \le n-1} \pi |t_i - t_{j+1}| \tanh(\pi |t_i - t_{j+1}|). \tag{3.5}$$

Exercise 1 (Properties of the Helgason-Fourier Transform).

(a) Prove part (2) of Theorem 1 above. Show that the hypothesis of K-invariance is necessary.
(b) Prove part (3) of Theorem 1.

Hint. (a) Use the power function identity (3.17) below. Note that the convolution property implies that $f * g = g * f$.

(5) *The K-Invariant Case of Theorem* 1. Suppose the function $f(Y)$ in (3.1) is K-invariant; i.e., $f(Y[k]) = f(Y)$ for all $Y \in \mathcal{P}_n$ and $k \in K = O(n)$. Then the Helgason-Fourier transform (3.1) is really only a function of the s-variable and we will write

$$\mathcal{H}f(s, k) = \hat{f}(s), \tag{3.6}$$

for K-invariant functions f, as in (2.30) of Section 4.2.3. Let $h_s(Y)$ denote the spherical function defined as in formula (2.28) of Section 4.2.3; i.e.,

$$h_s(Y) = \int_{\bar{k} \in K/M} p_s(Y[k]) \, d\bar{k}, \tag{3.7}$$

with $d\bar{k}$ normalized as in (3.2). Then we see from formulas (1.37), (1.42), and (1.43) of Section 4.1.5 that

$$\hat{f}(s) = b_n \int_{a_i > a_{i+1}} f(a) \overline{h_s(a)} J(a) \prod da_j,$$

where

$$a = \begin{pmatrix} a_1 & & 0 \\ & \ddots & \\ 0 & & a_n \end{pmatrix}, \quad b_n = \pi^{(n^2+n)/4} \prod_{j=1}^{n} \Gamma(j/2)^{-1}, \tag{3.8}$$

and

$$J(a) = \prod_{j=1}^{n} a_j^{-(n+1)/2} \prod_{1 \le i < j \le n} (a_i - a_j).$$

The transform (3.8) is often called the "spherical transform" (see Helgason [4, p. 449]).

Exercise 2. Prove that the spherical function $h_s(Y)$ is bounded when $\text{Re } s = -\rho$. Note that in the r-variables of (3.3), we are just saying that the spherical function is bounded when the powers r_i are purely imaginary. Helgason and Johnson [1] obtain a more precise result (see Helgason [4, pp. 458–466]). In fact, Helgason shows (loc. cit.) that the spherical functions are bounded on a tube domain where $\hat{f}(s)$ is holomorphic for K-invariant f. And Helgason [4, p. 480] shows, using the Riemann-Lebesgue lemma, that for $r = it$, the spherical function approaches 0 as $\|t\| \to \infty$. (See Lebedev [1, p. 191] for the case $G = SL(2, \mathbb{R})$, when h_s is $P_{-1/2+it}$.)

4.3. Harmonic Analysis on $\mathscr{P}_n$ in Polar Coordinates

Hints. Another reference is Gangolli [2]. You can use the fact that a holomorphic function of $s \in \mathbb{C}^n$ which is bounded on a region must be bounded on the convex hull of that region. The exercise requires use of the functional equations of the spherical function. Recall (3.3) relating the r- and s-variables. The permutation of r_i and r_{i+1} corresponds in the s-variables to the map from s to s' given by:

$$s'_i = -1 - s_i,$$
$$s'_{i\pm 1} = s_{i\pm 1} + s_i - \tfrac{1}{2},$$
$$s'_j = s_j, \quad j \neq i, i \pm 1.$$

So $\operatorname{Re} s_i = 0$ corresponds to $\operatorname{Re} s'_i = -1$. The convex hull contains $\operatorname{Re} s_i = -\tfrac{1}{2}$.

Exercise 3. Suppose that f is a K-invariant function on $\mathscr{P}_n$ of the sort considered in part (1) of Theorem 1; i.e., assume that $f(a) = 0$, for $a \in A$ such that the diagonal entries a_j satisfy

$$\Sigma \log a_j^2 > R^2.$$

This means that the geodesic distance $d(I, I[a])$ is greater than R. Prove that for every G-invariant differential operator $L \in D(\mathscr{P}_n)$ such that the eigenvalue polynomial $\lambda_L(s)$ does not vanish; i.e., $L p_s = \lambda_L(s) p_s$, with $\lambda_L(s) \neq 0$, there is a positive constant C such that

$$|\hat{f}(s)| \leq C \exp(\|u\| R) |\lambda_L(s)|^{-1}.$$

Here

$$u \in \mathbb{C}^n, \quad u_j = \operatorname{Re}(s_j + \cdots + s_n), \quad \|u\| = (\Sigma u_j^2)^{1/2}.$$

Note that we can find operators L to make $\lambda_L(s)$ have arbitrarily high degree. This result is the converse of the Paley-Wiener theorem proved by Helgason [4, pp. 450–454].

Hint. Note that

$$|p_s(a)| \leq \prod_{j=1}^n \exp(u_j \log a_j).$$

Use Iwasawa coordinates to write $\hat{f}(s)$ as a composition of a Mellin transform over A and the Harish transform

$$F_f(I[a]) = \alpha(a)^{1/2} \int_{n \in N} f(I[an]) \, dn$$

where $\alpha(a)$ is the Jacobian of Iwasawa coordinates as defined in (1.37) of Section 4.1.5. If is not hard to see that $d(I, I[a]) > R$ implies that $F_f(I[a]) = 0$. For you need only note that

$$d(I, I[a]) \leq d(I, I[an]) \quad \text{for all } a \in A, n \in N.$$

When we restrict everything to the determinant one surface, so that $|a| = 1$, we will write in (3.18) $\alpha(a) = p_{2\rho}(I[a])$, where $p_j = \frac{1}{2}, j = 1, \ldots, n - 1$. In this notation we have:

$$\int_{Y \in \mathcal{SP}_n} \overline{p_s(Y)} f(Y) \, dY = 2^n \int_{\substack{a \in A \\ |a|=1}} \overline{p_{s+\rho}(I[a])} F_f(a) \, da.$$

Note that if for $j = 1, \ldots, n - 1$, $\operatorname{Re} s_j = -\frac{1}{2}$, as in the Fourier inversion formula, then $\operatorname{Re}(s + \rho)_j = 0, j = 1, \ldots, n - 1$.

Exercise 4. Show that with f as in Exercise 3, we have

$$\int_{\operatorname{Re} s = -\rho} \hat{f}(s) |c_n(s)|^{-2} \, ds < \infty.$$

Hint. You need a bound on $|c_n(s)|^{-2}$ when $\operatorname{Re} s = -\rho$. For this, use formula (3.5).

4.3.2. Beginning of the Discussion of Part (1) of Theorem 1—Steps 1 and 2

Now we start to contemplate how one might prove somthing like part (1) of Theorem 1.

Step 1 (Pulling Out the Determinant). Write $Y = v^{1/n} W$ with $v > 0$ and $W \in \mathcal{SP}_n$, which is the determinant one surface in $\mathcal{P}_n$. Then by Exercise 23 of Section 4.1.3, we can normalize measures so that

$$d\mu_n(Y) = v^{-1} \, dv \, dW, \tag{3.9}$$

where dW is an $SL(n, \mathbb{R})$-invariant measure on $\mathcal{SP}_n$. It follows that the Helgason-Fourier transform can be rewritten as:

$$\mathcal{H}f(s, k) = \int_{v > 0} v^{\bar{r}} \int_{W \in \mathcal{SP}_n} f(v^{1/n} W) \overline{p_s(W[k])} \, dW \, v^{-1} \, dv, \tag{3.10}$$

$$\text{where } r = n^{-1} \sum_{j=1}^{n} j s_j.$$

It suffices to assume that $f(v^{1/n} W) = f_1(v) f_2(W)$. Then

$$\mathcal{H}f(s, k) = Mf_1(\bar{r}) \mathcal{H}^0 f_2(s, k), \tag{3.11}$$

where Mf_1 is the ordinary Mellin transform of f_1 (see Section 1.4, Vol. I) and $\mathcal{H}^0 f_2$ denotes *the Helgason-Fourier transform on the determinant one surface*:

$$\mathcal{H}^0 f_2(s, k) = \int_{W \in \mathcal{SP}_n} f_2(W) \overline{p_s(W[k])} \, dW. \tag{3.12}$$

4.3. Harmonic Analysis on $\mathscr{P}_n$ in Polar Coordinates

It follows from the ordinary Mellin inversion formula that the Harish-Chandra c-function does not depend on the variable s_n. Moreover, we need $\operatorname{Re} r = 0$ in the final inversion formula, and thus using formula (3.10):

$$\operatorname{Re} s_n = -n^{-1} \sum_{j=1}^{n-1} \operatorname{Re}(js_j).$$

Since we shall show that we need $\operatorname{Re} s_j = -\frac{1}{2}$, it follows that we need:

$$\operatorname{Re} s_n = (n-1)/4. \tag{3.13}$$

For the rest of proof of part (1) of Theorem 1 we will replace $\mathscr{P}_n$ by $\mathscr{SP}_n$ and G will denote $SL(n, \mathbb{R})$.

Step 2 (Helgason's Reduction to the Case that f Is K-Invariant). We base the following analysis on Helgason [1, pp. 60–61]. The idea is to imitate the proof of the Euclidean Fourier inversion formula (Theorem 1 of Section 1.2, Volume I). Let $\mathscr{T}$ denote the *inverse transform* defined for nice functions:

$$F: \mathbb{C}^{n-1} \times (K/M) \to \mathbb{C}$$

and for $Y \in \mathscr{SP}_n$ by:

$$\mathscr{T}F(Y) = \int_{\operatorname{Re} s = -\rho} \int_{\bar{k} \in K/M} F(s, \bar{k}) p_s(Y[k]) |c_n(s)|^{-2} \, d\bar{k} \, ds. \tag{3.14}$$

Lemma 1 (Two Properties of the Helgason-Fourier Transform and Its Inverse Transform).

(1) *Let $\mathscr{T}$ be defined by (3.14) and choose $\rho = (\frac{1}{2}, \ldots, \frac{1}{2}) \in \mathbb{C}^{n-1}$. For $x \in G$ and $f: \mathscr{SP}_n \to \mathbb{C}$, define*

$$f^x(W) = f(W[x]), \quad \text{for } W \in \mathscr{SP}_n.$$

Then $\mathscr{TH}^0$ commutes with the action of G; i.e.,

$$\mathscr{TH}^0(f^x) = (\mathscr{TH}^0 f)^x, \quad \text{for all } x \in G = SL(n, \mathbb{R}).$$

(2) *If f and g are infinitely differentiable functions with compact support on $\mathscr{SP}_n$, then*

$$\int_{\mathscr{SP}_n} f(W) \overline{\mathscr{TH}^0 g(W)} \, dW = \int_{\mathscr{SP}_n} (\mathscr{TH}^0 f)(W) \overline{g(W)} \, dW;$$

i.e., $\mathscr{TH}^0$ is self-adjoint.

PROOF.

(1) First we need to show that $\mathscr{TH}^0$ makes sense for functions f which are smooth and compactly supported. This follows from the estimate in Exercise 3 in the preceding subsection.

The main fact needed to prove part (1) is a certain identity satisfied by the power function $p_s(Y)$. This identity (3.17) provides a kind of substitute for the following property of exponentials:

$$\exp(x + y) = \exp(x)\exp(y), \quad \text{for } x, y \in \mathbb{R}.$$

To describe this identity, we need to recall the Iwasawa decomposition (see Exercise 20 of Section 4.1.3) of $x \in G = GL(n, \mathbb{R})$ into

$$x = K(x)A(x)N(x),$$

with $K(x) \in K = O(n)$, $A(x) \in A$, the group of positive diagonal matrices in G, and $N(x) \in N$, the (nilpotent or unipotent) group of upper triangular elements of G with ones on the diagonal. An element $x \in G$ acts on the boundary element $\bar{k} = kM$, with M defined to be the group of diagonal matrices lying in K, according to the formula:

$$x(\bar{k}) = K(xk)M \tag{3.15}$$

(see formula (1.20) of Section 4.1.3). Here $K(xk)$ denotes the K-part of xk in the Iwasawa decomposition.

It will help to use the following *notation for the power function*:

$$p_s(Y[k]) = p_s(Y, kM) = p_s(Y, \bar{k}). \tag{3.16}$$

Then we have the *power function identity*:

$$p_s(Y, \bar{k}) = p_s(Y[x], x^{-1}(\bar{k}))p_s(I[x^{-1}], \bar{k}), \tag{3.17}$$

for all $x \in G$, $Y \in \mathscr{SP}_n$, $\bar{k} \in K/M$. To prove formula (3.17), let $x^{-1}k = k_1 a_1 n_1$, with $k_1 \in K$, $a_1 \in A$, $n_1 \in N$. Clearly

$$p_s(Y[k]) = p_s(Y[xx^{-1}k]) = p_s(Y[xk_1])p_s(I[a_1]),$$

which gives the desired identity.

To prove part (1) of Lemma 1, note that for $x \in G$:

$$\mathscr{TH}^0(f^x)(Y) = \int\int\int f(W[x])p_{\bar{s}}(W[k])p_s(Y[k])|c_n(s)|^{-2} \, dW \, d\bar{k} \, ds$$

$$= \int\int\int f(V)p_{\bar{s}}(V[x^{-1}k])p_s(Y[k])|c_n(s)|^{-2} \, dV \, d\bar{k} \, ds$$

$$= \int\int\int f(V)p_{\bar{s}}(V, x^{-1}(\bar{k}))p_{\bar{s}}(I[x^{-1}], \bar{k})p_s(Y[k])|c_n(s)|^{-2} \, dV \, d\bar{k} \, ds$$

$$= \int\int\int f(V)p_{\bar{s}}(V, \bar{k})p_{\bar{s}}(I[x^{-1}], x(\bar{k}))p_s(Y, x(\bar{k}))$$

$$\times \alpha^{-1}(A(xk))|c_n(s)|^{-2} \, dV \, d\bar{k} \, ds.$$

The integrals are over $\operatorname{Re} s = -\rho$, $\bar{k} \in K/M$, and $W \in \mathscr{SP}_n$. We have first changed variables via $V = W[x]$, then used (3.17), and finally replaced $\bar{k}$ by

4.3. Harmonic Analysis on $\mathscr{P}_n$ in Polar Coordinates

$x(\bar{k})$, using Exercise 22 of Section 4.1.3 to give us the Jacobian of this change of variables, which is:

$$\alpha(A(xk)) = p_{2\rho}(I[x], \bar{k}) = p_{-2\rho}(I[x^{-1}], x(\bar{k})) = \prod_{i=1}^{n} a_i^{n-2i+1} \quad (3.18)$$

with $\rho \in \mathbb{C}^{n-1}$ given by formula (3.20) below.

Now choose

$$s = -\rho + i\lambda \quad \text{so that } \bar{s} + 2\rho = -s. \quad (3.19)$$

It follows that

$$\mathscr{T}\mathscr{H}^0(f^x)(Y) = \iiint f(V) p_{\bar{s}}(V, \bar{k}) p_{\bar{s}+2\rho}(I[x^{-1}], x(\bar{k}))$$

$$\times p_s(Y, x(\bar{k})) |c_n(s)|^{-2} dV \, d\bar{k} \, ds$$

$$= \iiint f(V) p_{\bar{s}}(V, \bar{k}) p_s(Y[x], \bar{k}) |c_n(s)|^{-2} dV \, d\bar{k} \, ds$$

$$= (\mathscr{T}\mathscr{H}^0 f)^x(Y).$$

Here we used formulas (3.17)–(3.19) to see that

$$p_{\bar{s}+2\rho}(I[x^{-1}], x(\bar{k})) p_s(Y, x(\bar{k})) = p_s(Y[x], \bar{k}).$$

Now we want to compute ρ in formula (3.18). Exercise 20 of Section 4.1.3 shows that if

$$a = \begin{pmatrix} a_1 & & 0 \\ & \ddots & \\ 0 & & a_n \end{pmatrix} \quad \text{with } |a| = 1,$$

then

$$\alpha(a) = \prod_{j=1}^{n} a_j^{n-2j+1} = (a_1 \cdots a_{n-1})^{n-1} \prod_{j \neq 1}^{n-1} a_j^{n-2j+1} = \prod_{j=1}^{n-1} a_j^{2(n-j)}.$$

Thus if $|a| = 1$ we find that

$$\alpha(a) = \prod_{j=1}^{n-1} a_j^{2(n-j)} = \prod_{j=1}^{n-1} (a_1 \cdots a_j)^{4\rho_j} = \prod_{j=1}^{n-1} a_j^{4(\rho_j + \cdots + \rho_{n-1})}.$$

It follows that $4\rho_j = 2(n - j - n + (j + 1))$ and therefore

$$\rho_j = 1/2 \; (j = 1, \ldots, n - 1). \quad (3.20)$$

(2) To prove part (2) of Lemma 1, note that, assuming we can change the order of integration, we have

$$\int_{\mathscr{S}\mathscr{P}_n} f(W) \overline{\mathscr{T} F(W)} \, dW = \int_{\mathrm{Re}\, s = -\rho} \int_{\bar{k} \in K/M} \mathscr{H}^0 f(s, k) \overline{F(s, \bar{k})} |c_n(s)|^{-2} \, d\bar{k} \, ds.$$

It follows that

$$\int_{\mathscr{SP}_n} f(W)\overline{\mathscr{TH}^0 g(W)}\, dW = \int_{\mathrm{Re}\, s=-\rho} \int_{\bar{k}\in K/M} \mathscr{H}^0 f(s,k)\overline{\mathscr{H}^0 g(s,k)} |c_n(s)|^{-2}\, d\bar{k}\, ds,$$

and therefore the operator $\mathscr{TH}^0$ is self-adjoint, completing the proof of Lemma 1. □

Exercise 5. Show that the interchange of integration orders is legal in the proof of part (2) of Lemma 1.

In order to finish our discussion of Step 2 in the proof of part (1) of Theorem 1, note that it suffices to prove the inversion formula for any Dirac sequence of functions $g_m : \mathscr{SP}_n \to \mathbb{C}$ approaching δ_I, the Dirac delta function at the identity matrix in $\mathscr{SP}_n$. One can always take such a Dirac sequence to be K-invariant. To see that the general inversion formula follows from that for such a Dirac sequence, one proceeds as in the analogous argument for $\mathbb{R}^n$ (which can be found in the proof of Theorem 1 in Section 1.2, Vol. I). For part (2) of Lemma 2 implies that if $\mathscr{TH}^0 g_m = g_m$, then

$$\int f g_m = \int (\mathscr{TH}^0 f) g_m,$$

assuming the functions g_m are real-valued. Now the left hand side of the above equality will approach $f(I)$, while the right hand side will approach $\mathscr{TH}^0 f(I)$, as m goes to infinity. Moreover, part (1) of Lemma 1 implies that it suffices to prove that $f(Y) = \mathscr{TH}^0 f(Y)$ at the point $Y = I$, or any other fixed point of $\mathscr{SP}_n$.

Remarks.

(1) Of course it is easy to believe that it should not really matter what point in G/K is chosen at which to try to prove the inversion formula for the Helgason-Fourier transform, since the result is an eigenfunction expansion for differential operators that commute with the action of G on the symmetric space G/K.
(2) We will use our results from Step 2 to push the support of the function in the inversion formula out to infinity. By this we mean out towards the boundary of the symmetric space where asymptotic expansions and ordinary Mellin inversion take over.
(3) The "lines" of integration in the inversion formula (part 1) of Theorem 1 are $\mathrm{Re}\, s_j = -\frac{1}{2}$, for $j = 1, \ldots, n-1$ and $\mathrm{Re}\, s_n = (n-1)/4$. If we change variables according to formula (3.3), then these lines correspond to $\mathrm{Re}\, r_j = 0, j = 1, \ldots, n$. These lines are certainly fixed by the $n!$ permutations of the variables r_j which represent the functional equations of the spherical functions (see part 4 of Theorem 3 in Section 4.2.3).

4.3. Harmonic Analysis on $\mathscr{P}_n$ in Polar Coordinates

Note that when $\operatorname{Re} r_j = 0$ the eigenvalues of the Laplacian on $\mathscr{P}_n$ are negative, since the eigenvalues are:

$$r_1^2 + \cdots + r_n^2 + \frac{n - n^3}{48}$$

(see Exercise 12 of Section 4.2.1).

The Harish-Chandra c-function in part (1) of Theorem 1 is quickly seen to have poles along the domain of integration for the inverse transform; i.e., when $r_i = r_{j+1}$. See formula (3.4) of the preceding subsection.

All of these phenomena occurred in the inversion formula for the Helgason-Fourier transform on the Poincaré upper half plane (see Section 3.2 of Volume I; e.g., on page 142).

4.3.3. End of the Discussion of Part (1) of Theorem 1—Steps 3 and 4

Step 3 (Asymptotics and Functional Equations). This part of the proof of part (1) of Theorem 1 was obtained by Harish-Chandra [2, Vol. II, pp. 409–539], who gave the general theory for G a semi-simple real Lie group, and by Bhanu-Murthy [1], who made Harish-Chandra's results explicit when $G = SL(n, \mathbb{R})$. Gindikin and Karpelevic [1] compute the spectral measure explicitly for general G. Helgason [4, pp. 425–466] gives a very detailed treatment of this last step in our discussion. We will stick to a heuristic version of the argument. See also Varadarajan [3].

Suppose that $f(Y) = f(Y[k])$ for all Y in $\mathscr{P}_n$ and $k \in K = O(n)$. The Helgason-Fourier Transform of f is obtained in formula (3.8) in the preceding subsection. If we pull out the determinant in formula (3.8), as in Step 1 of the proof, then we must change variables via

$$a = \begin{pmatrix} a_1 & & 0 \\ & \ddots & \\ 0 & & a_n \end{pmatrix} = v^{1/n} \begin{pmatrix} u_1 & & 0 \\ & \ddots & \\ 0 & & u_n \end{pmatrix}, \quad \text{with } v = |a| \text{ and } u_1 \cdots u_n = 1.$$

Then formula (3.8) becomes:

$$\hat{f}(s) = b_n \int_{v>0} \int_{u_i > u_{i+1}} f(v, u) v^{r-1} \prod_{i<j}(u_i - u_j) \overline{h_s(u)} \, dv \, du, \qquad (3.21)$$

where $f(a) = f(v, u)$,

$$du = \prod_{i=1}^{n-1} u_i^{-1} \, du, \qquad (3.22)$$

and r is defined by formula (3.10).

According to Step 1, we need to show that if we are given a function f on $\mathcal{SP}_n$ which is $O(n)$-invariant and if u and a denote positive diagonal matrices of determinant one, then:

$$f(u) = (2\pi i)\omega_n b_n \int_{\operatorname{Re} s=-\rho} \int_{a_i > a_{i+1}} f(a) \prod_{i<j} (a_i - a_j)\overline{h_s(a)}\, da\, h_s(u)|c_n(s)|^{-2}\, ds, \tag{3.23}$$

with da as in formula (3.22), b_n as in (3.8), and ω_n as in part (1) of Theorem 1. Suppose now that a_i/a_{i-1} is near zero for all $i = 2, \ldots, n$. Then

$$\prod_{i<j}(a_i - a_j) = \prod_{i=1}^{n-1} a_i^{n-i} \prod_{i<j}(1 - a_j/a_i),$$

which approaches

$$\prod_{i=1}^{n-1} a_i^{n-i}, \qquad \text{as } a_i/a_{i-1} \to 0,\, i = 2, \ldots, n. \tag{3.24}$$

Assume next that the function $f(a)$ is supported on positive diagonal matrices a with a_i/a_{i-1} near 0, for all $i = 2, \ldots, n$. Then the inversion formula of Harish-Chandra can be obtained from an asymptotics/functional equations principle similar to that seen in Section 3.2 of Vol. I, if we can show that the spherical functions have *asymptotic expansions* of the form:

$$h_s(a) \sim c_n(s) p_s(a), \qquad \text{if } a_j/a_{j-1} \to 0. \tag{3.25}$$

Here the argument a of the spherical function is a diagonal determinant one matrix with jth diagonal entry a_j and the parameter $s \in \mathbb{C}^{n-1}$ is fixed with $\operatorname{Re} s_j$ sufficiently large. In fact, $\operatorname{Re} s_j$ must be so large that s_j is outside of the domain of integration in the inversion transform. For such a parameter s, the term on the right in (3.25) should be replaced by a sum of $n!$ terms coming from the functional equations of the spherical functions. This is proved by Harish-Chandra [loc. cit.]. See also Helgason [4] and Varadarajan [3].

According to formulas (3.23)–(3.25), if $f(Y)$, $Y \in \mathcal{SP}_n$, is K-invariant and supported on $a \in A$ with a_j/a_{j-1} near 0 for all $j = 2, \ldots, n$, the *inversion formula* of part (1) of Theorem 1 would look approximately like:

$$f(u) = n!(2\pi i)\omega_n b_n \int_{\substack{s \in \mathbb{C}^{n-1} \\ \operatorname{Re} s = -\rho}} \int_{\substack{a \in A \\ a_i/a_{i-1} \text{ near } 0}} f(a) \prod_{i=1}^{n-1} a_i^{n-i}\overline{p_s(a)}\, da\, p_s(u)\, ds, \tag{3.26}$$

for $u \in A$ with u_j/u_{j-1} near 0, $j = 2, \ldots, n$. The spectral measure in part (1) of Theorem 1 is thus seen to be chosen to cancel out the term $\overline{c_n(s)}c_n(s)$ coming from the asymptotic formula (3.25) for the spherical function. The extra $n!$ comes from the $n!$ functional equations of the spherical function, which replaces the right-hand side of (3.25) by a sum of similar terms, where the sum is over the group of permutations acting on the r-variables for the power functions. Here one needs to note the orthogonality of the different exponentials that are summed over the permutation group.

4.3. Harmonic Analysis on $\mathscr{P}_n$ in Polar Coordinates

Clearly (3.26) is the same as

$$f(u) = n!(2\pi i)\omega_n b_n \int_{\substack{s \in \mathbb{C}^{n-1} \\ \operatorname{Re} s = -\rho}} \int_{\substack{a \in A \\ a_i/a_{i-1} \text{ near } 0}} f(a) \prod_{i=1}^{n-1} a_i^{\overline{s_i + \cdots + s_{n-1}} + n - i - 1} \, da_i \qquad (3.27)$$
$$\times \prod_{i=1}^{n-1} u_i^{s_i + \cdots + s_{n-1}} \, ds.$$

Then ordinary Mellin inversion and formula (3.8) imply that we must choose ω_n as in part (1) of Theorem 1. For if $\operatorname{Re} s_j = -\frac{1}{2}$, we find that the exponent of a_j is $e_j - 1$ while that of u_j is $-e_j$, which is just what is required for ordinary Mellin inversion. To give a rigorous justification of this argument would require us to delve into the proof of the Paley-Wiener theorem on $\mathscr{P}_n$. We shall not do this here but see Helgason [4, pp. 450–454]. The miracle is that one needs only the main term in the asymptotic expansion of the spherical function. This is very similar to the standard miracle of Paley-Wiener theory, as well as the result of Lemma 1 in Section 3.7 of Volume I. And this miracle seems believable, recalling the orthogonality of the distinct power functions.

Anyway, we are accordingly reduced to the computation of the Harish-Chandra c-function in formula (3.25).

Step 4 (Computation of Harish-Chandra's c-Function). We must compute the coefficient in the main term of the asymptotic formula (3.25) for the spherical function. This computation is most easily accomplished by changing variables from K/M to $\bar{N}$ in Harish-Chandra's integral formula (3.7) for the spherical function. Here $\bar{N}$ is the (nilpotent or unipotent) subgroup of $G = SL(n, \mathbb{R})$ consisting of all lower triangular matrices with one's on the diagonal. Harish-Chandra [2, Vol. II, p. 455] shows how to produce this change of variables as a consequence of the Bruhat decomposition to be discussed later. And Bhanu-Murthy [1] carried out the computation of the c-function explicitly for $G = SL(n, \mathbb{R})$. See also Helgason [4, pp. 198, 434–448]

Lemma 2. *If $P = MAN$, then P is the minimal parabolic subgroup of $G = SL(n, \mathbb{R})$ consisting of all upper triangular matrices and $P\bar{N}$ is an open subset of G with lower dimensional complement. Thus we can realize the boundary $B = K/M$ as $\bar{N}$ as far as integration is concerned, obtaining the integral formula for the change of variables from the boundary B to $\bar{N}$:*

$$\int_B f(b) \, db = \kappa \int_{\bar{N}} f(\bar{n}(M)) p_{-2\rho}(I[\bar{n}]) \, d\bar{n}, \quad \text{where } \kappa^{-1} = \int_{\bar{N}} p_{-2\rho}(I[\bar{n}]) \, d\bar{n}.$$

Here $\bar{n}(M)$ denotes the result of letting $\bar{n}$ act on the coset M in $B = K/M$. If $K(\bar{n}) = k_1$ is the K-component in the Iwasawa decomposition of $\bar{n}$, then

$$\bar{n} = k_1 a_1 n_1, \quad k_1 \in K, \quad a_1 \in A, \quad n_1 \in N,$$

and

$$\bar{n}(M) = K(\bar{n})M = k_1 M.$$

The measure $d\bar{n}$ is defined by:

$$d\bar{n} = \prod_{i>j} dx_{ij}, \quad \text{if } \bar{n} = \begin{pmatrix} 1 & & & 0 \\ & 1 & & \\ & & \ddots & \\ x_{ij} & & & 1 \end{pmatrix}.$$

PROOF. (Cf. Helgason [4, p. 198].) To see the first statement of the lemma, multiply the matrices below.

$$\begin{pmatrix} v_{11} & v_{12} & \cdots & v_{1,n-1} & v_{1n} \\ \vdots & \vdots & & \vdots & \vdots \\ 0 & 0 & \cdots & v_{n-1,n-1} & v_{n-1,n} \\ 0 & 0 & \cdots & 0 & v_{nn} \end{pmatrix} \begin{pmatrix} 1 & 0 & & 0 & 0 \\ \vdots & \vdots & & \vdots & \vdots \\ u_{n-1,1} & u_{n-1,2} & \cdots & 1 & 0 \\ u_{n1} & u_{n2} & \cdots & u_{n,n-1} & 1 \end{pmatrix}$$

$$= \begin{pmatrix} v_{11} + v_{12}u_{21} + \cdots + v_{1n}u_{n1} & \cdots & v_{1,n-1} + v_{1n}u_{n,n-1} & v_{1n} \\ v_{22}u_{21} + \cdots + v_{2n}u_{n1} & \cdots & v_{2,n-1} + v_{2n}u_{n,n-1} & v_{2n} \\ \vdots & & \vdots & \vdots \\ v_{n-1,n-1}u_{n-1,1} + v_{n-1,n}u_{n1} & \cdots & v_{n-1,n-1} + v_{n-1,n}u_{n,n-1} & v_{n-1,n} \\ v_{nn}u_{n1} & \cdots & v_{nn}u_{n,n-1} & v_{nn} \end{pmatrix}.$$

This is an element of $G = SL(n, \mathbb{R})$ such that the lower right $k \times k$ corner matrix is non-singular for all $k = 1, 2, \ldots, n$. Thus $P\bar{N}$ is indeed an open subset of G with lower dimensional complement.

Exercise 6. Prove these last two statements.

Hint. We can multiply block matrices as follows:

$$(0 \quad I)\begin{pmatrix} F & H \\ 0 & G \end{pmatrix}\begin{pmatrix} A & 0 \\ C & B \end{pmatrix}\begin{pmatrix} 0 \\ I \end{pmatrix} = GB.$$

Thus we can define a mapping

$$\psi : \bar{N} \to K/M$$
$$\bar{n} \mapsto \bar{n}(M) = K(\bar{n})M.$$

And we can use this mapping to identify $\bar{N}$ and the boundary $B = K/M \cong G/P$, with $P = MAN$ (recalling formula (1.20) of Section 4.1.3).

Next we seek the Jacobian $J(\bar{n})$ of the mapping ψ; i.e.,

$$\int_{B=K/M} f(b)\, db = \int_{\bar{N}} f(n(M)) J(\bar{n})\, d\bar{n}. \tag{3.28}$$

Now the integral formula for the action of G on B (see Exercise 22 of Section 4.1.3 and formula (3.18) above) gives:

$$\int_{b=\bar{k}\in K/M} f(\bar{n}_0(b))\alpha^{-1}(A(\bar{n}_0 k))\, d\bar{k} = \int_{b\in K/M} f(b)\, db \tag{3.29}$$

4.3. Harmonic Analysis on $\mathscr{P}_n$ in Polar Coordinates

with $\alpha(A(\bar{n}_0 k)) = p_{2\rho}(I[\bar{n}_0 k])$, for $\rho \in \mathbb{C}^{n-1}$ as in (3.20). Combining (3.28) and (3.29) yields:

$$\int_{\bar{n}\in \bar{N}} f(\bar{n}_0 \bar{n}(M))\alpha^{-1}(A(\bar{n}_0 K(\bar{n})))J(\bar{n})\,d\bar{n} = \int_{\bar{n}\in \bar{N}} f(\bar{n}_0 \bar{n}(M))J(\bar{n}_0 \bar{n})\,d\bar{n},$$

$$J(\bar{n}_0 \bar{n}) = \alpha^{-1}(A(\bar{n}_0 K(\bar{n})))J(\bar{n}).$$

Set $\bar{n} = I$ to obtain

$$J(\bar{n}_0) = p_{-2\rho}(I[\bar{n}_0])J(I).$$

The constant $J(I)$ is determined by demanding that the total volume of the boundary be one. This completes the proof of Lemma 2. □

If we apply Lemma 2 to the integral formula (3.7) for the spherical function, we find the *second integral formula for the spherical function*:

$$h_s(I[a]) = \kappa \int_{\bar{N}} p_s(I[a], \bar{n}(M)) p_{-2\rho}(I[\bar{n}])\,d\bar{n} \qquad (3.30)$$

with the constant κ given in Lemma 2.

To discover the asymptotics of (3.30), we need the *second power function identity*:

$$p_s(I[a], \bar{n}(M)) = p_{2s}(a)p_s(I[\bar{n}^a])p_{-s}(I[\bar{n}]), \qquad \text{with } \bar{n}^a = a\bar{n}a^{-1}. \quad (3.31)$$

To prove this, write $\bar{n} = k_1 a_1 n_1$, with $k_1 \in K$, $a_1 \in A$, $n_1 \in N$. Then

$$\bar{n}^a = ak_1 a_1 n_1 a^{-1} = ak_1 a_1 a^{-1}(an_1 a^{-1}).$$

Since $(an_1 a^{-1}) \in N$, it follows from the definition of the power function that

$$p_s(I[\bar{n}^a]) = p_s(I[a], \bar{n}(M))p_s(I[\bar{n}])p_s(I[a^{-1}]).$$

This implies (3.31).

Combining (3.30) and (3.31) gives the *third integral formula for the spherical function*:

$$h_s(I[a]) = \kappa p_{2s}(a) \int_{\bar{N}} p_s(I[\bar{n}^a])p_{-s}(I[\bar{n}])p_{-2\rho}(I[\bar{n}])\,d\bar{n}, \qquad (3.32)$$

with κ as in Lemma 2.

For Re s_j all sufficiently large, we can let a_j/a_{j-1} approach zero inside the integral in (3.32). If

$$\bar{n} = \begin{pmatrix} 1 & & 0 \\ & \ddots & \\ x_{ij} & & 1 \end{pmatrix} \quad \text{and} \quad \bar{n}^a = \begin{pmatrix} 1 & & 0 \\ & \ddots & \\ y_{ij} & & 1 \end{pmatrix},$$

then $y_{ij} = a_i x_{ij} a_j^{-1}$, for $i > j$. It follows that if a_i/a_{i-1} approaches zero, then y_{ij} approaches zero if $i > j$.

Thus for Re s_j sufficiently large, $j = 1, \ldots, n - 1$, as $a_j/a_{j-1} \to 0, j = 2, \ldots, n$, we see that formula (3.32) approaches

$$\kappa p_{2s}(a) \int_{\bar{N}} p_{-(s+2\rho)}(I[\bar{n}]) \, d\bar{n}.$$

This means that the *Harish-Chandra c-function* is:

$$c_n(s) = \kappa \int_{\bar{N}} p_{-(s+2\rho)}(I[\bar{n}]) \, d\bar{n}. \tag{3.33}$$

Now we want to use mathematical induction to evaluate $c_n(s)$. First, note that when $n = 2$, we have $\alpha(a) = a_1^2$ and $\rho = 1/2$. Thus we find that $c_2(s)$ is evaluated as follows in terms of beta functions (see Lebedev [1, p. 13]);

$$c_2(s) = \kappa \int_{x \in \mathbb{R}} (1 + x^2)^{-(s+1)} \, dx = \kappa B\left(\frac{1}{2}, s + \frac{1}{2}\right) = B\left(\frac{1}{2}, s + \frac{1}{2}\right) \bigg/ B\left(\frac{1}{2}, \frac{1}{2}\right).$$

In the general case we define

$$b_n(s) = \int_{\bar{N}} p_{-s}(I[\bar{n}]) \, d\bar{n}. \tag{3.34}$$

This is a special case of the Whittaker function in Section 4.2.2. Clearly $c_n(s) = b_n(s + 2\rho)/b_n(2\rho)$. We need to relate b_n with b_{n-1}. To do this, write the element $\bar{n}$ in $\bar{N}$ as:

$$\bar{n} = \begin{pmatrix} \bar{m} & 0 \\ {}^t x & 1 \end{pmatrix},$$

for $x \in \mathbb{R}^{n-1}$, $\bar{m}$ in the group $\bar{N}$ for $SL(n - 1, \mathbb{R})$, a group that we shall denote $\bar{N}_{n-1}$. Then

$$p_s(I[\bar{n}]) = p_s({}^t\bar{m}\bar{m} + x\,{}^tx),$$

and it follows that:

$$b_n(s) = \int_{\bar{m} \in \bar{N}_{n-1}} \int_{x \in \mathbb{R}^{n-1}} p_{-s}({}^t\bar{m}\bar{m} + x\,{}^tx) \, dx \, d\bar{m}.$$

Now write ${}^t\bar{m}\bar{m} = I[t]$, for t upper triangular with positive diagonal. Make the change of variables $x = {}^tt u$ in the last formula for $b_n(s)$. This gives:

$$b_n(s) = b_{n-1}(s_1, \ldots, s_{n-2}) b'_n(s_1, \ldots, s_{n-1}),$$

where

$$b'_n(s) = \int_{u \in \mathbb{R}^{n-1}} p_{-s}(I + u\,{}^tu) \, du. \tag{3.35}$$

To evaluate this last integral. write $u = (v, w)$ with $v \in \mathbb{R}^{n-2}$ and $w \in \mathbb{R}$. Also define $s = (r, s_{n-1})$, with $r \in \mathbb{C}^{n-2}$. Note that

4.3. Harmonic Analysis on $\mathcal{P}_n$ in Polar Coordinates

$$|I + u\,{}^t u| = 1 + {}^t uu = 1 + u_1^2 + \cdots + u_{n-1}^2.$$

This is an easy consequence of the spectral theorem, since $u\,{}^t u$ is an $(n-1) \times (n-1)$ matrix with only one non-zero eigenvalue. And that eigenvalue is the square of the norm of the vector u. By these remarks

$$b'_n(s) = \int_{v \in \mathbb{R}^{n-2}} p_{-r}(I + v\,{}^t v) \int_{w \in \mathbb{R}} (|I + v\,{}^t v| + w^2)^{-s_{n-1}} \, dw \, dv.$$

Next change variables via $w = |I + v\,{}^t v|^{1/2} y$ and obtain:

$$b'_n(s) = b'_{n-1}\left(s_1, \ldots, s_{n-3}, s_{n-2} + s_{n-1} - \frac{1}{2}\right) B\left(\frac{1}{2}, s_{n-1} - \frac{1}{2}\right)$$

$$= \prod_{j=1}^{n-1} B\left(\frac{1}{2}, s_j + s_{j+1} + \cdots + s_{n-1} - \frac{n-j}{2}\right).$$

It follows from this that

$$b_n(s) = \prod_{1 \le i \le j \le n-1} B\left(\frac{1}{2}, s_i + s_{i+1} + \cdots + s_j - \frac{j-i+1}{2}\right),$$

which quickly leads to the formula in part (1) of Theorem 1. Hopefully we have given the reader enough insight into the proof of Theorem 1 to find the theorem believable. Note that $\operatorname{Re} s_j > -1/2$, $j = 1, \ldots, n-1$, is required for the absolute convergence of the integral in the beta functions. Thus we find that the lines of integration in Theorem 1 are outside the region of absolute convergence and care must be taken because of this.

Remark. The function b'_n defined by formula (3.35) is a k-Bessel function of singular argument (see formula (2.20) in Section 4.2.2). The function $b_n(s)$ in (3.34) is a Whittaker function of singular argument (see the end of Section 4.2.2). One can also consider these functions to be analogues of the beta function (cf. Gindikin [1]).

A Short Table of Helgason Transforms for $\mathcal{P}_n$

$f(Y)$	$\hat{f}(\bar{s}) = \mathcal{H}f(\bar{s}, I) = \int_{\mathcal{P}_n} f(Y) p_s(Y) \, d\mu_n(Y)$	
$\exp[-\operatorname{Tr}(X^{-1}Y)], \quad X \in \mathcal{P}_n$	$p_s(X)\Gamma_n(s)$,	§4.2.1, formula (2.4)
$\exp[-\operatorname{Tr}(VY + WY^{-1})]$	$K_n(s|V, W)$,	§4.2.2, formula (2.21)
$\exp[-\operatorname{Tr}(XY)]p_{r-s-(n+1)/2}(I + Y)$	$\Psi_n(s, r; X)$,	§4.2.2
$K_n(r|I, Y)$	$\Gamma_n(s)\Gamma_n(s + r)$,	§4.2.2, Exercise 18

Exercise 7. Check the preceding table.

This completes our discussion of part (1) of Theorem 1. As we said earlier, we will not prove part (4). And we relegated parts (2) and (3) of Theorem 1 to Exercise 1.

The Helgason-Fourier transform of a non K-invariant function involves a variable $k \in K$. But one can use the Fourier inversion formula for K itself (see Chapter 2) to replace functions on K with functions of $\pi \in \hat{K}$, which is the set of equivalence classes of irreducible unitary representations of K. Thus we could replace $\mathscr{H}f(s, k)$, $k \in K$, with a matrix valued transform

$$\mathscr{H}f(s, \pi), \quad \pi \in \hat{K}.$$

4.3.4. Applications

In Section 3.2 of Volume I we used harmonic analysis on the Poincaré upper half plane to obtain a central limit theorem for rotation-invariant random variables on that space. Here we aim to generalize the result to $\mathscr{P}_n$, at least in the case $n = 3$. We will find that there is a significant difference between the situation for $GL(3, \mathbb{R})$ and that for $SL(2, \mathbb{R})$. The limiting density is not the same as the fundamental solution of the heat equation (see Exercise 8 below).

Of course, there have been many applications of analysis on $\mathscr{P}_n$ to multivariate statistics. See Sections 4.1.6 and 4.2.4, or Anderson [1], Farrell [1, 2], James [1–3], Morrison [1], and Muirhead [1]. Here we aim to show that it is possible to discuss a K-invariant central limit theorem on $\mathscr{P}_3$ using harmonic analysis on $\mathscr{P}_3$. The discussion of the central limit theorem will use analogous methods to those of Cramér [1], Dym and McKean [1], and Feller [1] (or our Section 3.2, Vol. I).

Our methods are special to the case of limit theorems for groups on which one can do harmonic analysis. Methods based on martingales, semigroups, stochastic difference or differential equations can produce more general results. But the methods of harmonic analysis can give more detailed information.

There are many papers on central limit theorems for Lie groups as we mentioned already on page 156 of Vol. I. We should add here the paper of Bougerol [1].

Exercise 8 (The Heat Equation on $\mathscr{P}_n$). Suppose that f is a K-invariant function on $\mathscr{P}_n$ which is continuous with compact support. We seek a solution $u(Y, t)$, $Y \in \mathscr{P}_n$, $t > 0$, to the heat equation:

$$\begin{cases} u_t = \Delta u, \quad \Delta = \mathrm{Tr}((Y \partial/\partial Y)^2) = \text{Laplacian on } \mathscr{P}_n, \\ u(Y, 0) = f(Y). \end{cases}$$

Show that the solution is $u(Y, t) = G_t * f$, with convolution as in formula (1.24) of Section 4.1.3 and show that the fundamental solution G_t (or normal density) is given by:

4.3. Harmonic Analysis on $\mathcal{P}_n$ in Polar Coordinates

$$G_t(Y) = \omega_n \int_{\operatorname{Re} s = -\rho} \exp[\lambda_2(s)t] h_s(Y) |c_n(s)|^{-2} ds,$$

where $\Delta p_s = \lambda_2(s) p_s$ (see Exercise 12 of Section 4.2.1), ω_n, ρ are defined in Theorem 1. Here $h_s(Y)$ denotes the spherical function in formula (3.7).

Hints. (See Gangolli [1, pp. 108ff.].) Imitate the method used to solve the heat equation in the non-Euclidean upper half plane in Section 3.2, Vol. I. Note that G_t approaches the Dirac delta distribution at the identity, as t approaches 0 from above, by the same argument that proved formula (3.35) in Section 3.2, Vol. I. Note that G_t cannot have compact support, since it is a solution of a parabolic partial differential equation and thus an analytic function of Y. Therefore we really need an extension of Theorem 1 to Schwartz functions on $\mathcal{P}_n$ in order to do this exercise rigorously.

Remark. It would also be interesting and useful to consider other partial differential equations on $\mathcal{P}_n$. For example, Helgason [1] and [4, pp. 342–343] investigates the wave equation on symmetric spaces and extensions of Huygen's principle.

We consider *random variables* Y in $\mathcal{P}_n$ with *density* $f(Y)$ in $L^1(\mathcal{P}_n, d\mu_n), f \geq 0$, where $d\mu_n$ denotes the G-invariant measure on $\mathcal{P}_n$. Then if S is a measurable subset of $\mathcal{P}_n$, the *probability* that the random variable Y, with density f, is in S is

$$P(Y \in S) = \int_{Y \in S} f(Y) d\mu_n(Y) = \int_{I[g] \in S} f(I[g]) dg,$$

where dg denotes Haar measure on $G = GL(n, \mathbb{R})$. There are many possible analogues of the mean and the standard deviation as is the case for $\mathbb{R}^n$. We will return to this subject below (see (3.37) and (3.38)).

Here *we consider only K-invariant* random variables Y; i.e., we will *always* assume that the density function $f = f_Y$ satisfies:

$$f(Y[k]) = f(Y), \qquad \text{for all } Y \in \mathcal{P}_n \text{ and } k \in K = O(n).$$

It will often be helpful to identify such a function f on $\mathcal{P}_n$ with a *K-bi-invariant* function on the group G via

$$f(I[g]) = f(g), \qquad \text{for all } g \in G.$$

The *composition* $Y_1 \circ Y_2$ of two K-invariant random variables Y_1 and Y_2 on $\mathcal{P}_n$ is defined to be that coming from multiplication of the corresponding group elements. If Y_j has density function $f_j, j = 1, 2$, then $Y_1 \circ Y_2$ has density the *convolution* $f_1 * f_2$, assuming that Y_1 and Y_2 are independent:

$$(f_1 * f_2)(x) = \int_G f_1(y) f_2(y^{-1} x) dy. \tag{3.36}$$

To see this, note that

$$P(Y_1 \circ Y_2 \in S) = \iint_{g_1 g_2 \in S} f_1(g_1) f_2(g_2) \, dg_1 \, dg_2$$

$$= \int_{g_1 \in G} f_1(g_1) \int_{w \in S} f_2(g_1^{-1} w) \, dw \, dg_1$$

$$= \int_{w \in S} (f_1 * f_2)(w) \, dw.$$

If we now seek to imitate our discussion from Vol. I, pp. 156–162, we find that the result for $\mathscr{P}_n$ is somewhat different from that for the Poincaré upper half plane.

Let $\{Y_v\}_{v \geq 1}$ be a sequence of independent K-invariant random variables on $\mathscr{P}_n$, each having the same density function $f(Y)$. We will assume the vanishing of the means with respect to the $h_j = \log a_j$, where a_j denotes the jth eigenvalue of Y, and we will also assume that the covariance matrix with respect to the h_j is the identity; i.e., we assume via the change to polar coordinates that the following integral formulas hold:

$$b_n \int_{H \in \mathfrak{a}} h_j f(\exp H) J(\exp H) \, dh = 0, \qquad j = 1, \ldots, n; \qquad (3.37)$$

$$b_n \int_{H \in \mathfrak{a}} h_i h_j f(\exp H) J(\exp H) \, dh = \delta_{ij}, \qquad 1 \leq i, j \leq n. \qquad (3.38)$$

Here

$$\mathfrak{a} = \left\{ H = \begin{pmatrix} h_1 & & 0 \\ & \ddots & \\ 0 & & h_n \end{pmatrix} \middle| \, h_j \in \mathbb{R} \right\}, \qquad dh = \prod_{j=1}^{n} dh_j.$$

Note that $\mathfrak{a}$ is the tangent space to A at the identity (cf. (2.35) in Section 4.2.3). The constant b_n and the Jacobian $J(\mathfrak{a})$ are given in formula (3.8) of Section 4.3.1.

Consider the composition $S_v = Y_1 \circ \cdots \circ Y_v$ which was defined in the paragraph preceding (3.36). We will normalize S_v as follows. Let $h_j = \log a_j$, where a_j denotes the jth eigenvalue of Y. Then normalize by replacing h_j by $v^{-1/2} h_j$. Call the resulting random variable $S_v^{\#}$. The characteristic function of the normalized random variable is:

$$\varphi_{S_v^{\#}}(s) = \left\{ b_n \int_{H \in \mathfrak{a}} f(\exp H) h_s(\exp(v^{-1/2} H)) J(\exp H) \, dH \right\}^v. \qquad (3.39)$$

Here we have used the convolution property of the Helgason-Fourier transform from Theorem 1.

Now let us assume that $n = 3$ and use (2.55) to see that, as v approaches infinity, the term inside the parentheses in (3.39) is asymptotic to:

4.3. Harmonic Analysis on $\mathscr{P}_n$ in Polar Coordinates

$$b_n \left\{ \int_{H \in \mathfrak{a}} f(\exp H) J(\exp H) dH + \int_{H \in \mathfrak{a}} \sum_{j=1}^{3} h_j f(\exp H) J(\exp H) dH \frac{1}{3\sqrt{v}} \sum_{j=1}^{3} r_j \right.$$

$$+ \int_{H \in \mathfrak{a}} \sum_{j=1}^{3} h_j^2 f(\exp H) J(\exp H) dH \frac{1}{30v} \left(3 \sum_{j=1}^{3} r_j^2 + 2 \sum_{1 \le i < j \le 3} r_i r_j - 1 \right)$$

$$+ \left. \int_{H \in \mathfrak{a}} \sum_{1 \le i < j \le 3} h_i h_j f(\exp H) J(\exp H) dH \frac{p(r)}{30v} \right\}.$$

Now (3.37) and (3.38) imply that as v approaches infinity:

$$\varphi_{S_v^\#}(s) \sim \left\{ 1 + \frac{1}{v} \frac{3[r_1^2 + r_2^2 + r_3^2] + 2[r_1 r_2 + r_1 r_3 + r_2 r_3] - 1}{10} \right\}^v$$

$$\sim \exp\left\{ \frac{3[r_1^2 + r_2^2 + r_3^2] + 2[r_1 r_2 + r_1 r_3 + r_2 r_3] - 1}{10} \right\}.$$

Here $s \in \mathbb{C}^3$ is the function of $r \in \mathbb{C}^3$ specified in (3.3). Recalling the formula for the eigenvalue of the Laplacian in Exercise 12 of Section 4.2.1, we see that the limit characteristic function does not appear to be related in a simple way to

$$\exp\{t(r_1^2 + r_2^2 + r_3^2 - 1/2)\},$$

the Fourier transform of the fundamental solution of the heat equation.

By the convolution theorem we see that the limit density is

$$\exp(1/20) G_{3/10} * F_{1/5},$$

where F_t is the function on $\mathscr{P}_3$ whose Helgason-Fourier transform is

$$\exp\{t(r_1 r_2 + r_1 r_3 + r_2 r_3)\}.$$

Here G_t is the fundamental solution of the heat equation from Exercise 8 above.

Theorem 2 (The Central Limit Theorem for $\mathscr{P}_3$). *Suppose that $\{Y_n\}_{n \ge 1}$ is a sequence of independent, SO(3)-invariant random variables in $\mathscr{P}_3$, each having the same density function $f(Y)$. And suppose that the density satisfies (3.37) and (3.38). Let $S_n = Y_1 \circ \cdots \circ Y_n$ be normalized as in (3.39). The normalized variable has density function $f_n^\#$. Then for measurable sets S in $\mathscr{P}_n$ we have, as $n \to \infty$:*

$$\int_S f_n^\#(Y) d\mu(Y) \sim e^{1/20} \int_S G_{3/10} * F_{1/5}(Y) d\mu(Y).$$

Here G_t is the fundamental solution of the heat equation from Exercise 8 and its Helgason-Fourier transform is

$$\hat{G}_t(s(r)) = \exp\{t(r_1^2 + r_2^2 + r_3^2 - \tfrac{1}{2})\}$$

while F_t has Helgason-Fourier transform:

$$\hat{F}_t(s(r)) = \exp\{t(r_1r_2 + r_1r_3 + r_2r_3)\}.$$

Note that we have reparametrized the Helgason-Fourier transform using the change of variables (3.3) from s-variables to r-variables.

PROOF. We need only to argue that the limiting behavior of densities mirrors that of their Fourier transforms. To see this, recall the inversion and Plancherel formulas from Theorem 1. Let β be an infinitely differentiable function with compact support on $\mathscr{P}_3$. Let $d\sigma(r)$ denote the spectral measure in part (1) of Theorem 1, using the r-variables from (3.3) rather than the s-variables of Theorem 1. Then, by the dominated convergence theorem, we have:

$$\lim_{n \to \infty} \int_{\mathscr{P}_3} f_n^\#(Y)\beta(Y)\,d\mu(Y) = \lim_{n \to \infty} \int_{\operatorname{Re} r = 0} \hat{f}_n^\#(s(r))\hat{\beta}(s(r))\,d\sigma(r)$$

$$= \int_{\operatorname{Re} r=0} \exp\left\{\frac{3}{10}(r_1^2+r_2^2+r_3^2) + \frac{1}{5}(r_1r_2+r_1r_3+r_2r_3) - \frac{1}{10}\right\} \hat{\beta}(s(r))\,d\sigma(r)$$

$$= \exp\left(\frac{1}{20}\right) \int_{Y \in \mathscr{P}_3} (G_{3/10} * F_{1/5})(Y)\beta(Y)\,d\mu.$$

Here we are using the fact (proved by Helgason [11, p. 458] that spherical functions are bounded on the lines of integration for the inverse transform and thus if f is in $L^1(\mathscr{P}_3, d\mu)$, then $\hat{f}(s(r))$ is bounded for $\operatorname{Re} r = 0$. Next let β approximate the indicator function of a set in $\mathscr{P}_3$ to complete the proof. □

Another reference for central limit theorems on Lie groups is the volume edited by Cohen, Kesten, and Newman [1]. It would be useful to compare our results here with the limit theorem of Oseledec discussed by several authors in this volume. See also Watkins [1]. And finally it would be nice to generalize Theorem 2 to the case $\mathscr{P}_n$. This has been done by D. St. P. Richards.

4.3.5. Other Directions in the Labyrinth

Our discussion of Fourier inversion on $\mathscr{P}_n$ mainly followed the path of Helgason [1]. At this point, the reader might like to travel some other paths. Varadarajan's introduction to Harish-Chandra [2] (the collected works) gives a good historical introduction to the representation-theoretic road to harmonic analysis, as it was travelled by Harish-Chandra and others. One part of the route involves *orbital integrals*:

$$\int_{G/T} f(xtx^{-1})\,dx, \quad \text{for t in a maximal abelian subgroup T of G.}$$

4.3. Harmonic Analysis on $\mathscr{P}_n$ in Polar Coordinates

Weyl [1] already made great use of these integrals in his development of the theory of representations of compact Lie groups. See also Broecker and tom Dieck [1] or Helgason [4]. Gelfand and Graev [1] use such an approach for complex groups such as $SL(n, \mathbb{C})$. In particular they utilize formulas for the residues of certain integrals of M. Riesz type defined by:

$$R(s) = \int_{x \in \mathbb{R}^m, Q[x] \geq 0} f(x) Q[x]^s \, dx, \quad \text{for } s \in \mathbb{C},$$

where Q is a symmetric matrix in $\mathbb{R}^{m \times m}$ and $f\colon \mathbb{R}^m \to \mathbb{C}$ is sufficiently differentiable. The residue formulas involve certain differential operators. This leads to a version of Fourier analysis of $f\colon G \to \mathbb{C}$ which is often called the Plancherel formula, for groups G like $SL(n, \mathbb{C})$, $U(n)$, at first, and then also real groups. One writes $f(e)$, $e = $ the identity of the group, as a differential operator applied to an orbital integral.

Michelle Vergne [1] provides a view of harmonic analysis on $G = SL(2, \mathbb{R})$ and other groups, which is close to that of Kirillov for nilpotent groups. Again the main direction is given by the *orbit method* which is used to classify representations according to orbits of the Adjoint action of G on the dual of its Lie algebra. For matrix groups, the Adjoint is conjugation.

Ehrenpreis and Mautner [1] give an interesting discussion of Fourier analysis on $SL(2, \mathbb{R})$ and $SL(2, \mathbb{R})/\Gamma$ from the point of view of "classical" analysis (after Laurent Schwartz), including a Riemann-Lebesgue lemma and a readable discussion of the Schwartz space.

Flensted-Jensen [1] gives some relations between analysis on symmetric spaces like $GL(n, \mathbb{R})/O(n)$ and $GL(n, \mathbb{C})/U(n)$. Analysis is much easier on the latter space. Healy [1] studies relations between harmonic analysis on $GL(2, \mathbb{C})/U(2)$ and that on $SU(2)$.

Becky Herb and Joe Wolf [1] note that Harish-Chandra's Plancherel formula was not proved for all real connected semisimple Lie groups; e.g. the universal covers of groups like $SL(2, \mathbb{R})$. They obtain the Plancherel formula for all real semigroups using different methods from Harish-Chandra. Once more, orbital integrals play a key role.

Orbital integrals are also of fundamental importance in the Selberg trace formula. See the conference volume edited by Hejhal, Sarnak and Terras [1] for many papers on that subject; e.g., that of Arthur, Herb and Sally.

When $G = SL(2, \mathbb{R})$, for example, the Plancherel formula involves a series as well as an integral (see Lang [3]). Why doesn't this happen for $G/K \cong H$? Or for $\mathscr{P}_n$? Equivalently, one wonders why there are no square-integrable eigenfunctions of the G-invariant differential operators on $\mathscr{P}_n$? One answer to this question comes from thinking about discrete subgroups Γ of $GL(n, \mathbb{R})$ or $SL(n, \mathbb{R})$. If there is a nonzero function f in $L^2(\mathscr{P}_n)$, such that $Lf = \lambda f$ for all L in $D(\mathscr{P}_n)$, it follows that $f \in L^2(\mathscr{P}_n/\Gamma)$ for *all* discrete subgroups Γ of $GL(n, \mathbb{R})$. This is absurd.

Furstenberg [2] defines a *boundary* M for a Lie group G to be a compact space such that there is a continuous G action $(g, x) \mapsto gx$ taking $G \times M$ into

M such that the action is

(i) associative: $(g_1 g_2)x = g_1(g_2 x)$;
(ii) transitive: for each x, y in M there is a g in G so that $gx = y$; and such that
(iii) for each probability measure π on M there is a sequence of elements g_n of G such that $g_n \pi$ converges to a point measure on M.

A *maximal boundary* $B(G)$ has the property that for any boundary M of G there is a map from $B(G)$ to M preserving the G actions. Furstenberg [loc. cit.] proves that a maximal boundary for $G = SL(n, \mathbb{R})$ is the boundary G/MAN appearing in Theorem 1 of this section. Here MAN consists of the upper triangular matrices in G and we can identify G/MAN with K/M by the Iwasawa decomposition of G. Furstenberg's result is actually more general and he goes on to show that Poisson's integral formula for bounded harmonic functions can be generalized using the maximal boundary. We will discuss this further in Chapter 5.

Exercise 9 (Boundaries of $G = SL(n, \mathbb{R})$).

(a) Show that if $G = SL(n, \mathbb{R})$ and MAN is the group of upper triangular matrices of determinant one, then the maximal boundary G/MAN can be identified with the *flag manifold* F_n of $n-1$ tuples $(V_1, V_2, \ldots, V_{n-1})$ where V_i denotes an i dimensional vector subspace of $\mathbb{R}^n$ and $V_1 \subset V_2 \subset \cdots \subset V_{n-1}$. The action of G on F_n is the obvious one defined via $gV_i = \{gx | x \in V_i\}$.
(b) Define $G_{i,n-1}$ to be the *Grassmann variety* of i dimensional subspaces of $\mathbb{R}^n$. Show that the mapping $(V_1, \ldots, V_{n-1}) \mapsto V_i$ sends F_n onto $G_{i,n-1}$ and preserves the G-actions. Thus $G_{i,n-1}$ is also a boundary of G. In particular, the projective space $G_{1,n-1} = \mathbb{P}^{n-1}$ is a boundary of G.
(c) Let $P(i, n-i)$ denote the parabolic subgroup of G consisting of matrices with block from

$$\begin{pmatrix} A & B \\ 0 & C \end{pmatrix}, \quad A \in GL(i, \mathbb{R}), \quad C \in GL(n-i, \mathbb{R}).$$

Show that we can identify the Grassmann variety $G_{i,n-1}$ of part (b) with $G/P(i, n-i)$.

4.4. Fundamental Domains for $\mathscr{P}_n/GL(n, \mathbb{Z})$

Seit meiner ersten Studienzeit war mir Minkowski der beste und zuverlässigste Freunde, der an mir hing mit der ganzen ihm eigenen Tiefe und Treue. Unsere Wissenschaft, die uns das liebste war, hatte uns zusammengeführt; sie erschien uns wie ein blühender Garten; in diesem Garten gibt es geebnete Wege, auf denen man mühelos geniesst, indem man sich umschaut, zumal an der Seite eines Gleichempfindenden. Gern suchten wir aber auch verborgene Pfade auf und entdeckten manche neue, uns schön dünkende Aussicht, und wenn der eine dem andern sie zeigte und wir sie gemeinsam bewunderten, war unsere Freude vollkommen.*

From Hilbert's speech in memory of Minkowski (see Minkowski [1, Vol. I, XXX]).

4.4.1. Introduction

In this section we study the action of the discrete group $GL(n, \mathbb{Z})$ consisting of $n \times n$ matrices with integer entries and determinant ± 1 on the space $\mathscr{P}_n$ of positive matrices. A *fundamental domain* D for $\mathscr{P}_n/GL(n, \mathbb{Z})$ is a subset of $\mathscr{P}_n$ which behaves like the quotient space $\mathscr{P}_n/GL(n, \mathbb{Z})$, at least up to boundary identifications. The fundamental domains for $\mathscr{P}_n/GL(n, \mathbb{Z})$ are much more difficult to visualize than those for $SL(2, \mathbb{Z})\backslash H$ which were considered in Section 3.3 of Volume I, since $\mathscr{P}_n$ is a subset of $n(n + 1)/2$-dimensional Euclidean space. Thus the smallest dimension for a picture of such a fundamental domain (for $n \geq 3$) would be six. If we consider only the determinant one surface $\mathscr{SP}_n/GL(n, \mathbb{Z})$, this reduces the dimension by one, making our picture 5-dimensional. We will include some pictures of projections of points in a fundamental domain for $\mathscr{SP}_3/GL(3, \mathbb{Z})$ in Section 4.4.3.

Much of this section is due to Minkowski, who was the first to describe a fundamental domain for $GL(n, \mathbb{Z})$ (see Section 4.4.2). We will discuss another fundamental domain—that of Grenier [1] in Section 4.4.3. The latter domain has the advantage of looking more like the one for $SL(2, \mathbb{Z})$ which we used in Section 3.3 of Volume I. There are indeed many unusual flowers in these higher dimensional gardens. The names of those who have cultivated these flowers include: Gauss, Hermite, Minkowski, Voronoi, Siegel, Weyl, Weil, Satake, Baily, Borel, Serre, Harish-Chandra, Mostow, Tamagawa, Mumford, Delone, Ryskov, ...

The reader may be wondering why one would want to wander about in these higher dimensional gardens. As we mentioned in Section 4.1.0, our main motivation is the desire to study some *relatives of Riemann's zeta function*. We will see that we can generalize Riemann's method of analytic continuation of the Riemann zeta function—a method used in Theorem 1 of Section 1.4 in

---

* Since my first days as a student, Minkowski, with his typical depth and faith, was my best and most reliable friend. Our beloved science had brought us together; it seemed to us like a blooming garden; in this garden there were smooth (well tended) paths that one enjoyed effortlessly while looking around, especially at the side of someone with the same feelings. But we also liked to seek out the hidden paths and discovered several new views which were beautiful in our opinion and when one of us showed them to the other and we both admired them, our joy was complete.

Volume I. This method involves taking a Mellin transform of a theta function. If $X \in \mathscr{P}_n$, and $Y \in \mathscr{P}_m$, for $1 \leq m \leq n$, we define the *theta function* by:

$$\theta(Y, X) = \sum_{A \in \mathbb{Z}^{n \times m}} \exp\{-\pi \operatorname{Tr}(X[A]Y)\}. \tag{4.1}$$

This theta function is related to a zeta function generalizing Epstein's zeta function from Section 1.4 of Volume I as well as the zeta function introduced in formula (1.4) of Section 4.1.0. The zeta function in question is called *Koecher's zeta function* because it was first studied by Koecher [1] and it is defined by:

$$Z_{m,n-m}(X, s) = \sum_{\substack{A \in \mathbb{Z}^{n \times m}/GL(m, \mathbb{Z}) \\ \operatorname{rank} A = m}} |X[A]|^{-s}, \quad \text{if } \operatorname{Re} s > \frac{n}{2}. \tag{4.2}$$

Here the sum is over $n \times m$ integral matrices A of rank m running through a complete set of representatives for the equivalence relation

$$A \sim B \text{ iff } A = BU \text{ for some } U \in GL(m, \mathbb{Z}).$$

Note that if $m = 1$, the theta function (4.1) is just that considered in Exercise 6 of Section 1.4 of Volume I, and in this case, Koecher's zeta function reduces to the Epstein zeta function defined in Section 1.4 of Volume I. When $n = m$, Koecher's zeta function is the function in formula (1.4) of Section 4.1.0 and we will prove that in this case it is a product of Riemann zeta functions (see Lemma 7 below)

$$Z_{n,0}(X, s) = |X|^{-s} \prod_{j=0}^{n-1} \zeta(2s - j). \tag{4.3}$$

In fact $Z_{n,0}(I, s)$ is the analogue of the Dedekind zeta function (considered in Section 1.4 of Volume I) for the simple algebra of all $n \times n$ rational matrices.

In order to imitate the proof of the analytic continuation of the Epstein zeta function given in Section 1.4 of Volume I, we need to Mellin transform the theta function (4.1). The Mellin transform used here is not a transform over all Y in $\mathscr{P}_m$, but instead over $\mathscr{P}_m/GL(m, \mathbb{Z})$. This is necessary because $\theta(Y[U], X) = \theta(Y, X)$ for all $U \in GL(m, \mathbb{Z})$. Explicitly, the *Mellin transform* is:

$$\int_{\mathscr{P}_m/GL(m, \mathbb{Z})} |Y|^s \theta_m(Y, X) \, d\mu_m(Y) = 2\pi^{-ms} \Gamma_m(0, \ldots, 0, s) Z_{m, n-m}(X, s). \tag{4.4}$$

Here Γ_m denotes the gamma function defined by (2.4) in Section 4.1.2 and θ_m denotes the partial sum of (4.1) over all $A \in \mathbb{Z}^{n \times m}$ such that the rank of A is m. Here we always assume that $1 \leq m \leq n$.

This kind of example motivates the search for analogues of *Hecke's correspondence* (see Section 3.6 of Volume I) which would relate Siegel modular forms such as the theta function in (4.1) with Dirichlet series of several variables. One needs more variables than the one complex variable s appearing in (4.4) in order to invert the matrix Mellin transform. This inversion was used by Kaori Imai [1] in the case of cuspidal Siegel modular forms of genus 2 (i.e., forms for $Sp(2, \mathbb{Z})$) to generalize Hecke's correspondence. Her results say that there is a dictionary which translates between the languages:

$$\begin{array}{c} \text{Siegel modular forms} \\ \text{of genus 2 for } Sp(2, \mathbb{Z}) \end{array} \Leftrightarrow \begin{array}{c} \text{Dirichlet series "twisted" by} \\ \text{automorphic forms for } GL(2, \mathbb{Z}) \\ \text{with functional equations.} \end{array}$$

The "→" can be found in Maass [2, Section 16] for Siegel modular forms of arbitrary genus, in fact. The converse correspondence "←" is proved for cusp forms by Imai [1] using the Roelcke-Selberg-Mellin inversion formula on $\mathscr{P}_2/GL(2,\mathbb{Z})$. See also Chapter 5, as well as Maass [4] and Roelcke [1]. Recently Weissauer [1] has extended the converse result to congruence subgroups of $Sp(n, \mathbb{Z})$ for all n.

The main goal of this chapter is to present some of the ideas necessary for harmonic analysis on $\mathscr{P}_n/GL(n, \mathbb{Z})$, from the same point of view that worked in the preceding section for $\mathscr{P}_n$ itself. The theory is not yet in its final form, however. But it is this goal that motivates our detailed study of the fundamental domain.

Jacquet, Piatetski-Shapiro, and Shalika [1] have shown that the adelic version of the Hecke converse theorem for $GL(n)$ does not require "twists" by automorphic forms for $GL(n-1)$, but only those for $GL(m)$, $m \leq n-2$. Such a converse theorem can be used to show, for example, that zeta and L-functions for totally real cubic number fields correspond to cusp forms for the adelized version of $GL(3)$, i.e., cusp forms for congruence subgroups of $GL(3)$. Thus when one sees L-functions with the right gamma factors in their functional equations, one expects to find corresponding cusp forms for $GL(n)$. But, in general, one must also have functional equations for L-functions "twisted" by automorphic forms for $GL(m)$, $m \leq n - 2$. Making use of the Rankin-Selberg convolution, which leads to L-functions with an Euler product that indicates the presence of an automorphic form for $GL(3)$, Gelbart and Jacquet [1] obtain a lifting of automorphic forms from $GL(2)$ to $GL(3)$.

The aforementioned results are part of a vast program of Langlands and many coworkers which gives a theory of L-functions attached to adelic irreducible automorphic representations of reductive groups over global fields. This theory is surveyed by Borel in Borel and Casselman [1, Vol. II, pp. 27–61] and by Gelbart [2]. Langlands has attached L-functions to an automorphic representation of the adelic $GL(n)$ by defining an Euler product. The p-factor for a given prime p is obtained by making use of the eigenvalues of the Hecke operators acting on an automorphic form to determine an $n \times n$ matrix A_p which is diagonal (see Gelbart [2, pp. 200–203]). Then the Euler factor is:

$$\det(I - A_p Np)^{-1},$$

when the representation is "unramified" at the prime ideal p. Langlands made a conjecture about his L-functions which generalizes the Artin reciprocity law in the theory of abelian extensions of number fields. This conjecture of Langlands would imply the Artin conjecture that the Artin L-functions are entire (excluding cases which are obviously not entire; e.g., when the character is trivial), since the Langlands L-function is entire for any nontrivial cuspidal representation of $GL(n)$. Attempts to prove the Artin conjecture this way have indeed made progress in the case of degree 2 representations of the Galois group of the extension (see Langlands [3], Tunnell [1], [2]). This progress involves the "twisted" Selberg trace formula and "base change."

At first sight, the Langlands L-function defined by an Euler product sounds rather different from an L-function defined by a Dirichlet series or a Mellin transform over a fundamental domain for $GL(n, \mathbb{Z})$. However, as we saw in part (5) of Theorem 4, Section 3.6, in Volume I, Hecke L-functions can be defined in either way, if the corresponding automorphic form is an eigenform for all the Hecke operators. We will find that an analogous result holds for $GL(n, \mathbb{Z})$ in Theorem 2 of Section 4.5. Thus we will study L-functions using Mellin transforms over $\mathscr{P}_n/GL(n, \mathbb{Z})$; and these L-functions will indeed have Euler products when the corresponding automorphic form is an eigenfunction of all the Hecke operators for $GL(n, \mathbb{Z})$. Of course, these Mellin transforms can also be used to study the Eisenstein series generalizing Koecher's zeta function (4.2). Such Eisenstein series need not have Euler products, except in certain special cases, such as that of (4.3), where the Euler product comes from that for the Riemann zeta function. Bump [1] provides more connections between the adelic point of view and the Dirichlet series point of view.

If you are not interested in these L-functions for $GL(n, \mathbb{Z})$, there are still lots of reasons to study fundamental domains for $\mathscr{P}_n/GL(n, \mathbb{Z})$. We listed some of these at the beginning of Section 4.1.0. Let's go into more detail here.

The embedding used by Hecke to relate zeta functions of algebraic number fields with Epstein zeta functions (see Theorem 2 of Section 1.4) leads one to suspect that explicit fundamental domains for $\mathscr{P}_n/GL(n, \mathbb{Z})$ should lead to *explicit algorithms for the computation of class numbers and units of number fields*. This was indeed the case for imaginary quadratic fields (see Exercise 5 of Section 3.3). The units in a number field are connected with a certain fundamental domain in a Euclidean space (see Step 4 in the proof of Theorem 2 in Section 1.4, Vol. I). The units and class number also influence the fundamental domains for $SL(2, O_K)$, $O_K =$ the ring of integers of a number field K—groups to be considered in the next chapter. In many ways, $\mathscr{P}_n/GL(n, \mathbb{Z})$ is the prototype for all fundamental domains.

Another related issue is that of the closed geodesics in $SL(2, \mathbb{Z})\backslash H$ corresponding to hyperbolic elements of $SL(2, \mathbb{Z})$. Such a geodesic corresponds to an element z in a real quadratic number field—z being fixed by the hyperbolic matrix γ. Here γ in $SL(2, \mathbb{Z})$ is called *hyperbolic* if the eigenvalues of γ are distinct, real, and different from 1 or -1. If

$$\sigma \begin{pmatrix} \varepsilon & 0 \\ 0 & 1/\varepsilon \end{pmatrix} \sigma^{-1} = \gamma, \quad \text{for } \sigma \in SL(2, \mathbb{R}),$$

then the geodesic fixed by γ is the image of the positive y-axis under σ. Then

4.4. Fundamental Domains for $\mathscr{P}_n/GL(n, \mathbb{Z})$

the eigenvalue ε of γ is a unit in a real quadratic field and the columns of σ are eigenvectors of γ. The periodic continued fraction expansions of these quadratic numbers z come from the translations and inversions needed to map the half circle connecting z and its conjugate into the fundamental domain for $SL(2, \mathbb{Z})$ (see Exercise 20 of Section 3.7). You might wonder how z and its conjugate z' over $\mathbb{Q}$ relate to ε and its conjugate ε^{-1}. It is not hard to see that

$$\{\varepsilon, \varepsilon^{-1}\} = \{cz + d, cz' + d\}.$$

For if

$$\gamma = \begin{pmatrix} a & b \\ c & d \end{pmatrix}, \quad \text{then} \quad \gamma\begin{pmatrix} z \\ 1 \end{pmatrix} = \begin{pmatrix} az + b \\ cz + d \end{pmatrix}$$

and it follows that because $\gamma z = (az + b)/(cz + d) = z$, we have

$$\gamma\begin{pmatrix} z & z' \\ 1 & 1 \end{pmatrix} = \begin{pmatrix} z & z' \\ 1 & 1 \end{pmatrix}\begin{pmatrix} cz + d & 0 \\ 0 & cz' + d \end{pmatrix}.$$

One wonders whether $GL(n, \mathbb{Z})$-analogues of the preceding remarks would lead to periodic algorithms for the approximation of elements of a totally real number field of degree n. There is a long history of the search for a generalization of the theorem that a real number is quadratic if and only if its continued fraction expansion is periodic. Minkowski [1, Vol. I, pp. 357–371]) gives an algorithm which is periodic in some cases. There are many other algorithms generalizing continued fractions, but none seems to be completely satisfactory.

There is a generalization to $GL(n, \mathbb{Z})$ of the relation between units in real quadratic fields and closed geodesics in $SL(2, \mathbb{Z}) \backslash H$ (see D. Wallace [2] for related results). A hyperbolic element γ in $GL(n, \mathbb{Z})$ is one with distinct real eigenvalues none of which are equal to ± 1. Thus γ has eigenvalues which are units in a totally real number field of degree n. If for $\sigma \in GL(n, \mathbb{R})$, we have $\sigma\gamma\sigma^{-1}$ is diagonal with jth diagonal entry ε_j, then the following totally geodesic submanifold G_a is fixed by γ:

$$\bigcup_{a \in \mathbb{R}^n} G_a[\sigma],$$

where

$$G_a = \left\{ \begin{pmatrix} e^{a_1 t} & & 0 \\ & \ddots & \\ 0 & & e^{a_n t} \end{pmatrix} \middle| t \in \mathbb{R} \right\}.$$

Other references for continued fraction type algorithms are Brentjes [1], Ferguson and Forcade [1]. There are many applications of higher dimensional continued fraction algorithms in coding and elsewhere (see Lagarias and Odlyzko [1]). Related references are: Ash, Mumford, Rapoport and Tai [1], Barrucand, Williams, and Baniuk [1], Cusick and Schoenfeld [1], Delone and Faddeev [1], Hirzebruch [1], and Williams and Broere [1].

It is also of interest to number theorists that the Euclidean volume of the subset of matrices Y in Minkowski's fundamental domain for $\mathscr{P}_n/GL(n,\mathbb{Z})$ such that $|Y| \leq 1$ involves a product of *Riemann zeta functions at odd as well as even integer arguments* (see Section 4.4.4). For recall from the discussion in Section 1.4 of Volume I (and Exercise 7 of Section 3.5 in Vol. I) that Euler found a nice formula for values of zeta at positive even integers, but no one has managed a similar result for odd integers. Siegel used formulas (4.3) and (4.4) above to prove Minkowski's formula for this volume (see Siegel [1, Vol. I, pp. 459–468 and Vol. III, pp. 328–333]). Weil [2, Vol. I, p. 561] notes:

> Siegel était arrivé à Princeton en 1940; pendant tout mon séjour aux Etats-Unis, je l'avais vu souvent. Depuis longtemps, avec juste raison, il attachait une grande importance au calcul du volume des domaines fondamentaux pour les sous-groupes arithmétiques des groupes simples; il avail consacré à ce sujet, inauguré autrefois par Minkowski, plusieurs mémoires importants. A ce propos il s'était vivement intéressé à la formule générale de Gauss-Bonnet, d'où pouvait résulter, du moins pour les sous-groupes à quotient compact, une détermination topologique des volumes en question. Je crois même me souvenir qu'il avait cru un jour tirer de là des conclusions au sujet de valeurs de $\zeta(n)$ pour n impair > 1, et s'était donné quelque mal pour les vérifier numériquement, avant de s'apercevoir qu'il s'agissait d'un cas où la courbure de Gauss-Bonnet est nulle.*

To bring up a different and quite old question from number theory, define the representation numbers $A_Y(m)$ for *the number of representations of an integer m in the form $m = Y[a]$ for a positive definite quadratic form Y in $\mathscr{P}_n$ with integer coefficients and an integral vector $a \in \mathbb{Z}^n$*. We discussed some of this at the end of Section 3.4 in Volume I. Gauss treated the cases $n = 2, 3$. The case that $Y = I_n$ is the $n \times n$ identity matrix has received special attention. For example, in 1829, Jacobi proved that

$$A_{I_4}(n) = 8 \sum_{0 < d|n} d, \quad \text{if } n \text{ is odd.}$$

One can view the left hand side of the equality as the Fourier coefficient of a theta function of weight 2 and the right hand side as the Fourier coefficient of an Eisenstein series of weight 2.

In 1883 when Minkowski was 17, he and Smith split a prize for proofs of Eisenstein's formula for the mass of a genus of quadratic forms (see Minkowski [1, Vol. I, pp. 157–202] or Hancock [1]). Siegel developed a vast extension

---

* Siegel arrived at Princeton in 1940; during my entire stay in the United States, I saw him often. For a long time, rightly, he attached a great importance to the calculation of the volume of the fundamental domain for arithmetic subgroups of simple groups; he had devoted several important papers to this subject which had been begun long before by Minkowski. In this regard he was keenly interested in the general Gauss-Bonnet formula, from which could result a topological characterization of the volume in question, at least for subgroups with compact quotient. I even believe that I remember that he once thought that he had derived conclusions from that on the subject of the values of $\zeta(n)$ for n odd > 1, and had taken some trouble to verify this numerically, before realizing that it was a question of a case where the Gauss-Bonnet curvature is zero.

4.4. Fundamental Domains for $\mathscr{P}_n/GL(n, \mathbb{Z})$

of these results in the 1930's (see Siegel [1, Vol. I, pp. 326–405, 410–443, 469–548; Vol. II, pp. 1–7, 20–40] and Milnor and Husemoller [1]). The general result can be viewed as an identity between Siegel modular forms. See also Freitag [1, pp. 285–297]. There is a brief exposition of Siegel's work and related developments in Cassels [1, pp. 374–388].

These studies of quadratic forms require a knowledge of the fundamental domain for $\mathscr{P}_n/GL(n, \mathbb{Z})$. For the fundamental domain for the Siegel modular group cannot be understood without first understanding the fundamental domain for $GL(n, \mathbb{Z})$, as we shall see in Chapter 5.

In the 1960's Tamagawa, Weil, Ono, and Kneser obtained an adelic version of Siegel's results on quadratic forms. Some references are the article of Kneser in Cassels and Fröhlich [1, pp. 250–265], the articles of Mars in Borel and Mostow [1, pp. 133–142], and Weil [2, Vol. III pp. 1–157].

Fundamental domains for groups like $GL(n, \mathbb{Z})$ are not just of interest to number theorists. They also provide food for thought to those interested in *geometry and topology*. Ash, Mumford, Tai, and Rappoport [1] have obtained smooth compactifications of such fundamental domains. This would allow one to use the Riemann-Roch theorem and other methods from geometry to compute dimensions of spaces of modular forms. These smooth compactifications are obtained explicitly using ideas of Minkowski and Voronoi, as well as the theory of toroidal embeddings. References include: Baily and Borel [1], Borel and Serre [1], Chai [1], Mostow and Tamagawa [1], Namikawa [1], Satake [2,3], and Yamazaki [1]. References related to the computation of cohomology of arithmetic groups are: Ash [1], Ash, Grayson and Green [1], Borel [1], Borel and Serre [1], Borel and Wallach [1], Schwermer [1], [2], Serre [1], Soulé [1], [2].

There are many places in *physics* where automorphic forms for $GL(n, \mathbb{Z})$ and $Sp(n, \mathbb{Z})$ have popped up. Of course, it should not be surprising to find that abelian integrals and thus Riemann theta functions such as (4.1) above should bear solutions to partial differential equations as their fruit. For example, classical theta functions such as those discussed in Section 3.4 of Volume I appear in the solutions by Euler, Lagrange, and Poisson of two special cases of the problem of describing the motion of a solid body rotating about a fixed point. The third known case of this problem was solved by Sonya Kovalevsky [1] using Siegel modular forms (Riemann theta functions). She was awarded the Prix Bordin for this work in 1888. Evidently no less a mathematician than Picard told Kovalevsky in 1886 that he was skeptical that theta functions for $Sp(n, \mathbb{Z})$ "can be useful in the integration of certain differential equations" (see Dubrovin, Matveev, and Novikov [1]). But ninety years later, in the paper of Dubrovin et al [loc. cit.] theta functions are used to solve the Korteweg-deVries partial differential equation arising in the theory of solitons. For related papers and some short articles on Kovalevsky's life see the volume edited by Linda Keen [1]. The books of Cooke [1] and Koblitz [1] give more detailed discussions.

The theta functions for $Sp(n, \mathbb{Z})$ are also intrinsic to Siegel's work on quadratic forms mentioned above. This work has recently been connected with quantum mechanics via the Segal-Shale-Weil representation. References are Lion and Vergne [1], Shale [1], Wallach [3], and Weil [2, Vol. 3, pp. 1–157]. See also the book by Mumford [1].

Finding *densest lattice packings of spheres* in $\mathbb{R}^n$ is a part of Hilbert's eighteenth problem (see Cassels [1], Davenport [1], Milnor [1], Rogers [1], Siegel [2] [3], Sloane [1,2]), and Thompson [1]). A *lattice L* in $\mathbb{R}^n$ is a subgroup of the additive group of $\mathbb{R}^n$ of the form:

$$L = \mathbb{Z}v_1 \oplus \mathbb{Z}v_2 \oplus \cdots \oplus \mathbb{Z}v_n, \tag{4.5}$$

where the vectors $v_1, v_2, \ldots, v_n$ form a vector space basis of $\mathbb{R}^n$. It can be shown that this is equivalent to saying that L is a discrete subgroup of $\mathbb{R}^n$ such that $\mathbb{R}^n/L$ is compact; i.e. a discrete cocompact subgroup of $\mathbb{R}^n$. For a proof of this last remark see Siegel [2, pp. 9–12]. The problem of finding the densest lattice packings of spheres in $\mathbb{R}^n$ is that of finding a lattice L such that if non-overlapping open spheres of equal radii are centered at each point of L, the largest possible volume is filled up. This sphere packing problem goes back to a book review that Gauss wrote in 1831.

There is an *identification between lattices L* in $\mathbb{R}^n$ *and positive matrices Y* in a fundamental domain for $\mathscr{P}_n/GL(n, \mathbb{Z})$ which is made as follows. Suppose we are given a lattice L as in formula (4.5). Define the positive matrix $Y(L)$ in $\mathscr{P}_n$ by:

$$Y(L) = I[v], \quad \text{for } v = (v_1 v_2 \cdots v_n) \in \mathbb{R}^n. \tag{4.6}$$

Since the lattice L remains the same upon change of $\mathbb{Z}$-basis, which amounts to replacing v by $v\gamma$, for some $\gamma \in GL(n, \mathbb{Z})$, we must consider $Y(L)$ as an equivalence class in $\mathscr{P}_n/GL(n, \mathbb{Z})$.

Using the identification (4.6), the problem of finding the lattice L giving the densest packing of spheres of equal radius r with centers at points in L turns out to equivalent to the problem of choosing Y in $\mathscr{P}_n$ to maximize the *minimum over the integer lattice*:

$$m_Y = \min\{Y[a] | a \in \mathbb{Z}^n - 0\}. \tag{4.7}$$

To see this, note first that if $a \in \mathbb{Z}^n$, and we set

$$w = a_1 v_1 + \cdots + a_n v_n,$$

then $Y[a] = {}^t w w$ is the square of the distance from the lattice point $w \in L$ to the origin. Next note that the spheres must not intersect, which means that one should take them to have radius equal to one-half the minimum distance of any lattice point from the origin. This means that the radius r must be chosen to be $\frac{1}{2}(m_Y)^{1/2}$.

The density of space occupied by spheres of radius r centered at points in the lattice L is:

$$d_L = \lim_{X \to \infty} \frac{v_n(r) \cdot \#(L \cap (\text{cube of volume } X))}{X}, \tag{4.8}$$

4.4. Fundamental Domains for $\mathscr{P}_n/GL(n,\mathbb{Z})$

with $v_n(r)$ being the volume of the sphere of radius r in $\mathbb{R}^n$. Now the number of points of L in a cube of volume X is easily seen to be asymptotic to $|Y|^{1/2} X$ as X approaches infinity. Therefore

$$d_L = r^n v_n(1)|Y|^{-1/2} = \left(\frac{m_Y}{4}\right)^{n/2} v_n(1)|Y|^{-1/2},$$

and

$$v_n(1) = \frac{\pi^{n/2}}{\Gamma(1+n/2)}. \tag{4.9}$$

Note that the density d_L is unchanged if we multiply the $\mathbb{Z}$-basis of L by a constant c (or equivalently if we multiply the corresponding matrix $Y(L)$ by c^2) for then r is multiplied by c and $|Y|^{-1/2}$ is multiplied by $1/c$. The fact that the density d_L must be less than or equal to one gives the *Minkowski upper bound for the minimum* m_Y:

$$m_Y \leq c_n |Y|^{1/n}, \quad \text{with } c_n = \frac{4}{\pi}\Gamma(1+n/2)^{2/n} \sim \frac{2n}{\pi e}, \quad \text{as } n \to \infty. \tag{4.10}$$

The asymptotic behavior of c_n comes from Stirling's asymptotic formula for the gamma function (see Lebedev [1]).

We will give another (and more detailed) view of (4.10) later in this section as a consequence of Lemma 1 below. Blichfeldt showed in 1914 that the constant c_n can be halved. This is equivalent to showing that the density d_L cannot exceed about $2^{-.5n}$, for large n. Kabatiansky and Levenshtein [1] have shown that for large n the density cannot exceed about $2^{-.599n}$. This leads to an upper bound on m_Y of about

$$\frac{n}{\pi e} 2^{-.198n}, \quad \text{for large } n.$$

In Section 4.4.5 we will consider Minkowski's result that there exist Y in $\mathscr{P}_n$ such that the minimum satisfies

$$m_Y > \frac{n}{2\pi e}|Y|^{1/n}, \quad \text{for large } n$$

(see Corollary 1 to Proposition 2). This translates to the statement that there are lattice packings of $\mathbb{R}^n$ whose density is greater than 2^{-n} for large n. Minkowski's result does not, however, give a construction for these lattices. Some work on finding dense lattice packings explicitly is surveyed by Sloane [1]. For example, it is noted that Barnes and Sloane have constructed lattice packings in dimensions up to 100,000 with density roughly $2^{-1.25n}$. But Sloane notes that Minkowski's theorem guarantees that there exist packings that are $10^{4,000}$ times denser. See also Rush [1] and Rush and Sloane [1].

It may surprise the reader to learn that it is still an open question whether the densest lattice packing in $\mathbb{R}^3$ actually gives the densest not necessarily lattice centered packing of spheres in $\mathbb{R}^3$. See Sigrist [1] for a short survey on sphere packing. Sigrist notes the following quotes on the problem:

H.S.M. Coxeter: "It is conceivable that some irregular packing might be still denser."

C.A. Rogers: "Many mathematicians believe, and all physicists know, that the density cannot exceed $\pi/\sqrt{18}$."

See Figures 4.1 and 4.2 for the densest lattice packings in the plane and 3-space. It is known that the densest lattice packing in the plane also gives the densest packing lattice or not. But, as we've said, this is not known for 3-space.

Exercise 1.

(a) Prove formula (4.9) for the volume of the unit sphere in $\mathbb{R}^n$.
(b) Fill in the details in the rest of the discussion of formula (4.10) above.

Hints. (a) Note that

$$\int_{\mathbb{R}^n} \exp(-{}^t xx)\, dx = \pi^{n/2}.$$

Suppose now that w_n denotes the surface area of the unit sphere in $\mathbb{R}^n$; i.e.,

$$w_n = \text{surface area } \{x \in \mathbb{R}^n | {}^t xx = 1\}.$$

Use polar coordinates on the preceding integral to show that

$$\pi^{n/2} = w_n \tfrac{1}{2} \Gamma(n/2).$$

On the other hand, polar coordinates can be used to show that

$$v_n(1) = w_n/n.$$

This says that the volume of the unit sphere in $\mathbb{R}^n$ gets much smaller than the surface area as n goes to infinity. In fact, both w_n and $v_n(1)$ approach zero—a fact that we will use later in this section (see Corollary 1 to Theorem 4 in Section 4.4.4). Hamming [1, Ch. 9, 10] gives an interesting paradox related to these facts as well as applications to information theory. We'll consider this paradox in Section 4.4.4.

Denote by L_n, a lattice giving the densest lattice packing of spheres of equal radii in $\mathbb{R}^n$. For $n \leq 5$, the lattice L_n was determined by Korkine and Zolotareff. For $n \leq 8$, L_n was found by Blichfeldt. The lattice L_2 is often called the regular hexagonal lattice because the *Voronoi polyhedron*, which is the set of points in $\mathbb{R}^2$ lying as close to the origin as any lattice point, is a regular hexagon (see Figure 4.1). The lattice L_2 has $\mathbb{Z}$-basis $v_1 = (1, \sqrt{3})$, $v_2 = (2, 0)$ and thus corresponds to the positive matrix

$$2\begin{pmatrix} 2 & 1 \\ 1 & 2 \end{pmatrix}. \qquad \text{And } d_L = \frac{\pi}{2\sqrt{3}} \cong .9068.$$

4.4. Fundamental Domains for $\mathscr{P}_n/GL(n, \mathbb{Z})$

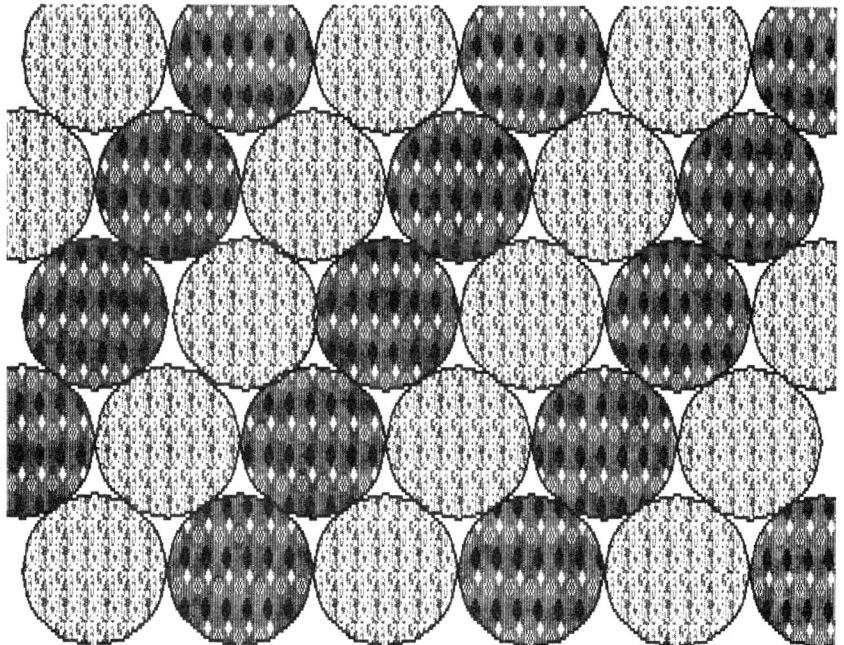

Figure 4.1. Densest lattice packing of circles of equal radii in the plane.

Figure 4.2. Part of the face-centered cubic lattice packing in 3-space. (From Sigrist [1].)

The lattice L_3 is the face centered cubic lattice pictured in the crystallography discussion in Section 1.4 of Volume I. See also Figure 4.2. This lattice occurs in crystals of gold, silver, and aluminum, for example. It has $\mathbb{Z}$-basis $v_1 = (1, 1, 0)$, $v_2 = (1, 0, 1)$, $v_3 = (0, 1, 1)$ and thus corresponds to the positive matrix:

$$\begin{pmatrix} 2 & 1 & 1 \\ 1 & 2 & 1 \\ 1 & 1 & 2 \end{pmatrix}. \quad \text{Here } d_L = \frac{\pi}{\sqrt{18}} \cong .7404.$$

The Voronoi polyhedron for L_3 is a rhombic dodecahedron (a solid bounded by 12 rhombuses). The lattices L_n, for $n \leq 8$, can be shown to correspond to root systems for the simple Lie groups $A_2, A_3, D_4, D_5, E_6, E_7, E_8$. See Chapter 5 for a discussion of root systems and see Milnor [1, p. 502] or Milnor and Husemoller [1] for a description of how to get Y out of the Dynkin diagram for the Lie group. See Thompson [1, Appendix 1] for a table of densest known sphere packings.

We know from Section 3.4 of Volume I that there are connections between sphere packings and *coding theory* (see Sloane [1], [2]). And Shannon found that the problem of finding densest sphere packings in spherical space has applications to information theory (see Van der Waerden [3]). Note that one can consider various non-Euclidean analogues of the sphere-packing problem (see Fejes Tóth [1]). Moreover, the work on codes has led to the discovery of the Leech lattice in $\mathbb{R}^{24}$ which gives rise to many of the densest known lattice packings as well as some new simple groups. Thompson [1] provides a survey of the connection.

Dyson [1] discusses some of these stories about the interplay between dense lattice packings, codes, and simple groups in an article about unfashionable mathematics. We quote:

> Roughly speaking, unfashionable mathematics consists of those parts of mathematics which were declared by the mandarins of Bourbaki not to be mathematics. A number of very beautiful mathematical discoveries fall into this category. To be mathematics according to Bourbaki, an idea should be general, abstract, coherent, and connected by clear logical relationships with the rest of mathematics. Excluded from mathematics are particular facts, concrete objects which just happen to exist for no identifiable reason, things which a mathematician would call accidental or sporadic. Unfashionable mathematics is mainly concerned with things of accidental beauty, special functions, particular number fields, exceptional algebras, sporadic finite groups. It is among these unorganized and undisciplined parts of mathematics that I would advise you to look for the next revolution in physics.

Analysis on the fundamental domain $\mathscr{P}_n/GL(n, \mathbb{Z})$ can also be applied to the problem of finding the best lattice of points in $\mathbb{R}^n$ to use for *numerical integration* (see Ryskov [1]).

Explicit fundamental domains for $\mathscr{P}_n/GL(n, \mathbb{Z})$ are pertinent to the problem of finding subgroups of $GL(n, \mathbb{Z})$ fixing some Y in $\mathscr{P}_n$, a problem which is of interest in *crystallography* and was thus solved long ago for $n \leq 3$. The cases $n = 4$ and 5 have been solved by Dade and Ryskov (see Ryskov [1] for the references).

This completes our list of reasons for beginning the study of the fundamental domain $\mathscr{P}_n/GL(n, \mathbb{Z})$. The main references for this section are: Borel [1,2], Borel's article in Borel and Mostow [1, pp. 20–25], Cassels [1], [2], Davenport [1], Delone and Ryskov [1], Freitag [1], Grenier [1], [2], Grenier, Gordon and Terras [1], Hancock [1], Humphreys [1], O.-H. Keller [1], Maass [2], Minkowski [1], Raghunathan [1], Rogers [1], Ryskov [1], Ryskov and Baranovskii [1], Schwarzenberger [1], Séminaire Cartan [1], Siegel [2, 3], Van der Waerden [1], [2], [3], Weil [2, Vol. I, pp. 339–358], [3], Weyl [1, Vol. III, pp. 719–757, Vol. IV, pp. 46–96]. Some of the earlier references are: Gauss [1, Vol. I, p. 188], Hermite [1, Vol. I, pp. 94–164], Korkine and Zolotareff [1], Lagrange [1, Vol. III, pp. 693–758], Seeber [1], and Voronoi [1], [2].

4.4.2. Minkowski's Fundamental Domain

Before describing Minkowski's fundamental domain for $\mathscr{P}_n/GL(n, \mathbb{Z})$, we need to retrace Minkowski's steps and consider his most fundamental results in the geometry of numbers. These results have immediate applications in the very foundations of algebraic number theory. In general, they are useful when one wants to know whether some inequality has a solution in integers. Here we are interested in the size of the minimum of a quadratic form Y over the integer lattice; i.e., in the size of m_Y defined by (4.7) for $Y \in \mathscr{P}_n$. We have already given one approach to this problem, which led to the inequality (4.10). Now let us consider another approach. Define the *ellipsoid* in $\mathbb{R}^n$ associated to the positive matrix $Y \in \mathscr{P}_n$:

$$S_Y(t) = \{x \in \mathbb{R}^n | Y[x] < t\}, \quad \text{for } t > 0. \tag{4.11}$$

This is a *convex* set; i.e., if $x, y \in S_Y(t)$ and $a \in [0, 1]$, then

$$ax + (1 - a)y \in S_Y(t).$$

Exercise 2.

(a) Show that $S_Y(t)$ defined in (4.11) is a convex set. Show also that its closure is compact. Why do we call it an ellipsoid?
(b) Show that the volume of $S_Y(t)$ is $|Y|^{-1/2} v_n(t^{1/2})$, where $v_n(t^{1/2})$ is the volume of the sphere of radius $t^{1/2}$, obtained using formula (4.9).

Minkowski used the fundamental facts below to see that

$$S_Y(t) \cap \mathbb{Z}^n \neq \{0\}, \quad \text{if Vol } S_Y(t) > 2^n.$$

This means that

$$t > \frac{4}{\pi} \Gamma\left(1 + \frac{n}{2}\right)^{2/n} |Y|^{1/n}$$

implies that there exists an $a \in \mathbb{Z}^n - 0$ such that $Y[a] < t$. The inequality (4.10) for m_Y follows from this result.

Lemma 1 (Minkowski's Fundamental Lemma in the Geometry of Numbers).

(1) *Suppose that S is a Lebesgue measurable set in $\mathbb{R}^n$ with $\text{Vol}(S) > 1$. Then there are two points x, y in S such that $0 \neq x - y \in \mathbb{Z}^n$.*
(2) *Let S be a Lebesgue measurable subset of $\mathbb{R}^n$ which is symmetric with respect to the origin (i.e., $x \in S$ implies $-x \in S$) and convex. If, in addition, $\text{Vol}(S) > 2^n$, then $S \cap \mathbb{Z}^n \neq \{0\}$.*

PROOF.

(1) (From Weil [2, p. 36].) One has the following integral formula (as a special case of formula (1.6) from Section 4.1.0):

$$\int_{\mathbb{R}^n} f(x)\, dx = \int_{[0,1]^n} \sum_{a \in \mathbb{Z}^n} f(x + a)\, dx.$$

Let f be the characteristic function of S; i.e., $f(x) = 1$ if $x \in S$ and 0 otherwise. If the conclusion of part (1) of the lemma were false, the inner sum over $\mathbb{Z}^n$ on the right hand side of this integral formula would be less than or equal to one for all x in $[0, 1]^n$. This gives a contradiction to the hypothesis that $\text{Vol}(S) > 1$.

Another way to see part (1) is to translate S to the unit cube $[0, 1]^n$ by elements of $\mathbb{Z}^n$. If there were no overlap among these translates, the volume of S would be less than one. See Figure 4.3.

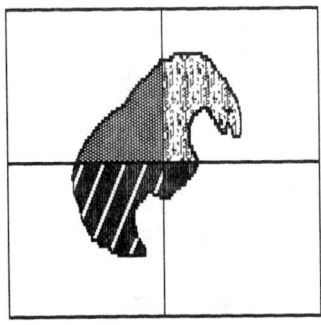

Figure 4.3. Picture proof of part 1 of Minkowski's fundamental lemma in 2 dimensions. Each square is a unit square. The 4 parts of the big square are translated to the square on the lower right by integral translations. There is an overlap.

4.4. Fundamental Domains for $\mathscr{P}_n/GL(n, \mathbb{Z})$

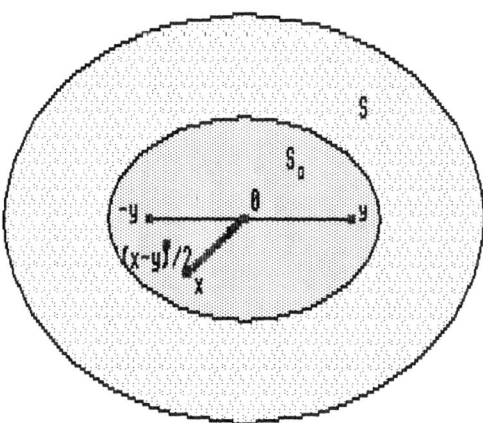

Figure 4.4. Picture proof of part 2 of Minkowski's fundamental lemma in 2 dimensions.

(2) The proof of this part of the lemma is illustrated in Figure 4.4. More explicitly, let us define S_0 to be the set $\frac{1}{2}S$ consisting of vectors of the form $\frac{1}{2}x$, for $x \in S$. Then $\text{Vol}(S_0) > 1$ by hypothesis and thus, by part (1) of this lemma, there are points x, y in S_0 such that $x - y \in \mathbb{Z}^n - 0$. It follows that $\frac{1}{2}(x - y)$ lies in S_0 by the convexity and symmetry of S. So $x - y$ lies in S. This completes the proof of Lemma 1. □

Minkowski's fundamental lemma leads quickly to the finiteness of the class number of an algebraic number field K as well as to Dirichlet's unit theorem giving the structure of the group of units in the ring of integers of K (see Section 1.4 of Volume I and the references mentioned there). A lower bound on the absolute value of the discriminant of K is also a consequence.

Next we need a lemma.

Lemma 2 (Vectors That Can Be First Columns of Elements of $GL(n, \mathbb{Z})$. A vector $a = {}^t(a_1, \ldots, a_n) \in \mathbb{Z}^n$ can be made the first column of a matrix in $GL(n, \mathbb{Z})$ if and only if the greatest common divisor g.c.d. $(a_1, \ldots, a_n)$ equals one.

PROOF.
$\Rightarrow$

This direction is clear upon expanding the determinant of A by its first column a.
$\Leftarrow$

Suppose that $a \in \mathbb{Z}^n$ has g.c.d. $(a_1, \ldots, a_n) = 1$. We need to obtain a matrix U in $GL(n, \mathbb{Z})$ such that $Ua = {}^t(1, 0, \ldots, 0)$. For then a is the first column of U^{-1}.

Our goal can be attained by multiplying our column vector a on the left by combinations of matrices giving rise to the elementary row operations on the vector. These *elementary row operations* are:

(i) changing the order of rows,
(ii) multiplying any row by $+1$ or -1,
(iii) adding an integral multiple of any row to any other row.

Using operation (ii), we can assume that all the entries in the given column vector a are non-negative. Operation (i) allows us to assume that a_1 is the smallest non-zero entry of the vector a. We can use operation (iii) to replace any non-zero entry a_j, for $j > 1$, by its remainder upon division by a_1. In this way, we can cause all the entries $a_2, \ldots, a_n$ to lie in the interval $[0, a_1)$. Either all the a_j are zero for $j \geq 2$, or there is a smallest positive entry $a_j, j \geq 2$. Use operation (i) to put this a_j in the first row and continue in this way. The final result is a vector ${}^t(m, 0, \ldots, 0)$, with $m > 0$. Thus $a = mU^{-1}\,{}^t(1, 0, \ldots, 0)$. Since the greatest common divisor of the entries in the vector a is one, it follows that $m = 1$ and the proof of Lemma 2 is complete. □

Note. The preceding lemma has even been established in an analogous situation in which $\mathbb{Z}$ is replaced by $F[X_1, \ldots, X_n]$ for fields F, proving a conjecture of Serre (see Bass [1]).

Our main goal is to produce a *fundamental domain* for $\mathscr{P}_n/GL(n, \mathbb{Z})$ and study its properties. That is, we want to find a subset D of $\mathscr{P}_n$ such that

(i) $$\mathscr{P}_n = \bigcup_{A \in GL(n, \mathbb{Z})} D[A]$$

and

(ii) $Y, W \in D$, with $W = Y[A]$ for some $A \in GL(n, \mathbb{Z})$, with $A \neq \pm I$ implies that Y must lie in the boundary ∂D of D.

Given $Y \in \mathscr{P}_n$, we will also want a procedure for finding a matrix $A \in GL(n, \mathbb{Z})$ such that $Y[A]$ lies in the fundamental domain D. Such a procedure is called a *reduction algorithm*. We will also call the fundamental domain a *set of reduced matrices*.

In this section we will discuss Minkowski's fundamental domain which is a convex, closed subset of $\mathscr{P}_n$, bounded by a finite set of hyperplanes through the origin. The domain was found by Minkowski who was motivated by the work of mathematicians such as Lagrange, Gauss, Seeber, and Hermite. Minkowski's fundamental domain $\mathscr{M}_n$ is defined as follows:

$$\mathscr{M}_n = \left\{ Y = (y_{ij}) \in \mathscr{P}_n \,\middle|\, \begin{array}{l} Y[a] \geq y_{kk}, \text{ if } a \in \mathbb{Z}^n, \text{ g.c.d. } (a_k, \ldots, a_n) = 1 \\ y_{k, k+1} \geq 0, \text{ for all } k \end{array} \right\}. \quad (4.12)$$

The domain in (4.12) appears to be bounded by an infinite number of hyperplanes. We will show in Theorem 1 that a finite number of inequalities actually suffices to give the region.

4.4. Fundamental Domains for $\mathscr{P}_n/GL(n,\mathbb{Z})$

By Lemma 2, it is easily seen that $\mathscr{M}_n$ has the alternative definition:

$$\mathscr{M}_n = \left\{ Y \in \mathscr{P}_n \,\middle|\, \begin{array}{l} Y[a] \geq y_{kk} \text{ if } (e_1 e_2 \cdots e_{k-1} a * \cdots *) \in GL(n,\mathbb{Z}) \\ y_{k,k+1} \geq 0, \text{ for all } k \end{array} \right\}. \quad (4.13)$$

Here e_k denotes the kth element in the standard basis for $\mathbb{R}^n$; i.e., e_k is a vector with all its entries 0 except for the kth entry which is 1. To see (4.13), note that Lemma 2 implies that the condition on the vector a in (4.13) is equivalent to asking that the greatest common divisor $(a_k, \ldots, a_n) = 1$.

Exercise 3.

(a) Prove that

$$\mathscr{M}_2 = \{Y \in \mathscr{P}_n | 0 \leq 2y_{12} \leq y_{11} \leq y_{22}\}. \quad (4.14)$$

(b) Note that restricting to the determinant one surface

$$\mathscr{SM}_2 = \{W \in \mathscr{M}_2 | |W| = 1\}$$

does not change the form of the inequalities in (4.14). Now recall the mapping from Exercise 9 of Section 3.1 in Volume I which identifies $\mathscr{SP}_2$ and the Poincaré upper half plane H:

$$W_z = \begin{pmatrix} y^{-1} & 0 \\ 0 & y \end{pmatrix} \begin{bmatrix} 1 & -x \\ 0 & 1 \end{bmatrix}, \quad \text{for } z = x + iy \in H.$$

Show that $W_z \in \mathscr{SM}_2$ is equivalent to the inclusion of z in the left half of the fundamental domain for $SL(2,\mathbb{Z}) \backslash H$ which is pictured in Figure 3.11 of Section 3.3 of Volume I. It is not the whole fundamental domain because $GL(2,\mathbb{Z})/SL(2,\mathbb{Z})$ has order 2 and the non-trivial coset comes from the matrix

$$\begin{pmatrix} -1 & 0 \\ 0 & 1 \end{pmatrix}.$$

Hints. To see that $\mathscr{M}_2$ is actually given by (4.14), you must note that if e_k, $k = 1, 2$, denotes the standard basis of $\mathbb{R}^2$, as usual, then

$$y_{11} \geq y_{22} \text{ comes from } Y[e_2] = y_{22} \geq y_{11}$$

since the g.c.d. of the entries of e_2 is 1. Similarly

$$y_{11} \geq 2y_{12} \text{ comes from } Y\begin{bmatrix} 1 \\ -1 \end{bmatrix} = y_{11} - 2y_{12} + y_{22} \geq y_{22}.$$

To see that it suffices to use only the inequalities in (4.13) coming from the vectors $a = e_2$ and ${}^t(-1, 1)$, you need to use the fact that for any vector $a \in \mathbb{Z}^n$ we have for Y in the set of (4.14):

$$Y[a] \geq y_{11}(a_1^2 - |a_1 a_2| + a_2^2) = y_{11}\{(|a_1| - |a_2|)^2 + |a_1 a_2|\}.$$

This is greater or equal to y_{11} if g.c.d. $(a_1, a_2) = 1$. And if $a_2 = \pm 1$, we see that if Y satisfies the inequalities in (4.14):

$$Y[a] = y_{11}a_1^2 + 2y_{12}a_1a_2 + y_{22}a_2^2 \geq y_{22}.$$

Exercise 4 (Successive Minima). Show that $Y \in \mathcal{M}_2$ is equivalent to saying that Y satisfies the following inequalities:

$$y_{12} \geq 0;$$

$$Y\begin{bmatrix} 1 \\ 0 \end{bmatrix} = y_{11} = m_Y = \min\{Y[a] | a \in \mathbb{Z}^2 - 0\} = N_1;$$

$$Y\begin{bmatrix} 0 \\ 1 \end{bmatrix} = y_{22} = \min\left\{Y[a] \middle| a \in \mathbb{Z}^2, \begin{vmatrix} 1 & a_1 \\ 0 & a_2 \end{vmatrix} \neq 0\right\} = N_2.$$

Here we call N_1 and N_2 the 1st and 2nd successive minima of Y respectively.

Similar ideas to those of Exercise 4 work for $GL(n, \mathbb{Z})$ with $n = 3$, as was found by Seeber [1] in 1831 and Gauss [1, Vol. II, p. 188]. That is, $Y \in \mathcal{M}_3$ is equivalent to requiring that y_{kk} be the minimum of the values $Y[a]$, $a \in \mathbb{Z}^3$ such that a is linearly independent of the standard basis vectors $e_1, \ldots, e_{k-1}$, for $k = 1, 2, 3$ as well as requiring $y_{k,k+1} \geq 0$ as usual.

For $n \geq 5$ reduction by successive minima is not possible because successive minima do not necessarily occur at vectors a_k which give matrices $(a_1 a_2 \cdots a_n)$ in $GL(n, \mathbb{Z})$.

We note here that Minkowski was probably inspired not only by the work of Gauss and Seeber, but also by that of Hermite [1, Vol. I, pp. 94–164]. However, with Minkowski, the theory progressed by a quantum leap. Minkowski [1, Vol. II, pp. 51–100] proved all of the following theorem as well as Theorem 3 in Section 4.4.4. His proofs have been rewritten by many eminent mathematicians. See the references mentioned earlier for alternative treatments.

Theorem 1 (The Minkowski Fundamental Domain). *Let $\mathcal{M}_n$ denote the Minkowski fundamental domain defined by formula (4.12). This domain has the following properties.*

(1) *For any Y in $\mathcal{P}_n$, there exists a matrix A in $GL(n, \mathbb{Z})$ such that $Y[A]$ lies in the Minkowski domain $\mathcal{M}_n$.*
(2) *Only a finite number of inequalities are necessary in the definition of $\mathcal{M}_n$. Thus $\mathcal{M}_n$ is a convex cone through the origin bounded by a finite number of hyperplanes.*
(3) *If Y and $Y[A]$ both lie in the Minkowski domain $\mathcal{M}_n$, and A is an element of $GL(n, \mathbb{Z})$ distinct from $\pm I$, then Y must lie on the boundary $\partial \mathcal{M}_n$. Moreover, $\mathcal{M}_n$ is bounded by a finite number of images $\mathcal{M}_n[A]$, for A in $GL(n, \mathbb{Z})$. That is $\mathcal{M}_n \cap (\mathcal{M}_n[A]) \neq \emptyset$, for only finitely many $A \in GL(n, \mathbb{Z})$.*

4.4. Fundamental Domains for $\mathscr{P}_n/GL(n, \mathbb{Z})$

(4) *When* $n = 2, 3, 4$, *we have*

$$\mathscr{M}_n = \{Y \in \mathscr{P}_n | y_{k,k+1} \geq 0;\ Y[a] \geq y_{kk}, \text{if } a_k = 1 \text{ and } a_j = 0 \text{ or } \pm 1, \text{for all } k\}.$$

The fundamental domains $\mathscr{M}_5$ *and* $\mathscr{M}_6$ *were determined explicitly by Minkowski* [1, Vol. I, pp. 145–148, 154, 218] *and* $\mathscr{M}_7$ *was found by Tammela* [1]. *These domains require more inequalities.*

PROOF.

(1) Suppose $Y \in \mathscr{P}_n$ is given. We need to find some $A \in GL(n, \mathbb{Z})$ such that $Y[A] \in \mathscr{M}_n$. To find A we locate the columns $a^{(j)}$ of $A = (A^{(1)} \cdots A^{(n)})$ as follows. First choose $a^{(1)}$ so that

$$Y[a^{(1)}] = \min\{Y[a] | a \in \mathbb{Z}^n - 0\}.$$

Such a vector $a^{(1)} \in \mathbb{Z}^n - 0$ must exist because if c is the smallest eigenvalue of Y, we have the inequality:

$$cI[x] \leq Y[x], \quad \text{for all } x \in \mathbb{R}^n.$$

This implies the finiteness of the set of $a \in \mathbb{Z}^n$ such that $Y[a]$ is less than any given bound.

Now if $a^{(1)}$ minimizes $Y[a^{(1)}]$, it follows that the greatest common divisor of the entries of $a^{(1)}$ must be one. By Lemma 2, then $a^{(1)}$ is the first column of a matrix in $GL(n, \mathbb{Z})$. So there exists a vector $b \in \mathbb{Z}^n$ such that the $n \times 2$ matrix $(a^{(1)}b)$ is the first two columns of a matrix in $GL(n, \mathbb{Z})$. This means that there exists some $a^{(2)}$ in $\mathbb{Z}^n$ such that $(a^{(1)}a^{(2)})$ can be completed to a matrix in $GL(n, \mathbb{Z})$ and $Y[a^{(2)}]$ is minimal.

Continue inductively to obtain the matrix $A = (a^{(1)}a^{(2)} \cdots a^{(n)})$ in $GL(n, \mathbb{Z})$. The column $a^{(k)}$ in $\mathbb{Z}^n$ is defined by requiring that the $n \times k$ matrix $(a^{(1)} \cdots a^{(k)})$ can be completed to a matrix in $GL(n, \mathbb{Z})$ and that

$$Y[a^{(k)}] = \min\{Y[b] | (a^{(1)} \cdots a^{(k-1)}b * \cdots *) \in GL(n, \mathbb{Z})\}.$$

And lastly we require that

$${}^t a^{(k-1)} Y a^{(k)} \geq 0, \quad \text{for all } k = 2, \ldots, n.$$

This last requirement is possible, since we can always multiply $a^{(k)}$ by -1, if necessary.

Now we must show that if $A = (a^{(1)} \cdots a^{(n)})$ is constructed as above, then $W = Y[A]$ must lie in the Minkowski domain $\mathscr{M}_n$. To see this, note that (4.13) is an equally good definition of the Minkowski domain. That is, W is Minkowski reduced (assuming $w_{k,k+1} \geq 0$) if

$$W[c] \geq w_{kk} \quad \text{when} \quad (e_1 \cdots e_{k-1} c * \cdots *) \in GL(n, \mathbb{Z}).$$

Here, as usual e_j denotes the standard basis vector in $\mathbb{R}^n$, with all entries 0 but the jth which is 1. To complete the proof that $W \in \mathscr{M}_n$, observe that

$$A(e_1 \cdots e_{k-1} c * \cdots *) = (a^{(1)} \cdots a^{(k-1)} d * \cdots *),$$

if $Ac = d$. And
$$W[c] = Y[d] \geq Y[a^{(k)}] = w_{kk}, \qquad (*)$$
by the construction of A. This concludes the proof that $W \in \mathcal{M}_n$. Conversely, note that if W is in $\mathcal{M}_n$, then we also have the inequality (*).

(4) The case $n = 2$ was Exercise 3. Suppose now that $n = 2, 3,$ or 4 and set
$$\mathcal{M}_n^* = \{Y \in \mathcal{P}_n | y_{k-1,k} \geq 0, Y[a] \geq y_{kk}, \text{if } a_k = 1, a_j = 0 \text{ or } \pm 1, \text{for } j \neq k, 1 \leq k \leq n\}.$$
Clearly $\mathcal{M}_n \subset \mathcal{M}_n^*$. To show the reverse inclusion, we need to prove that $Y \in \mathcal{M}_n^*$ and $m \in \mathbb{Z}^n$ with g.c.d. $(m_k, \ldots, m_n) = 1$ imply that $Y[m] \geq y_{kk}$. We can assume by induction and changing signs that all the m_j are positive.

Among the numbers $m_1, \ldots, m_n$, let m_t be the last occurrence of the minimum. Define vectors u and w in $\mathbb{R}^n$ by:
$$u_i = \begin{cases} m_t, & \text{when } i \neq t, \\ 0, & i = t; \end{cases}$$
and $w_i = 1$, for all i. Then we have the following equality:
$$Y[m] - Y[m - u] = m_t^2(Y[w] - y_{tt}) + 2 \sum_{i \neq t} (m_i - m_t) m_t \sum_{j \neq t} y_{ij}. \quad (4.15)$$
Now $Y \in \mathcal{M}_n^*$ implies that $Y[w] \geq y_{tt}$, and if $n \leq 4$, $Y \in \mathcal{M}_n^*$ implies that
$$\sum_{j \neq t} y_{ij} \geq 0, \quad \text{if } i \neq t. \qquad (4.16)$$

Exercise 5.

(a) Prove formula (4.15).
(b) Prove formula (4.16) if $n \leq 4$. Show that the inequality fails if $n \geq 5$.

It follows from (4.15) and (4.16) that $Y[m] \geq Y[m - u]$. Since $m - u$ has smaller entries than m, the proof of part (4) of Theorem 1 is completed by induction on the norm of m.

Before proving parts (2) and (3) of Theorem 1, we need a proposition, among other things.

Proposition 1 (Other Properties of Minkowski's Fundamental Domain).

(a) *If $Y \in \mathcal{M}_n$, then the entries of Y satisfy the following inequalities:*
$$y_{11} \leq y_{22} \leq \cdots \leq y_{nn}$$
and
$$|y_{ij}| \leq y_{ii}/2, \quad \text{if } 1 \leq i < j \leq n.$$

(b) *If $Y \in \mathcal{M}_n$, Y satisfies the inequality:*
$$k_n y_{11} \cdots y_{nn} \leq |Y| \leq y_{11} \cdots y_{nn},$$

4.4. Fundamental Domains for $\mathscr{P}_n/GL(n, \mathbb{Z})$

where k_n is a positive constant (depending only on n and not on Y). The right hand inequality actually holds for any matrix Y in $\mathscr{P}_n$.

(c) If $Y \in \mathscr{M}_n$, set

$$Y_0 = \begin{pmatrix} y_{11} & & 0 \\ & \ddots & \\ 0 & & y_{nn} \end{pmatrix}.$$

Then there is a positive constant γ_n such that

$$\gamma_n^{-1} Y_0[x] \le Y[x] \le \gamma_n Y_0[x], \quad \text{for all } x \in \mathbb{R}^n.$$

The constant γ_n depends only on n and not on Y or x.

(d) Suppose $Y \in \mathscr{P}_n$ has Iwasawa decomposition $Y = D[T]$ given by

$$D = \begin{pmatrix} d_1 & & 0 \\ & \ddots & \\ 0 & & d_n \end{pmatrix} \quad \text{and} \quad T = \begin{pmatrix} 1 & & t_{ij} \\ & \ddots & \\ 0 & & 1 \end{pmatrix}, \quad \text{with } d_j > 0,$$

then

$$d_i/d_{i+1} \le \kappa_n \quad \text{and} \quad |t_{ij}| \le \kappa_n,$$

for a positive constant κ_n depending only on n and not on Y.

PROOF OF PROPOSITION 1.

(a) Let e_i denote the standard basis vectors of $\mathbb{R}^n$, $i = 1, \ldots, n$. From formula (4.12) defining $\mathscr{M}_n$, we have

$$y_{ii} = Y[e_i] \le Y[e_{i+1}] = y_{i+1,i+1}$$

and

$$y_{ii} \pm 2y_{ij} + y_{jj} = Y[e_i \pm e_j] \ge Y[e_j] = y_{jj}, \quad \text{if } 1 \le i < j \le n.$$

The inequalities in part (a) are an easy consequence.

(b) The inequality $|Y| \le y_{11} \cdots y_{nn}$, for $Y \in \mathscr{P}_n$, is easily proved by induction on n. To do so, use the partial Iwasawa decomposition:

$$Y = \begin{pmatrix} V & 0 \\ 0 & w \end{pmatrix} \begin{bmatrix} I & q \\ 0 & 1 \end{bmatrix}, \quad \text{for } V \in \mathscr{P}_{n-1}, w > 0, q \in \mathbb{R}^{n-1}.$$

Then $w = y_{nn} - V^{-1}[q]$ which implies that $w \le y_{nn}$. It follows that

$$|Y| = |V|w \le |V|y_{nn}.$$

Note also that V is the upper left hand corner of the matrix Y to complete the proof of the right hand inequality in (b).

We will have to work harder to prove the left hand inequality in (b) (cf. Freitag [1], Siegel [2, pp. 44–46], Maass [2, pp. 124–127], and Minkowski [1, Vol. II, pp. 63–67]). We shall follow an approach of Van der Waerden [2]. Let $N_1, \ldots, N_n$ denote the n successive minima of $Y \in \mathscr{P}_n$; i.e.,

$$N_1 = Y[a_1] = m_Y = \min\{Y[a] | a \in \mathbb{Z}^n - 0\}, \tag{4.17}$$

and given $N_i = Y[a_i]$, for $i = 1, \ldots, k-1$,

$$N_k = \min\{Y[a] | a \in \mathbb{Z}^n; a_1, \ldots, a_{k-1}, a \text{ linearly independent}\}.$$

It is clear that $Y \in \mathcal{M}_n$ implies that $N_k \leq y_{kk}$, for $k = 1, \ldots, n$ and $N_1 = y_{11}$. We need two inequalities for $Y \in \mathcal{M}_n$:

$$N_1 N_2 \cdots N_n \leq c_n^n |Y|, \quad \text{with } c_n = \frac{4}{\pi}\Gamma\left(1 + \frac{n}{2}\right)^{2/n} \text{ as in (4.10),} \tag{4.18}$$

$$y_{kk} \leq \delta_k N_k, \quad \text{where } \delta_k = 1, \text{ for } k \leq 4 \text{ and } \delta_k = \left(\frac{5}{4}\right)^{k-4}, k \geq 4. \tag{4.19}$$

These inequalities will be proved in Lemma 3. They imply that the constant k_n in part (b) is given by the following expression:

$$k_n = \begin{cases} \left(\frac{\pi}{4}\right)^n \Gamma\left(1 + \frac{n}{2}\right)^{-2}, & \text{if } n \leq 4, \\ \left(\frac{\pi}{4}\right)^n \left(\frac{4}{5}\right)^{p_n} \Gamma\left(1 + \frac{n}{2}\right)^{-2}, & \text{if } n \geq 4, \end{cases} \tag{4.20}$$

with

$$p_n = \frac{(n-3)(n-4)}{2}.$$

This completes the proof of (b).

(c) The inequality to be proved says that the eigenvalues of $W = Y[Y_0^{-1/2}]$ are bounded above and below by constants independent of Y. Call these eigenvalues $\rho_1, \ldots, \rho_n$. Since

$$\rho_1 + \cdots + \rho_n = \text{Tr}(W) = n,$$

it is clear that $\rho_j < n$, for all j.

Using part (b), one has

$$\rho_1 \cdots \rho_n = |W| = |Y|/(y_{11} \cdots y_{nn}) \geq k_n > 0.$$

It follows that $\rho_j > k_n n^{1-n}$, for all j. This completes the proof of part (c).

(d) First we need to recall Exercise 13 of Section 4.1, which gave formulas for d_i and t_{ij} in terms of the entries of the matrix Y:

$$d_i = |Y_i|/|Y_{i-1}| \quad \text{and} \quad t_{ij} = Y(1, \ldots, i-1, i | 1, \ldots, i-1, j).$$

Here Y_i denotes the upper left hand $i \times i$ corner of the matrix Y and $Y(*|*)$ stands for the subdeterminant of Y obtained by using the indicated rows and columns of Y.

Now to prove part (d), use part (b) to see that since $Y_i \in \mathcal{M}_i$, as is demonstrated in Exercise 6, we have the following inequalities:

$$\frac{d_i}{d_{i+1}} = \frac{|Y_i||Y_i|}{|Y_{i-1}||Y_{i+1}|} \leq \frac{(y_{11} \cdots y_{ii})^2}{k_{i-1}k_{i+1}(y_{11} \cdots y_{i-1,i-1})^2 y_{ii}y_{i+1,i+1}} \leq \frac{1}{k_{i-1}k_{i+1}}$$

4.4. Fundamental Domains for $\mathscr{P}_n/GL(n, \mathbb{Z})$

and

$$t_{ij} = \frac{Y(1,\ldots,i-1,i|1,\ldots,i-1,j)}{|Y_i|} \leq \frac{i! \, y_{11} \cdots y_{ii}}{2k_i y_{11} \cdots y_{ii}} = \frac{i!}{2k_i}.$$

This completes the proof of Proposition 1. □

Exercise 6. Prove that if Y lies in the Minkowski domain $\mathscr{M}_n$ and Y_i denotes the upper left hand $i \times i$ corner of Y, then Y_i lies in the Minkowski domain $\mathscr{M}_i$.

Before returning to the proof of Theorem 1, we need to prove the lemma which was used in the proof of part (b) of Proposition 1.

Lemma 3 (More Inequalities for Reduced Matrices).

(i) Minkowski's Inequality for the Product of the Successive Minima.
   Let the successive minima N_k of $Y \in \mathscr{M}_n$ be defined by (4.17). Then

$$N_1 N_2 \cdots N_n \leq c_n^n |Y|,$$

where

$$c_n = \frac{4}{\pi} \Gamma\left(1 + \frac{n}{2}\right)^{2/n},$$

as defined in (4.10).

(ii) Mahler's Inequality Relating the Diagonal Entries of a Minkowski-Reduced Matrix and the Successive Minima.
   Suppose that $Y \in \mathscr{M}_n$. Then

$$y_{kk} \leq \delta_k N_k, \quad k = 1, 2, \ldots, n,$$

where $\delta_k = 1$ for $k \leq 4$ and $\delta_k = (5/4)^{k-4}$, for $k \geq 4$.

PROOF OF LEMMA 3.

(i) (Cf. Minkowski [1, Section 51] and Van der Waerden [1].) Suppose that

$$N_k = Y[a_k] \quad \text{with} \quad A = (a_1 \cdots a_n) \in \mathbb{Z}^{n \times n} \text{ having rank } n.$$

And let $Y[A] = D[T]$ be the Iwasawa decomposition with:

$$D = \begin{pmatrix} d_1 & & 0 \\ & \ddots & \\ 0 & & d_n \end{pmatrix}, \quad d_i > 0,$$

and (4.21)

$$T = \begin{pmatrix} 1 & & t_{ij} \\ & \ddots & \\ 0 & & 1 \end{pmatrix}.$$

Form the matrix

$$Y^{\#} = \begin{pmatrix} d_1/N_1 & & 0 \\ & \ddots & \\ 0 & & d_n/N_n \end{pmatrix} [T].$$

Part (i) will follow from formula (4.10) if we can show that if $m_{Y^{\#}}$ is the minimum of $Y^{\#}[a]$ over $a \in \mathbb{Z}^n - 0$, then $m_{Y^{\#}}$ is greater than or equal to one.

To see this, let

$$T = {}^t(\xi_1 \cdots \xi_n) \qquad \text{with the } \xi_j \text{ being column vectors in } \mathbb{R}^n. \qquad (4.22)$$

It follows that if x is a column vector in $\mathbb{Z}^n - 0$, then

$$Y^{\#}[x] = \frac{d_1}{N_1}({}^t\xi_1 x)^2 + \cdots + \frac{d_n}{N_n}({}^t\xi_n x)^2.$$

Let ${}^t\xi_k x$ be the last of the ${}^t\xi_j x$'s that is not zero. Then because

$${}^t\xi_k = {}^t(0, \ldots, 1, t_{k,k+1}, \ldots, t_{kn}),$$

we know that x must be linearly independent of $e_1, \ldots, e_{k-1}$. It follows that

$$Y^{\#}[x] \geq \frac{1}{N_k}(d_1({}^t\xi_1 x)^2 + \cdots + d_n({}^t\xi_n x)^2) = \frac{1}{N_k} Y[x] \geq 1,$$

by the definition of N_k.

(ii) (Cf. Mahler [1], Weyl [1, Vol. III, pp. 719–757], Remak [1], and Van der Waerden [1].) Clearly $\delta_1 = 1$. We will show that

$$\delta_k = \max(1, \tfrac{1}{4}\delta_1 + \cdots + \tfrac{1}{4}\delta_{k-1} + \tfrac{1}{4}),$$

which leads quickly to the desired formula for δ_k. This is Remak's formula for the constants involved.

Thus we want to use mathematical induction to show that $y_{kk} \leq \delta_k N_k$, with δ_k given by the preceding formula. Suppose that $Y[a] = N_k$ with

$$a = \begin{pmatrix} a_1 \\ ma_2 \end{pmatrix}, \qquad m \in \mathbb{Z}^+, \qquad a_1 \in \mathbb{Z}^{k-1}, \qquad a_2 \in \mathbb{Z}^{n-k+1}, \qquad \text{g.c.d. } a_2 = 1.$$

Write

$$Y = \begin{pmatrix} Y_1 & 0 \\ 0 & W \end{pmatrix} \begin{bmatrix} I_{k-1} & B \\ 0 & I_{n-k+1} \end{bmatrix}.$$

Since $m = 1$ implies that $N_k \geq y_{kk}$, we may assume that $m \geq 2$. Then

$$N_k = Y[a] = Y_1[a_1 + Bma_2] + W[ma_2] \geq 4W[a_2],$$

which implies

$$W[a_2] \leq \tfrac{1}{4} N_k. \qquad (*)$$

And

$$y_{kk} \leq Y \begin{bmatrix} c \\ a_2 \end{bmatrix} = Y_1[c + Ba_2] + W[a_2].$$

4.4. Fundamental Domains for $\mathscr{P}_n/GL(n,\mathbb{Z})$

Write $Y = D[T]$ with D, T as in (4.21) and (4.22). Then we can choose $c \in \mathbb{Z}^{k-1}$ so that

$$Y_1[c + Ba_2] \le \tfrac{1}{4}(d_1 + \cdots + d_{k-1}). \qquad (**)$$

To see this, let $x = c + Ba_2$ and note that, using (4.22), $Y_1[x]$ equals

$$Y_1[x] = \sum_{j=1}^{k-1} d_j \,{}^t\xi_j x_j, \qquad (***)$$

for

$${}^t\xi_j x_j = x_j + t_{j,j+1} x_{j+1} + \cdots + t_{j,k-1} x_{k-1}.$$

We can choose the entries of $c \in \mathbb{Z}^{k-1}$ so that all the terms ${}^t\xi_j x_j$ in the formula $(***)$ are forced to lie in the interval $[-\tfrac{1}{2}, +\tfrac{1}{2}]$. Thus we obtain $(**)$.

It follows from the inequalities $(*)$ and $(**)$ that:

$$y_{kk} \le Y\begin{bmatrix} c \\ a_2 \end{bmatrix} \le \tfrac{1}{4}(d_1 + \cdots + d_{k-1}) + \tfrac{1}{4} N_k$$

$$\le \tfrac{1}{4}(y_{11} + \cdots + y_{k-1,k-1}) + \tfrac{1}{4} N_k$$

$$\le \tfrac{1}{4}(\delta_1 N_1 + \cdots + \delta_{k-1} N_{k-1} + N_k),$$

by induction, since $d_i \le y_{ii}$ (using the analogous formula to $(***)$ for $Y[x]$ with $x = e_n$). This completes the proof of (ii). $\square$

We continue the intermission in the proof of Theorem 1 with a discussion of the Iwasawa coordinates (4.21) and certain domains related directly to these coordinates. In the next subsection we will make even greater use of these coordinates in obtaining a fundamental domain.

We define a *Siegel set* of matrices to be given by the following expression, assuming that u and v are positive numbers and $Y = D[T]$ as in (4.21):

$$\mathscr{S}_{u,v} = \left\{ Y \in \mathscr{P}_n \,\middle|\, \frac{d_i}{d_{i+1}} \le v, |t_{ij}| \le u, \text{ all } i,j \right\}. \qquad (4.23)$$

Siegel [2, p. 49] makes great use of these sets. The name "Siegel set" appears for example in Borel's article in Borel and Mostow [1, pp. 20–25] and Borel [1]. And Borel defines a *fundamental set* to be a subset $\mathscr{S}$ of $\mathscr{P}_n$ possessing the following two properties:

(a) $$\bigcup_{A \in GL(n,\mathbb{Z})} \mathscr{S}[A] = \mathscr{P}_n$$

and

(b) $\quad \mathscr{S} \cap \mathscr{S}[A] \ne \varnothing, \qquad$ for only finitely many $A \in GL(n,\mathbb{Z})$.

One can show that the Siegel set $\mathscr{S}_{v,u}$ is a fundamental set when $v = \frac{4}{3}$ and $u = \frac{1}{2}$ (see Exercise 7 below and Borel [1, p. 34] who uses the Bruhat decomposition to prove property (b)). But note that, for example, when $n = 2$, if one restricts to the determinant one surface, a Siegel set corresponds to a rectangle in the upper half plane and is thus *not a fundamental domain*.

Exercise 7.

(a) Show that $\mathscr{S}_{v,u}$ satisfies property (a) of a fundamental set if $v \geq \frac{4}{3}$ and $u \geq \frac{1}{2}$.

(b) Use part (a) to show that the Euclidean volume of
$$\{Y \in \mathscr{S}_{v,u} | |Y| \leq 1\}$$
is finite.

Hint. See Borel [1, p. 14] or Raghunathan [1, pp. 160–161].

Part (d) of Proposition 1 shows that $\mathscr{M}_n \subset \mathscr{S}_{v,u}$ for some large constants v and u. In order to obtain a reverse inclusion, we need to replace $\mathscr{M}_n$ by a related set whose definition does not involve the inequalities $y_{k,k+1} \geq 0$. To do this, again motivated by Proposition 1, we define a set also considered by Siegel [2, p. 49]:

$$\mathscr{R}_a = \left\{ Y \in \mathscr{P}_n \left| \frac{y_{kk}}{y_{k+1,k+1}} < a, |y_{ij}| < a y_{ii}, y_{11} \cdots y_{nn} < a|Y|, \text{ all } i,j \right. \right\}. \quad (4.24)$$

The proof of part (d) of Proposition 1 shows that given $a > 0$, there are positive numbers v and u such that $\mathscr{R}_a \subset \mathscr{S}_{v,u}$. One can also show that given v and u, there is a positive number a depending on v and u such that $\mathscr{S}_{v,u} \subset \mathscr{R}_a$. To see this, suppose that $Y \in \mathscr{S}_{v,u}$ with Iwasawa coordinates given by (4.21); that is:

$$y_{kj} = d_k t_{kj} + \sum_{h=1}^{k-1} d_h t_{hk} t_{hj}, \quad \text{for } 1 \leq k \leq j \leq n. \quad (4.25)$$

We shall assume that $v > 1$. Clearly:

$$|y_{kj}| \leq \{u + v^{k-1} u^2 (k-1)\} d_k \leq \eta y_{kk}, \quad \text{if } \eta = u + v^{n-1} u^2 (n-1), \quad (4.26)$$

since $d_k \leq y_{kk}$ (as is seen by setting $k = j$ in (4.25)). Also

$$y_{kk} \leq \eta v y_{k+1,k+1} \quad \text{and} \quad y_{11} \cdots y_{nn} \leq \eta^n |Y|.$$

This completes the proof that $\mathscr{S}_{v,u} \subset \mathscr{R}_a$, for some $a = a(v,u)$.

Exercise 8 (Some Finiteness Results).

(a) Show that there are only finitely many integral matrices of determinant d in Minkowski's fundamental domain $\mathscr{M}_n$.

(b) Let $S \in \mathscr{R}_a$ as defined in (4.24) and suppose that S has the block matrix decomposition:

4.4. Fundamental Domains for $\mathscr{P}_n/GL(n, \mathbb{Z})$

$$S = \begin{pmatrix} S_1 & S_{12} \\ {}^t S_{12} & S_2 \end{pmatrix}.$$

Show that the matrix

$$S_1^{-1} S_{12}$$

has all of its elements bounded in absolute value by a constant depending only on n and a.

(c) Show that $\mathscr{M}_n \subset \mathscr{R}_a$ for some value of a. Can one prove that $\mathscr{R}_a$ is contained in a finite union of images of $\mathscr{M}_n$ under $GL(n, \mathbb{Z})$, for sufficiently small a?

Hints.
(a) See Freitag [1, p. 36]. Use the boundedness of the product of the diagonal entries in such an integral matrix.
(b) See Siegel [2, p. 51]. Use the Iwasawa decomposition:

$$S = D[T] = \begin{pmatrix} D_1 & 0 \\ 0 & D_2 \end{pmatrix} \begin{bmatrix} T_1 & T_{12} \\ 0 & T_2 \end{bmatrix},$$

and write the matrix of interest in terms of the D_i, T_i, T_{12}.

Now finally we turn again to the proof of Theorem 1.

PROOF OF PART (2) OF THEOREM 1. (See Minkowski [1, Vol. II, pp. 67–68] or Hancock [1, Vol. II, pp. 787–788].) We know from (4.25) and (4.26) that if $Y \in \mathscr{M}_n$ and Y has the Iwasawa decomposition $Y = D[T]$ given in (4.21), then

$$y_{ii} \geq d_i \geq \lambda_i y_{ii}, \qquad (4.27)$$

with λ_i positive and independent of Y. Moreover the t_{ij} are bounded. As in (4.22) we set $T = {}^t(\xi_1, \ldots, \xi_n)$. Then, if $a \in \mathbb{Z}^n$, define

$$\zeta_j(a) = {}^t\xi_j a = a_j + t_{j,j+1} a_{j+1} + \cdots + t_{jn} a_n, \qquad (4.28)$$

and note that

$$Y[a] = \sum_{j=1}^n d_j \zeta_j^2(a).$$

It is easily seen that a *finite* set of $a \in \mathbb{Z}^n$ (independent of $Y \in \mathscr{M}_n$) satisfy the following inequalities (see Exercise 9) with $\zeta_j = \zeta_j(a)$ in (4.28):

$$\lambda_1 \zeta_1^2 < \frac{1}{4}, \qquad \lambda_2 \zeta_2^2 < \frac{2}{4}, \qquad \ldots, \qquad \lambda_n \zeta_n^2 < \frac{n}{4}. \qquad (4.29)$$

Exercise 9.

(a) Prove that the $a \in \mathbb{Z}^n$ satisfying (4.29) form a finite set independent of the particular $Y \in \mathscr{M}_n$.

(b) Compute the λ_i in formula (4.27) as a function of v and u such that $\mathcal{M}_n \subset \mathcal{S}_{v,u}$, for the Siegel set defined in (4.23). Show that we can take λ_i to be greater than or equal to λ_{i+1} for all $i = 1, 2, \ldots, n-1$.

Hint. (a) The triangular nature of the transformation $\zeta_j(a)$ in (4.28) allows one to bound the entries of a recursively.

Now we prove that to put Y in $\mathcal{M}_n$ we need only the inequalities $Y[a] \geq y_{kk}$ with $a \in \mathbb{Z}^n$ determined by the inequalities (4.29) and assuming that a has the property that g.c.d. $(a_k, \ldots, a_n) = 1$. This list of vectors a includes $a = e_j$ and $a = e_i \pm e_j$. We know that these inequalities are all that are needed to obtain formula (4.27) which is:

$$y_{hh} \geq d_h \geq \lambda_n y_{hh}.$$

Now suppose that $b \in \mathbb{Z}^n$ with g.c.d. $(b_j, \ldots, b_n) = 1$. In order that b not satisfy (4.29), there must exist an index $h \leq n$ such that

$$\lambda_h \zeta_h^2(b) \geq h/4.$$

Let $h \leq n$ denote the largest such index. So b satisfies the inequalities in (4.29) except for those corresponding to the indices $j = 1, \ldots, h$. Then, since $d_h \geq \lambda_h y_{hh}$, we have

$$\begin{aligned} Y[b] &\geq d_h \zeta_h^2(b) + \cdots + d_n \zeta_n^2(b) \\ &\geq y_{hh}\frac{h}{4} + d_{h+1}\zeta_{h+1}^2(b) + \cdots + d_n \zeta_n^2(b). \end{aligned} \quad (4.30)$$

Here if $h = n$, the last $n - h$ terms will be nonexistent, of course.

Now form

$$b_1^*, \ldots, b_h^* \text{ such that } |\zeta_j(b^*)| \leq 1/2,$$

by subtracting integers from $b_1, \ldots, b_h$. And set

$$b_{h+1}^* = b_{h+1}, \ldots, b_n^* = b_n.$$

Now

$$\text{g.c.d. } (b_j^*, \ldots, b_n^*) = 1.$$

Therefore, making use of (4.30), we have:

$$\begin{aligned} Y[b^*] &\leq \tfrac{1}{4}(d_1 + \cdots + d_h) + d_{h+1}\zeta_{h+1}^2(b) + \cdots + d_n \zeta_n^2(b) \\ &\leq \tfrac{1}{4}(y_{11} + \cdots + y_{hh}) + d_{h+1}\zeta_{h+1}^2(b) + \cdots + d_n \zeta_n^2(b) \\ &\leq \tfrac{1}{4}h y_{hh} + d_{h+1}\zeta_{h+1}^2(b) + \cdots + d_n \zeta_n^2(b) \leq Y[b]. \end{aligned}$$

And

$$Y[b] \geq Y[b^*] \geq y_{jj}$$

4.4. Fundamental Domains for $\mathscr{P}_n/GL(n, \mathbb{Z})$

since b^* satisfies all the inequalities in (4.29). This completes the proof of part (2) of Theorem 1, as it is now clear that $\mathscr{M}_n$ is a convex cone through the origin which is bounded by a finite number of hyperplanes through the origin.

Exercise 10. In the definition (4.12) of $\mathscr{M}_n$, call those inequalities *boundary inequalities* which occur with equality. These boundary inequalities imply all the other inequalities. Prove this.

Hint. Suppose that $Y_0 \in \mathscr{M}_n$ and Y_0 satisfies only strict inequalities. Let $Y_1 \notin \mathscr{M}_n$ satisfy all the boundary inequalities. Consider the line segment joining Y_0 and Y_1; i.e., the points $Y_t = (1 - t)Y_0 + tY_1$, for $0 \leq t \leq 1$. Let

$$u = \text{l.u.b.} \{t \in [0, 1] | Y_t \in \mathscr{M}_n\}.$$

Show that Y_u must satisfy a boundary inequality and then deduce a contradiction, using the linearity of the inequalities in the Y's. Note that we use the finiteness of the number of inequalities needed to define $\mathscr{M}_n$ here, but see Weyl [1, Vol. III, p. 743] for a modification and a reformulation of Minkowski's proof of part (2) of Theorem 1.

PROOF OF PART (3) OF THEOREM 1. It is easy to see that $Y, W = Y[A] \in \mathscr{M}_n$, for $A \in GL(n, \mathbb{Z})$, $A \neq \pm I$, implies that Y lies in the boundary of $\mathscr{M}_n$. For suppose that A is diagonal, with diagonal entries $a_1, \ldots, a_n$. Then there must be a first sign change, say between a_k and a_{k+1}. It follows that $w_{k,k+1} = -y_{k,k+1}$. Then $w_{k,k+1} \geq 0$ and $y_{k,k+1} \geq 0$ imply that $y_{k,k+1} = 0$ and thus Y must lie on the boundary of $\mathscr{M}_n$.

If A is not diagonal, there is a first column of A, say a_k, such that $a_k \neq \pm e_k$, for e_k the standard kth basis vector for $\mathbb{R}^n$. Then the kth column of A^{-1} has the same property. Call it b_k. Suppose that

$$a_k = {}^t(\alpha_1, \ldots, \alpha_n) \quad \text{and} \quad b_k = {}^t(\beta_1, \ldots, \beta_n).$$

Then

$$\text{g.c.d.}(\alpha_k, \ldots, \alpha_n) = 1 = \text{g.c.d.}(\beta_k, \ldots, \beta_n).$$

So it follows that

$$y_{kk} \leq Y[a_k] = w_{kk} \quad \text{and} \quad w_{kk} \leq W[b_k] = y_{kk}.$$

Thus we must have $y_{kk} = Y[a_k]$ and Y lies on the boundary of $\mathscr{M}_n$.

To complete the proof of part (3) of Theorem 1, we need to show that Y and $Y[A]$ both lie in $\mathscr{M}_n$ for at most a finite number of $A \in GL(n, \mathbb{Z})$. Again we shall follow Minkowski [1, Vol. II, p. 70] or Hancock [1, Vol. II, pp. 790–794].

Suppose that Y and $W = Y[A]$ both lie in $\mathscr{M}_n$ for $A \in GL(n, \mathbb{Z})$. Write $A = (a_1 \cdots a_n)$ with $a_j \in \mathbb{Z}^n$. Then $Y[a_j] = w_{jj}$. Suppose that a_{hk} is the last non-zero entry of a_k. Then using (4.27) and Exercise 9(b), we find that:

$$w_{kk} \geq y_{hh} \geq d_h \geq \lambda_h y_{hh} \geq \lambda_n y_{hh}.$$

Claim. For each j, we have $w_{jj} \geq \lambda_n y_{jj}$ and $y_{jj} \geq \lambda_n w_{jj}$.

PROOF OF CLAIM. Suppose that $w_{jj} < \lambda_n y_{jj}$ for some j. Then
$$w_{11} \leq w_{22} \leq \cdots \leq w_{jj} < \lambda_n y_{jj} \leq \lambda_n y_{j+1,j+1} \leq \cdots \leq \lambda_n y_{nn}.$$
So no quantity a_{hk} ($h = j, j+1, \ldots, n; k = 1, \ldots, j$) can be the last not equal to zero. This means that the numbers a_{hk}, $h = j, \ldots, n; k = 1, \ldots, j$, must all be zero which implies that the determinant of A is zero, a contradiction, proving the claim.

If $w_{kk} \geq \lambda_n y_{k+1,k+1}$, for $k = 1, \ldots, n-1$, then we can easily bound the elements of a_k. For then we have:
$$y_{11} \geq \lambda_n w_{11} \geq \lambda_n^2 y_{22} \geq \lambda_n^3 w_{22} \geq \cdots$$
$$\geq \lambda_n^{2k-1} w_{kk} \geq \lambda_n^{2k} y_{k+1,k+1} \geq \cdots.$$
It follows that $w_{kk}/y_{11} \leq \lambda_n^{1-2k}$. We know that there is a positive constant c such that
$$cy_{11} {}^t a_k a_k \leq c Y_0[a_k] \leq Y[a_k] = w_{kk},$$
from part (c) of Proposition 1. Thus a_k has bounded norm.

Next suppose that j is the largest index such that
$$w_{j-1,j-1} < \lambda_n y_{jj}.$$
The considerations in the proof of the claim show that
$$A = \begin{pmatrix} A_1 & A_{12} \\ 0 & A_2 \end{pmatrix} \quad \text{with } A_1 \in GL(j-1, \mathbb{Z}).$$
Induction says that A_1 has bounded entries, since
$$Y[A] = \begin{pmatrix} Y_1 & * \\ * & * \end{pmatrix} \begin{bmatrix} A_1 & A_{12} \\ 0 & A_2 \end{bmatrix} = \begin{pmatrix} Y_1[A_1] & * \\ * & * \end{pmatrix}.$$
The hypothesis on j implies that if $j < n$, then
$$w_{kk} \geq \lambda_n y_{k+1,k+1}, \qquad \text{for } k = j, \ldots, n-1.$$
Thus we have for $k = j, \ldots, n$, the inequality:
$$y_{jj} \geq \lambda_n^{2k-2j} y_{kk} \geq \lambda_n^{2k-2j+1} w_{kk}.$$
It follows that if $k \geq j$ and $Y = D[T]$ with D and T as usual in the Iwasawa decomposition (4.21), then if $\zeta_i(x)$ is defined by (4.28), for $x \in \mathbb{R}^n$, we have the following inequality (making use of (4.27)):
$$w_{kk} = Y[a_k] = \sum_{i=1}^{n} d_i \zeta_i^2(a) \geq \lambda_n y_{jj}(\zeta_j^2(a_k) + \cdots + \zeta_n^2(a_k)).$$
Therefore
$$\lambda_n^{-(2k-2j+2)} \geq \zeta_j^2(a_k) + \cdots + \zeta_n^2(a_k), \qquad \text{for } k = j, \ldots, n,$$

4.4. Fundamental Domains for $\mathscr{P}_n/GL(n,\mathbb{Z})$

which implies that the matrix A_2 has bounded entries. To see this, argue as in part (a) of Exercise 9.

It only remains to show that the matrix A_{12} has bounded entries. Since $W \in \mathscr{M}_n$, we know that for $k = j, \ldots, n$, $x_i \in \mathbb{Z}$, we have

$$W[x] \geq w_{kk}, \quad \text{if } x = {}^t(x_1, \ldots, x_{j-1}, \delta_{kj}, \ldots, \delta_{kn}).$$

Here δ_{ki} denotes the Kronecker delta; i.e., it is 1 if $k = i$ and 0 otherwise. Our inequality means that

$$Y[Ax] \geq Y[a_k] \quad \text{with} \quad Ax = {}^t(x_1^*, \ldots, x_{j-1}^*, a_{jk}, \ldots, a_{nk}) = a^*,$$

for an arbitrary vector $(x_1^*, \ldots, x_{j-1}^*) \in \mathbb{Z}^{j-1}$. We can choose this arbitrary vector to insure that $|\zeta_i(a^*)| \leq \frac{1}{2}$, for $i = 1, \ldots, j-1$. This implies that

$$d_1 \zeta_1^2(a^*) + \cdots + d_n \zeta_n^2(a^*) \geq d_1 \zeta_1^2(a_k) + \cdots + d_n \zeta_n^2(a_k)$$

and

$$\tfrac{1}{4}(d_1 + \cdots + d_{j-1}) \geq d_1 \zeta_1^2(a_k) + \cdots + d_{j-1} \zeta_{j-1}^2(a_k).$$

Thus $\zeta_i^2(a_k)$ is bounded for $i = 1, \ldots, j-1$, which implies that the entries of A_{12} are bounded, completing the proof of part (3) of Theorem 1, and thus the entire proof of the theorem at last. □

Thus we have finished the discussion of the basic properties of the Minkowski fundamental domain $\mathscr{M}_n$ for $GL(n, \mathbb{Z})$. We tried to follow Minkowski's own reasoning in many places, despite the fact that we found it rather tortuous. In the next section we consider another sort of fundamental domain for the general linear group. The discussion will be much simpler. And in the last section we will look at the formula for the Euclidean volume of the subset of $\mathscr{M}_n$ consisting of all matrices of determinant less than or equal to one. Before this, let's consider a few miscellaneous geometric questions related to $\mathscr{M}_n$.

Exercise 11 (Edge Forms). Minkowski defined an *edge form* Q in $\mathscr{M}_n$ to be a reduced form such that $Q = Y + W$ with $Y, W \in \mathscr{M}_n$ implies that Y is a positive scalar multiple of W. Show that an edge form must be on an edge of $\mathscr{M}_n$. Define an equivalence relation between edge forms Y, W by saying that Y is equivalent to W if there is a positive real number c such that $Y = cW$. Show that there are only finitely many equivalence classes of edge forms. Show finally that every Y in $\mathscr{M}_n$ has an expression as a linear combination of edge forms with non-negative scalars.

Hint. See Minkowski [1, Vol. II, p. 69] or Hancock [1, Vol. II, pp. 790–791].

Exercise 12 (The Determinant One Surface).

(i) Suppose that Y and W lie in the determinant one surface $\mathscr{SP}_n$ and $Y \neq cW$ for $c = \pm 1$. If $t \in (0, 1)$, then $tY + (1 - t)W$ is in $\mathscr{P}_n$ with determinant greater than one.

(ii) Show that the determinant one surface in $\mathcal{M}_n$ is everywhere convex as seen from the origin; i.e., the surface lies on the side of the tangent plane away from the origin.

Hint. See Minkowski [1, Vol. II, p. 73] or Hancock [1, Vol. II, pp. 794–797]. Use the fact that there is a $k \in O(n)$ such that both $Y[k]$ and $W[k]$ are diagonal. You will also need the inequality below which holds for positive real numbers $a_1, \ldots, a_n, b_1, \ldots, b_n$:

$$\sqrt{\prod_{i=1}^{n}(a_i + b_i)} \geq \sqrt{\prod_{i=1}^{n} a_i} + \sqrt{\prod_{i=1}^{n} b_i},$$

with equality only if $a_i/b_i = a_{i+1}/b_{i+1}$, for all $i = 1, \ldots, n-1$.

Extreme forms $Y \in \mathcal{P}_n$ are defined to be forms for which the first minimum $m_Y|Y|^{-1/n}$ takes on a local maximum value. Recall that the problem of finding the densest lattice packings of spheres in Euclidean space seeks to find the Y with global maximum value of $m_Y|Y|^{-1/n}$. Korkine and Zolotareff [1] found these forms in low dimensions ($n = 4, 5$). They used a reduction theory which makes use of the Iwasawa decomposition (see Ryskov and Baranovskii [1]). Minkowski [1, Vol. II, p. 76] shows that an extreme form of $\mathcal{M}_n$ must be an edge form. The proof uses Exercise 12 on the convexity of the determinant one surface in the fundamental domain. Minkowski finds representatives of the classes of positive edge forms with $m_Y = 2$ for $n = 2, 3, 4$ [loc. cit., p. 79] and claims that they are all extreme. Elsewhere [loc. cit., p. 218] he considers the cases $n = 5$ and 6. The inequivalent positive edge forms with $m_Y = 2$, for $n = 2, 3, 4$, are:

$$\begin{pmatrix} 2 & 1 \\ 1 & 2 \end{pmatrix} \quad \begin{pmatrix} 2 & 1 & 1 \\ 1 & 2 & 1 \\ 1 & 1 & 2 \end{pmatrix} \quad \begin{pmatrix} 2 & 1 & 1 & 1 \\ 1 & 2 & 1 & 1 \\ 1 & 1 & 2 & 1 \\ 1 & 1 & 1 & 2 \end{pmatrix} \quad \begin{pmatrix} 2 & 0 & 0 & 1 \\ 0 & 2 & 0 & 1 \\ 0 & 0 & 2 & 1 \\ 1 & 1 & 1 & 2 \end{pmatrix}.$$

Korkine and Zolotareff [1] also proved the following results.

(i) If f_n denotes the number of distinct vectors $a \in \mathbb{Z}^n - 0$ such that $Y[a] = m_Y$ for an extreme form Y, then

$$n(n+1)/2 \leq f_n \leq (3^n - 1)/2.$$

(ii) The representations of the minimum m_Y of an extreme form Y determine Y completely, up to multiplication by a positive scalar.

Voronoi [1] gives an algorithm for finding all extreme forms and worked out the cases $n = 3, 4, 5$. Barnes [1] carried out the computation in the case $n = 6$. Voronoi's methods have also turned out to be useful in resolving the singularities of compactifications of the fundamental domain (see Ash, Mum-

4.4. Fundamental Domains for $\mathscr{P}_n/GL(n,\mathbb{Z})$

ford, Rapoport, and Tai [1, pp. 145–150] and Namikawa [1, pp. 85–112]). Important in this work is the *Voronoi map* which is defined to be

$$V\colon \mathbb{R}^n \to \text{Closure}\,(\mathscr{P}_n)$$
$$x \mapsto V(x) = x\,{}^t x.$$

Note that $V(x)$ is positive semi-definite, since $V(x)[a] = ({}^t va)^2 \geq 0$ for any $a \in \mathbb{R}^n$. The *Voronoi points* are the points x in $\mathbb{Z}^n$ with relatively prime coordinates. The *Voronoi cell* or *polyhedron* $\Pi(n)$ is the closure of the convex hull of the set of $V(x)$ such that x is a Voronoi point. See Ryskov and Baranovskii [1, pp. 40ff.] for more details on Voronoi's theory.

Exercise 13 (Hermite-Mahler). A subset of Minkowski's fundamental domain has compact closure if m_Y (the first minimum defined by (4.17)) is bounded from below and the determinant $|Y|$ is bounded from above. In the language of lattices, this says that the minimum separation of the lattice points is bounded from below and the volume of the fundamental parallelepiped is bounded from above.

Hint. Use parts (a) and (b) of Proposition 1.

The theory of smooth compactifications involves finding $GL(n,\mathbb{Z})$-invariant cone decompositions of the closure of $\mathscr{P}_n$. For $n \leq 3$, this comes from the following exercise.

Exercise 14. Define the *fundamental cone* by

$$C_0 = \left\{ X \in \mathbb{R}^{n \times n} \,\middle|\, {}^t X = X,\ x_{ij} \leq 0,\ \text{all } i,j,\ \sum_{i,j=1}^{n} x_{ij} \geq 0 \right\}.$$

Show that the closure of $\mathscr{P}_n$ is a union of images $C_0[A]$ over A in $GL(n,\mathbb{Z})$, when $n \leq 3$. According to Mumford: "This illustrates the interesting fact that only in 4 space or higher do lattice packing problems and related geometry of numbers problems get interesting" (see Ash, Mumford, Rapoport and Tai [1, p. 146]).

Exercise 15.

(a) Use Theorem 1 to show that $GL(n,\mathbb{Z})$ is finitely generated. In fact, show that the A in $GL(n,\mathbb{Z})$ such that $\mathscr{M}_n \cap (\mathscr{M}_n[A]) \neq \emptyset$ generate $GL(n,\mathbb{Z})$.
(b) Prove that $GL(n,\mathbb{Z})$ is generated by the matrices which give the elementary row and column operations; i.e., by matrices of the following 3 forms:
   (i) diagonal with ± 1 as the entries,
   (ii) permutation matrices,
   (iii) upper triangular matrices with ones on the diagonal and all elements above the diagonal equal to zero except 1 which is equal to 1.

Minkowski [loc. cit., p. 95] showed that if $H(d)$ is the number of integral $Y \in \mathscr{P}_n$ of determinant d and inequivalent modulo $GL(n, \mathbb{Z})$, then

$$\lim_{x \to \infty} \left(x^{-(n+1)/2} \sum_{d=1}^{x} H(d) \right) = \text{Euclidean Volume } \{Y \in \mathscr{M}_n | |Y| \leq 1\}.$$

Note that $H(d)$ is finite by Exercise 8.

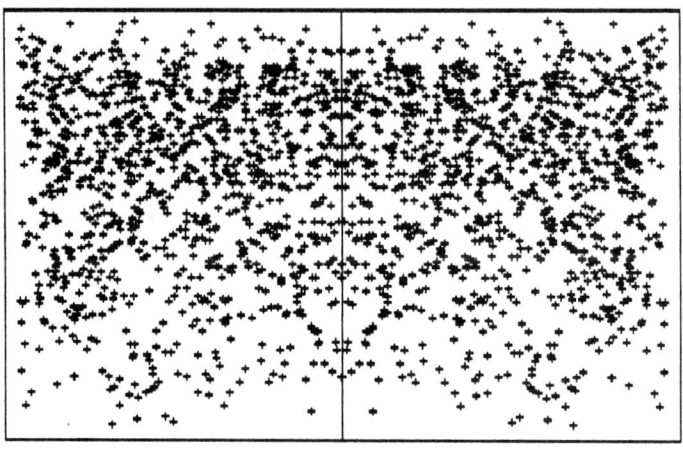

$GL(3)$ ink blot test.

4.4.3. Grenier's Fundamental Domain

Here we consider a fundamental domain and reduction algorithm discussed by Grenier [1], [2] and pictured for $SL(3, \mathbb{Z})$ in Grenier, Gordon, and Terras [1]. Grenier's method is analogous to that used by Siegel for $Sp(n, \mathbb{Z})$ in Siegel [1, Vol. II, pp. 105–113, 298–317] (see also Maass [2] and Gottschling [1]).

We need to consider another fundamental domain than Minkowski's for the reasons that follow. Recall that Minkowski's fundamental domain for $GL(3, \mathbb{Z})$ is:

$$\mathscr{M}_3 = \left\{ Y = (y_{ij}) \in \mathscr{P}_3 \left| \begin{array}{l} y_{11} \leq y_{22} \leq y_{33}, 0 \leq y_{12} \leq \tfrac{1}{2} y_{11} \\ 0 \leq y_{23} \leq \tfrac{1}{2} y_{22}, |y_{13}| \leq \tfrac{1}{2} y_{11} \\ Y[e] \geq y_{33}, e = \pm e_1 \pm e_2 \pm e_3 \end{array} \right. \right\}.$$

Here the e_j are the standard basis vectors for $\mathbb{R}^3$. Suppose Y has the Iwasawa decomposition:

$$Y = \begin{pmatrix} y_1 & 0 & 0 \\ 0 & y_2 & 0 \\ 0 & 0 & y_3 \end{pmatrix} \begin{bmatrix} 1 & x_1 & x_2 \\ 0 & 1 & x_3 \\ 0 & 0 & 1 \end{bmatrix}$$

$$= \begin{pmatrix} y_1 & y_1 x_1 & y_1 x_2 \\ y_1 x_1 & y_1 x_1^2 + y_2 & y_1 x_1 x_2 + y_2 x_3 \\ y_1 x_2 & y_1 x_1 x_2 + y_2 x_3 & y_1 x_2^2 + y_2 x_3^2 + y_3 \end{pmatrix}. \tag{4.31}$$

4.4. Fundamental Domains for $\mathscr{P}_n/GL(n, \mathbb{Z})$

Thus if Y lies in Minkowski's fundamental domain, we find that:

$$0 \leq x_1 \leq \tfrac{1}{2}, \quad |x_2| \leq \tfrac{1}{2}, \quad 0 \leq y_1 x_1 x_2 + y_2 x_3 \leq \tfrac{1}{2}(y_1 x_1^2 + y_2).$$

If follows that:

$$-\tfrac{1}{4} y_1/y_2 \leq x_3 \leq \tfrac{1}{2} - (\tfrac{3}{8}) y_1/y_2. \tag{4.32}$$

This means that Minkowski's fundamental domain has only an *approximate box shape* as Y approaches the boundary; i.e., as $y_1/y_2 \to 0$. We prefer an *exact box shape*; that is, we prefer to see the inequality $0 \leq x_3 \leq \tfrac{1}{2}$, particularly when computing the integrals of Eisenstein series over truncated fundamental domains (a necessary prelude to generalizing the Selberg trace formula to $SL(3, \mathbb{Z})$).

Grenier [1], [2] describes a fundamental domain for $GL(n, \mathbb{Z})$ which makes essential use of the Iwasawa coordinates. Moreover, Grenier's domain has an exact box shape at the boundary. And Grenier gives a reduction algorithm to move Y in $\mathscr{P}_n$ into this fundamental domain via a "highest point method." Other fundamental domains and reduction algorithms are considered by Korkine and Zolotareff [1], Novikova [11], Ryskov [2], Ryskov and Baranovskii [1], Siegel [1, Vol. II, pp. 105–113, 298–317] and A.B. Venkov [1]. We will say a little more about some of these other methods at the end of the section.

Before defining Grenier's fundamental domain, we need to fix a set of partial Iwasawa coordinates for $Y \in \mathscr{P}_n$ given by

$$Y = \begin{pmatrix} v & 0 \\ 0 & W \end{pmatrix} \begin{bmatrix} 1 & {}^t x \\ 0 & I_{n-1} \end{bmatrix}, \quad \text{for } v > 0, \, Y \in \mathscr{P}_{n-1}, \, x \in \mathbb{R}^{n-1}. \tag{4.33}$$

Next let us consider how the action of a matrix in $GL(n, \mathbb{R})$ affects the v-coordinate. Suppose

$$M = \begin{pmatrix} a & {}^t b \\ c & D \end{pmatrix}, \quad \text{for } a \in \mathbb{Z}, \quad b, c \in \mathbb{Z}^{n-1}, \, D \in \mathbb{Z}^{(n-1) \times (n-1)}.$$

If $Y[M] = Y^*$ has partial Iwasawa coordinates v^*, W^*, and x^*, then we find that

$$v^* = v[a + {}^t xc] + W[c].$$

We want to think of the coordinate v as the reciprocal of the height of Y. This agrees with the idea of height in the case $n = 2$ which was used in Section 3.4 of Volume I to put a point into the fundamental domain for $SL(2, \mathbb{Z})$ by the highest point method. Thus we want Y to have coordinate v such that

$$v \leq v[a + {}^t xc] + W[c]$$

for any a, c forming the 1st column of a matrix in $GL(n, \mathbb{Z})$. It is thus natural to make the following definition.

Grenier's fundamental domain for $GL(n, \mathbb{Z})$ is the set $\mathscr{F}_n$ of Y in $\mathscr{P}_n$ satisfying:

(1) $v \leq v[a + {}^t xc] + W[c]$, for $a \in \mathbb{Z}, c \in \mathbb{Z}^{n-1} - 0$, and

$$M = \begin{pmatrix} a & {}^tb \\ c & D \end{pmatrix} \in GL(n, \mathbb{Z});$$

(2) $W \in \mathscr{F}_{n-1}$, the fundamental domain for $GL(n-1, \mathbb{Z})$;
(3) $0 \leq x_1 \leq \frac{1}{2}, |x_i| \leq \frac{1}{2}$, for $i = 2, \ldots, n-1$.

Note that for $n = 2$, this fundamental domain is just the same as Minkowski's. But for $n \geq 3$, it differs. In particular, it puts the x-variables into a "box" shape (see Exercise 17 below). The fundamental domain $\mathscr{F}_n$ was considered by Hermite (see Cassels [2, p. 259]). But Cassels notes that it doesn't appear that anyone (before Grenier) managed to show that the domain is a reasonable one to consider since it is defined by inequalities more similar to those that worked in the case $n = 2$ than those giving Minkowski's domain.

We take the discussion of the following results from Grenier [1], [2].

Theorem 2. $\mathscr{F}_n$ *is a fundamental domain for* $GL(n, \mathbb{Z})$.

Before proving this result we need a lemma.

Lemma 4. *If Y is in Grenier's fundamental domain $\mathscr{F}_n$ and v, W are as in (4.33) with v the upper left hand entry of Y (and the inverse height of Y), then if w_j is the jth diagonal entry in W, we have the inequality:*

$$w_{jj} \geq \frac{3v}{4}.$$

PROOF. If Y is in $\mathscr{F}_n$, then we know that $v \leq v[a + {}^txc] + W[c]$ for any a, c that can be made the first column of a matrix in $GL(n, \mathbb{Z})$. Take $a = 0$ and $c = e_j$, the jth element of the standard basis of $\mathbb{R}^{n-1}$. Then

$$v \leq vx_j^2 + w_{jj} \leq \tfrac{1}{4}v + w_{jj}.$$

The result follows. □

PROOF OF THEOREM 2. We need to show two things:

(1) for any $Y \in \mathscr{P}_n$, there is a matrix $M \in GL(n, \mathbb{Z})$ such that $Y[M] \in \mathscr{F}_n$;
(2) for Y and $Y[M]$ both in $\mathscr{F}_n$ with $M \neq \pm I$, Y must be on the boundary of $\mathscr{F}_n$.

We proceed by induction on n. The case $n = 2$ is already done. So assume that $\mathscr{F}_{n-1}$ is a fundamental domain for $GL(n-1, \mathbb{Z})$.

To prove (1), suppose $Y \in \mathscr{P}_n$ is given with partial Iwasawa decomposition (4.33). If $M \in GL(n, \mathbb{Z})$ has first column given as usual by ${}^t(ac)$, for $a \in \mathbb{Z}$, then $Y[M]$ has its upper left corner $v^* = v[a + {}^txc] + W[c]$. It is easily seen that there are only finitely many a and c forming the first column of a matrix M in $GL(n, \mathbb{Z})$ such that v^* stays less than any given bound. Thus we may choose a and c so that v^* is minimal.

4.4. Fundamental Domains for $\mathscr{P}_n/GL(n, \mathbb{Z})$

Then locate D^* in $GL(n-1, \mathbb{Z})$ so that $W[D^*] \in \mathscr{F}_{n-1}$ by the induction hypothesis. This means that we want to set

$$M_1 = M \begin{pmatrix} 1 & 0 \\ 0 & D^* \end{pmatrix} \in GL(n, \mathbb{Z}),$$

in order to say that $Y[M_1]$ satisfies the first two prerequisites for being a member of $\mathscr{F}_n$. Next we can put the x-coordinates of $Y[M_1]$ in the desired intervals by acting on it by:

$$N = \begin{pmatrix} \pm 1 & b^* \\ 0 & I \end{pmatrix}.$$

Thus $Y[M_1 N] \in \mathscr{F}_n$ and we're done.

To prove part (2), we use induction again. The case $n = 2$ is done. Now assume that the case of $\mathscr{F}_{n-1}$ is proved. And suppose that both Y and $Y^* = Y[M]$ lie in $\mathscr{F}_n$ for $M \in GL(n, \mathbb{Z})$. We know that in terms of the inverse heights, since $Y^*[M^{-1}] = Y$:

$$v \leq v[a + {}^t xc] + W[c] = v^* \quad \text{and similarly} \quad v^* \leq v.$$

Thus we see that $v = v^* = v[a + {}^t xc] + W[c]$. If $c \neq 0$, this equality puts Y on the boundary of $\mathscr{F}_n$ because it is close to something outside of $\mathscr{F}_n$. If $c = 0$, the equality means that $v = va^2$ and thus $a = \pm 1$ and

$$M = \begin{pmatrix} \pm 1 & {}^t b \\ 0 & D \end{pmatrix} \quad \text{and} \quad D \in GL(n-1, \mathbb{Z}).$$

Then $W^* = W[D]$ and W both lie in $\mathscr{F}_{n-1}$ and the induction hypothesis implies that W lies on the boundary of $\mathscr{F}_{n-1}$ and thus Y is on the boundary of $\mathscr{F}_n$ unless $D = \pm I$. In the latter case, by looking at the effect of M on the x's, we see that:

$$x \text{ and } \pm x \pm b \in [-\tfrac{1}{2}, \tfrac{1}{2}]^{n-1}, \quad \text{with } x_1, \pm x_1 \pm b_1 \geq 0.$$

Since $b \in \mathbb{Z}^{n-1}$, we see that either:

$$x_i = \pm \tfrac{1}{2} \text{ and } b_i = \pm 1, \quad i = 2, \ldots, n-1; \quad x_1 = \tfrac{1}{2}, \quad b_1 = 1$$

or

$$x_i \neq \pm \tfrac{1}{2} \text{ and } b_i = 0, \quad i = 1, \ldots, n-1.$$

If all the $b_i = 0$ and if $M \neq \pm I$, it follows that $x_1 = 0$ and Y must lie on the boundary of $\mathscr{F}_n$. □

Theorem 3 (Grenier). $\mathscr{F}_n$ *has a finite number of boundary inequalities, which can be explicitly given for small values of n.*

PROOF. Once more we use induction on n. We already know the case $n = 2$ and now we assume the result for $\mathscr{F}_{n-1}$. We need to show that only a finite number of inequalities of the form

$$v \leq v[a + {}^txc] + W[c], \quad \text{for all } a, c \text{ forming the first column of a matrix in } GL(n, \mathbb{Z})$$

are necessary to place Y in $\mathscr{F}_n$. Using the partial Iwasawa coordinate decomposition (4.33), we know that $W \in \mathscr{F}_{n-1}$ and so we can write

$$W = \begin{pmatrix} v' & 0 \\ 0 & W' \end{pmatrix} \begin{bmatrix} 1 & {}^tx' \\ 0 & I_{n-2} \end{bmatrix}, \quad v' > 0, \quad W' \in \mathscr{F}_{n-1}, \quad x' \in [-\tfrac{1}{2}, \tfrac{1}{2}]^{n-1}.$$

By the induction hypothesis there are only finitely many inequalities:

$$v' \leq v'[a' + {}^tx'c'] + W'[c']$$

which must be considered. Why?

And clearly

$$W[c] = v'[c_1 + {}^tx'c'] + W'[c'] \quad \text{with} \quad {}^tc' = (c_2, \ldots, c_n).$$

So we now have a finite number of vectors c. Given the bounds on the x_i, this leads to a finite number of a.

To describe what is happening more explicitly, continue the partial Iwasawa decompositions until you reach the full Iwasawa decomposition and obtain:

$$v[a + {}^txc] + W[c] = v(a + {}^txc)^2 + v'(c_1 + {}^tx'c')^2 + \cdots$$
$$+ v^{(n-2)}(c_{n-2} + x^{(n-2)}c_{n-1})^2 + v^{(n-1)}c_{n-1}^2.$$

By repeated application of Lemma 4, we know that

$$v^{(j)} \geq \kappa^j v, \quad \text{for } \kappa = \tfrac{3}{4}.$$

Thus

$$v[a + {}^txc] + W[c] \geq v\{(a + {}^txc)^2 + \kappa(c_1 + {}^tx'c')^2 + \cdots$$
$$+ \kappa^{n-2}(c_{n-2} + x^{(n-2)}c_{n-1})^2 + \kappa^{n-1}c_{n-1}^2\}.$$

This means that we need only consider the a and c such that

$$(a + {}^txc)^2 + \kappa(c_1 + {}^tx'c')^2 + \cdots + \kappa^{n-1}c_{n-1}^2 \leq 1.$$

Since the x_j's are bounded by $\tfrac{1}{2}$ in absolute value, this bounds the a's and c's. □

One can go on to determine the exact list of inequalities for small values of n. See Grenier [1], [2] for the cases $n \leq 5$.

Exercise 16. Show that on the determinant one surface, Grenier's fundamental domain $\mathscr{S}\mathscr{F}_3$ for $SL(3, \mathbb{Z})$ is the set of $Y \in \mathscr{S}\mathscr{P}_3$ with partial Iwasawa coordinates having

$$W = \begin{pmatrix} w & 0 \\ 0 & 1/w \end{pmatrix} \begin{bmatrix} 1 & x_3 \\ 0 & 1 \end{bmatrix},$$

4.4. Fundamental Domains for $\mathscr{P}_n/GL(n,\mathbb{Z})$

$$Y = Y(v,w,x) = \begin{pmatrix} v & 0 \\ 0 & v^{-1/2}W \end{pmatrix} \begin{bmatrix} 1 & (x_1, x_2) \\ 0 & I_2 \end{bmatrix}, \quad x = (x_1, x_2, x_3),$$

satisfying the following inequalities:

$0 \le x_1 \le \frac{1}{2}, \quad |x_2| \le \frac{1}{2}, \quad 0 \le x_3 \le \frac{1}{2},$

$1 \le w^{-2} + x_3^2,$

$v \le v(a + {}^txc)^2 + v^{-1/2} W[c] \quad \text{for} \quad \begin{cases} a=0, & {}^tc=(1,0),(0,1),(1,-1), \\ a=1, & {}^tc=(-1,1). \end{cases}$

Hint. See Grenier [1], [2] or Gordon, Grenier, and Terras [1]. Note that $GL(3,\mathbb{Z})/SL(3,\mathbb{Z})$ has order 2 and the nontrivial coset is represented by $-I$ which does nothing to an element $Y \in \mathscr{SP}_3$, so the fundamental domain for $SL(3,\mathbb{Z})$ is the same as that for $GL(3,\mathbb{Z})$.

Setting $t^{-2} = v^{-3/2}w$, the explicit inequalities for $\mathscr{SF}_3$ are:

(i) $1 \le (1 - x_1 + x_2)^2 + t^{-2}\{(1 - x_3)^2 + w^{-2}\}$
(ii) $1 \le (x_1 - x_2)^2 + t^{-2}\{(1 - x_3)^2 + w^{-2}\}$
(iii) $1 \le x_1^2 + t^{-2}$
(iv) $1 \le x_2^2 + t^{-2}(x_3^2 + w^{-2})$
(v) $1 \le x_3^2 + w^{-2}$
(vi) $0 \le x_1 \le \frac{1}{2}$
(vii) $0 \le x_3 \le \frac{1}{2}$
(viii) $-\frac{1}{2} \le x_2 \le \frac{1}{2}.$

(4.34)

Note that inequalities (iii) (v), (vi), and (vii) say that $x_3 + iw^{-1}$ and $x_1 + it^{-1}$ lie in the fundamental domain for $GL(2,\mathbb{Z})\backslash H$ which was given in Exercise 1 of Section 3.3 of Volume I.

Exercise 17.

(a) Show that $w \le 1$ and $t \le 1$ for $t^{-2} = v^{3/2}w$, plus inequalities (vi)–(viii) from (4.34) imply that the point $Y = Y(v, w, x)$ with Iwasawa coordinates given in Exercise 16 must lie in the fundamental domain $\mathscr{SF}_3$. That is, in this case, we do not need all of the inequalities (4.34). This shows that Grenier's fundamental domain does have an exact box shape at infinity, unlike Minkowski's fundamental domain (cf. (4.32)). Note that, in general,

$$Y \in \mathscr{SF}_3 \text{ implies that } w \text{ and } t \text{ are both } \le 2/\sqrt{3}.$$

(b) In the case of $GL(3,\mathbb{Z})$, compare the Grenier fundamental domain with the Siegel set $\mathscr{S}_{u,v}$ defined in formula (4.23) of Section 4.4.2. In particular, show that for the 3 × 3 case we have:

$$\mathscr{SF}_3 \subset \mathscr{S}_{1/2, 4/3} \cap \mathscr{SP}_3$$

and

$$\mathscr{S}_{1/2,1} \cap \mathscr{SP}_3 \cap \{Y = Y(v,w,x) | x_1, x_3 \geq 0\} \subset \mathscr{SF}_3.$$

This is a similar situation to that which occurred for $GL(2,\mathbb{Z})$.

Boundary identifications for the fundamental domain $\mathscr{SF}_3$ come from completing the '(ac) in Exercise 16 to matrices in $SL(3,\mathbb{Z})$, a process which can be carried out as follows:

$$T1 = \begin{pmatrix} 1 & 1 & 0 \\ 0 & 1 & 0 \\ 0 & 0 & 1 \end{pmatrix}, \quad T2 = \begin{pmatrix} 1 & 0 & 1 \\ 0 & 1 & 0 \\ 0 & 0 & 1 \end{pmatrix}, \quad T3 = \begin{pmatrix} 1 & 0 & 0 \\ 0 & 1 & 1 \\ 0 & 0 & 1 \end{pmatrix},$$

$$S1 = \begin{pmatrix} 0 & 0 & 1 \\ 1 & 0 & 0 \\ 0 & 1 & 0 \end{pmatrix}, \quad S2 = \begin{pmatrix} 0 & 1 & 0 \\ 0 & 0 & 1 \\ 1 & 0 & 0 \end{pmatrix}, \quad S3 = \begin{pmatrix} 0 & 1 & 0 \\ 1 & 0 & -1 \\ -1 & 0 & 0 \end{pmatrix},$$

$$S4 = \begin{pmatrix} 1 & 0 & 0 \\ -1 & 1 & 0 \\ 1 & 0 & 1 \end{pmatrix}, \quad S5 = \begin{pmatrix} 1 & 0 & 0 \\ 0 & 0 & 1 \\ 0 & -1 & 0 \end{pmatrix},$$

$$U1 = \begin{pmatrix} -1 & 0 & 0 \\ 0 & -1 & 0 \\ 0 & 0 & 1 \end{pmatrix}, \quad U2 = \begin{pmatrix} 1 & 0 & 0 \\ 0 & -1 & 0 \\ 0 & 0 & 1 \end{pmatrix}.$$

Note. This gives more than enough generators for $SL(3,\mathbb{Z})/\pm I$, but we do not appear to be able to get rid of any of the inequalities in (4.34) (cf. Exercise 15).

Grenier's reduction algorithm is a "highest point method" where the height of Y is $1/v$, for $v = $ the entry y_{11}. The algorithm goes as follows:

Step I. Set $S0 = I$ and pick j to minimize the v-coordinate of $Y[Sj]$, for $j = 0, 1, 2, 3, 4$. Then replace Y by $Y[Sj]$.

Step II. Let $W(Y)$ denote the element of $\mathscr{SP}_2$ defined by the equations in Exercise 16. Put $W(Y)[\delta]$ in $\mathscr{SF}_2$ using $\delta \in GL(2,\mathbb{Z})$. That is, we make w and x_3 satisfy inequalities (v) and (vii) in (4.34). Replace Y by $Y[\gamma]$ for

$$\gamma = \begin{pmatrix} \pm 1 & 0 \\ 0 & \delta \end{pmatrix} \in SL(3,\mathbb{Z}).$$

Here $\gamma = S5$, $U1$, or $(T3)^n$ for some $n \in \mathbb{Z}$.

Step III. Translate the x_1, x_2-coordinates of Y in Exercise 16 by applying $\gamma = (Tj)^p$ to Y, for $p = [\frac{1}{2} - x_j]$, $j = 1, 2$. Here $[x]$ denotes the greatest integer $\leq x$.

Step IV. Make $x_1 \geq 0$ by replacing Y by $Y[U2]$, if necessary.

4.4. Fundamental Domains for $\mathcal{P}_n/GL(n, \mathbb{Z})$

Keep performing Steps I–IV until the process stabilizes. Jeff Stopple suggested that we use this last test; i.e., see whether the process has repeated itself to stop the program. This idea is useful since it allows us not to test all the inequalities at each step, as some might be tempted to do. On the other hand one might worry that the program would get into an infinite loop. This does not happen if one is careful in writing the code. However, one must be rather cautious because there can be great loss of precision due to subtraction and division. Thus we found that we had to use double precision when performing the algorithm in BASIC on the UCSD VAX.

Exercise 18. Write a program to carry out Grenier's reduction algorithm.

Note that in obtaining the matrices Sj, $j = 1, 2, 3, 4, 5$, we completed the matrices $^t(ac)$ from Exercise 11 to 3×3 matrices in $SL(3, \mathbb{Z})$. This can be done in a number of ways—each differing by matrices of the form

$$\begin{pmatrix} 1 & {}^t q \\ 0 & R \end{pmatrix}$$

with $q \in \mathbb{Z}^2$ and $R \in \mathbb{Z}^{2 \times 2}$. The choice of Sj will affect the reduction algorithm, but not the final result that the algorithm does send a point into the fundamental domain.

Let us now consider the results of some computer experiments we did with Dan Gordon and Doug Grenier which were published in Gordon, Grenier, and Terras [1]. Our aim was understanding what $\mathcal{SF}_3$ looks like.

First recall that we saw in Figure 3.30 from Section 3.5 and in Exercise 18 in Section 3.6 of Volume I that one can use Hecke operators to help us visualize the fundamental domain for $SL(2, \mathbb{Z})$. Now we would like to do something similar to visualize Grenier's fundamental domain $\mathcal{SF}_3$. Before attempting that, let us look at Figure 4.5 which is a picture of the standard fundamental domain for $SL(2, \mathbb{Z})$ from Figure 3.11 of Volume 1 transformed by sending y to $v = 1/y$. Then look at Figure 4.6 which consists of images of the point $z = 1.4i$ under matrices from the Hecke operator for the prime $p = 997$, again making the transformation $v = 1/y$ and mapping x versus v. Figure 4.6 was constructed exactly as Figure 3.30 in Volume 1 except that we graphed the x, $v = 1/y$ coordinates of the points rather than applying the Cayley transform to ship the points into the unit disk.

Since $\mathcal{SF}_3$ is 5-dimensional, we will take the easy way out and look at graphs of 2 coordinates from $(v, w, x1, x2, x3)$. So there are 10 possible graphs. The most interesting is that of (w, v) showing the shape of the *cuspidal region*, where v or w approaches 0.

We quickly see that from (4.34), as in Exercise 12, that

$$w \leq 2(3^{-1/2}) \cong 1.154701 \text{ and also } t \leq 2(3^{-1/2}). \tag{4.35}$$

Therefore

$$v \leq 4/3 \cong 1.333333.$$

154   IV. The Space $\mathscr{P}_n$ of Positive $n \times n$ Matrices

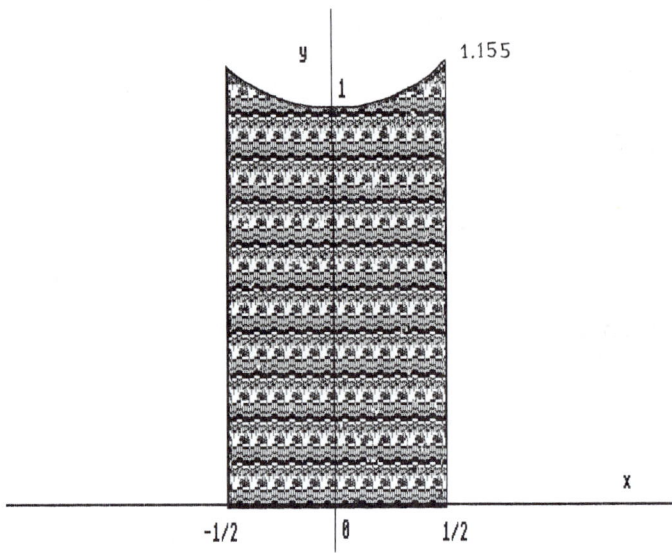

Figure 4.5. The standard fundamental domain $\mathscr{SF}_2$ for $SL(2, \mathbb{Z})$ transformed by $v = 1/y$.

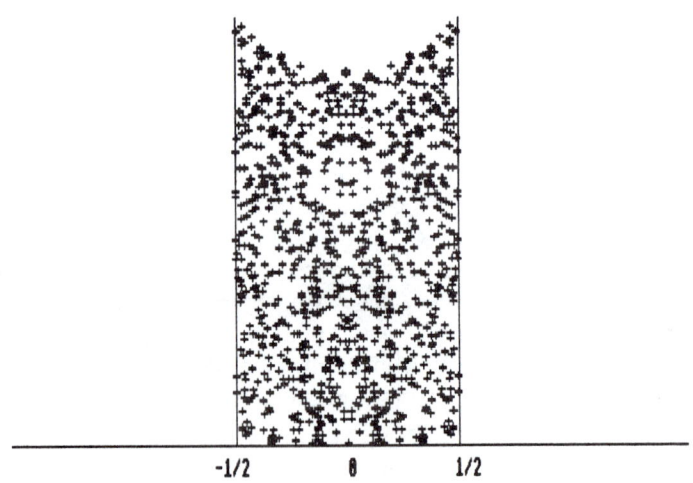

Figure 4.6. Images of Hecke points $(z_0 + j)/p$, $0 \leq j \leq p - 1$, in the fundamental domain for $SL(2, \mathbb{Z})$, with $z_0 = 1.4i$ and $p = 997$.

4.4. Fundamental Domains for $\mathscr{P}_n/GL(n, \mathbb{Z})$

Hecke operators for $\Gamma_3 = SL(3, \mathbb{Z})$ will be discussed in Section 4.5. Here we need only consider the simplest aspects of the theory. For $\Gamma_3 = GL(3, \mathbb{Z})$, $f: \mathscr{SP}_3/\Gamma_3 \to \mathbb{C}$ and $m \in \mathbb{Z}^+$, define *the Hecke operator* T_m by:

$$T_m f(Y) = \sum_{A \in M_m/\Gamma_3} f(Y[A]^0) \tag{4.36}$$

where

$$M_m = \{A \in \mathbb{Z}^{3 \times 3} | \det A = m\}$$

and

$$Y^0 = (\det Y)^{-1/3} Y \in \mathscr{SP}_3.$$

It is easily seen (as in Lemma 6 which follows) that one can take representatives of $M_3(m)/\Gamma_3$ of the form

$$\begin{pmatrix} d_1 & d_{12} & d_{13} \\ 0 & d_2 & d_{23} \\ 0 & 0 & d_3 \end{pmatrix}, \quad d_i > 0, \quad \prod_{i=1}^{3} d_i = m, \quad 0 \le d_{ij} < d_i. \tag{4.37}$$

Maass [5] studied Hecke operators for the Siegel modular group $Sp(n, \mathbb{Z})$ in 1951. We are imitating his version of the theory. It is a theory which is basic to the study of automorphic forms on higher rank symmetric spaces G/K and it connects with many questions in representation theory, p-adic group theory, combinatorics, and number theory. Applications of Hecke operators to numerical integration on spheres are given by Lubotzky, Phillips, and Sarnak [1].

It is not hard to see that the Hecke operators for $SL(3, \mathbb{Z})$ have the following properties:

(i) $T_n T_m = T_{mn}$, if g.c.d. $(m, n) = 1$;

for p = prime

(ii) $\sum_{r \ge 0} T_{p^r} X^r = (I - T_p X + [(T_p)^2 - T_{p^2}] X^2 - p^3 X^3)^{-1}.$

It follows that L-functions associated to eigenforms f of the Hecke operators must have Euler products. We will discuss all these things in Section 4.5.

Here we graph points from the operator T_p, p = prime. We use only the matrices

$$M(p; a, b) = p^{-1/3} \begin{pmatrix} p & a & b \\ 0 & 1 & 0 \\ 0 & 0 & 1 \end{pmatrix}, \quad 0 \le a, b \le p - 1. \tag{4.38}$$

The other matrices in T_p do not appear to be necessary. *Moreover, we will restrict b so that $b \equiv 5a + 163 \pmod{p}$.* This will restrict the number of points to p rather than p^2.

Figures 4.7–4.13 show plots of pairs of Iwasawa-type coordinates of Γ_3-images of Hecke points in the fundamental domain $\mathscr{SF}_3$ or in the union of fundamental domains obtained by letting the x-coordinates run between $-\frac{1}{2}$ and $+\frac{1}{2}$. This allows us to produce more pleasing symmetric pictures than those in Grenier, Gordon, and Terras [1]. By "Hecke points" we mean points of the form

$$Y_0[M(p;a,b)], \quad 0 \le a, b \le p-1, \tag{4.39}$$

for $M(p; a, b)$ as in (4.38), $p = 3001$, and fixed Y_0 with its Iwasawa coordinates, as in Exercise 16, given by:

$$(v, w, x_1, x_2, x_3) = (.7815, .6534, .2123, .0786, .3312).$$

In Figure 4.7, the graph shows t versus w, where $t = v^{3/4}w^{-1/2}$. Note that t is the square root of the quotient of the first two diagonal entries in the Iwasawa decomposition of Y. The coordinate w is the square root of the quotient of the last two diagonal entries in the Iwasawa decomposition of Y. And we have seen in (4.34) that t and w play a more similar role than v and w, the variables we graphed in Gordon, Grenier, and Terras [1].

Figures 4.8 and 4.9 show plots of the coordinates (x_1, t) and (x_3, w) of Hecke points for the prime $p = 3001$. These are the variables in copies of the fundamental domain of $SL(2, \mathbb{Z})$ in the Poincaré upper half plane. The figures *do* give a good approximation to Figure 4.6, as expected.

Figures 4.10 and 4.11 give plots of $(x1, x2)$ and $(x1, x3)$, respectively. The plots look like randomly placed points in $[-\frac{1}{2}, \frac{1}{2}]^2$.

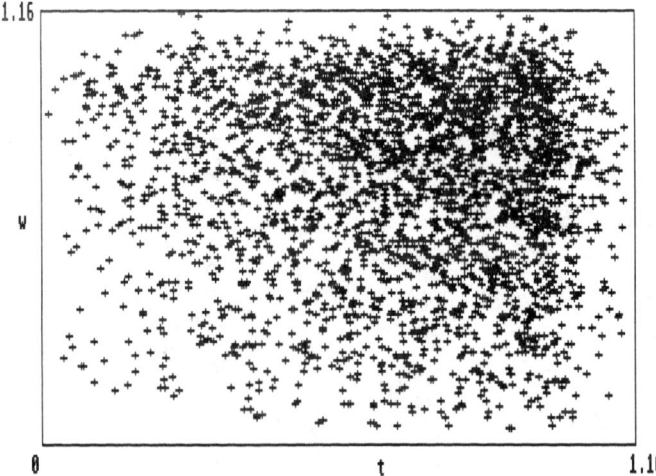

Figure 4.7. (t, w) coordinates of $SL(3, \mathbb{Z})$-images of Hecke points $Y_0[M(p;a,b)]$ in the fundamental domain $\mathscr{SF}_3$ for $SL(3, \mathbb{Z})$ using the notation (4.38), with Y_0 defined in (4.39) and $p = 3001$. Here $b \equiv 5a + 163 \pmod p$.

4.4. Fundamental Domains for $\mathscr{P}_n/GL(n,\mathbb{Z})$

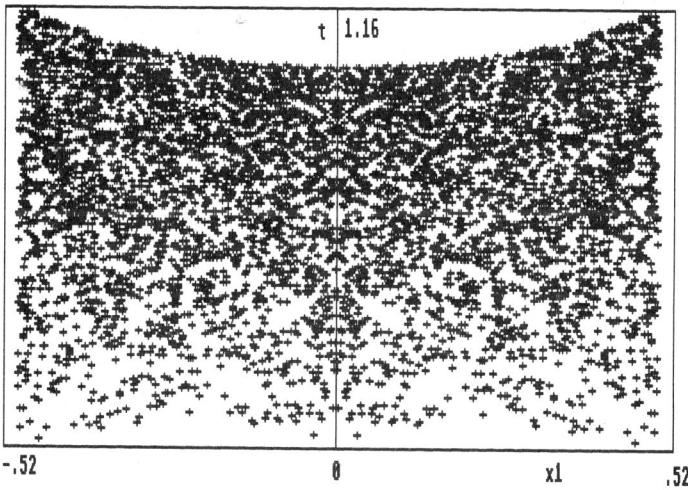

Figure 4.8. (x_1, t) coordinates of $SL(3,\mathbb{Z})$-images of Hecke points $Y_0[M(p;a,b)]$ in the union of the projection of the fundamental domain $\mathscr{SF}_3$ for $SL(3,\mathbb{Z})$ and its mirror image under the reflection across the t-axis. We use the notation (4.38), with Y_0 defined in (4.39) and $p = 3001$. Here $b \equiv 5a + 163 \pmod{p}$.

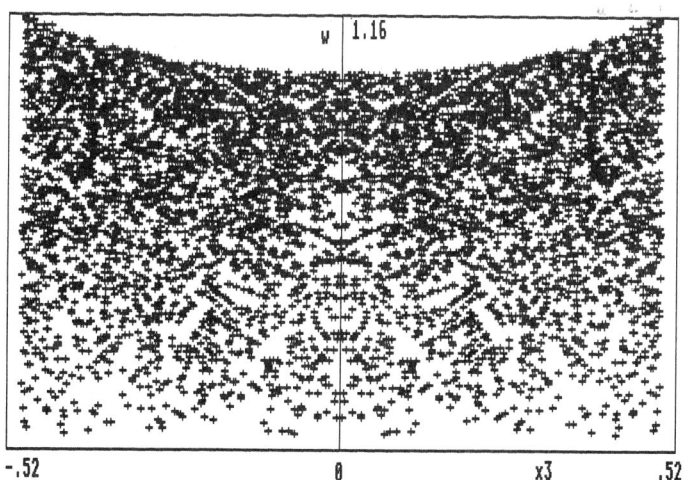

Figure 4.9. (x_3, w) coordinates of $SL(3,\mathbb{Z})$-images of Hecke points $Y_0[M(p;a,b)]$ in the union of the projection of the fundamental domain $\mathscr{SF}_3$ for $SL(3,\mathbb{Z})$ and its mirror image under reflection across the w-axis. We use the notation (4.38), with Y_0 defined in (4.39) and $p = 3001$. Here $b \equiv 5a + 163 \pmod{p}$.

158 IV. The Space $\mathscr{P}_n$ of Positive $n \times n$ Matrices

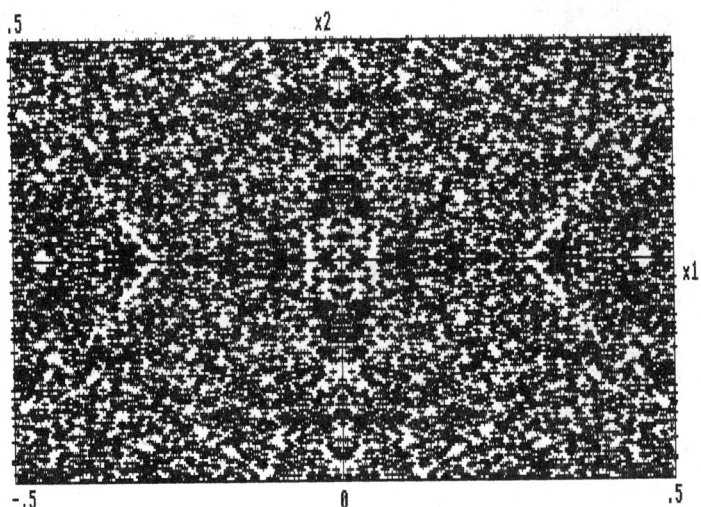

Figure 4.10. (x_1, x_2) coordinates of $SL(3, \mathbb{Z})$-images of Hecke points $Y_0[M(p; a, b)]$ in the union of the projection of the fundamental domain $\mathscr{SF}_3$ for $SL(3, \mathbb{Z})$ and its mirror image under reflection across the x_2-axis. We use the notation (4.38), with Y_0 defined in (4.39) and $p = 3001$. Here $b \equiv 5a + 163 \pmod{p}$.

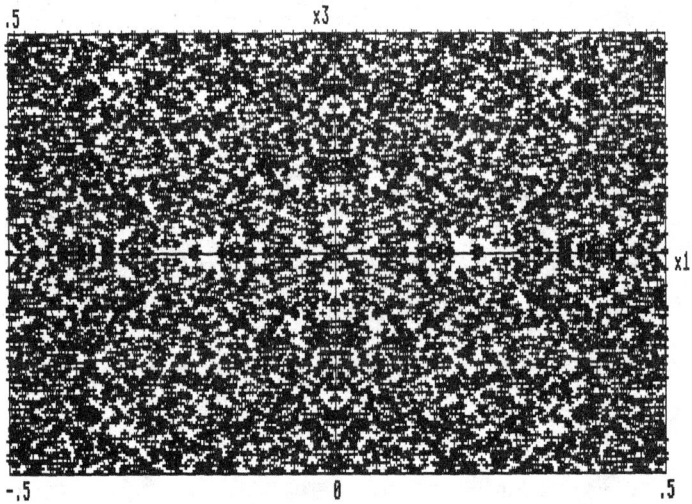

Figure 4.11. (x_1, x_3) coordinates of $SL(3, \mathbb{Z})$-images of Hecke points $Y_0[M(p; a, b)]$ in the union of the projection of the fundamental domain $\mathscr{SF}_3$ for $SL(3, \mathbb{Z})$ and its mirror images under reflections across the x_1 and x_3 axes. We use the notation (4.38), with Y_0 defined in (4.39) and $p = 3001$. Here $b \equiv 5a + 163 \pmod{p}$.

4.4. Fundamental Domains for $\mathscr{P}_n/GL(n,\mathbb{Z})$

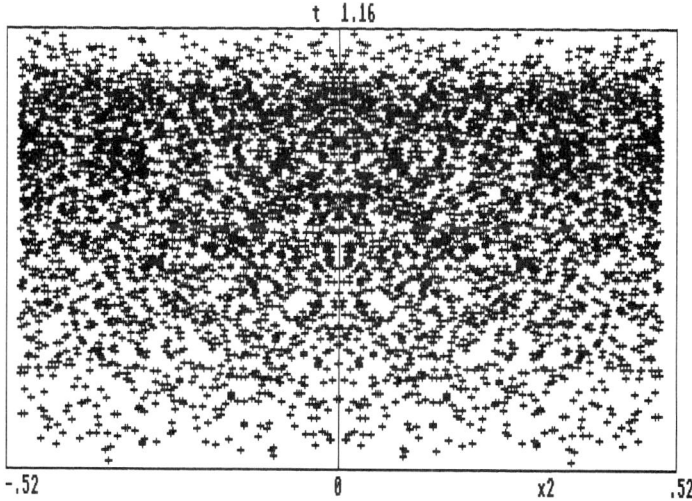

Figure 4.12. (x_2, t) coordinates of $SL(3,\mathbb{Z})$-images of Hecke points $Y_0[M(p;a,b)]$ in the fundamental domain $\mathscr{SF}_3$ for $SL(3,\mathbb{Z})$ using the notation (4.38), with Y_0 defined in (4.39) and $p = 3001$. Here $b \equiv 5a + 163 \pmod{p}$.

Figures 4.12, 4.13, and 4.14 are plots of (x_2, t), (x_3, t), and (x_2, w). The result should be compared with Figures 4.8 and 4.9. If we do so, we see that the top curves of Figures 4.12–4.14 cannot be those of Figures 4.8 and 4.9. The variables in Figures 4.12–4.14 are less closely related.

These figures should also be compared with those in Gordon, Grenier, and Terras [1] where, for example, the points were plotted as points. Here we are plotting small cross marks.

One might complain that our graphs still do not give a real 5-dimensional feeling for the fundamental domain. We hope to make "$\mathscr{SF}_3$ THE MOVIE" some day making use of motion and color. This would be a non-Euclidean analogue of Banchoff's movie of a rotating 4-dimensional cube. For you may view our region $\mathscr{SF}_3$ as a 5-dimensional non-Euclidean crystal. It would also be nice to produce a figure representing the tessellation of the 5-dimensional space $\mathscr{SP}_3$ corresponding to $SL(3,\mathbb{Z})$ images of $\mathscr{SF}_3$. These would be 5-dimensional analogues of pictures that inspired the artist M.C. Escher. Such graphs could be obtained by plotting images of geodesics under matrices generated by S_i, T_i, U_i appearing in Grenier's reduction algorithm.

Figure 4.6 was obtained from a UCSD zeta plotter, after computing the points on a VAX. Figures 4.7–4.14 were obtained with a Fujitsu printer using an Atari 1040ST with Snapshot and Degas Elite to process the points computed on the UCSD VAX computer.

There are various ways of understanding why the Hecke points should be dense in $\mathscr{SF}_3$. One could imitate an argument of Zagier using Eisenstein series

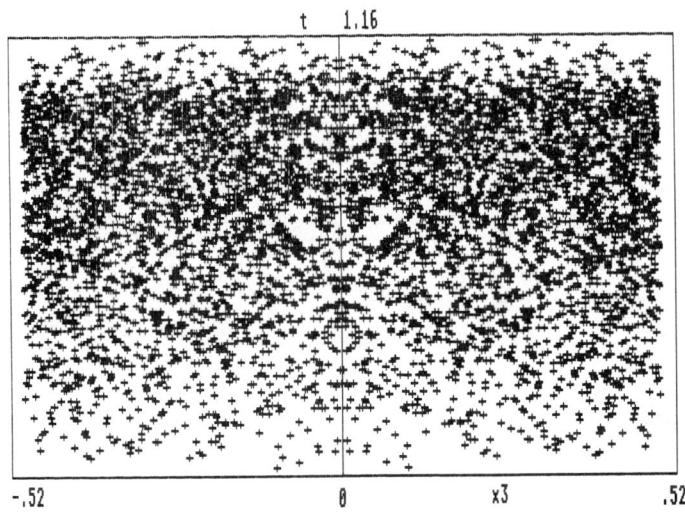

Figure 4.13. (x_3, t) coordinates of $SL(3, \mathbb{Z})$-images of Hecke points $Y_0[M(p;a,b)]$ in the union of the projection of the fundamental domain $\mathscr{SF}_3$ for $SL(3, \mathbb{Z})$ and its mirror image under reflection across the t-axis. We use the notation (4.38), with Y_0 defined in (4.39) and $p = 3001$. Here $b \equiv 5a + 163 \pmod{p}$.

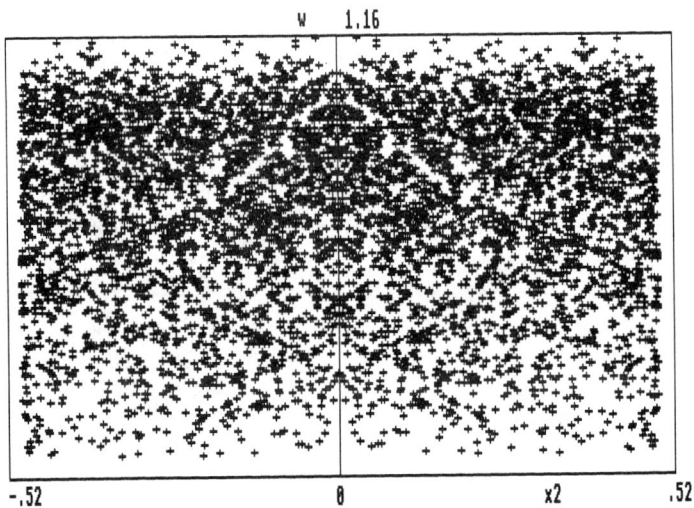

Figure 4.14. (x_2, w) coordinates of $SL(3, \mathbb{Z})$-images of Hecke points $Y_0[M(p;a,b)]$ in the fundamental domain $\mathscr{SF}_3$ for $SL(3, \mathbb{Z})$ using the notation (4.38), with Y_0 defined in (4.39) and $p = 3001$. Here $b \equiv 5a + 163 \pmod{p}$.

4.4. Fundamental Domains for $\mathcal{P}_n/GL(n,\mathbb{Z})$

(see page 248 of Volume I for the $SL(2,\mathbb{Z})$-version of the argument) to show that the image of a horocycle C_Y becomes dense in $\mathcal{SF}_3$ as Y approaches the boundary of $\mathcal{SP}_3$. Here, by a *horocycle* we mean the set:

$$C_Y = \{Y[n] \mid n \text{ is upper triangular with 1 on the diagonal}\}.$$

We will return to this question after studying Eisenstein series in the next section.

This result is also related to standard results in ergodic theory for connected noncompact simple Lie groups G with finite center (e.g., $G = SL(3,\mathbb{R})$) saying that if H is a closed non compact subgroup of G and Γ is an irreducible lattice (e.g. $\Gamma = SL(3,\mathbb{Z})$) then H acts ergodically on G/Γ. Here we are closest to looking at an equally spaced finite set of points in

$$H = \left\{ \begin{pmatrix} 1 & x & y \\ 0 & 1 & 0 \\ 0 & 0 & 1 \end{pmatrix} \Bigg| x, y \in \mathbb{R} \right\}.$$

For we are looking at points from T_p acting on a fixed $Y_0 \in \mathcal{SP}_3$ via

$$Y_0 \left[\begin{pmatrix} p & 0 & 0 \\ 0 & p & 0 \\ 0 & 0 & p \end{pmatrix} \begin{pmatrix} 1 & a/p & b/p \\ 0 & 1 & 0 \\ 0 & 0 & 1 \end{pmatrix} \right],$$

with $0 \le a, b \le p - 1$. For the ergodic theory result, see Zimmer [1, p. 19ff.].

Ultimately one would hope to be able to use the points $M(p;a,b)$ to generalize the results of Stark [2] given as Table 3.10 of Volume I. This will require programs for the computation of matrix argument K-Bessel or Whittaker functions.

At this point, there are various natural questions.

Questions.

(1) Exercise 15 gives generators of $GL(n,\mathbb{Z})$. What happened to Poincaré's generators and relations theorems (cf. Exercise 1 in Section 3.3 of Volume I) in this context?
(2) Is there some way of visualizing the tessellation of $\mathcal{SP}_n$ produced by writing

$$\mathcal{SP}_n = \bigcup_{\gamma \in \Gamma_n} \mathcal{SF}_n[\gamma]?$$

Perhaps we should take a hint from topology and look at retracts (cf. Ash [1]).
(3) In the classical case of $SL(2,\mathbb{Z})$, the reduction algorithm for putting a point $z \in H$ into the standard fundamental domain, using a sequence of translations and flips, is the same as the algorithm for finding a continued fraction expansion of a real number. Thus Grenier's algorithm for putting a matrix $Y \in \mathcal{SP}_n$ into $\mathcal{SF}_n$ by some combination of matrices from those listed after

Exercise 17 gives an analogue of a continued fraction algorithm. This should be compared with the continued fraction algorithms of Ferguson and Forcade [1] and other work mentioned in the introduction to Section 4.1.

(4) There is an analogue for $GL(n, \mathbb{Z})$ of the method of perpendicular bisectors which writes the fundamental domain for a discrete subgroup Γ of $GL(n, \mathbb{R})$ as follows, for a point $W \in \mathscr{P}_n$ such that $W \neq W[\gamma]$ if $\gamma \in \Gamma$ and $\gamma \neq \pm I$:

$$\{Y \in \mathscr{P}_n | d(Y, W) \leq d(Y, W[\gamma]), \text{ for all } \gamma \in \Gamma\}.$$

Here d denotes the distance obtained from the Riemannian structure. See Siegel [1, Vol. II, pp. 298–301]. Why can (and should) we choose the point W as stated?

B.A. Venkov [1] considers a related domain defined for a fixed $H \in \mathscr{P}_n$ by:

$$\{Y \in \mathscr{P}_n | \text{Tr}(YH) \leq \text{Tr}(YH[\gamma]) \text{ for all } \gamma \in \Gamma\}.$$

When the point H is such that $H = H[\gamma]$ implies $\gamma = \pm I$, then this Venkov domain is a fundamental domain for $GL(n, \mathbb{Z}) = \Gamma$. See also Ryskov [2].

Siegel [loc. cit., p. 310] notes: "The application of the general method ...[given above] would lead to a rather complicated shape of the frontier of [the fundamental domain] F." However, the method does lead to the standard fundamental domain for $SL(2, \mathbb{Z})$ if the point W in the Poincaré upper half plane is chosen to be $2i$, for example.

The question here is to compare all these domains with those of Minkowski and Grenier.

(5) One should relate our fundamental domain to that which would be obtained if one replaced $\Gamma = GL(n, \mathbb{Z})$ by integral matrices with arbitrary nonzero determinant. The question concerns the relationship between Minkowski reduction and reduction by successive minima. Or one could consider replacing Γ by $\Gamma \cap (A \Gamma A^{-1})$, where A is some integral matrix of positive determinant d. This has something to do with Hecke operators.

(6) What geodesics of $\mathscr{P}_n$, if any, induce dense geodesics on the fundamental domain for $\mathscr{P}_n/GL(n, \mathbb{Z})$?

Recall that in 1835 Jacobi showed that a line with irrational slope in $\mathbb{R}^2$ induces a densely wound line in the torus $\mathbb{R}^2/\mathbb{Z}^2$. Weyl [1] further developed the theory in 1916. Artin [3, pp. 499–504] showed in 1924 that almost all geodesics in the Poincaré upper half plane will induce densely wound lines in the standard fundamental domain for $SL(2, \mathbb{Z})$. See Figure 4.15. In fact, geodesics are typically dense in the unit tangent bundle for the $SL(2, \mathbb{Z})$ case. However, the geodesic flow is not ergodic on the unit tangent bundle in higher rank (cf. Mautner [2, pp. 419–421]). This still leaves open the question of density in the fundamental domain for the higher rank case. Perhaps it is more sensible to look for $n - 1$-dimensional totally geodesic submanifolds. See Zimmer [1, especially pp. 18–19].

4.4. Fundamental Domains for $\mathscr{P}_n/GL(n,\mathbb{Z})$

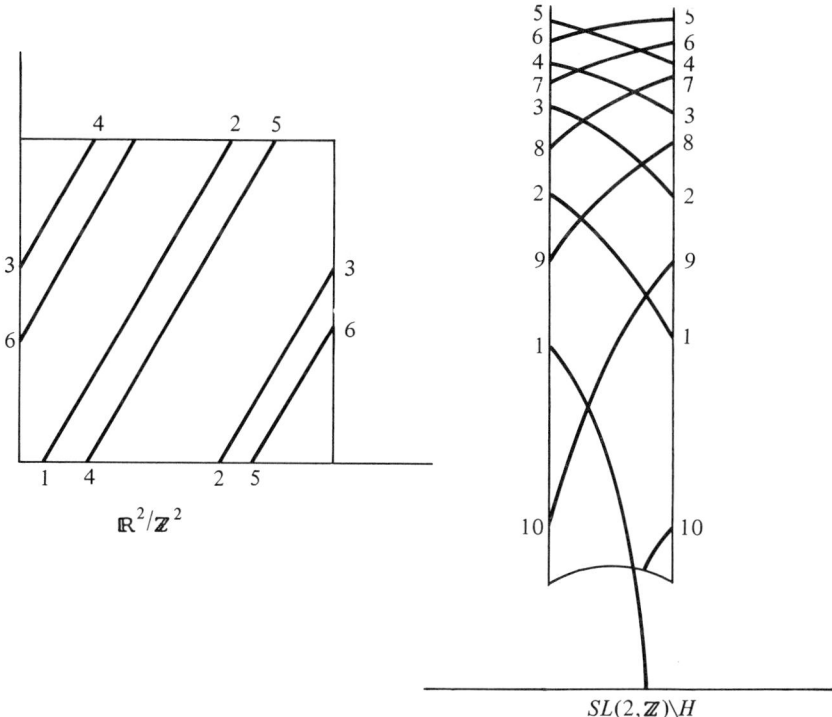

Figure 4.15. Geodesics in fundamental domains.

One can ask analogous questions about "horocycles" (i.e., conjugates of the group N of upper triangular matrices with ones on the diagonal). This sort of question was already considered in connection with our pictures of images of Hecke points in the fundamental domain $\mathscr{S}\mathscr{F}_3$.

The next exercise gives the finiteness of the volume of $\mathscr{S}\mathscr{F}_n$. In the next section we will obtain an exact formula.

Exercise 19 (Finiteness of the Volume of the Fundamental Domain in the Determinant One Surface).

(a) Show that, if $W \in \mathscr{S}\mathscr{P}_n$ and

$$W = \begin{pmatrix} u & 0 \\ 0 & u^{-(1/n-1)}V \end{pmatrix} \begin{bmatrix} 1 & q \\ 0 & I_{n-1} \end{bmatrix}, \quad \text{for } u > 0, V \in \mathscr{S}\mathscr{P}_{n-1}, q \in \mathbb{R}^{n-1},$$

then

$$dW = u^{(n/2)-1}\, du\, dV\, dq.$$

See Exercise 30 in §4.1.5.

(b) Use part (a) to show that
$$\text{Vol}(\mathscr{SF}_n) \leq \text{constant Vol}(\mathscr{SF}_{n-1}).$$
(c) Conclude that the volume of $SO(n)\backslash SL(n,\mathbb{R})/SL(n,\mathbb{Z})$ is finite.

4.4.4. Integration over Fundamental Domains

Next we turn to integral formulas on fundamental domains $\mathscr{D}$ for $\mathscr{P}_n/GL(n,\mathbb{Z})$. Here we will often take $\mathscr{D}$ to be the Minkowski domain $\mathscr{M}_n$. Define the determinant one surface in the fundamental domain to be $\mathscr{SD} = \mathscr{D} \cap \mathscr{SP}_n$. Recall Exercise 23 of Section 4.1.3 which gave *the relation between the $GL(n,\mathbb{R})$-invariant measure $d\mu_n$ on $\mathscr{P}_n$ and an $SL(n,\mathbb{R})$-invariant measure dW on the determinant one surface $\mathscr{SP}_n$*:

$$Y = t^{1/n}W, \quad Y \in \mathscr{P}_n, \quad t = |Y| > 0, \quad W \in \mathscr{SP}_n, \quad d\mu_n(Y) = t^{-1}\,dt\,dW, \quad (4.40)$$

where the ordinary Euclidean volume element dY on $\mathscr{P}_n$ is related to the invariant measure $d\mu_n(Y)$ by:

$$d\mu_n(Y) = |Y|^{-(n+1)/2}\,dY. \quad (4.41)$$

Exercise 20. Compute the Jacobian of the change of variables $Y = t^{1/n}W$, from $Y \in \mathscr{P}_n$ to $t > 0$ and $W \in \mathscr{SP}_n$, using all but one of the entries of $W = Y^0$ above or on the diagonal (e.g., leave out w_{nn}).

Answer.
$$|dY/d(t,W)| = (nt)^{-1}t^{(n+1)/2}.$$

Thus (4.40) normalizes measures "wrong" by throwing away the factor of $1/n$. This does not really matter.

If we set $G = SL(n,\mathbb{R})$, $\Gamma = SL(n,\mathbb{Z})$, the quotient space G/Γ has a G-invariant volume element $d\bar{g}$, which is unique up to a positive constant multiple (see Lang [1], Siegel [1, Vol. III, pp. 39–46], or Weil [1, pp. 42–45]). Therefore we can normalize $d\bar{g}$ to obtain:

$$\int_{\mathscr{SP}_n/SL(n,\mathbb{Z})} f(W)\,dW = \int_{G/\Gamma} f({}^t gg)\,dg. \quad (4.42)$$

Our first goal is to compute the volume of the fundamental domain in the determinant one surface. We know that this volume is finite by Exercise 19 of the preceding section.

Lemma 5. *The Euclidean volume of the set of matrices in $\mathscr{M}_n$ having determinant less than or equal to one is related to the $SL(n,\mathbb{R})$-invariant volume of $\mathscr{SM}_n$*

4.4. Fundamental Domains for $\mathscr{P}_n/GL(n,\mathbb{Z})$

(obtained using the measure dW in (4.40)) as follows:

$$\text{Euclidean Vol}\{Y \in \mathscr{M}_n \mid |Y| \le 1\} = 2\,\text{Vol}(\mathscr{SM}_n)/(n+1).$$

PROOF. By formulas (4.40) and (4.41), we have:

$$\int_{\substack{|Y| \le 1 \\ Y \in \mathscr{M}_n}} dY = \int_{t=0}^{1} t^{(n-1)/2}\, dt\, \text{Vol}(\mathscr{SM}_n).$$

This clearly gives the stated formula. □

Our plan is to determine $\text{Vol}(\mathscr{SM}_n)$ using an inductive procedure which derives from work of Minkowski [1, Vol. II, pp. 80–94], Siegel [1, Vol. III, pp. 39–46], and Weil [2, Vol. I, pp. 339–358]. Weil writes [loc. cit., p. 56] that he was able to use his simplification of Siegel's work on this subject to calculate the Tamagawa number, which gives the adelic formulation of Siegel's main theorem on quadratic forms. The determination of $\text{Vol}(\mathscr{SM}_n)$ is closely related to the following proposition.

Proposition 2 (Siegel's Integral Formula in the Geometry of Numbers). *Let $G = SL(n,\mathbb{R})$, $\Gamma = SL(n,\mathbb{Z})$, and $f: \mathbb{R}^n \to \mathbb{C}$ be an integrable function. Then we have the following equalities:*

$$\frac{1}{\text{Vol}(G/\Gamma)} \int_{\bar{g} \in G/\Gamma} \sum_{a \in \mathbb{Z}^n - 0} f(ga)\, d\bar{g} = \int_{\mathbb{R}^n} f(x)\, dx,$$

$$\frac{\zeta(n)}{\text{Vol}(G/\Gamma)} \int_{\bar{g} \in G/\Gamma} \sum_{\substack{a \in \mathbb{Z}^n - 0 \\ \text{g.c.d.}(a) = 1}} f(ga)\, d\bar{g} = \int_{\mathbb{R}^n} f(x)\, dx.$$

Here dx denotes Lebesgue measure on $\mathbb{R}^n$, $d\bar{g}$ is a G-invariant measure on G/Γ, the vectors $a \in \mathbb{Z}^n - 0$ are column vectors, and ga denotes the column vector that results from multiplying a by the $n \times n$ matrix g.

PROOF (Weil [loc. cit.]). The main idea is to use the following integration formula which holds for a unimodular locally compact topological group G with closed unimodular subgroup G_1. Here "unimodular" means that right and left Haar measures coincide. The integral formula in question is:

$$\int_{G/G_1} \int_{G_1} f(gg_1)\, dg_1\, d\bar{g} = c \int_G f(g)\, dg, \qquad (4.43)$$

where c is a positive constant, dg and dg_1 are Haar measures on G and G_1, respectively, $d\bar{g}$ is a G-invariant measure on G/G_1. References for this result are Helgason [2], [4], Lang [1], Weil [1, p. 45]. Formula (4.43) can be extended to non unimodular G and G_1 provided that the modular functions of G and G_1 are equal on G_1.

Two applications of formula (4.43) are required to prove Siegel's integral formula. There are, in fact, two quotients in Siegel's integral formula. The

obvious quotient is G/Γ and the other is $\mathbb{R}^n - 0 \cong G/H$, where H is the subgroup:
$$H = \{g \in G | g e_1 = e_1\}, \tag{4.44}$$
and e_1 is the standard unit basis vector in $\mathbb{R}^n$; i.e. $e_1 = {}^t(1, 0, \ldots, 0)$. Note that the elements of H have the form:
$$\begin{pmatrix} 1 & * \\ 0 & * \end{pmatrix}.$$

The mapping that identifies G/H with $\mathbb{R}^n - 0$ is:
$$G/H \to \mathbb{R}^n - 0$$
$$gH \mapsto g e_1 = \text{the first column of } g.$$

Now let $\gamma = H \cap \Gamma$. Suppose that $f: G/H \to \mathbb{C}$ satisfies the hypotheses of the proposition. Then
$$c \int_{G/H} f(x) \, dx = \int_{G/\Gamma} \int_{\Gamma/\gamma} f(g y) \, d\bar{y} \, d\bar{g}, \tag{4.45}$$
for some positive constant c (independent of f). To see this, note that
$$\int_{G/H} \int_{H/\gamma} f(gh) \, d\bar{h} \, d\bar{g} = c_1 \int_{G/\gamma} f(g) \, dg,$$
$$\int_{G/\Gamma} \int_{\Gamma/\gamma} f(ga) \, d\bar{a} \, d\bar{g} = c_2 \int_{G/\gamma} f(g) \, dg,$$
for some positive constants c_1 and c_2.

Next observe that from Lemma 2,
$$\Gamma/\gamma = \{a \in \mathbb{Z}^n | \text{g.c.d. } a = 1\}. \tag{4.46}$$
So (4.45) says that:
$$c \int_{\mathbb{R}^n} f(x) \, dx = \int_{G/\Gamma} \sum_{\substack{a \in \mathbb{Z}^n \\ \text{g.c.d. } (a)=1}} f(ga) \, d\bar{g}.$$
This implies, by change of variables, that if $t > 0$, we have:
$$ct^{-n} \int_{\mathbb{R}^n} f(x) \, dx = \int_{G/\Gamma} \sum_{\substack{a \in \mathbb{Z}^n \\ \text{g.c.d. } (a)=1}} f(tga) \, d\bar{g}.$$
Now sum over $t = 1, 2, 3, \ldots$ and obtain:
$$c\zeta(n) \int_{\mathbb{R}^n} f(x) \, dx = \int_{G/\Gamma} \sum_{a \in \mathbb{Z}^n - 0} f(ga) \, d\bar{g}. \tag{4.47}$$

The proof of Siegel's integral formula is completed by showing that $c\zeta(n) = \text{Vol}(G/\Gamma)$. Weil's proof of this fact uses the Poisson summation formula (see

4.4. Fundamental Domains for $\mathscr{P}_n/GL(n, \mathbb{Z})$

Theorem 2 of Section 1.3 in Volume I). Let $c^* = c\zeta(n)$ and $V = \text{Vol}(G/\Gamma)$. From (4.47) it follows that:

$$Vf(0) + c^* \int_{\mathbb{R}^n} f(x)\,dx = \int_{G/\Gamma} \sum_{a \in \mathbb{Z}^n} f(ga)\,d\bar{g}. \tag{4.48}$$

And Poisson tells us that for $g \in G$:

$$\sum_{a \in \mathbb{Z}^n} f(ga) = \sum_{a \in \mathbb{Z}^n} \hat{f}({}^t g^{-1} a),$$

where $\hat{f}$ denotes the Fourier transform of f over $\mathbb{R}^n$. Note that

$$\hat{f}(0) = \int_{\mathbb{R}^n} f(y)\,dy.$$

Therefore

$$Vf(0) + c^* \hat{f}(0)$$

$$= \int_{G/\Gamma} \sum_{a \in \mathbb{Z}^n} \hat{f}({}^t g^{-1} a)\,d\bar{g} = \int_{G/\Gamma} \sum_{a \in \mathbb{Z}^n} \hat{f}(ga)\,d\bar{g}. \tag{4.49}$$

Replace f by $\hat{f}$ in formula (4.48) or formula (4.49) to find that

$$V\hat{f}(0) + c^* f(0) = Vf(0) + c^* \hat{f}(0),$$

which says that $(V - c^*)(f(0) - \hat{f}(0)) = 0$. Since we can easily find a function $f(x)$ such that $f(0) \neq \hat{f}(0)$, it follows that $V = c^*$, and we're finished with the proof of Siegel's integral formula. □

Corollary. *Suppose that* $f: \mathbb{R}^+ \to \mathbb{C}$ *is suitably chosen for convergence. Then*

$$\frac{1}{\text{Vol}(\mathscr{S}\mathscr{M}_n)} \int_{\mathscr{S}\mathscr{M}_n} \sum_{a \in \mathbb{Z}^n - 0} f(W[a])\,dW = \int_{\mathbb{R}^n} f({}^t xx)\,dx.$$

PROOF. This corollary follows immediately from (4.42) and Proposition 2 in the case n is odd, since then $\mathscr{S}\mathscr{P}_n/SL(n, \mathbb{Z}) = \mathscr{S}\mathscr{P}_n/GL(n, \mathbb{Z}) = \mathscr{S}\mathscr{M}_n$ because $GL(n, \mathbb{Z})/SL(n, \mathbb{Z})$ has representatives $I, -I$, both having no effect on $W \in \mathscr{S}\mathscr{P}_n$. However, one has to make a more complicated argument when n is even. In that case suppose that I, γ represent $GL(n, \mathbb{Z})/SL(n, \mathbb{Z})$. Then note that

$$\mathscr{S}\mathscr{M}_n \cup \mathscr{S}\mathscr{M}_n[\gamma]$$

is a fundamental domain for $SL(n, \mathbb{Z})$. Thus, for even n,

$$\text{Vol}(\mathscr{S}\mathscr{P}_n/SL(n, \mathbb{Z})) = 2\,\text{Vol}(\mathscr{S}\mathscr{P}_n/GL(n, \mathbb{Z})).$$

And, setting $\Gamma^0 = SL(n, \mathbb{Z})$ and $\Gamma = GL(n, \mathbb{Z})$, we have

$$\int_{\mathscr{S}\mathscr{P}_n/\Gamma^0} \sum_{a \in \mathbb{Z}^n - 0} f(W[a])\,dW = 2 \int_{\mathscr{S}\mathscr{P}_n/\Gamma} \sum_{a \in \mathbb{Z}^n - 0} f(W[a])\,dW.$$

The reason for this is the fact that

$$\sum_{a \in \mathbb{Z}^n - 0} f(W[a]) = \sum_{a \in \mathbb{Z}^n - 0} f(W[\gamma a])$$

for any $\gamma \in GL(n, \mathbb{Z})$. □

There are many applications of the integral formulas in Proposition 2 and their generalizations. For example, they give integral tests for the convergence of Eisenstein series. They also imply the existence of quadratic forms with large minima; i.e., the existence of dense lattice packings. This result is usually called the Minkowski-Hlawka theorem in the geometry of numbers. But before discussing these applications, let us compute the exact volume of $\mathscr{S}\mathscr{M}_n$.

Theorem 4 (Volume of the Fundamental Domain).

(1) *Using the normalization of measures, given in* (4.40), *we have*

$$\operatorname{Vol}(\mathscr{S}\mathscr{M}_n) = \prod_{k=2}^{n} \Lambda(k/2), \qquad \Lambda(s) = \pi^{-s} \Gamma(s) \zeta(2s).$$

(2) *Euclidean Volume* $\{Y \in \mathscr{M}_n \mid |Y| \le 1\} = 2 \operatorname{Vol}(\mathscr{S}\mathscr{M}_n)/(n+1)$.

PROOF. Note that part (2) is an easy consequence of Lemma 5.

We proceed to prove part (1). If the function f in the corollary to Proposition 2 is radial, then from the formula for the surface area of the unit sphere in Exercise 1 of Section 4.4.1, we have the following sequence of equalities for suitable $f: \mathbb{R}^+ \to \mathbb{C}$:

$$\frac{\pi^{n/2}}{\Gamma(n/2)} \int_{r>0} f(r) r^{n/2-1} \, dr = \int_{\mathbb{R}^n} f({}^t x x) \, dx$$

$$= \frac{\zeta(n)}{\operatorname{Vol}(\mathscr{S}\mathscr{M}_n)} \sum_{\substack{a \in \mathbb{Z}^n \\ \text{g.c.d. }(a)=1}} \int_{\mathscr{S}\mathscr{M}_n} f(W[a]) \, dW$$

$$= \frac{\zeta(n)}{\operatorname{Vol}(\mathscr{S}\mathscr{M}_n)} \sum_{(a*) \in \Gamma/\Gamma \cap H} \int_{\mathscr{S}\mathscr{P}_n/\Gamma} f(W[a]) \, dW$$

if $\Gamma = GL(n, \mathbb{Z})$, and

$$H = \begin{pmatrix} 1 & * \\ 0 & * \end{pmatrix}.$$

Therefore if $f: \mathbb{R}^+ \to \mathbb{C}$ is suitably chosen for convergence, we have:

$$\frac{\pi^{n/2}}{\Gamma(n/2)} \int_{r>0} f(r) r^{(n/2)-1} \, dr = \frac{\zeta(n)}{\operatorname{Vol}(\mathscr{S}\mathscr{M}_n)} \int_{W=(w_{ij}) \in \mathscr{S}\mathscr{P}_n/\Gamma \cap H} f(w_{11}) \, dx. \quad (4.50)$$

Next we need some exercises.

4.4. Fundamental Domains for $\mathscr{P}_n/GL(n,\mathbb{Z})$

Exercise 21. Use formula (4.40) to show that if $f: \mathscr{SP}_n \to \mathbb{C}$ is integrable, then

$$\int_{W \in \mathscr{SP}_n} f(W)\, dW = \int_{\substack{Y \in \mathscr{P}_n \\ |Y| \le 1}} f(|Y|^{-1/n} Y)\, |Y|^{-(n-1)/2}\, dY,$$

where dY is as in (4.41).

Hint. Let $h(t^{1/n} W) = t\chi_{[0,1]}(t) f(W)$, for $t > 0$, $W \in \mathscr{SP}_n$ in formula (4.40), where

$$\chi_{[0,1]}(t) = \begin{cases} 1, & \text{if } t \in [0,1], \\ 0, & \text{otherwise.} \end{cases}$$

Exercise 22. Use the partial Iwasawa decomposition

$$Y = \begin{pmatrix} t & 0 \\ 0 & V \end{pmatrix} \begin{bmatrix} 1 & {}^t h \\ 0 & I \end{bmatrix}, \qquad t > 0, \qquad V \in \mathscr{P}_{n-1}, \qquad h \in \mathbb{R}^{n-1},$$

to obtain an explicit fundamental domain for $\mathscr{P}_n / \Gamma \cap H$, with H as it was defined just before formula (4.50).

It follows from Exercises 21 and 22 above and Exercise 19 of Section 4.1.3 that the integral appearing on the right hand side of formula (4.50) can be rewritten as:

$$\int_{\mathscr{SP}_n/\Gamma_n \cap H} f(w_{11})\, dW$$

$$= \int_{\substack{t|V| \le 1 \\ V \in \mathscr{P}_{n-1}/\Gamma_{n-1} \\ h \in [0,1]^{n-1}}} f((t|V|)^{-1/n} t)(t|V|)^{-(n-1)/2} t^{n-1}\, dV\, dt\, dh. \quad (4.51)$$

Upon setting $U = t^{1/(n-1)} V$, formula (4.51) becomes:

$$\int_{\mathscr{SP}_n/\Gamma_n \cap H} f(w_{11})\, dW = \int_{\substack{|U| \le 1 \\ U \in \mathscr{M}_{n-1} \\ 0 < t}} f(|U|^{-1/n} t) |U|^{(1-n)/2} t^{(n/2)-1}\, dt\, dU.$$

Therefore if we substitute $x = t|U|^{-1/n}$ and use formulas (4.40) and (4.41), we obtain:

$$\int_{\mathscr{SP}_n/\Gamma_n \cap H} f(w_{11})\, dW = \int_{\substack{|U| \le 1 \\ U \in \mathscr{M}_{n-1} \\ 0 < x}} f(x) |U|^{1-(n/2)} x^{(n/2)-1}\, dt\, dU$$

$$= \mathrm{Vol}(\mathscr{SM}_{n-1}) \int_{x > 0} f(x) x^{(n/2)-1}\, dx.$$

Thus we have proved:

$$\int_{\mathscr{SP}_n/\Gamma_n \cap H} f(w_{11})\, dW = \mathrm{Vol}(\mathscr{SM}_{n-1}) \int_{x > 0} f(x) x^{(n/2)-1}\, dt. \quad (4.52)$$

If $f(x)x^{-1+n/2}$ is positive and integrable over $(0, \infty)$, then (4.50) and (4.52) combine to give:

$$\text{Vol}(\mathscr{SM}_n) = \text{Vol}(\mathscr{SM}_{n-1})\pi^{-n/2}\Gamma(n/2)\zeta(n).$$

The theorem follows by induction, using the case $n = 2$ which was obtained in Chapter 3 of Volume I. □

Three corollaries of Siegel's integral formula (Proposition 2) can now be derived quite easily.

Corollary 1 (The Minkowski-Hlawka Theorem). *There is a matrix $Y \in \mathscr{P}_n$ such that the first minimum m_Y (defined in (4.7) of Section 4.4.1) satisfies:*

$$m_Y > d_n |Y|^{1/n}, \quad \text{with } d_n = n/(2\pi e),$$

for n sufficiently large.

PROOF. Consider a ball $B_n \subset \mathbb{R}^n$ of radius

$$k_n = \left(\frac{n}{2\pi e}\right)^{1/2}.$$

Then

$$\text{Vol}(B_n) = \left(\frac{n}{2\pi e}\right)^{n/2} \frac{\pi^{n/2}}{\Gamma(1 + n/2)}.$$

Stirling's formula implies that $\text{Vol}(B_n)$ tends to zero as n goes to infinity. This is rather surprising, since the radius is blowing up. See the note below for a related paradox.

Now let

$$\chi_{B_n} = \begin{cases} 1, & \text{for } x \in B_n, \\ 0, & \text{otherwise.} \end{cases}$$

If this function is plugged into Siegel's integral formula and $V = \text{Vol}(G/\Gamma)$, we obtain:

$$\text{Vol}(B_n) = \frac{1}{V} \int_{G/\Gamma} \sum_{a \in \mathbb{Z}^n - 0} \chi_{B_n}(ga) \, d\bar{g}.$$

When n is sufficiently large, it is clear that, since $\text{Vol}(B_n) \to 0$, we must have:

$$\sum_{a \in \mathbb{Z}^n - 0} \chi_{B_n}(ga) < 1$$

for some $g \in G = SL(n, \mathbb{R})$. This means that ${}^t gg[a] > k_n^2$ for some g and all a in $\mathbb{Z}^n - 0$. Set $Y = {}^t gg$ to finish the proof. □

Note. There is an interesting paradox associated with balls in high-dimensional spaces (cf. Hamming [1, pp. 168–170]). In n-dimensional space, con-

4.4. Fundamental Domains for $\mathscr{P}_n/GL(n,\mathbb{Z})$

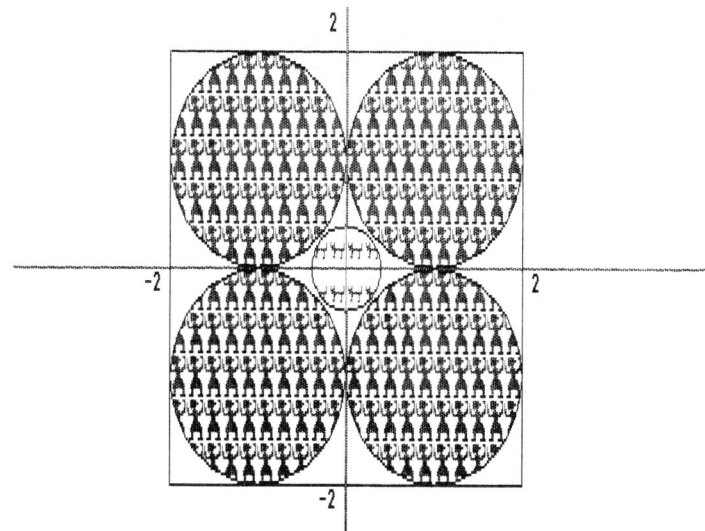

Figure 4.16. A paradox.

sider a hypercube having a side of length 4 and centered at the origin. Put 2^n unit spheres in each corner of this cube such that each sphere touches all its n neighboring spheres. The distance from the origin to the center of one of these spheres is $\sqrt{n}$. Thus we can put a sphere of radius $\sqrt{n} - 1$ inside all the unit spheres at the corners. See Figure 4.16 for the case $n = 2$. When $n \geq 10$, this inner sphere reaches outside the cube, since $\sqrt{10} - 1 > 2$. Weird.

Corollary 1 gives the existence of dense lattice packings and should be compared with (4.10) in Section 4.4.1. Our next corollary is a result that we already know, but the proof will be easily generalized to more complicated Eisenstein series.

Corollary 2 (Integral Test Proof of the Convergence of Epstein's Zeta Function). *The Epstein zeta function*

$$Z(Y,s) = \sum_{a \in \mathbb{Z}^n - 0} Y[a]^{-s}$$

converges absolutely for $\operatorname{Re} s > n/2$, $Y \in \mathscr{P}_n$.

PROOF. Siegel tells us that:

$$\frac{1}{\operatorname{Vol}(\mathscr{S}\mathscr{M}_n)} \int_{W \in \mathscr{S}\mathscr{M}_n} \sum_{\substack{a \in \mathbb{Z}^n - 0 \\ W[a] \geq 1}} W[a]^{-s} \, dW = \int_{\substack{x \in \mathbb{R}^n \\ {}^t xx \geq 1}} (xx)^{-s} \, dx.$$

The integral on the right is easily evaluated as

$$\pi^{n/2} \Gamma(n/2)^{-1} (s - n/2)^{-1},$$

if $\operatorname{Re} s > n/2$. Then Fubini's theorem says that the series being integrated converges for almost all W in $\mathscr{SM}_n$. The series differs from Epstein's zeta function $Z(W, s)$ by at most a finite number of terms. Thus $Z(W, x)$ converges for $\operatorname{Re} s > n/2$ and almost all W in $\mathscr{SM}_n$. In order to deduce the convergence of $Z(Y, s)$ for all Y in $\mathscr{P}_n$, note that there is a positive constant c such that

$$cI[a] \leq Y[a] \leq c^{-1}I[a], \quad \text{for all } a \text{ in } \mathbb{R}^n. \qquad \square$$

Corollary 3 (The Vanishing of Epstein's Zeta Function in $(0, n/2)$). *For all s with $0 < \operatorname{Re} s < n/2$*

$$\int_{\mathscr{SM}_n} Z(W, s)\, dW = 0.$$

It follows that for any $s \in (0, 1)$ there exist $Y \in \mathscr{P}_n$ such that $Z(Y, s) > 0$. Similarly there are $Y \in \mathscr{P}_n$ such that $Z(Y, s) < 0$, or such that $Z(Y, s) = 0$.

PROOF. Use the incomplete gamma expansion of Epstein's zeta function (Theorem 1 of Section 1.4 in Volume I):

$$2\pi^{-s}\Gamma(s)Z(Y, s)$$
$$= \frac{|Y|^{-1/2}}{s - \frac{n}{2}} - \frac{1}{s} + \sum_{a \in \mathbb{Z}^n - 0} \left\{ G(s, \pi Y[a]) + |Y|^{-1/2} G\left(\frac{n}{2} - s, \pi Y^{-1}[a]\right) \right\},$$

where the incomplete gamma function is:

$$G(s, a) = \int_{t \geq 1} t^{s-1} \exp(-at)\, dt, \quad \text{for } \operatorname{Re} a > 0.$$

Note that we can use this expansion to see that $Z(W, s)$ is integrable over the fundamental domain $\mathscr{SM}_n$ provided that $0 < \operatorname{Re} s < n/2$.

Now apply Siegel's integral formula to see that:

$$\frac{2}{\operatorname{Vol}(\mathscr{SM})} \int_{W \in \mathscr{SM}_n} \pi^{-s}\Gamma(s) Z(W, s)\, dW$$
$$= \frac{1}{s - \frac{n}{2}} - \frac{1}{s} + \int_{\mathbb{R}^n} \left\{ G(s, \pi\, {}^t x x) + G\left(\frac{n}{2} - s, \pi\, {}^t x x\right) \right\} dx.$$

Use the definition of $G(s, a)$ to write the integral over x in $\mathbb{R}^n$ on the right hand side of this equality as a double integral over x and t. Then make the change of variables $y = t^{1/2} x$ to see that if $\operatorname{Re} s < n/2$, then we have, for example,

$$\int_{x \in \mathbb{R}^n} G(s, \pi\, {}^t x x)\, dx = \frac{1}{\frac{n}{2} - s}.$$

This completes the proof. $\qquad \square$

When $n = 2$, Corollary 3 gives the orthogonality of the Eisenstein series and the constants in the spectral decomposition of the Laplacian on $L^2(H/SL(2, \mathbb{Z}))$

4.4. Fundamental Domains for $\mathscr{P}_n/GL(n, \mathbb{Z})$

(see Lemma 1 of Section 3.7 in Volume I). When $n = 2$ or 3 there are explicit criteria on Y which tell, for example, whether $Z(Y, (n-1)/2)$ is positive or negative (see Terras [9]). More general results, showing $Z(Y, nu/2) > 0$ if the first minimum $m_Y \leq nu/(2\pi e)$, for $u \in (0, 1)$ and n sufficiently large, can be found in Terras [10].

In our studies of Eisenstein series we will need more general integral formulas than the one given in Proposition 2. The following proposition gives an example which was stated by Siegel [1, Vol. III, p. 46].

Proposition 3 (Siegel's Integral Formula, Part II, A Generalization). *Let $G = SL(n, \mathbb{R})$, $\Gamma = SL(n, \mathbb{Z})$ and if $1 \leq k < n$, let $h: \mathbb{R}^{n \times k} \to \mathbb{C}$ be integrable. Then*

$$\text{Vol}(G/\Gamma)^{-1} \int_{G/\Gamma} \sum_{\substack{N \in \mathbb{Z}^{n \times k} \\ \text{rk}(N) = k}} h(gN)\, d\bar{g} = \int_{\mathbb{R}^{n \times k}} h(x)\, dx.$$

PROOF. We imitate the proof that we gave for Proposition 2. Let $H = H_k$ be the subgroup of G consisting of matrices of the form:

$$\begin{pmatrix} I_k & * \\ 0 & * \end{pmatrix},$$

and consider $f: G/H \to \mathbb{C}$. Then as before, we have the integral formula:

$$\int_{G/H} f(g)\, d\bar{g} = c \int_{G/\Gamma} \int_{a \in \Gamma/\Gamma \cap H} f(ga)\, d\bar{g}.$$

Note that

$$G/H \cong \{x \in \mathbb{R}^{n \times k} \mid \text{rank } x = k\}.$$

The complement of the set on the right in $\mathbb{R}^{n \times k}$ has measure 0. Note also that

$$\Gamma/\Gamma \cap H \cong \{A_1 \in \mathbb{Z}^{n \times k} \mid A_1 \text{ can be completed to a matrix in } \Gamma\}.$$

It follows that we have the formula below upon replacing $f(x)$ by $f(xB)$ with B in $\mathbb{Z}^{k \times k}$:

$$\int_{G/\Gamma} \sum_{(A*) \in \Gamma} f(gAB)\, d\bar{g} = c|B|^{-n} \int_{\mathbb{R}^{n \times k}} f(x)\, dx. \tag{4.53}$$

Now sum over B in $\mathbb{Z}^{n \times k}$ of rank k modulo $GL(k, \mathbb{Z})$; i.e., B in a complete set of representatives for the equivalence relation:

$$B \sim C \quad \text{iff} \quad B = CU, \quad \text{for some } U \text{ in } GL(k, \mathbb{Z}).$$

The matrices B are the right g.c.d.'s of the matrices AB essentially. Siegel has described the theory of matrix g.c.d.'s and other such concepts in his long paper on the theory of quadratic forms in Vol. I of his collected works (see Siegel [1, Vol I, pp. 331–332]). The generalization of concepts from number theory (such as that of class number) to simple algebras (like $\mathbb{Q}^{n\times n}$) is described in Deuring [1] and the end of Weil [5] to some extent.

We need several lemmas to continue the proof. First we want to show that for $\Gamma_k = GL(k, \mathbb{Z})$:

$$\sum_{A \in SL(n,\mathbb{Z})/H_k} \sum_{B \in \mathbb{Z}^{k\times k}/\Gamma_k} f(gA\,{}^tB) = \sum_{N \in \mathbb{Z}^{n\times k}\,\text{rk}\,k} f(gN). \tag{4.54}$$

Here H_k is the subgroup of $G = SL(n, \mathbb{R})$ consisting of matrices with block form

$$\begin{pmatrix} I_k & * \\ 0 & * \end{pmatrix}.$$

To prove (4.54) we need the following lemma.

Lemma 6 (A Decomposition for $n \times k$ Integral Rank k Matrices). *If $1 \leq k < n$ and $N \in \mathbb{Z}^{n\times k}$ has rank k, then N has the unique expression:*

$N = A\,{}^tB$, *with* $(A*)$ *in* $SL(n, \mathbb{Z})/H_k$, A *in* $\mathbb{Z}^{n\times k}$; B *in* $\mathbb{Z}^{k\times k}/GL(k, \mathbb{Z})$, $\text{rk}\,B = k$.

Here H_k is as above.

PROOF.

(a) *Existence of the Decomposition.*

Since $GL(n, \mathbb{Z})$ consists of matrices generated by those corresponding to elementary row and column operations (as described in the proof of Lemma 2 in Section 4.1), there is a diagonal integral $k \times k$ matrix D such that

$$N = U\begin{pmatrix} D \\ 0 \end{pmatrix}V, \quad \text{for } U \text{ in } GL(n, \mathbb{Z}), V \text{ in } GL(k, \mathbb{Z}).$$

Set $U = (A*)$, with A in $\mathbb{Z}^{n\times k}$. Note that by changing V we can put U in $SL(n, \mathbb{Z})$. Then

$$N = ADV \quad \text{and we can set} \quad {}^tB = DV.$$

Then reduce $B \bmod GL(k, \mathbb{Z})$ on the right and modify A to preserve the equality $N = A\,{}^tB$. This proves the existence of the decomposition.

(b) *Uniqueness of the Decomposition.*

Suppose that

$$A\,{}^tB = A'\,{}^tB'.$$

Then let $U = (A*)$, $U' = (A'*)$ be matrices in $SL(n, \mathbb{Z})$. It follows that

$$U\begin{pmatrix} {}^tB \\ 0 \end{pmatrix} = \begin{pmatrix} {}^tB' \\ 0 \end{pmatrix}.$$

Now write

4.4. Fundamental Domains for $\mathscr{P}_n/GL(n, \mathbb{Z})$

$$U'^{-1}U = \begin{pmatrix} P & Q \\ R & S \end{pmatrix}.$$

It follows that

$$P\,{}^tB = {}^tB' \quad \text{and} \quad R\,{}^tB = 0.$$

Since B is invertible, $R = 0$. So P is in $GL(k, \mathbb{Z})$ and then $B = B'$ so that $P = I$. But then

$$U'^{-1}U \text{ is in the subgroup } H_k$$

and then $(A'*)$ and $(A*)$ are equivalent modulo H_k. This completes the proof. $\square$

Next we want to prove a result which was mentioned already in Section 4.4.1, namely the factorization formula (4.3) for the zeta function of the simple algebra of $n \times n$ rational matrices.

Lemma 7 (A Factorization of the Analogue of the Dedekind Zeta Function for the Simple Algebra of All $n \times n$ Rational Matrices). *If $\operatorname{Re} s > k$, we have the following factorization of the matrix analogue of Riemann's zeta function into a product of ordinary Riemann zeta functions:*

$$\zeta_{\mathbb{Q}^{k \times k}}(s) = \sum_{B \in \mathbb{Z}^{k \times k} \text{ rk } k/\Gamma} |B|^{-s} = \prod_{j=0}^{k-1} \zeta(s-j).$$

Here $\Gamma = GL(k, \mathbb{Z})$.

PROOF. We need a system of representatives for the sum over the $k \times k$ integral matrices B having rank k modulo $GL(k, \mathbb{Z})$. That is we need a complete set of representatives for the equivalence relation:

$$B_1 \sim B_2 \quad \text{iff} \quad B_1 = B_2 g \quad \text{for } g \text{ in } GL(k, \mathbb{Z}). \tag{4.55}$$

The complete set of representatives we choose is the following set of upper triangular matrices:

$$\begin{pmatrix} d_{11} & & d_{ij} \\ & \ddots & \\ 0 & & d_{kk} \end{pmatrix}, \tag{4.56}$$

where $d_{ij} \in \mathbb{Z}$, for all i, j, $d_{ii} > 0$, for all i, and $0 \le d_{ij} < d_{ii}$.

To prove this, note that if B in $\mathbb{Z}^{k \times k}$ has rank k, then there exists a matrix U in $GL(k, \mathbb{Z})$ such that

$$BU = \begin{pmatrix} a & c \\ 0 & D \end{pmatrix}, \quad \text{with } a \in \mathbb{Z}^+, D \in \mathbb{Z}^{(k-1) \times (k-1)}.$$

To see this, you must solve $k - 1$ homogeneous linear equations in k unknowns. These equations can be solved with relatively prime integers.

Those integers can then be made the first column of a matrix in $GL(k, \mathbb{Z})$ by Lemma 2 from Section 4.4.1. Moreover, we can insure that c is reduced modulo a. Induction finishes the proof that we can choose the representatives for B as given above. And it is not hard to see that, in fact, any pair of distinct upper triangular matrices from (4.56) are inequivalent modulo $GL(n, \mathbb{Z})$ in the sense of (4.55).

But then

$$\zeta_{\mathbb{Q}^{k \times k}}(s) = \sum_{B \in \mathbb{Z}^{k \times k} \text{ rk } k/\Gamma} |B|^{-s} = \sum_{d_{ii} > 0} \prod_{i=1}^{k} d_{ii}^{k-i-s}.$$

The term on the right in this last formula is indeed the product of Riemann zeta functions and Lemma 6 is proved. □

Note. Formula (4.56) for a complete set of representatives of the equivalence relation defined on the rank k matrices in $\mathbb{Z}^{k \times k}$ by right multiplication by matrices in $GL(n, \mathbb{Z})$ is the generalization to $k \times k$ matrices of formula (3.87) of Section 3.6 of Volume I, a formula which was important in the theory of Hecke operators. We will use formula (4.56) again when we study Hecke operators for $GL(n)$ in the next section.

LAST STEP IN THE PROOF OF PROPOSITION 3. First note that Lemma 6 implies (4.54) and then (4.53) and (4.54) give:

$$\int_{G/\Gamma} \sum_{N \in \mathbb{Z}^{n \times k} \text{ rk } k} f(gN) \, d\bar{g} = c^* \int_{\mathbb{R}^{n \times k}} f(x) \, dx, \qquad (4.57)$$

where

$$c^* = \zeta_{\mathbb{Q}^{k \times k}}(n)c, \qquad V = \text{Vol}(G/\Gamma), \quad \text{and } \Gamma = SL(n, \mathbb{Z}).$$

Our problem is to determine the constant in Siegel's integral formula by showing that $c^* = V$. It is natural to try to imitate Weil's argument from the proof of Proposition 2. If we do that, our formula (4.57) plus Poisson summation yields:

$$(c^* - V)(\hat{f}(0) - f(0)) = \sum_{j=1}^{k-1} \int_{G/\Gamma} \sum_{N \in \mathbb{Z}^{n \times k} \text{ rk } j} (\hat{f}(gN) - f(gN)) \, dg.$$

The claim is that this is zero. But it is not at all obvious from staring at the right hand side—unless you see something I don't. Perhaps this will be a useful result when we try to analytically continue Eisenstein series using Riemann's method of theta functions and find that terms of lower rank give divergent integrals. Anyway that fact makes one worry slightly about this whole procedure, doesn't it? So, it seems better to try to find a new argument which does not use Poisson summation.

Let us imitate the argument given by Siegel in his original paper for the case $k = 1$. Note that by the definition of the integral:

4.4. Fundamental Domains for $\mathscr{P}_n/GL(n,\mathbb{Z})$

$$\int_{\mathbb{R}^{n\times k}} f(x)\,dx = \lim_{h\to 0+}\left\{h^{nk}\sum_{\substack{N\in\mathbb{Z}^{n\times k}\\ \operatorname{rk} k}} f(hgN)\right\}.$$

To see this, think of what happens if you multiply the points N in the lattice $\mathbb{Z}^{n\times k}$ by a small positive number h. You will get points in a grid with a very small mesh such that each individual hypercube has volume h^{nk}.

Integrate the preceding formula over $\bar{g}$ in G/Γ and obtain:

$$V\int_{\mathbb{R}^{n\times k}} f(x)\,dx = \lim_{h\to 0+}\left\{h^{nk}\int_{G/\Gamma}\sum_{\substack{N\in\mathbb{Z}^{n\times k}\\ \operatorname{rk} k}} f(hgN)\,d\bar{g}\right\}$$

$$= \lim_{h\to 0+} h^{nk}c^*\int_{\mathbb{R}^{n\times k}} f(hx)\,dx, \quad \text{using (4.57),}$$

$$= \lim_{h\to 0+} c^*\int_{\mathbb{R}^{n\times k}} f(x)\,dx, \quad \text{as } |d(hx)/dx| = h^{nk}.$$

This completes the proof that $c^* = V$ and thus finishes (at last) the proof of Siegel's 2nd integral formula. □

Next we combine the integral formulas of Siegel (Proposition 3) and Wishart (which was formula (2.57) in Section 4.2.4).

Corollary (Siegel and Wishart).

$$\operatorname{Vol}(\mathscr{S}\mathscr{P}_n/GL(n,\mathbb{Z}))^{-1}\int_{\mathscr{S}\mathscr{P}_n/GL(n,\mathbb{Z})} \sum_{\substack{N\in\mathbb{Z}^{n\times k}\\ \operatorname{rank}(N)=k}} f(W[N])\,dW$$

$$= c_{n,k}\int_{\mathscr{P}_k} f(Y)|Y|^{n/2}\,d\mu_k.$$

Here $d\mu_k$ is the invariant volume element on $\mathscr{P}_k$ and $c_{n,k}$ is the constant in Wishart's integral formula; i.e.,

$$c_{n,k} = \prod_{j=n-k+1}^{n} \pi^{j/2}\Gamma(j/2)^{-1}.$$

PROOF. Use an argument similar to that given in the corollary to Proposition 2. □

Exercise 23 (A Generalization of the Minkowski-Hlawka Theorem). For $Y\in \mathscr{P}_n$ define the following generalization of the first minimum m_Y in (4.7) if $1 \le k \le n$ by:

$$m_{Y,k} = \min\{|Y[A]|\,|\, A\in\mathbb{Z}^{n\times k},\ \operatorname{rank} A = k\}.$$

Prove that if k is fixed $1 \le k < n$, n is sufficiently large (depending on k), and $r < (n/(2\pi e))^k$, there exists a matrix $Y\in\mathscr{P}_n$ such that

$$m_{Y,k} > r|Y|^{k/n}.$$

Hint. Imitate the proof of Corollary 1 of Proposition 2. More details can be found in Terras [1, pp. 478–479].

Exercise 24 (Convergence of Koecher's Zeta Function by an Integral Test). Show that Koecher's zeta function $Z_{k,n-k}(Y,s)$ defined by formula (4.2) in Section 4.4.1 will converge absolutely for $\text{Re } s > n/2$. Imitate the proof of Corollary 2 of Proposition 2.

Hint. The only new idea that is required is the following. There is a positive constant c such that if $I[A]$ is in Minkowski's fundamental domain $\mathcal{M}_k$, then $|Y[A]| \leq c|I[A]|$.

Similarly, there is a positive constant c^* such that if $Y[A] \in \mathcal{M}_k$ then $|Y[A]| \geq c^*|I[A]|$. Here c and c^* depend on Y and not on A. You also need to know that the set of matrices $A \in \mathbb{Z}^{n \times k}$ modulo $GL(k, \mathbb{Z})$ such that $|I[A]| \leq 1$ is finite. This follows from Exercise 8(a) in Section 4.4.2.

Koecher's zeta function is another Eisenstein series for the general linear group and thus its analytic continuation is of interest to us. We will consider that problem later and find that our integral formulas are useful in this regard. The analytic continuation of Eisenstein series for $GL(n)$ is much harder than that of Epstein's zeta function.

But we will want to consider more general Eisenstein series than Koecher's zeta function. Such Eisenstein series are associated to parabolic subgroups of $GL(n)$. Suppose that

$$n = n_1 + \cdots + n_q, \quad \text{with } n_i \in \mathbb{Z}^+,$$

is a *partition* of n. Then the (standard) *parabolic subgroup* $P = P(n_1, \ldots, n_q)$ of $GL(n)$ is defined to be the group of matrices U with block form:

$$\begin{pmatrix} U_1 & & * \\ & \ddots & \\ 0 & & U_q \end{pmatrix}, \quad \text{with } U_j \text{ in } GL(n_j). \tag{4.58}$$

Koecher's zeta function is an Eisenstein series associated to a maximal parabolic subgroup (the case $q = 2$). When $n = 2$ there is only one such standard parabolic subgroup, but for general n there are many such subgroups—as many as there are partitions of n, a number denoted by $p(n)$. The partition function $p(n)$ has been much studied by number theorists. It is a very rapidly increasing function of n. Some examples are:

$$p(10) = 42, \quad p(100) = 190{,}569{,}292, \quad p(200) = 3{,}972{,}999{,}029{,}388.$$

These are asymptotic and exact formulas for $p(n)$ when n is large, thanks to the work of Rademacher, Hardy, Littlewood, and Ramanujan, as well as the fact that $p(n)$ is the nth Fourier coefficient of a modular form of weight $-\tfrac{1}{2}$, namely $\eta(z)^{-1}$ from formula (3.56) in Section 3.4 of Volume I.

We will develop one version of the general integral formula in some exercises.

4.4. Fundamental Domains for $\mathscr{P}_n/GL(n, \mathbb{Z})$

Exercise 25 (Siegel's Integral Formula for a Maximal Parabolic Subgroup $P(k, n - k)$).

(a) Suppose that $P = P(k, n - k)$ is the maximal parabolic subgroup defined in (4.58) and that $f: \mathscr{P}_k/\Gamma_k \to \mathbb{C}$, with $\Gamma_k = GL(k, \mathbb{Z})$. Show that

$$\int_{\mathscr{S}\mathscr{M}_n} \sum_{\substack{(A*) \in \Gamma_n/P \\ A \in \mathbb{Z}^{n \times k}}} f(W[A]) \, dW = \text{Vol}(\mathscr{S}\mathscr{M}_{n-k}) \int_{\mathscr{M}_k} f(X) |X|^{n/2} \, d\mu_k(X).$$

(b) Assuming the necessary integrability conditions on the following functions:

$$f_1: \mathscr{P}_k/\Gamma_k \to \mathbb{C}, \qquad f_2: \mathbb{R}^+ \to \mathbb{C},$$

show that if $f(Y) = f_1(Y_1) f_2(|Y|)$ for

$$Y = \begin{pmatrix} Y_1 & * \\ * & * \end{pmatrix}, \qquad \text{for } Y_1 \in \mathscr{P}_k,$$

$$\int_{\mathscr{M}_n} \sum_{\substack{(A*) \in \Gamma_n/P \\ A \in \mathbb{Z}^{n \times k}}} f(Y[A]) \, d\mu_n(Y)$$

$$= \text{Vol}(\mathscr{S}\mathscr{M}_{n-k}) \int_{t>0} f_2(t) t^{-k/2 - 1} \, dt \int_{\mathscr{M}_k} f_1(X) |X|^{n/2} \, d\mu_k(X).$$

Hints.
(a) Summing over $(A*) \in GL(n, \mathbb{Z})/P(k, n - k)$ is the same as summing over $A \mod GL(k, \mathbb{Z})$, where $A \in \mathbb{Z}^{n \times k}$ and A fits into a matrix in $GL(n, \mathbb{Z})$. So we find from similar arguments to those that gave us the corollary to Proposition 3 and Lemma 7 that:

$$\int_{\mathscr{S}\mathscr{M}_n} \sum_{\substack{(A*) \in \Gamma_n/P \\ A \in \mathbb{Z}^{n \times k}}} f(W[A]) \, dW = c \int_{\mathscr{M}_k} f(X) |X|^{n/2} \, d\mu_k(X).$$

Here

$$c = c_{n,k} \text{Vol}(\mathscr{S}\mathscr{M}_m) \prod_{j=0}^{k-1} \zeta(n - j)^{-1},$$

where $c_{n,k}$ is the constant in Wishart's integral formula (see the corollary to Proposition 3). Now the formula for $\text{Vol}(\mathscr{S}\mathscr{M}_n)$ in Theorem 2 and the formula for $c_{n,k}$ in the corollary to Proposition 3 finish this part of the exercise.
(b) Start with part (1) of the exercise and replace $f(X)$ by $f_2(t) f_1(t^{1/n} X)$. Then integrate with respect to $t^{-1} dt$ over $t > 0$ to get the result.

Exercise 26 (An Integral Formula for an Arbitrary Parabolic Subgroup). Suppose that

$$f(Y) = \prod_{j=1}^{q} f_j(|Y_j|), \qquad \text{where } Y_j \in \mathscr{P}_{N_j}, \qquad N_j = n_1 + \cdots + n_j,$$

and

$$Y = \begin{pmatrix} Y_j & * \\ * & * \end{pmatrix}.$$

Prove that if $\Gamma_n = GL(n, \mathbb{Z})$ and $P = P(n_1, \ldots, n_q)$, then

$$\int_{\mathcal{M}_n} \sum_{A \in \Gamma_n/P} f(Y[A]) \, d\mu_n(Y)$$

$$= V_{n_q} \int_{t_q > 0} f_q(t_q) t_q^{(-k/2)-1} \, dt_q \prod_{j=1}^{q-1} V_{n_j} \int_{t_j > 0} f_j(t_j) t_j^{(e_j/2)-1} \, dt_j,$$

where $V_i = \text{Vol}(\mathcal{SM}_{n_i})$, $k = N_{q-1}$, and $e_j = n_j + n_{j+1}$.

Hint. The case $q = 2$ comes from part (b) of Exercise 25. To prove the general result, use induction on q and write $A \in GL(n, \mathbb{Z})/P$, as

$A = BC$, with $B = (B_1 *) \in GL(n, \mathbb{Z})/Q$, $B_1 \in \mathbb{Z}^{n \times k}$, $Q = P(k, n_q)$, $k = N_{q-1}$,

$$C = \begin{pmatrix} D & * \\ 0 & * \end{pmatrix} \in Q/P, \quad D \in GL(k, \mathbb{Z})/P^*, \quad P^* = P(n_1, \ldots, n_{q-1}).$$

Note that if $A = (A_1 *)$ with $A_1 \in \mathbb{Z}^{n \times k}$, then $A_1 = B_1 D$. Thus

$$\int_{\mathcal{M}_n} \sum_{A \in \Gamma_n/P} f(Y[A]) \, d\mu_n(Y)$$

$$= V_{n_q} \int_{t_q > 0} f_q(t_q) t_q^{(-k/2)-1} \, dt_q \int_{\mathcal{M}_k} f_{q-1}(|Y|) |Y|^{n/2}$$

$$\times \sum_{D \in \Gamma_k/P^*} \prod_{j=1}^{q-1} f_j(|(Y[D])_j|) \, d\mu_k(Y).$$

The proof is completed by induction.

Exercise 27 (Another Integral Formula for Arbitrary Parabolic Subgroups). Suppose that Y has the partial Iwasawa decomposition:

$$Y = \begin{pmatrix} V_1 & & 0 \\ & \ddots & \\ 0 & & V_q \end{pmatrix} \left[\begin{pmatrix} I_{n_1} & & R_{ij} \\ & \ddots & \\ 0 & & I_{n_q} \end{pmatrix} \right], \quad \text{with } V_j \in \mathcal{P}_{n_j}, R_{ij} \in \mathbb{R}^{n_i \times n_j},$$

and consider a function $g(Y) = h(V_1, \ldots, V_q)$, satisfying suitable integrability conditions. Show that if $P = P(n_1, \ldots, n_q)$ and $\Gamma = GL(n, \mathbb{Z})$, then:

$$\int_{\mathcal{M}_n} \sum_{A \in \Gamma/P} g(Y[A]) \, d\mu_n(Y)$$

$$= \prod_{j=1}^{q} \text{Vol}(\mathcal{SM}_{n_j}) \int_{V_j \in \mathcal{M}_{n_j}} h(V_1, \ldots, V_q) |V_j|^{f_j} \, d\mu_{n_j}(V_j),$$

where $f_j = (n - N_j - N_{j-1})/2$.

Hint. Actually you should not try to compute the constant in this formula until later. Up to the computation of the constant, the formula follows from:

$$\int_{\mathscr{P}_n/\Gamma} \sum_{\Gamma/P} = c \int_{\mathscr{P}_n/P}$$

and the Jacobian of the partial Iwasawa decomposition (see Maass [2, pp. 149–150] or Varadarajan [2, p. 293]):

$$d\mu_n(Y) = \prod_{j=1}^{q} |V_j|^{f_j} d\mu_{n_j}(V_j) \prod_{1 \le i \le k \le q} dR_{ik},$$

where f_j is as given in the problem and dR_{ij} is ordinary Lebesgue measure on $n_i \times n_j$ matrix space.

The preceding exercises are based on what can be viewed as an analogue of the integral formula involved in the Rankin-Selberg method (i.e., formula (3.97) in Section 3.6 of Volume I).

4.5. Automorphic Forms for $GL(n, \mathbb{Z})$ and Harmonic Analysis on $\mathscr{P}_n/GL(n, \mathbb{Z})$

... and the manuscript was becoming an albatross about my neck. There were two possibilities: to forget about it completely, or to publish it as it stood; and I preferred the second.

From Langlands [1, Preface]

4.5.1. Analytic Continuation of Eisenstein Series by the Method of Inserting Larger Parabolic Subgroups

In order to do harmonic analysis on $\mathscr{P}_n/GL(n, \mathbb{Z})$, as in the case $n = 2$, we need to study automorphic forms for $\Gamma_n = GL(n, \mathbb{Z})$ and, in particular, obtain the analytic continuation of the Eisenstein series which form the continuous spectrum of the $GL(n, \mathbb{R})$-invariant differential operators on the fundamental domain. Selberg [1], [3], [4] had already noticed this in the 1950's and the methods we will develop are probably similar to the unpublished methods of Selberg. We will discuss two of Selberg's methods in this and the next subsection. These methods are also discussed by Maass [2]. There is a third unpublished method of Selberg which makes more use of functional analysis (see Wong [1]). For other points of view, see the books by Langlands [1] and Osborne and Warner [1] (cf. Langlands [2]) which discuss the analytic continuation of more general Eisenstein series.

We define an *automorphic form v* for $GL(n, \mathbb{Z}) = \Gamma$ to be a function $v \colon \mathscr{P}_n \to \mathbb{C}$ such that:

(1) v is an eigenfunction of all the invariant differential
operators L in $D(\mathscr{P}_n)$; i.e., $Lv = \lambda_L v$, for some eigenvalue λ_L;
(2) v is Γ-invariant; i.e., $v(Y[A]) = v(Y)$ for all $Y \in \mathscr{P}_n$, $A \in \Gamma$;   (5.1)
(3) v has at most polynomial growth at infinity; i.e.,

$$|v(Y)| \leq C|p_s(Y)| \qquad \text{for some } s \in \mathbb{C}^n \text{ and } C > 0.$$

We shall use the notation $\mathscr{A}(\Gamma, \lambda)$ for the space of automorphic forms for a given eigenvalue system λ.

Definition (5.1) is clearly a generalization of the concept of Maass wave form which appeared in Section 3.5 of Volume I. Maass considers these automorphic forms in [2, Section 10] and he calls them "grossencharacters." That name can be explained by the fact that Hecke grossencharacters play the same role in harmonic analysis for $GL(2)$ over a number field that forms in $\mathscr{A}(\Gamma, \lambda)$ play for harmonic analysis on $\mathscr{P}_n/\Gamma$ (see Hecke [1, pp. 215–234, 249–287], Jacquet and Langlands [1], Stark [1], and Weil [4]).

Motivated by the study of representations of semi-simple Lie groups, Harish-Chandra has given a much more general definition of automorphic form (see Borel's lecture in Borel and Mostow [1, pp. 199–210]). This definition includes (5.1) above as well as the concept of Siegel modular form for $Sp(n, \mathbb{Z})$ which was introduced by C.L. Siegel in his work on quadratic forms (see Chapter 5).

Another motivation for the study of automorphic forms for $GL(n)$ is the need to study various kinds of L-functions with many gamma factors in their functional equations. For L-functions corresponding to automorphic forms for $GL(n)$ will indeed have lots of gamma factors and will have an Euler product if they are eigenforms for the algebra of Hecke operators for $GL(n)$. See Section 4.5.2 which follows or Bump [1]. An adelic treatment of this subject is part of the Langlands theory (see Gelbart [2], Godement and Jacquet [1], or Jacquet, Piatetski-Shapiro, and Shalika [1]). Langlands has conjectured that there is a reciprocity law generalizing the Artin reciprocity law which says that each Artin L-function corresponding to an n-dimensional representation of a Galois group of an extension of number fields is the L-function for some automorphic representation of $GL(n)$. See Langlands [3], [4], [5], [7], Arthur [3], Casselman [1], and Gelbart [2]. In fact, Artin L-functions do have functional equations involving multiple gamma functions as well as Euler products and this certainly gives good evidence for Langlands' conjecture. There are other sorts of L-functions with Euler products and multiple gamma factors in their functional equations—the analogues of the Rankin-Selberg L-functions studied in formula (3.95) of Volume I. These Rankin-Selberg type L-functions have applications to the problem to prove a Ramanujan-Petersson conjecture for cusp forms for $GL(n, \mathbb{Z})$ (cf. Vol. I, formula (3.72) for the case $n = 2$). Some references are: Bump and Friedberg [1],

4.5. Automorphic Forms for $GL(n,\mathbb{Z})$ and Harmonic Analysis on $\mathscr{P}_n/GL(n,\mathbb{Z})$

Elliott, Moreno and Shahidi [1], Friedberg [1], Jacquet and Shalika [1], Jacquet, Piatetski-Shapiro and Shalika [2], Moreno and Shahidi [1], [2], Novodvorsky and Piatetski-Shapiro [1], Piatetski-Shapiro [1], Shahidi [1].

In the last section we needed to go backwards in time to commune with Minkowski. The present section unfortunately demands a time machine that will carry us into the future. Lacking this item, the section will be incomplete.

References for this section include Arthur [1–3], Ash [1], Baily [1], Borel and Casselman [1], Borel and Mostow [1], Bump [1], Casselman [1], Flicker [1], Gelbart [2], Gelbart and Jacquet [1], Gelfand, Graev, and Piatetski-Shapiro [1], Godement [3], [4], Godement and Jacquet [1], Harish-Chandra [1], Hejhal, Sarnak and Terras [1], Jacquet [1], Jacquet, Piatetski-Shapiro, and Shalika [1], Jacquet and Shalika [1–2], Kazhdan and Patterson [1], Langlands [1–7], Maass [2], [5], Ramanathan [1,2], Selberg [1], [3–4], Tamagawa [1–3], A.B. Venkov [1], and Dorothy Wallace (Andreoli) [1–6].

In earlier sections, we saw how to build up the eigenfunctions of the G-invariant differential operators in $D(\mathscr{P}_n)$ by integrating power functions over orthogonal, abelian and nilpotent subgroups of $GL(n,\mathbb{R})$, as in formulas (2.20), (2.26), and (2.28) of Section 4.2. The powers s in $p_s(Y)$ provide a way of indexing the eigenvalues of an invariant differential operator $L \in D(\mathscr{P}_n)$ via $Lp_s(Y) = \lambda_L(s)p_s(Y)$. We shall use this sort of indexing when we speak of the dimensionality of the spectrum components. For inversion of the Helgason-Fourier transform on $\mathscr{P}_n$, the spectrum needed was n-dimensional (see Theorem 1 of Section 4.3.1). The inverse transform required integration over a product of n lines: $\mathrm{Re}\,s_j = -\frac{1}{2}$, $j = 1, 2, \ldots, n-1$, and $\mathrm{Re}\,s_n = (n-1)/4$. We shall see that life is much more complicated in $\mathscr{P}_n/GL(n,\mathbb{Z})$, since there are also discrete and lower dimensional spectra. However, the basic method of constructing $GL(n,\mathbb{Z})$-invariant eigenfunctions in the highest dimensional part of the spectrum is analogous to the construction of spherical and K-Bessel functions. That is, one must sum power functions over $GL(n,\mathbb{Z})$ modulo a parabolic subgroup. But it is not so simple for general n as it was for $n = 2$ in Section 3.7 of Volume I. In the following discussion, we will sometimes consider only the case of $GL(3,\mathbb{Z})$ in order to simplify the formulas.

Before defining Eisenstein series, we need to consider another sort of automorphic form for $GL(n,\mathbb{Z})$—the cusp form. A *cusp form* is an automorphic form $f \in \mathscr{A}(\Gamma, \lambda)$, $\Gamma = GL(n,\mathbb{Z})$, with the property that for any k with $1 \leq k \leq n - 1$, we have:

$$\int_{X \in (\mathbb{R}/\mathbb{Z})^{k \times (n-k)}} f\left(Y \begin{bmatrix} I & X \\ 0 & I \end{bmatrix}\right) dX = 0, \quad \text{for all } Y \in \mathscr{P}_n.$$

This just signifies the vanishing of the constant terms in a bunch of Fourier expansions of $f(Y)$ as a periodic function of the X-variable in partial Iwasawa coordinates (see page 213 of Volume I). If we knew enough about Fourier expansions for $GL(n)$, we should be able to show that a cusp form is bounded in the fundamental domain. See the article of Borel and Jacquet in the volume of Borel and Casselman [1, p. 192] or see Harish-Chandra [1]. More information on Fourier expansions can be found in Section 4.5.3 or Bump [1].

The cusp forms and constants form the discrete spectrum of the $GL(n, \mathbb{R})$-invariant differential operators on the fundamental domain. It can be proved, using a method of Gelfand and Piatetski-Shapiro, that cusp forms exist (cf. Section 4.5.4 and Theorem 2 in Section 3.7 of Volume I or Godement's article in Borel and Mostow [1, pp. 225–234]). We will not be able to give any examples of cusp forms, just as we could not give any examples for the case $n = 2$ in Section 3.5 of Volume I. However, there are adelic examples of cusp forms belonging to congruence subgroups of $GL(3, \mathbb{Z})$ corresponding via generalizations of Hecke's correspondence to Hecke L-functions of cubic number fields (see Jacquet, Piatetski-Shapiro, and Shalika [1]). There are also cuspidal examples corresponding via Hecke theory to Rankin-Selberg L-functions for $GL(2, \mathbb{Z})$ (see Gelbart and Jacquet [1], Moreno and Shahidi [1]). Ash [1] and Ash, Grayson, and Green [1] compute cohomology of $SL(3, \mathbb{Z})$ using Hecke operators and methods of algebraic topology and differential geometry. They show the existence of cusp forms for $SL(3, \mathbb{Z})$ which come from the DeRham cohomology of the fundamental domain and are analogous to holomorphic automorphic forms of weight 2. See also Lee and Schwermer [1], Lee and Szczarba [1], Schwermer [1], [2], and Soulé [1], [2]. Donnelly [1] finds an upper bound for the dimension of the space of cusp forms.

We will not say much more about cusp forms. Most of our attention will be directed at Eisenstein series. We will consider the methods that Selberg and Maass used to continue these series. Before giving our first definition of Eisenstein series, we need to recall some notation concerning the determinant one surface $\mathcal{SP}_n$ in $\mathcal{P}_n$. If $Y \in \mathcal{P}_n$, we write

$$Y = t^{1/n} W, \qquad t = |Y| > 0, \qquad W = Y^0 \in \mathcal{SP}_n. \tag{5.2}$$

Clearly, one can define automorphic forms on $\mathcal{SP}_n/\Gamma_n$, $\Gamma_n = GL(n, \mathbb{Z})$, as in (5.1). We will denote these spaces $\mathcal{A}^0(\Gamma_n, \lambda)$ to indicate that the functions have domain the determinant one surface in $\mathcal{P}_n$. See Exercise 1 below for the relationship between the Laplacian on $\mathcal{P}_n$ and the Laplacian on $\mathcal{SP}_n$.

Now we give our first definition of Eisenstein series. Suppose that v is an automorphic form on a lower rank determinant one surface; i.e. let

$$v \in \mathcal{A}^0(GL(m, \mathbb{Z}), \lambda), \qquad 1 \leq m < n, \qquad \operatorname{Re} s > n/2 \quad \text{and} \quad Y \in \mathcal{P}_n.$$

Then we define the *Eisenstein series* $E_{m, n-m}(v, s | Y) = E(v, s | Y)$ with lower rank automorphic form v by:

$$E(v, s | Y) = \sum_{A = (A_1 *) \in \Gamma_n/P(m, n-m)} |Y[A_1]|^{-s} v(Y[A_1]^0). \tag{5.3}$$

Here $A_1 \in \mathbb{Z}^{n \times m}$, $\Gamma_n = GL(n, \mathbb{Z})$, and $P(m, n - m)$ is the parabolic subgroup defined in formula (4.58) of the preceding section.

Note that Exercise 24 of Section 4.4.4 says that the series in (5.3) converges whenever v is bounded. To see that the series in (5.3) converges when v is an

4.5. Automorphic Forms for $GL(n, \mathbb{Z})$ and Harmonic Analysis on $\mathscr{P}_n/GL(n, \mathbb{Z})$ 185

integrable function on the fundamental domain $\mathscr{S}\mathscr{M}_m \cong \mathscr{S}\mathscr{P}_m/GL(m, \mathbb{Z})$, use the integral formula of Exercise 25 in Section 4.4.4 and imitate Exercise 24 of that same section. We can thus compare the series (5.3) with the integral:

$$\int_{\mathscr{S}\mathscr{M}_n} \sum_{\substack{(A*) \in \Gamma_n/P \\ |Y[A]| \geq 1}} f(Y[A]) \, d\mu_n(Y)$$

$$= \text{Vol}(\mathscr{S}\mathscr{M}_{n-m}) \int_{X \in \mathscr{M}_m, |X| \geq 1} f(X) |X|^{n/2} \, d\mu_m(X),$$

with $f(X) = |X|^{-s} v(X^0)$. Then formula (4.40) in the preceding section says this last integral is:

$$\text{Vol}(\mathscr{S}\mathscr{M}_{n-m}) \int_{X \in \mathscr{S}\mathscr{M}_m} v(W) \, dW \int_{t \geq 1} t^{n/2 - s - 1} \, dt,$$

which converges for $\text{Re } s > n/2$, if v is integrable on the fundamental domain $\mathscr{S}\mathscr{M}_m$. One can obtain a similar domain of convergence assuming that v is bounded by a power function.

Exercise 1. Why is the Eisenstein series defined in (5.3) an eigenfunction of all the $GL(n, \mathbb{R})$-invariant differential operators in $D(\mathscr{P}_n)$? Compute the eigenvalue of the Laplacian acting on (5.3) as a function of s and of the eigenvalue of the Laplacian acting on v.

Hint. Maass [2, p. 73] gives formulas for the Laplacian in partial Iwasawa coordinates:

$$Y = \begin{pmatrix} F & 0 \\ 0 & H \end{pmatrix} \begin{bmatrix} I & X \\ 0 & I \end{bmatrix}, \quad F \in \mathscr{P}_m, \quad H \in \mathscr{P}_{n-m}, \quad X \in \mathbb{R}^{m \times (n-m)}.$$

Maass finds that for functions of the form $u(Y) = f(F)h(H)$ the invariant differential operators look like:

$$\text{Tr}\left(\left(Y \frac{\partial}{\partial Y}\right)^k\right) = \text{Tr}\left(\left(F \frac{\partial}{\partial F} + \frac{n-m}{2} I\right)^k\right) + \text{Tr}\left(\left(H \frac{\partial}{\partial H}\right)^k\right)$$

$$- \frac{1}{2} \sum_{j=1}^{k-1} \text{Tr}\left(\left(F \frac{\partial}{\partial F} + \frac{n-m}{2} I\right)^j\right) \text{Tr}\left(H \frac{\partial}{\partial H}\right)^{k-1-j}.$$

We proved the special case $m = 1$ in formula (2.14) of Section 4.2.1. Now in the case under consideration in this exercise our function is

$$u(Y) = |F|^{-s} v(F^0), \quad \text{where } F^0 = |F|^{-1/m} F \in \mathscr{S}\mathscr{P}_m.$$

So the only term of interest in the formula for the Laplacian in partial Iwasawa coordinates is the first term.

One must also relate the Laplacian on the space $\mathscr{P}_m$ and that on the determinant one surface $\mathscr{S}\mathscr{P}_m$. Writing $Y = t^{1/m} W$, for $Y \in \mathscr{P}_m$, $t = |Y| > 0$,

$W \in \mathcal{SP}_m$, one can show that $\Delta_Y = m(t\partial/\partial t)^2 + \Delta_W$, where Δ_W is the Laplacian on the determinant one surface induced by the arc length $ds_W^2 = \text{Tr}((W^{-1}\,dW)^2)$.

When the automorphic form v is identically equal to one, the Eisenstein series (5.3) is a quotient of Koecher zeta functions defined in formula (4.2) of the preceding section; i.e.,

$$Z_{m,n-m}(Y, s) = Z_{m,0}(I, s) E(1, s | Y). \tag{5.4}$$

To see this, we need the following lemma.

Lemma 1 (A Decomposition of the Matrices in the Sums Defining Eisenstein Series). *The quotient $\mathbb{Z}^{n \times m}$ rank m/Γ_n, which means the $n \times m$ rank m integral matrices in a complete set of matrices inequivalent under right multiplication by matrices in $\Gamma_m = GL(m, \mathbb{Z})$, for $1 \leq m < n$, can be represented by matrices $A = BC$, where*

$$B \in \mathbb{Z}^{n \times m}, \qquad (B*) \in GL(n, \mathbb{Z})/P(m, n - m),$$
$$C \in \mathbb{Z}^{m \times m} \text{ rank } m/GL(m, \mathbb{Z}).$$

PROOF. First note that by elementary divisor theory, there are matrices $U \in \Gamma_n$ and $V \in \Gamma_m$ such that

$$A = U \begin{pmatrix} D \\ 0 \end{pmatrix} V, \qquad \text{with } D \text{ diagonal } m \times m \text{ and nonsingular}.$$

Suppose that $U = (U_1 *)$ with $U_1 \in \mathbb{Z}^{n \times m}$. Then $A = U_1 DV$ and U_1 may be taken modulo $GL(m, \mathbb{Z})$, by throwing the difference into DV. The existence of the stated decomposition follows quickly.

To see the uniqueness of the decomposition, suppose

$$B_2 C_2 = B_1 C_1 W, \qquad \text{for some } W \in \Gamma_n$$

with

$$V_i = (B_i *) \in \Gamma_n/P(m, n - m) \quad \text{and} \quad C_i \in \mathbb{Z}^{m \times m} \text{ rank } m/\Gamma_m, \; i = 1, 2.$$

Then

$$V_2 \begin{pmatrix} C_2 \\ 0 \end{pmatrix} = V_1 \begin{pmatrix} C_1 W \\ 0 \end{pmatrix} \text{ implies } V_1^{-1} V_2 \begin{pmatrix} C_2 \\ 0 \end{pmatrix} = \begin{pmatrix} C_1 W \\ 0 \end{pmatrix}.$$

Let

$$Z = V_1^{-1} V_2 \in \Gamma_n \quad \text{and} \quad Z = \begin{pmatrix} X & * \\ Y & * \end{pmatrix},$$

so that $XC_2 = C_1 W$, $YC_2 = 0$, which means that Y must vanish. Then the inequivalence of V_2 and V_1 modulo $P(m, n - m)$ implies that $X = I$. The inequivalence of C_1 and C_2 implies that $C_2 = C_1$ and $W = I$, to complete the proof of the lemma. □

4.5. Automorphic Forms for $GL(n, \mathbb{Z})$ and Harmonic Analysis on $\mathscr{P}_n/GL(n, \mathbb{Z})$

Formula (5.4) follows immediately from Lemma 1. After we have studied Hecke operators for $GL(n, \mathbb{Z})$, we will be able to prove a similar result for a general Eisenstein series of the form (5.3) (see formula (5.35) below). Recall that formula (4.3) (proved in Lemma 7 of the previous section) expressed Koecher's zeta function $Z_{m,0}(I, s)$ as a product of Riemann zeta functions. Thus formula (5.4) is quite analogous to the formula in part (c) of Exercise 1 in Section 3.5 of Volume I—a result which writes the Eisenstein series for $GL(2, \mathbb{Z})$ as a quotient of Epstein's zeta function divided by Riemann's zeta function.

Exercise 2. In formula (5.3) for the Eisenstein series $E_{m,n-m}(s, \varphi | Y)$, let $n = 3$, $m = 2$, and $\varphi \in \mathscr{A}^0(GL(2, \mathbb{Z}), \lambda = u(u - 1))$; i.e., φ is a Maass wave form for $GL(2, \mathbb{Z})$ as in Section 3.5 of Volume I. Suppose, in particular, that φ is an Eisenstein series: $\varphi(z) = E_u(z)$, $z \in H$, defined in Equation (3.66) of Volume I. Recall the standard identification of the upper half plane H with $\mathscr{PP}_2$ given in Exercise 1 of Section 3.5 of Volume I. Then show that

$$E(E_u, s | Y) = \sum_{\substack{(C_i*) = C \in \Gamma_3/P_{(3)} \\ C_i \in \mathbb{Z}^{3 \times i}}} Y[C_1]^{-u} | Y[C_2]|^{-s+u/2},$$

where $P_{(3)} = P(1, 1, 1)$ is the minimal parabolic subgroup of $\Gamma_3 = GL(3, \mathbb{Z})$ consisting of all upper triangular matrices with ± 1 on the diagonal (see formula (4.58) of the preceding section). Use an integral test argument similar to the ones given after formula (5.3) above and Exercise 26 of the preceding section to see that the series on the right hand side of this formula converges for $\operatorname{Re} u > 1$ and $\operatorname{Re}(s - u/2) > 1$. Note also that E_u is integrable if $0 < \operatorname{Re} u < 1$.

Hint. We can write the sum over $\Gamma_3/P(1, 1, 1)$ in the right hand side of the formula to be demonstrated as a double sum over $\Gamma_3/P(2, 1)$ and over $P(2, 1)/P(1, 1, 1)$. The latter sum can be identified as a sum over matrices of block form:

$$\begin{pmatrix} \Gamma_2/P(1, 1) & 0 \\ 0 & \pm 1 \end{pmatrix}.$$

This shows that the right hand side of the formula looks like:

$$\sum_{(A_1*) \in \Gamma_3/P(2, 1)} |Y[A_1]|^{-s+u/2} \sum_{(b*) \in \Gamma_2/P(1, 1)} Y[A_1][b]^{-u},$$

which is $E(E_u, s | Y)$.

Motivated somewhat by the preceding exercise, we define *Selberg's Eisenstein Series for a Parabolic Subgroup* $P = P(n_1, \ldots, n_q)$ to be a function of $Y \in \mathscr{P}_n$ and $s \in \mathbb{C}^q$ given by:

$$E_P(s | Y) = \sum_{\substack{(A_j*) = A \in \Gamma_n/P \\ A_j \in \mathbb{Z}^{n \times N_j}}} \prod_{j=1}^{q} |Y[A_j]|^{-s_j} \qquad (5.5)$$

with $N_j = n_1 + \cdots + n_j$. We will also write
$$E_{n_1,\ldots,n_q}(s|Y) = E_P(s|Y).$$
The integral test coming from Exercises 26 and 27 of the preceding section generalizes to show that the series above converges absolutely for
$$\operatorname{Re} s_j > (n_j + n_{j+1})/2, \quad j = 1, 2, \ldots, q-1.$$

Exercise 3. Prove the last statement about the region of convergence of the series (5.5).

Since Selberg proved all the basic properties of the functions (5.5), there is good reason for calling $E_P(s|Y)$ "Selberg's Eisenstein series." Some authors (e.g. Maass [2] and Christian [2]) call (5.5) "Selberg's zeta function." This is confusing since there is another function which has been given that name (namely, the zeta function in formula (3.150) of Volume I).

It is also possible to create Eisenstein series involving automorphic forms $v_j \in \mathscr{A}^0(GL(n_j, \mathbb{Z}), \lambda_j)$ by summing a function $f(Y)$ defined by
$$f(Y) = \prod_{j=1}^{q} |Q_j|^{r_j} v(Q_j^0),$$
where
$$Y = \begin{pmatrix} Q_1 & 0 & \cdots & 0 \\ 0 & Q_2 & \cdots & 0 \\ \vdots & & \ddots & \vdots \\ 0 & \cdots & & Q_q \end{pmatrix} \begin{bmatrix} I_{n_1} & * & \cdots & * \\ 0 & I_{n_2} & \cdots & * \\ \vdots & & \ddots & \vdots \\ 0 & \cdots & & I_{n_q} \end{bmatrix},$$
and the Q_j are positive $n_j \times n_j$ matrices. The *Eisenstein series* is then:
$$E_P(f|Y) = \sum_{\gamma \in \Gamma_n/P} f(Y[\gamma]). \tag{5.6}$$

Here $\Gamma_n = GL(n, \mathbb{Z})$ and $P = P(n_1, \ldots, n_q)$. Our integral tests will show that this series converges for $\operatorname{Re} r_j$ sufficiently large, assuming that the automorphic forms v_j are bounded by power functions or integrable over the fundamental domain for $GL(n_j, \mathbb{Z})$. In the present subsection we will be studying the special case that all of the v_j are identically 1.

We need to study the analytic continuation, functional equations, residues, Fourier expansions, etc., of these Eisenstein series for $GL(n, \mathbb{Z})$. We will be interested in *two methods for analytic continuation of Eisenstein series.*

The first method is that of inserting larger parabolic subgroups between $\Gamma_n = GL(n, \mathbb{Z})$ and $P(n_1, \ldots, n_q)$. An example of this method has already appeared in Exercise 1. The method is also discussed by Maass [2, pp. 275–278]. Maass attributes the idea to Selberg, who announced results of the sort we shall discuss in several places (see Selberg [1], [3], [4]). The method was also used by Langlands [1, Appendix I]. Other references are Terras [6], [7].

4.5. Automorphic Forms for $GL(n,\mathbb{Z})$ and Harmonic Analysis on $\mathscr{P}_n/GL(n,\mathbb{Z})$

A *second method for continuing Eisenstein series is the method of theta functions* which goes back to Riemann. Many complications occur, as we shall see in the next subsection. Again, this method was developed by Selberg, who did not publish his proofs. Selberg did explain this method to various people (including the present author when she was a graduate student in 1969). It is developed in some detail in Maass [2, Section 16] and Terras [6, 7]. The main idea that eliminates the exploding integral is that of making use of differential operators chosen to annihilate the singular terms in the theta series. See also Siegel [1, Vol. III, pp. 328–333].

Other methods of obtaining analytic continuations of Eisenstein series are explained in Harish-Chandra [1], Kubota [1], Langlands [1], and Wong [1]. These methods are more function-theoretic and apply in a more general context. The method of Wong [1] was outlined by Selberg in a talk with the author and Carlos Moreno in 1984. These methods make use of the Fourier expansion of the Eisenstein series, a topic which we will discuss in a later subsection. There are also adelic methods (see Jacquet's talk in Borel and Casselman [1, Vol. II, pp. 83–84] and Helen Strassberg [1]). And there is a method which uses Eisenstein series for the Siegel modular group (see Arakawa [1]). See also Christian [2] for the analytic continuation of Eisenstein series for congruence subgroups. And Diehl [1] obtains a relation between Eisenstein series for $GL(n,\mathbb{Z})$ and $Sp(n,\mathbb{Z})$ by methods similar to those of Lemma 2 and Exercise 9 below.

Now we begin the discussion of the method of inserting larger parabolic subgroups to continue Selberg's Eisenstein series (5.5). We are restricting our attention here to Eisenstein series with the maximal number of complex variables and no lower rank automorphic forms. We shall use the notation $P_{(n)}$ for the minimal parabolic subgroup $P(1,\ldots,1)$ of $\Gamma_n = GL(n,\mathbb{Z})$; i.e., $P_{(n)}$ consists of all upper triangular matrices in Γ_n. And we define, for $Y \in \mathscr{P}_n, s \in \mathbb{C}^n$, Selberg's Eisenstein Series Associated to $P_{(n)}$ by:

$$E_{(n)}(s|Y) = \sum_{\gamma \in \Gamma_n/P_{(n)}} p_{-s}(Y[\gamma]), \qquad \text{if } \operatorname{Re} s_j > 1, j = 1, \ldots, n-1. \quad (5.7)$$

First we generalize Exercise 1.

Lemma 2 (Relations Between $E_{(n)}$ and $E_{(2)}$). *When* $\operatorname{Re} s_k > 1$, $k = 1, 2, \ldots, n-1$, *we have the following formula for the Selberg Eisenstein series defined in (5.7), for each* $i = 1, 2, \ldots, n-1$:

$$E_{(n)}(s|Y) = \sum_{\substack{V \in \Gamma_n/P_i^* \\ V = (V_k*), V_k \in \mathbb{Z}^{n \times k}}} E_{(2)}(s_i|T) \prod_{\substack{j=1 \\ j \neq i}}^n |Y[V_j]|^{-s_j},$$

if $P_i^* = P(1,\ldots,1,2,1,\ldots,1)$ *with the 2 in the ith position. Here* $R \in \mathscr{P}_2$ *is defined by the partial Iwasawa decomposition given below (see Exercise 11 of Section 4.1.2):*

$$Y[V_{i+1}] = \begin{pmatrix} Y[V_{i-1}] & 0 \\ 0 & R \end{pmatrix} \begin{bmatrix} I_{i-1} & Q \\ 0 & I_2 \end{bmatrix}, \qquad i = 2, \ldots, n-1;$$

and
$$Y[V_2] = R, \quad \text{when } i = 1.$$

And $T \in \mathscr{P}_2$ is defined by:
$$T = |Y[V_{i-1}]|R, \quad \text{if } i = 2, \ldots, n-1;$$
$$T = R = Y[V_2], \quad \text{if } i = 1.$$

Then the determinant of T is given by:
$$|T| = \begin{cases} |Y[V_{i-1}]||Y[V_{i+1}]|, & \text{if } i = 2, \ldots, n-2; \\ |Y[V_2]|, & \text{if } i = 1; \\ |Y[V_{n-2}]||Y|, & \text{if } i = n-1. \end{cases}$$

PROOF. Observe that $U \in \Gamma_n/P_{(n)}$ can be expressed uniquely as $U = VW$, with $V \in \Gamma_n/P_i^*$ and $W \in P_i^*/P_{(n)}$. Moreover, W can be chosen to have the form:
$$W = \begin{pmatrix} I_{i-1} & 0 & 0 \\ 0 & W^* & 0 \\ 0 & 0 & I_{n-i-1} \end{pmatrix}, \quad W^* \in \Gamma_2/P_{(2)}. \tag{5.8}$$

Here i always denotes our fixed index. If $i = 1$, the top row and first column in (5.8) are not present and if $i = n-1$, the bottom row and last column are absent. Next we write:
$$V = (V_k *), \quad V_k \in \mathbb{Z}^{n \times k}, \quad k = 1, 2, \ldots, n;$$
$$V_{i+1} = (V_{i-1} V_i^*), \quad V_i^* \in \mathbb{Z}^{n \times 2}. \tag{5.9}$$

Then, according to Exercise 4 below, we have:
$$|Y[(VW)_j]| = |Y[V_j]|, \quad \text{for } W \text{ as in (5.8)}, j \neq i; \tag{5.10}$$
and
$$|Y[(VW)_i]| = R[W_1^*]|Y[V_{i-1}]|, \tag{5.11}$$
where R is defined in the lemma being proved and $W_1^* \in \mathbb{Z}^2$ is defined by
$$W^* = (W_1^* *), \quad \text{for } W^* \text{ as in (5.8)}. \tag{5.12}$$

The lemma follows immediately. □

Exercise 4. Use the definitions in Lemma 2.
(a) Show that a complete set of representatives for $W \in P_{(n)}/P_i^*$ can be expressed in the form (5.8).
(b) Prove formula (5.10).
(c) Prove formula (5.11).

Hints. You can do part (a) by multiplying matrices in the appropriate block form. Part (b) follows from the remark that

4.5. Automorphic Forms for $GL(n, \mathbb{Z})$ and Harmonic Analysis on $\mathscr{P}_n/GL(n,\mathbb{Z})$

$$Y\left[(A \ B)\begin{pmatrix} C \\ 0 \end{pmatrix}\right] = Y[AC], \quad \text{for } A \in \mathbb{R}^{n \times k}, B \in \mathbb{R}^{n \times (n-k)}, C \in \mathbb{R}^{k \times k},$$

$$0 \in \mathbb{R}^{(n-k) \times k},$$

where 0 denotes a matrix of zeros. Part (c) is proved using partial Iwasawa decompositions to obtain:

$$Y[(VW)_i] = Y\left[V_{i+1}\begin{pmatrix} I_{i-1} & 0 \\ 0 & W_1^* \end{pmatrix}\right]$$

$$= \begin{pmatrix} Y[V_{i-1}] & 0 \\ 0 & R \end{pmatrix}\left[\begin{pmatrix} I_{i-1} & Q \\ 0 & I_2 \end{pmatrix}\begin{pmatrix} I_{i-1} & 0 \\ 0 & W_1^* \end{pmatrix}\right]$$

$$= \begin{pmatrix} Y[V_{i-1}] & 0 \\ 0 & R[W_1^*] \end{pmatrix}\begin{bmatrix} I_{i-1} & X \\ 0 & 1 \end{bmatrix}.$$

It is now possible to begin the process of analytic continuation of the Selberg Eisenstein series $E_{(n)}$ to a meromorphic function of $s \in \mathbb{C}^n$ by utilizing the analytic continuation of $E_{(2)}$ which was obtained in Section 3.5 of Volume I, using Theorem 1 of Section 1.4 in Volume I.

Lemma 3 (First Step in the Analytic Continuation of Selberg's Eisenstein Series with n Complex Variables). *If $E_{(n)}$ denotes the Selberg Eisenstein series defined in (5.7), set*

$$\Lambda(s) = \pi^{-s}\Gamma(s)\zeta(2s) \quad \text{and} \quad \Lambda_i(s|Y) = 2\Lambda(s_i)E_{(n)}(s|Y),$$

for $i = 1, 2, \ldots, n-1$. Then $\Lambda_i(s|Y)$ can be analytically continued to the region D_i pictured in Figure 4.17. Moreover $\Lambda_i(s|Y)$ satisfies the functional equation: $\Lambda_i(s|Y) = \Lambda_i(s'|Y)$, where

$$s_j' = \begin{cases} s_j, & j \neq i, i \pm 1, \\ 1 - s_i, & j = i, \\ s_{i\pm 1} + s_i - \frac{1}{2}, & j = i \pm 1. \end{cases}$$

The only poles of Λ_i in the region D_i occur when $s_i = 0$ or 1. Moreover,

$$E_{(n)}(s|Y)|_{s_i=0} = E_{P_i^*}(s_1, \ldots, s_{i-1}, s_{i+1}, \ldots, s_n|Y),$$

with E_P denoting, as usual, the Eisenstein series associated to the parabolic subgroup $P = P_i^*$ of Lemma 2. And if $s_j^* = s_j$ for $j \neq i \pm 1$, $s_{i\pm 1}^* = s_{i\pm 1} + \frac{1}{2}$, then

$$\text{Res } E_{(n)}(s|Y)|_{s_i=1} = \frac{E_{P_i^*}(s_1^*, \ldots, s_{i-1}^*, s_{i+1}^*, \ldots, s_n^*|Y)}{2\Lambda_i(1)}.$$

Here $2\Lambda_i(1) = \pi/3$, which is the volume of the fundamental domain for $SL(2, \mathbb{Z})$ acting on the Poincaré upper half plane.

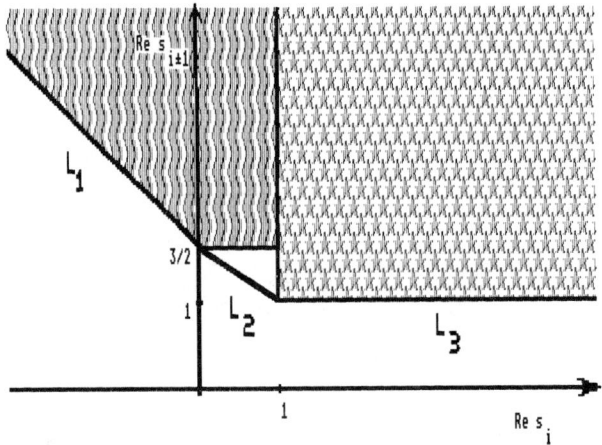

Figure 4.17. Real parts of s_i and $s_{i\pm 1}$ for the region D_i of analytic continuation of the Selberg Eisenstein series $E_{(n)}$. The region shaded with wavy lines or stars represents a projection of the region:

$$D_i = \{s \in \mathbb{C}^n | \operatorname{Re} s_i \geq 0, \operatorname{Re} s_j \geq \tfrac{3}{2}, j \neq i\}$$
$$\cup \{s \in \mathbb{C}^n | \operatorname{Re} s_i \leq 0, \operatorname{Re} s_j \geq \tfrac{3}{2}, j \neq i, i \pm 1, \operatorname{Re}(s_{i\pm 1} + s_i) \geq \tfrac{3}{2}\}.$$

The original region of analyticity for $E_{(n)}$ is that which is shaded with stars. We can enlarge the region D_i to that which lies above the lines L_1, L_2, L_3, but this would lead to more complicated equations. Here the line L_1 has the equation $\operatorname{Re}(s_{i+1} + s_i) = \tfrac{3}{2}$.

PROOF. We can write

$$E_{(2)}(s_i | T) = |T|^{-s_i/2} E_{(2)}(s_i | T^0),$$

where

$$T^0 = |T|^{-1/2} T \in \mathscr{SP}_2.$$

Substitute the incomplete gamma expansion of $E_{(2)}(s_i | T^0)$ from Theorem 1 of Section 1.4 in Volume I into the expression for $E_{(n)}$ given in Lemma 2. This leads to the formula

$$\Lambda_i(s | Y) = \frac{E_{P_i^*}(s^* | Y)}{s_i - 1} - \frac{E_{P_i^*}(s^* | Y)}{s_i} + S_3 + S_4, \qquad (5.13)$$

where $E_{P_i^*}(s^* | Y) = E_{P_i^*}(s_1^*, \ldots, s_{i-1}^*, s_{i+1}^*, \ldots, s_n^*)$,

$$s_i^* = \begin{cases} s_j, & j \neq i \pm 1, \\ s_j + s_i/2, & j = i \pm 1, \end{cases}$$

and

4.5. Automorphic Forms for $GL(n, \mathbb{Z})$ and Harmonic Analysis on $\mathcal{P}_n/GL(n, \mathbb{Z})$

$$S_3 = \sum_{V \in \Gamma_n/P_i^*} \prod_{j \neq i} |Y[V_j]|^{-s_j^*} \int_1^\infty x^{-s_i} \sum_{a \in \mathbb{Z}^2 - 0} \exp\{-\pi x (T^0)^{-1}[a]\} \, dx,$$

$$S_4 = \sum_{V \in \Gamma_n/P_i^*} \prod_{j \neq i} |Y[V_j]|^{-s_j^*} \int_1^\infty x^{s_i - 1} \sum_{a \in \mathbb{Z}^2 - 0} \exp\{-\pi x T^0[a]\} \, dx.$$

For any positive ε, we can easily bound the terms S_3 and S_4 in formula (5.13) by $\Lambda_i(\rho | Y)$, where

$$\rho_j = \begin{cases} \operatorname{Re} s_j, & \text{for } j \neq i \pm 1, \text{ or } i, \\ \operatorname{Re} s_j + \frac{1}{2}\{\operatorname{Re} s_i - |\operatorname{Re} s_i| - (1 + \varepsilon)\}, & \text{for } j = i \pm 1, \\ |\operatorname{Re} s_i| + 1 + \varepsilon, & \text{for } j = i. \end{cases}$$

Here we are using the fact that

$$(T^0)^{-1} \begin{bmatrix} a_1 \\ a_2 \end{bmatrix} = T^0 \begin{bmatrix} -a_2 \\ a_1 \end{bmatrix},$$

which implies that the sums over $a \in \mathbb{Z}^2 - 0$ in S_3 and S_4 are the same.

It follows that $E_{(n)}$ can be analytically continued to the region D_i with the indicated functional equation, poles, residues, and behavior at $s_i = 0$. Note that the transformation $s \to s'$, which appears in the functional equation, maps the region in Figure 4.17 above the lines L_1, L_2, L_3, into itself. In particular, $s \to s'$ takes the line L_1 to L_3 and fixes L_2. □

Note that the formulas for values of Eisenstein series at $s_i = 0$ are much simpler than the formulas for the residues at $s_i = 1$. The moral of this may be that one should attempt to study the spectral decomposition in the region where the Dirichlet series for $E_{(n)}$ does not converge. This is not even what we did in the case $n = 2$. But it has often been noticed by number theorists that zeta and L-functions have simpler behavior to the left of the line fixed by the functional equation. For example, the Riemann zeta function has rational values at negative odd integers (cf. Exercise 7 of Section 3.5 of Volume I), while the values at positive even integers are more complicated, involving powers of π.

Note also that the zeros of $\Lambda(s_i)$ can produce poles of

$$E_{(n)}(s | Y) = \Lambda_i(s | Y)/2\Lambda(s_i).$$

And the zeros of the Riemann zeta function are thus of interest to us here. Actually they will lie to the left of the line $\operatorname{Re} s_i = \frac{1}{2}$, which is the line of interest for our continuation to reach the spectrum of the $GL(n, \mathbb{R})$-invariant differential operators. This happens because $\Lambda(s_i)$ involves $\zeta(2s_i)$ not $\zeta(s_i)$.

Exercise 5. Prove all the statements made in the proof of Lemma 3.

Hint. You can use Theorem 2.5.10 in Hörmander [1], which says that every function holomorphic on a connected tube in $\mathbb{C}^n$, $n \geq 2$, can be continued to a function holomorphic on the convex hull of the connected tube. A *tube* $\Omega \subset \mathbb{C}^n$ has the form $\Omega = \{z \in \mathbb{C}^n | \operatorname{Re} z \in \omega\}$ for some set ω in $\mathbb{R}^n$.

Selberg noticed that the situation in Lemma 3 is clarified by introducing *new variables* $z \in \mathbb{C}^n$ (cf. Proposition 1 and formula (2.9) of Section 4.2.1):

$$s_j = z_{j+1} - z_j + \tfrac{1}{2}, \qquad j \le n, \qquad s_n = -z_n + \tfrac{1}{2}, \tag{5.14}$$
$$z_j = -(s_n + s_{n-1} + \cdots + s_j) + (n - j + 1)/2.$$

Note that the z-variables are closely related to the r-variables in part (4) of Theorem 3 in Section 4.2.3.

Exercise 6.

(a) Show that if $z(s)$ is given by formula (5.14) and if $s \to s'$ denotes the transformation appearing in the functional equation of $E_{(n)}$ in Lemma 3, then $z' = z(s')$ is the transformation σ_i of the z-variables which permutes z_i and z_{i+1}, while leaving the rest of the z_j fixed for $j \ne i, i + 1$.

(b) Relate the z-variables of (5.14) with the r-variables in part (4) of Theorem 3 in Section 4.2.3.

Hint. For example, $z'_{i-1} =$

$$-(s_n + s_{n-1} + \cdots + (s_{i+1} + s_i - \tfrac{1}{2}) + (1 - s_i) + (s_{i-1} + s_i - \tfrac{1}{2})) + \frac{n - i + 2}{2}$$

$$= -(s_n + \cdots + s_{i-1}) + \frac{n - i + 2}{2} = z_{i-1}.$$

The rest of the calculations are easier.

Exercise 7. Show that in terms of the z-variables the domain D_i in Figure 4.17 contains the following region, after setting $x_j = \operatorname{Re} z_j$:

$$R_i = \{z \in \mathbb{C}^n | x_{j+1} - x_j \ge 1, j \ne i, x_{i+1} - x_i \ge -\tfrac{1}{2}\}$$
$$\cup \{z \in \mathbb{C}^n | x_{j+1} - x_j \ge 1, j \ne i, i \pm 1, x_{i-2} - x_i \ge 1,$$
$$x_{i+1} - x_{i-1} \ge 1, x_i - x_{i+1} \ge -\tfrac{1}{2}\}.$$

Note that the second set in the union is the image of the first under the transformation σ_i of Exercise 6 which permutes z_i and z_{i+1}, leaving the rest of the variables fixed.

Now, as Selberg observed, we have the generators of the group of permutations of the variables $z_1, \ldots, z_n$ (i.e., the Weyl group of $GL(n)$). Thus we should be able to obtain $n!$ functional equations for the Eisenstein series $E_{(n)}$, if we include the identity $E_{(n)} = E_{(n)}$ as a functional equation. Moreover, we will be able to continue $E_{(n)}$ as a meromorphic function in the complex space $\mathbb{C}^n$.

4.5. Automorphic Forms for $GL(n, \mathbb{Z})$ and Harmonic Analysis on $\mathscr{P}_n/GL(n,\mathbb{Z})$

Theorem 1 (Selberg) (The Analytic Continuation and Functional Equations of Selberg's Eisenstein Series). *Let $\varphi(s) = \Gamma(s)\zeta(2s)$ and define (using (5.14)):*

$$\Xi(z|Y) = \pi^p E_{(n)}(z(s)|Y) \prod_{1 \leq i < j \leq n} \varphi(z_j - z_i + \tfrac{1}{2}),$$

if

$$p = -2 \sum_{j=1}^{n} j z_j.$$

Then

$$E_{(n)}(z(s)|Y) \prod_{1 \leq i < j \leq n} (z_j - z_i - \tfrac{1}{2})\zeta(2(z_j - z_i + \tfrac{1}{2}))$$

can be continued to a holomorphic function for all $z \in \mathbb{C}^n$. And $\Xi(z|Y)$ satisfies the $n!$ functional equations

$$\Xi(\sigma(z)|Y) = \Xi(z|Y)$$

for every permutation σ of n elements. Here

$$\sigma(z) = (z_{\sigma(1)}, \ldots, z_{\sigma(n)}) \quad \text{if} \quad z = (z_1, \ldots, z_n).$$

PROOF. Let us compute $\Xi(\sigma_i(z))$ for $\sigma_i = (i \quad i+1)$, the transposition of i and $i+1$. The power of π in $\Xi(\sigma_i(z))$ is:

$$p = -2 \sum_{j=1}^{n} j z_j + 2(z_{i+1} - z_i). \tag{5.15}$$

The product of the φ's in $\Xi(\sigma_i(z))$ is:

$$\frac{\varphi(z_i - z_{i+1} + \tfrac{1}{2})}{\varphi(z_{i+1} - z_i + \tfrac{1}{2})} \prod_{1 \leq i < j \leq n} \varphi(z_j - z_i + \tfrac{1}{2}). \tag{5.16}$$

Thus Lemma 3 completes the proof that $\Xi(\sigma(z)|Y) = \Xi(z|Y)$ for all σ of the form $\sigma = \sigma_i = (i \quad i+1)$. Since these permutations generate the symmetric group of permutations of n elements, it follows that $\Xi(z|Y)$ is invariant under all permutations of the entries of $z \in \mathbb{C}^n$. Thus, in fact, this function behaves like the spherical function $h_s(Y)$ (see Theorem 3 in Section 4.2.3).

Next we claim that $E_{(n)}$ can be continued as a meromorphic function in the region

$$B^* = \bigcup_{k=1}^{n} B_k, \tag{5.17}$$

where, writing $x_j = \text{Re } z_j$,

$$B_k = \{z \in \mathbb{C}^n | x_{j+1} - x_j > 2, j \neq k, k-1, x_{k+1} - x_{k-1} > 2\}. \tag{5.18}$$

Here, we simply drop inequalities that do not make sense; e.g., for $n = 3$, we have

$$B_1 = \{z \in \mathbb{C}^3 | x_3 - x_2 > 2\}.$$

To prove (5.17), first continue $E_{(n)}$ to B_1. To do this, note that if $z \in B_1$, there is a permutation σ of n elements such that $\sigma(z) \in R_j$, the region considered in Exercise 7. For suppose

$$x_j < x_1 \leq x_{j+1}, \quad \text{for some } j = 2, \ldots, n.$$

Then either $x_{j+1} - x_1 > 1$ or $x_1 - x_j > 1$, otherwise $x_{j+1} - x_j \leq 2$, contradicting the definition of B_1.

If $x_{j+1} - x_1 > 1$, then $z \in \sigma^{-1}(R_{j-1})$, where R_{j-1} is defined in Exercise 7 and

$$\sigma = \begin{pmatrix} 1 & 2 & \cdots & j-1 & j \\ 2 & 3 & \cdots & j & 1 \end{pmatrix} = (12)(23)\cdots(j-1\ j).$$

Here we use the standard notation for permutations:

$$\sigma = \begin{pmatrix} 1 & 2 & \cdots & n \\ \sigma(1) & \sigma(2) & \cdots & \sigma(n) \end{pmatrix}.$$

We did not list integers that are fixed. And $(i\ j)$ denotes the transposition that interchanges i and j, leaving the other integers fixed. To see that $z \in \sigma^{-1}(R_{j-1})$, note that the following inequalities say that σz with $\operatorname{Re} z_j = x_j$ lies in the first subset in the definition of R_{j-1} given in Exercise 7:

$$x_3 - x_2 \geq 1, \ldots, x_j - x_{j-1} \geq 1, x_1 - x_j \geq 0, x_{j+1} - x_1 \geq 1, \ldots, x_n - x_{n-1} \geq 1.$$

And, similarly, if $x_1 - x_j > 1$, then $z \in \sigma^{-1}(R_j)$, since the following inequalities say that σz with $\operatorname{Re} z_j = x_j$ lies in the first subset of R_j defined in Exercise 7:

$$x_3 - x_2 \geq 1, \ldots, x_j - x_{j-1} \geq 1, x_1 - x_j \geq 1, x_{j+1} - x_1 \geq 0, \ldots, x_n - x_{n-1} \geq 1.$$

The functional equations of $\Xi(z|Y)$ allow us to continue it to B_1. And

$$E_{(n)}(z|Y) \prod_{j=2}^{n} (z_j - z_1 - \tfrac{1}{2})\zeta(2(z_j - z_i + \tfrac{1}{2}))$$

is holomorphic in B_1. For example, the only poles of

$$\zeta(2(z_{j+1} - z_j) + 1)E_{(n)}(z|Y)$$

in the region R_j of Exercise 7 occur when $z_{j+1} - z_j = \tfrac{1}{2}$. This gives $z_{j+1} - z_1 = \tfrac{1}{2}$ in

$$\sigma^{-1}(R_j) = (j-1\ j)\cdots(23)(12)R_j = \sigma^{-1}(j\ j+1)R_j.$$

Now, in order to continue $E_{(n)}$ to the domain B_k, use the fact that

$$B_k = (k\ k-1)B_{k-1}.$$

So we can use the functional equations of $\Xi(z|Y)$ to continue $E_{(n)}$ to B_k. We find also that

$$E_{(n)}(z|Y) \prod_{1 \leq i < j \leq n} (z_j - z_i - \tfrac{1}{2})\zeta(2(z_j - z_i + \tfrac{1}{2}))$$

is holomorphic in the region B^*.

4.5. Automorphic Forms for $GL(n, \mathbb{Z})$ and Harmonic Analysis on $\mathcal{P}_n/GL(n, \mathbb{Z})$

To complete the analytic continuation of $E_{(n)}$, we need to use a theorem from several complex variables mentioned in Exercise 5. This insures that the function can always be continued to the convex hull of any region in which it is holomorphic (see Theorem 2.5.10 in Hörmander [2]). Thus it suffices to show that B^* is connected with convex hull $\mathbb{C}^n$. This is proved in Exercise 8. □

Exercise 8. Show that B^* defined in (5.17) is a connected set and then show that B^* has $\mathbb{C}^n$ as its convex hull.

Hint. You can show that n independent lines through the same point lie in the set
$$b^* = \{x \in \mathbb{R}^n | x_j = \operatorname{Re} z_j, z \in B^*\}.$$
You can take, for example, the lines
$$\lambda_i = \{x \in \mathbb{R}^n | x = 3(1, 2, \ldots, n) + te_i, t \in \mathbb{R}\},$$
with $e_i = (0, \ldots, 0, 1, 0, \ldots, 0)$, the ith element of the standard basis of $\mathbb{R}^n$. This follows from the fact that the definition of B_i does not restrict the ith coordinate at all.

This completes our discussion of the analytic continuation of $E_{(n)}(s|Y)$ by the method of inserting larger parabolic subgroups. It is possible to generalize our formula relating $E_{(n)}$ and $E_{(2)}$, as the following exercises show.

Exercise 9. (More General Decompositions Associated to Two Parabolics). Let $P = P(n_1, \ldots, n_q)$ be any parabolic subgroup of $\Gamma_n = GL(n, \mathbb{Z})$. Show that, in the region where the Dirichlet series for $E_{(n)}$ converges absolutely, we have:

$$E_{(n)}(s|Y) = \sum_{V \in \Gamma_n/P} \prod_{i=1}^{q} |Y[V_{N_i}]|^{-s_i} E_{(n_i)}(s_{N_{i-1}+1}, \ldots, s_{N_i-1}, 0 | T_i),$$

for $N_i = n_1 + \cdots + n_i$, $V = (V_j*)$, $V_j \in \mathbb{Z}^{n \times j}$, and $R_i \in \mathcal{P}_v$, $v = n_i$, is defined by the partial Iwasawa decomposition:

$$Y[V_{N_i}] = \begin{pmatrix} Y[V_{N_{i-1}}] & 0 \\ 0 & R_i \end{pmatrix} \begin{bmatrix} I_{N_{i-1}} & Q \\ 0 & I_{n_i} \end{bmatrix} \quad \text{and} \quad T_i = |Y[V_{N_{i-1}}]|R_i.$$

Hints (Terras [6]). Note that we can imitate the proof of Lemma 2. Write $U \in \Gamma_n/P_{(n)}$ uniquely as $U = VW$ with $V \in \Gamma_n/P$, $W \in P/P_{(n)}$:

$$W = \begin{pmatrix} W_1^* & & 0 \\ & \ddots & \\ 0 & & W_q^* \end{pmatrix}, \quad W_i \in \Gamma_{n_i}/P_{(n_i)}.$$

Then if $N_{i-1} < j < N_i$, we have
$$|Y[VW_j]| = |Y[V_{N_{i-1}}]||R_i[(W_i^*)_{j-N_{i-1}}]|.$$

Exercise 10. Extend Exercise 9 to relate the Eisenstein series E_P to E_{P^*} if $P^* \supset P$ are two parabolic subgroups of $GL(n, \mathbb{Z})$. Show that this implies that:

$$E_P(s_{N_1}, \ldots, s_{N_q} | Y)|_{s_{N_j} = 0, \text{ if } N_j \neq M_i} = E_{P^*}(s_{M_1}, \ldots, s_{M_{q^*}} | Y),$$

where

$$N_{i_k} = \sum_{\alpha=1}^{i_k} n_\alpha = M_k = \sum_{\beta=1}^{k} m_\beta,$$

if $P = P(n_1, \ldots, n_q)$ and $P^* = P(m_1, \ldots, m_{q^*})$.

Hint (Terras [6]). Use induction on q.

Exercise 11. Compute

$$\text{Res}_{s_j = 1, j = 1, \ldots, n} E_{(n)}(s | Y).$$

Answer.

$$2^{1-n} \text{Vol}(\mathscr{SP}_n / GL(n, \mathbb{Z}))^{-1}.$$

Exercise 11 is useful when one seeks to generalize Zagier's argument of Exercise 16 of Section 3.6, Volume I, in order to show that the Hecke points (as seen in Figures 4.7–4.14 of Section 4.4.3) actually become dense in the fundamental domain for $SL(3, \mathbb{Z})$ (actually a union of 2^2 copies of the fundamental domain in which all x-variables run between $-\frac{1}{2}$ and $+\frac{1}{2}$).

The analytic continuation given in Theorem 1 began with a very concrete formula for $E_{(2)}$—the incomplete gamma expansion found in Section 1.4 of Volume I. But the proof of Theorem 1 ended in a rather existential way—using the result from several complex variables which extends holomorphic functions of more than one complex variable in a connected tube domain to the convex hull of the domain.

4.5.2. Hecke Operators and Analytic Continuation of L-Functions Associated to Modular Forms by the Method of Theta Functions

Let us now begin the discussion of the second method of analytically continuing Eisenstein series—*the method of theta functions*. In order to relate Eisenstein series and theta functions, we need a generalization of formula (5.4), and thus a generalization of Lemma 1 in the preceding section. This will require a discussion of Hecke operators for $GL(n, \mathbb{Z})$. We will mostly restrict our attention here to Eisenstein series of the form (5.3), although the method can be generalized much further.

Suppose that $f: \mathscr{SP}_n / GL(n, \mathbb{Z}) \to \mathbb{C}$. Then for any positive integer m, the mth *Hecke operator* T_m is defined by:

4.5. Automorphic Forms for $GL(n, \mathbb{Z})$ and Harmonic Analysis on $\mathscr{P}_n/GL(n,\mathbb{Z})$

$$T_m f(Y) = \sum_{A \in \Omega_m} f(Y[A]^0), \quad \text{for } Y \in \mathscr{P}_n, Y^0 = |Y|^{-1/n} Y \in \mathscr{SP}_n. \quad (5.19)$$

Here Ω_m denotes any complete system of representatives for $M_m/GL(n,\mathbb{Z})$, where

$$M_m = \{A \in \mathbb{Z}^{n \times n} \mid |A| = m\}.$$

According to formula (4.56) in the proof of Lemma 7 in Section 4.4.4, we can take

$$\Omega_m = \left\{ \begin{pmatrix} d_1 & d_{12} & \cdots & d_{1n} \\ 0 & d_2 & \cdots & d_{2n} \\ \vdots & \vdots & & \vdots \\ 0 & 0 & \cdots & d_n \end{pmatrix} \middle| \begin{array}{l} d_j > 0, \quad j = 1, \ldots, n, \\ d_{ij} = 0, \quad \text{if } i > j, \\ 0 \le d_{ij} < d_i, \quad \text{if } i < j, \\ \prod_{i=1}^{m} d_i = m. \end{array} \right\}. \quad (5.20)$$

Note that the scalar matrices A cancel out when one computes $(Y[A]^0)$ in (5.19). Observe also that we have indeed generalized the Hecke operators given in formula (3.89) of Section 3.6, Volume I, although here we use a slightly different normalization in that we do not multiply by $m^{-1/2}$. We have already considered some of the history of Hecke operators in Section 3.6, Vol. I. Maass [5] studied Hecke operators for the Siegel modular group $Sp(n, \mathbb{Z})$. A good reference for the Hecke ring of a general group is Shimura [2, Ch. 3], where there is an exposition of work of Tamagawa [2] connecting Hecke operators with combinatorial results about lattices as well as p-adic convolution operators and a p-adic version of Selberg [1]. Hecke operators for $Sp(n, \mathbb{Z})$ and $SL(n, \mathbb{Z})$ are also considered by Andrianov [1–4] and Freitag [1]. Other references for the Hecke ring of $GL(n)$ over p-adic number fields are Macdonald [2] and Satake [1]. Another general reference is Krieg [2].

The Hecke operator (5.19) appears in many calculations associated with the general linear group. For example, set the function $f(Y) \equiv 1$ identically for all $Y \in \mathscr{SP}_n$. Then, from formulas (5.19) and (5.20), we have:

$$\sum_{m \ge 1} T_m f(Y) m^{-s} = \sum_{A \in \mathbb{Z}^{n \times n}/\Gamma_n} |A|^{-s} = Z_{n,0}\left(I, \frac{s}{2}\right) = \prod_{j=0}^{n-1} \zeta(s-j),$$

where $\Gamma_n = GL(n, \mathbb{Z})$ and $Z_{n,0}(Y, s)$ is Koecher's zeta function from formula (4.2) of Section 4.4.1. Solomon [1] considers generalizations of such results and connections with combinatorics. Operators like T_m are also intrinsic to formulas akin to (5.35) below connecting Eisenstein series such as (5.3) defined as sums over $GL(n, \mathbb{Z})$ and zeta functions defined as sums over rank m matrices in $\mathbb{Z}^{n \times m}$. These zeta functions are higher dimensional Mellin transforms of the non-singular terms in a theta function (see formula (5.4)) and we will be able to use a modification of Riemann's method of theta functions to obtain an analytic continuation of these zeta functions for $GL(n, \mathbb{Z})$. Thus it is formula

(5.35) and our search for analytic continuations of Eisenstein series that motivate our study of Hecke operators here.

The basic properties of Hecke operators for $GL(n, \mathbb{Z})$ are contained in the following theorem.

Theorem 2 (Hecke Operators for $GL(n, \mathbb{Z})$).

(1) The Hecke operator T_m maps automorphic forms $f \in \mathcal{A}^0 (GL(n, \mathbb{Z}), \lambda)$, as defined in (5.1) and just after (5.2) of the preceding section, to automorphic forms with the same eigenvalue system; i.e., $T_m f \in \mathcal{A}^0(GL(n, \mathbb{Z}), \lambda)$.

(2) The Hecke operator T_m is a Hermitian operator with respect to the inner product:

$$(f, g) = \int_{\mathcal{SP}_n/GL(n, \mathbb{Z})} f(W) \overline{g(W)} \, dW,$$

$dW = $ the $SL(n, \mathbb{R})$-invariant measure on $\mathcal{SP}_n$.

(3) The ring of Hecke operators is commutative and thus has a set of simultaneous eigenfunctions which span the space of all automorphic forms for $GL(n, \mathbb{Z})$.

(4) If $(k, m) = 1$, then $T_k T_m = T_{km}$. When the group is $GL(3, \mathbb{Z})$, one has the following formal power series in the indeterminate X for any prime p:

$$\sum_{r \geq 0} T_{p^r} X^r = (I - T_p X + [(T_p)^2 - T_{p^2}]X^2 - p^3 X^3)^{-1}.$$

(5) Suppose that $f \in \mathcal{A}^0(GL(n, \mathbb{Z}), \lambda)$ is an eigenfunction for all the Hecke operators; i.e., $T_m f = u_m f$, for some $u_m \in \mathbb{C}$. Form the Dirichlet series:

$$L_f(s) = \sum_{m \geq 1} u_m m^{-s}.$$

This series converges for $\operatorname{Re} s > n/4$ iff f is integrable on the fundamental domain $\mathcal{SM}_n$. Moreover, L_f can be analytically continued to a meromorphic function of s with functional equation:

$$\Lambda(f, s) = \pi^{-ns} L_{f^*}(2s) \Gamma_n(r(f, s)) = \Lambda(f^*, n/2 - s),$$

where $f^*(W) = f(W^{-1})$, $r = r(f, s) \in \mathbb{C}^n$ is defined via Proposition 2 of Section 4.2.3 and the formula:

$$\frac{\Omega(|Y|^s f(Y^0))}{|Y|^s f(Y^0)} = \frac{\Omega(p_r(Y))}{p_r(Y)}, \quad \text{for any } \Omega \in D(\mathcal{P}_n),$$

with $D(\mathcal{P}_n) = $ the algebra of $GL(n, \mathbb{R})$-invariant differential operators on $\mathcal{P}_n$ and $p_r(Y)$ the power function defined in (2.1) of Section 4.2.1. Here $\Gamma_n(r)$ is the gamma function defined in (2.4) of Section 4.2.1.

For $GL(3, \mathbb{Z})$ it follows that $L_f(s)$ has the Euler product:

$$L_f(s) = \prod_{p \text{ prime}} (1 - u_p p^{-s} + [(u_p)^2 - u_{p^2}] p^{-2s} - p^{3-3s})^{-1}.$$

4.5. Automorphic Forms for $GL(n, \mathbb{Z})$ and Harmonic Analysis on $\mathscr{P}_n/GL(n, \mathbb{Z})$

Remarks. The Euler product in part (5) of Theorem 2 should be compared with that obtained by Bump [1] using Fourier expansions of automorphic forms as sums of Whittaker functions.

Theorem 2 gives an analogue of much of Theorem 4 of Section 3.6 in Volume I. However, the converse result for part (5) is missing. Converse theorems have been obtained adelically for $GL(3)$ (see Jacquet, Piatetski-Shapiro, and Shalika [1]). One would expect to need more than one complex variable s in order to be able to invert the Mellin transform over $\mathscr{P}_n/\Gamma$ that leads to $L_f(s)$. Jacquet, Piatetski-Shapiro, and Shalika find that one needs to twist by automorphic forms for $GL(n-2, \mathbb{Z})$ for a $GL(n)$ converse theorem.

There are many other sorts of Hecke operators and attached Dirichlet series that produce results similar to those which we have stated here. See the references mentioned earlier for some examples.

We will see that the analytic continuation result in (5) is similar to that which we need for the Eisenstein series (5.3).

PROOF. Mostly we can imitate the arguments given by Maass [5] for the Siegel modular group $Sp(n, \mathbb{Z})$.

(1)
Exercise 12. Prove part (1) of Theorem 2.

Hint. Imitate the proof of the analogous result for $SL(2, \mathbb{Z})$ to be found in Section 3.6 of Volume I.

(2) One need only imitate the proof of the analogous result for $SL(2, \mathbb{Z})$ which was part (4) of Theorem 4 in Section 3.6 of Volume I. Set $\Gamma = GL(n, \mathbb{Z})$ and if $A \in M_m$, the set of matrices of determinant m, write $h_A(W) = f((W[A])^0)$. Note that h_A remains invariant under the *congruence subgroup*:

$$\Gamma(m) = \{B \in \Gamma | B \equiv I \pmod{m}\}. \tag{5.21}$$

For $B \in \Gamma(m)$ and $A \in M_m$ imply that $A^{-1}BA \in \Gamma$. To see this, observe that

$$A^{-1} = (1/m)\,{}^t(\mathrm{adj}\, A) \in (1/m)\mathbb{Z}^{n \times n}.$$

Thus $mA^{-1}BA \in \mathbb{Z}^{n \times n}$ and $mA^{-1}BA$ is congruent to $mA^{-1}A = mI$ and thus to 0 modulo m. So $A^{-1}BA$ is an integral matrix of determinant one.

Since the fundamental domain $\mathscr{SP}_n/\Gamma(m)$ consists of $[\Gamma : \Gamma(m)] = \#(\Gamma/\Gamma(m))$ copies of $\mathscr{SP}_n/\Gamma$ (see Exercise 13 below), one has the following equalities:

$$\begin{aligned}
(T_m f, g) &= \sum_{A \in \Omega_m} [\Gamma : \Gamma(m)]^{-1} \int_{W \in \mathscr{SP}_n/\Gamma(m)} f((W[A])^0)\overline{g(W)}\, dW \\
&= \sum_{A \in \Omega_m} [\Gamma : A^{-1}\Gamma(m)A]^{-1} \int_{X \in \mathscr{SP}_n/A^{-1}\Gamma(m)A} f(X)\overline{g(X[A^{-1}]^0)}\, dX \\
&= (f, T_m g).
\end{aligned}$$

The second equality is obtained by substituting $X = W[A]$ and noting that $[\Gamma : \Gamma(m)] = [\Gamma : A^{-1}\Gamma(m)A]$.

Exercise 13.

(a) Show that the index $[\Gamma : \Gamma(m)] = \#(\Gamma/\Gamma(m))$ is finite, for $\Gamma = GL(n,\mathbb{Z})$ and $\Gamma(m)$ as defined by (5.21).
(b) Prove part (3) of Theorem 2.

Hint. (b) This is proved by Shimura [1, Ch. 3] using the existence of the anti-automorphism $X \mapsto {}^t X$ of $GL(n,\mathbb{Z})$.

(3) See Exercise 13(b).
(4) To see that $T_k T_m = T_{km}$ if $(k,m) = 1$, one need only multiply the matrices below:

$$B = \begin{pmatrix} d_1 & d_{12} & \cdots & d_{1n} \\ 0 & d_2 & \cdots & d_{2n} \\ \vdots & \vdots & & \vdots \\ 0 & 0 & \cdots & d_n \end{pmatrix} \begin{pmatrix} c_1 & c_{12} & \cdots & c_{1n} \\ 0 & c_2 & \cdots & c_{2n} \\ \vdots & \vdots & & \vdots \\ 0 & 0 & \cdots & c_n \end{pmatrix}$$

$$= \begin{pmatrix} d_1 c_1 & d_1 c_{12} + c_2 d_{12} & \cdots & d_1 c_{1n} + d_{12} c_{2n} + \cdots + d_{1n} c_n \\ 0 & d_2 c_2 & \cdots & d_2 c_{2n} + \cdots + d_{2n} c_n \\ \vdots & \vdots & & \vdots \\ 0 & 0 & \cdots & d_n c_n \end{pmatrix}.$$

For if d_{ij} runs through a complete set of representatives mod d_i and c_{ij} runs through a complete set of representatives mod c_i, then consider the i,j-entry in the product above for $i < j$:

$$b_{ij} = d_i c_{ij} + d_{i,i+1} c_{i+1,j} + \cdots + d_{ij} c_j.$$

Inductively we can assume that the terms $d_{ij'}$ with $j' < j$ and $c_{i'j}$ with $i' > i$ are fixed. Thus what remains is

$$d_i c_{ij} + d_{ij} c_j + \text{a fixed number}.$$

This gives a complete set of representatives modulo $d_i c_j$.

Next we consider the proof of the formula which implies the Euler product for the L-function in part (5) which corresponds to an eigenform of the Hecke operators for $GL(3,\mathbb{Z})$. The proof which follows involves only matrix multiplication but clearly becomes more complicated for $GL(n,\mathbb{Z})$ with $n > 3$. See Exercises 14 and 15 for connections with other methods of obtaining such Euler product formulas as well as Shimura [2] and Freitag [1].

Observe that

$$T_k T_m f(Y) = \sum_{A \in \Omega_m} \sum_{B \in \Omega_k} f(Y[BA]^0).$$

4.5. Automorphic Forms for $GL(n, \mathbb{Z})$ and Harmonic Analysis on $\mathscr{P}_n/GL(n, \mathbb{Z})$

It will be helpful to set up the following notation. Suppose that S is a subset of M_m (the set of all $n \times n$ integral matrices of determinant m) and let $T(S)$ denote the operator:

$$T(S)f(Y) = \sum_{A \in S} f(Y[A]^0). \tag{5.22}$$

The formal power series identity in part (4) derives from the following two formulas, which are easily checked by multiplying the matrix representatives of the operators involved. The first formula is:

$$T_{p^r} T_p = T_{p^{r+1}} + T(S_1^r) + T(S_2^r), \tag{5.23}$$

where

$$S_1^r = \left\{ \begin{pmatrix} p^e & p(a_1 \bmod p^e) & a_2 \bmod p^e \\ 0 & p^{f+1} & a_3 \bmod p^{f+1} \\ 0 & 0 & p^g \end{pmatrix} \middle| e \geq 1; f, g \geq 0; e + f + g = r \right\}$$

$$S_2^r = \left\{ \begin{pmatrix} p^e & a_1 \bmod p^e & p(a_2 \bmod p^e) \\ 0 & p^f & p(a_3 \bmod p^f) \\ 0 & 0 & p^{g+1} \end{pmatrix} \middle| \begin{array}{l} e \geq 1 \text{ or } f \geq 1 \\ g \geq 0; e + f + g = r \end{array} \right\}.$$

The second formula is:

$$T_{p^r}[(T_p)^2 - T_{p^2}] = p^3 T_{p^{r-1}} + T_{p^{r+1}} T_p + T_{p^{r+2}}, \quad \text{for } r \geq 1. \tag{5.24}$$

To prove formula (5.23), use the following calculations:

$$\begin{pmatrix} p^e & a_1 \bmod p^e & a_2 \bmod p^e \\ 0 & p^f & a_3 \bmod p^f \\ 0 & 0 & p^g \end{pmatrix} \begin{pmatrix} p & b_1 \bmod p & b_2 \bmod p \\ 0 & 1 & 0 \\ 0 & 0 & 1 \end{pmatrix}$$

$$= \begin{pmatrix} p^{e+1} & c_1 \bmod p^{e+1} & c_2 \bmod p^{e+1} \\ 0 & p^f & c_3 \bmod p^f \\ 0 & 0 & p^g \end{pmatrix},$$

$$\begin{pmatrix} p^e & a_1 \bmod p^e & a_2 \bmod p^e \\ 0 & p^f & a_3 \bmod p^f \\ 0 & 0 & p^g \end{pmatrix} \begin{pmatrix} 1 & 0 & 0 \\ 0 & p & b_3 \bmod p \\ 0 & 0 & 1 \end{pmatrix}$$

$$= \begin{pmatrix} p^e & p(c_1 \bmod p^e) & c_2 \bmod p^e \\ 0 & p^{f+1} & c_3 \bmod p^{f+1} \\ 0 & 0 & p^g \end{pmatrix},$$

$$\begin{pmatrix} p^e & a_1 \bmod p^e & a_2 \bmod p^e \\ 0 & p^f & a_3 \bmod p^f \\ 0 & 0 & p^g \end{pmatrix} \begin{pmatrix} 1 & 0 & 0 \\ 0 & 1 & 0 \\ 0 & 0 & p \end{pmatrix} = \begin{pmatrix} p^e & c_1 \bmod p^e & p(c_2 \bmod p^e) \\ 0 & p^f & p(c_3 \bmod p^f) \\ 0 & 0 & p^{g+1} \end{pmatrix}.$$

The first set of matrices gives $T_{p^{r+1}}$, except for the $e + 1 = 0$ terms. The second set of matrices gives all of the $e = 0$ terms of $T_{p^{r+1}}$ except the terms with $e = 0$

and $f+1 = 0$, and it also gives S_1^r. The third set of matrices gives the $e = f = 0$ terms in $T_{p^{r+1}}$ plus S_2^r.

To prove (5.24), use formula (5.23) with $r = 1$ to see that

$$(T_p)^2 - T_{p^2} = T(R_1) + T(R_2) + T(R_3),$$

where

$$R_1 = \left\{ \begin{pmatrix} p & p(b_1 \bmod p) & b_2 \bmod p \\ 0 & p & b_3 \bmod p \\ 0 & 0 & 1 \end{pmatrix} \right\},$$

$$R_2 = \left\{ \begin{pmatrix} p & b_1 \bmod p & p(b_2 \bmod p) \\ 0 & 1 & 0 \\ 0 & 0 & p \end{pmatrix} \right\},$$

$$R_3 = \left\{ \begin{pmatrix} 1 & 0 & 0 \\ 0 & p & p(b_3 \bmod p) \\ 0 & 0 & p \end{pmatrix} \right\}.$$

Then compute the matrix products to find that $T_{p^r} T(R_j) = T(Q_j)$, where

$$Q_1 = \left\{ \begin{pmatrix} p^{e+1} & p(a_1 \bmod p^{e+1}) & a_2 \bmod p^{e+1} \\ 0 & p^{f+1} & a_3 \bmod p^{f+1} \\ 0 & 0 & p^g \end{pmatrix} \right\},$$

$$Q_2 = \left\{ \begin{pmatrix} p^{e+1} & a_1 \bmod p^{e+1} & p(a_2 \bmod p^{e+1}) \\ 0 & p^f & p(a_3 \bmod p^f) \\ 0 & 0 & p^{g+1} \end{pmatrix} \right\},$$

$$Q_3 = \left\{ \begin{pmatrix} p^e & p(a_1 \bmod p^e) & p(a_2 \bmod p^e) \\ 0 & p^{f+1} & p(a_3 \bmod p^{f+1}) \\ 0 & 0 & p^{g+1} \end{pmatrix} \right\}.$$

Now $T(Q_1)$ gives $T(S_1^{r+1})$ from (5.23). And $T(Q_2)$ gives the $e + 1 \neq 0$ part of $T(S_2^{r+1})$ in (5.23). The $e = 0$ part of $T(Q_3)$ gives the remainder of $T(S_2^{r+1})$. The $e \geq 1$ part of $T(Q_3)$ gives $p^3 T_{p^{r-1}}$, since

$$\begin{pmatrix} p^{e-1} & a_1 \bmod p^e & a_2 \bmod p^e \\ 0 & p^f & a_3 \bmod p^{f+1} \\ 0 & 0 & p^g \end{pmatrix}$$

$$= \begin{pmatrix} p^{e-1} & b_1 \bmod p^{e-1} + p^{e-1}(c_1 \bmod p) & b_2 \bmod p^{e-1} + p^{e-1}(c_2 \bmod p) \\ 0 & p^f & b_3 \bmod p^f + p^f(c_3 \bmod p) \\ 0 & 0 & p^g \end{pmatrix}$$

$$= \begin{pmatrix} p^{e-1} & b_1 \bmod p^{e-1} & b_2 \bmod p^{e-1} \\ 0 & p^f & b_3 \bmod p^f \\ 0 & 0 & p^g \end{pmatrix} \begin{pmatrix} 1 & c_1 \bmod p & c_2 \bmod p \\ 0 & 1 & c_3 \bmod p \\ 0 & 0 & 1 \end{pmatrix}.$$

4.5. Automorphic Forms for $GL(n, \mathbb{Z})$ and Harmonic Analysis on $\mathscr{P}_n/GL(n,\mathbb{Z})$

Exercise 14. Read Shimura [1, Ch. 3] for another discussion of Hecke operators. If $A \in \mathbb{Z}^{n \times n}$ has rank n, one considers the double coset decomposition:

$$\Gamma A \Gamma = \bigcup_{B \in S_A} B\Gamma \text{ (disjoint)}.$$

Here S_A is a set of representatives for $\Gamma A \Gamma / \Gamma$, with $\Gamma = GL(n, \mathbb{Z})$.

Then one can prove a result due to Tamagawa which says that for prime p:

$$\sum_{r \geq 0} T_{p^r} X^r = \left(\sum_{j=0}^{n} (-1)^j p^{j(j-1)/2} T(S_{A_j}) \right)^{-1},$$

where

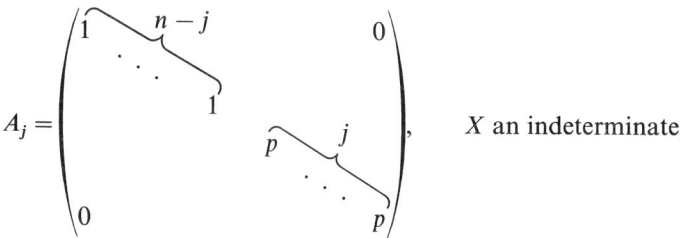 X an indeterminate,

and we use the notation $T(S_A)$ in (5.22). Show that when $\Gamma = GL(3, \mathbb{Z})$, Tamagawa's formal power series above is the same as ours in part (4) of Theorem 2. In particular, show that

$$(T_p)^2 - T_{p^2} = pT(S_{A_2}).$$

Exercise 15. Langlands [5] defines a Hecke operator T_A corresponding to a matrix $A \in \mathbb{Q}^{n \times n}$ with $|A| > 0$ by:

$$T_A f(Y) = \sum_{\gamma \in \Gamma / \Gamma \cap (A\Gamma A^{-1})} f((Y[\gamma A])^0),$$

for $f: \mathscr{SP}_n/\Gamma \to \mathbb{C}$, $\Gamma = GL(n, \mathbb{Z})$, where $Y^0 = |Y|^{-1/n} Y \in \mathscr{SP}_n$ for $Y \in \mathscr{P}_n$. Show that this definition agrees with that in Exercise 14.

Hint. Does

$$\sum_{\delta \in \Gamma A \Gamma / \Gamma} f(Y[\delta]^0) = \sum_{\gamma \in \Gamma / \Gamma \cap (A\Gamma A^{-1})} f(Y[\gamma A]^0)?$$

Bump [1] notes that these operators are no longer self adjoint.

Exercise 16. Check the Euler product in part (5) of Theorem 2 by letting f be the form that is identically one everywhere. Then

$$L_1(s) = \sum_{m \geq 1} T_m(1) m^{-s} = Z_{3,0}\left(I, \frac{s}{2}\right) = \prod_{j=0}^{2} \zeta(s-j),$$

where $Z_{3,0}(I, s/2)$ is Koecher's zeta function from (4.2) of Section 4.4.1 and the factorization into a product of Riemann zeta functions is formula (4.3) of

Section 4.4.1. Show that if you substitute the Euler product for Riemann's zeta function (see Exercise 4 of Section 1.4 in Volume I) into this formula for $L_1(s)$, you obtain the Euler product given in part (5) of Theorem 2.

Remarks on Part (5) *of Theorem* 2. We will prove part (5) in a slightly more general situation in order to obtain the analytic continuation of the Eisenstein series (5.3) simultaneously. The idea is to imitate Riemann's method of theta functions which gave the analytic continuation of Epstein's zeta function (see the proof of Theorem 1 of Section 1.4, Volume I). However the sailing is not so smooth here because when $k > 1$ there are many singular terms in the sum defining the theta function $\theta(Y, X)$, for $Y \in \mathscr{P}_n$, $X \in \mathscr{P}_k$, $1 \leq k \leq n$, from (4.1) in Section 4.4.1. These terms come from matrices $A \in \mathbb{Z}^{n \times k}$ of rank less than k. When $k = 1$ the only such term comes from the zero vector, but when $k > 1$ there are an infinite number of these terms to deal with.

This problem has led to gaps in many papers—gaps coming from the subtraction of divergent integrals. There was even a gap of this kind in Siegel's first paper on the computation of the volume of the fundamental domain $\mathscr{SM}_n$ (Siegel [1, Vol. I, pp. 459–468]). And such gaps appear in Koecher's paper [1] on the analytic continuation of his zeta function and other Dirichlet series associated to Siegel modular forms. And Käte Hey's thesis [1] on zeta functions of central simple algebras has a similar gap. Siegel makes the following remark at the beginning of his paper that fills the gap (Siegel [1, Vol. III, p. 328]): "Die Korrektur des Fehlschlusses ist dann keineswegs so einfach, wie man zunächst in Gedanken an die Renormalisierung in physikalischen Untersuchungen glauben möchte, und benötigt genauere Abschätzungen unendlicher Reihen."* We will not follow Siegel's method here, for it does not seem to give the complete analytic continuation (just the continuation beyond the first pole at $s = n/2$). Siegel's method does, however, allow the computation of the residue at $s = n/2$ of Koecher's zeta function $Z_{m,n-m}(Y,s)$ defined in formula (4.2) of Section 4.4.1. Another reference for Siegel's method is Maass [2, Section 16]. Instead we follow a method due to Selberg which makes use of invariant differential operators to annihilate the lower rank terms in the theta function, generalizing Exercises 9 and 10 of Section 1.4, Volume I. Other references for Selberg's method are Maass [2, Section 16] and Terras [1, 6, 8]. There are also other ways of accomplishing the analytic continuation, as we mentioned in Section 4.5.1, but we will not discuss them here.

(5) In order to prove part (5) and complete the proof of Theorem 2, we need the *theta function* from formula (4.1) of Section 4.4.1, with $1 \leq k \leq n$:

$$\theta(Y, X) = \sum_{A \in \mathbb{Z}^{n \times k}} \exp\{-\pi \operatorname{Tr}(Y[A]X)\}, \quad \text{for } Y \in \mathscr{P}_n \text{ and } X \in \mathscr{P}_k. \quad (5.25)$$

---

* The correction of the wrong deduction is then by no means so simple, as one should like to believe to begin with when thinking of renormalization in the physics literature, and it necessitated more exact estimates of infinite series.

4.5. Automorphic Forms for $GL(n, \mathbb{Z})$ and Harmonic Analysis on $\mathscr{P}_n/GL(n, \mathbb{Z})$

And we set

$$\theta_r(Y, X) = \sum_{\substack{A \in \mathbb{Z}^{n \times k} \\ \text{rank } A = r}} \exp\{-\pi \operatorname{Tr}(Y[A]X)\}. \tag{5.26}$$

Before proceeding with the proof, we must beg the reader to do the following exercises.

Exercise 17 (The Transformation Formula of the Theta Function). Show that if the theta function is defined by (5.25), then

$$\theta(Y, X) = |Y|^{-k/2}|X|^{-n/2}\theta(Y^{-1}, X^{-1}).$$

Hint. Imitate the proof of Exercise 6 in Section 1.4 of Volume I, using the Poisson summation formula.

Exercise 18 (A Gamma Integral Associated to an Automorphic Form for $GL(n)$). Suppose that f is an automorphic form on the determinant one surface; i.e., that $f \in \mathscr{A}^0(GL(n, \mathbb{Z}), \lambda)$, where

$$\frac{\Omega(f(Y^0)|Y|^s)}{f(Y^0)|Y|^s} = \frac{\Omega p_r(Y)}{p_r(Y)} = \lambda_\Omega(r) \quad \text{for } r = r(f, s) \in \mathbb{C}^n,$$

for all invariant differential operators $\Omega \in D(\mathscr{P}_n)$. Here $p_r(Y)$ is a power function as in formula (2.1) of Section 4.2.1. Use Proposition 4 of Section 4.2.3 and the definition of the gamma function Γ_k in formula (2.4) of Section 4.2.1 to prove that:

$$\int_{X \in \mathscr{P}_k} \exp\{-\operatorname{Tr}(Y^{-1}X)\} f(X^0)|X|^s d\mu_k(X) = \Gamma_k(r(f, s))f(Y^0)|Y|^s.$$

There is another discussion of this result in Maass [2, Section 7]. Related results were obtained in Exercise 4 of Section 4.2.1 and in formula (2.60) of Section 4.2.4.

We will write $\Gamma_k = GL(k, \mathbb{Z})$ and hope that the context will make it clear that this is not the gamma function. Defining

$$J(f, s|Y) = \int_{X \in \mathscr{M}_k} f(X^0)|X|^s \theta_k(Y, X) d\mu_k(X), \tag{5.27}$$

it follows from Exercise 18 that

$$J(f, s|Y) = 2\int_{\mathscr{P}_k} \sum_{\substack{A \in \mathbb{Z}^{n \times k} \\ \text{rank } k/\Gamma_k}} \exp\{-\pi \operatorname{Tr}(Y[A]X)\} f(X^0)|X|^s d\mu_k(X)$$

$$= 2\pi^{-ks} \sum_{\substack{A \in \mathbb{Z}^{n \times k} \\ \text{rank } k/\Gamma_k}} |Y[A]|^{-s} f^*(Y[A]^0) \Gamma_k(r(f, s)).$$

Here $f^*(W) = f(W^{-1})$. The factor 2 comes from the fact that

$$\bigcup_{A\in\Gamma_k} \mathcal{M}_k[A]$$

represents each point of $\mathcal{P}_n$ exactly twice (see part (3) of Theorem 1 in Section 4.4.2).

Now suppose that $f(W)$ is an eigenfunction of all the Hecke operators. Then

$$J(f,s|Y) = \begin{cases} 2\pi^{-ns}\Gamma_n(r(f,s))L_{f^*}(2s)|Y|^{-s}f^*(Y^0), & \text{if } k=n; \\ 2\pi^{-ks}\Gamma_k(r(f,s))L_{f^*}(2s)E_{k,n-k}(f^*,s|Y), & \text{if } 1 \le k < n. \end{cases} \quad (5.28)$$

The formula for $k=n$ is clear. That for $1 \le k < n$ will be proved in Corollary 1. Only the case $k=n$ is needed for Theorem 5. Here $E_{k,n-k}(f^*,s|Y)$ is the Eisenstein series of formula (5.3).

Exercise 19. Show that if f is an integrable function on $\mathcal{SM}_k$ satisfying the rest of the hypotheses of part (5) of Theorem 2, then the associated L function $L_f(2s)$ converges for $\operatorname{Re} s > n/2$.

Hint. Use a similar argument to that which was used to prove the convergence of the Eisenstein series (5.3).

Next we want to consider some differential operators on $\mathcal{P}_k$. As in Section 4.1.4, we write the matrix operator

$$\partial/\partial X = (\tfrac{1}{2}(1+\delta_{ij})\partial/\partial x_{ij}), \quad \text{if } X = (x_{ij}) \in \mathcal{P}_k. \quad (5.29)$$

See formula (1.28) in Section 4.1.4. Now define the *determinant operator*

$$\partial_X = \det(\partial/\partial X). \quad (5.30)$$

This is a departure from Section 4.1.4, in which we only considered traces of matrix operators. The property of ∂_X that endears it to us is:

$$\partial_X \exp[\operatorname{Tr}(XY)] = |Y|\exp[\operatorname{Tr}(XY)]. \quad (5.31)$$

This means that the operator ∂_X annihilates the singular terms $\theta_r(Y,X)$ in (5.26) when $r < k$. But ∂_X is not $GL(n,\mathbb{R})$-invariant. To obtain such an operator consider *Selberg's differential operators*:

$$D_a = |X|^a \partial_X |X|^{1-a}, \quad a \in \mathbb{R}. \quad (5.32)$$

Let $\sigma(X) = X^{-1}$ for $X \in \mathcal{P}_k$ and $D^\sigma f = D(f \circ \sigma) \circ \sigma^{-1}$, for a differential operator D and a function $f: \mathcal{P}_k \to \mathbb{C}$. We know from Theorem 1 of Section 4.2.1 that D^σ is the conjugate adjoint $\bar{D}^*$. This allows you to do part (b) of the following exercise.

Exercise 20.

(a) Show that D_a defined by (5.32) is indeed a $GL(k,\mathbb{R})$-invariant differential operator on $\mathcal{P}_k$; i.e., $D_a \in D(\mathcal{P}_k)$.

4.5. Automorphic Forms for $GL(n, \mathbb{Z})$ and Harmonic Analysis on $\mathscr{P}_n/GL(n, \mathbb{Z})$

(b) Show that if D_a is as in (5.32) and if $\sigma(X) = X^{-1}$, then
$$D_a^\sigma = D_a^* = (-1)^k D_{a^*}, \qquad \text{where } a^* = 1 - a + (k+1)/2.$$

Hint. For part (b), use integration by parts to find the adjoint of D_a.

Exercise 21.

(a) Let D_a be as defined in (5.32). Set $D = D_a D_1$ for $a = (k - n + 1)/2$. Show that
$$(D_1 D_a)^\sigma |X|^{n/2} = |X|^{n/2} D_a D_1.$$

(b) Use part (a) to show that if we write $\theta(Y, X) = \theta^Y(X)$, then
$$D\theta^Y(X) = |Y|^{-k/2} |X|^{-n/2} (D\theta^{Y^{-1}})(X^{-1}).$$

Exercise 21 shows that the differentiated theta function $D\theta^Y(X)$ satisfies the same transformation formula as $\theta^Y(X)$ itself. By formula (5.31) however, the differentiated theta function is missing all the lower rank terms; i.e.,
$$D\theta^Y(X) = D\theta_k^Y(X).$$

Exercise 22. Let D be as defined in Exercise 21 and $\lambda_{D^*}(r(f, s))$ as defined in Exercise 18. Show that
$$\lambda_{D^*}(r(f, s)) = \lambda_{D^*}\left(r\left(f^*, \frac{n}{2} - s\right)\right),$$
with $f^*(X) = f(X^{-1})$ and $D^* = $ the adjoint of D.

It is now possible to complete the proof of part (5) of Theorem 2 by writing:
$$J(f, s | Y) \lambda_{D^*}(r(f, s)) = \int_{\mathscr{M}_k} D\theta^Y(X) |X|^s f(X^\circ) \, d\mu_k(X).$$

Break the fundamental domain into two parts according to whether the determinant is greater than one or not. This gives:
$$J(f, s | Y) \lambda_{D^*}(r(f, s)) = \int_{X \in \mathscr{M}_k, |X| \geq 1} + \int_{X \in \mathscr{M}_k, |X| \leq 1}.$$

In the second integral replace X by X^{-1} to see that:
$$J(f, s | Y) \lambda_{D^*}(r(f, s))$$
$$= \int_{\substack{X \in \mathscr{M}_k \\ |X| \geq 1}} \left(|X|^s f(X^\circ) D\theta^Y(X) + |Y|^{-k/2} |X|^{n/2 - s} f^*(X^\circ) D\theta^{Y^{-1}}(X) \right) d\mu_k(X).$$

It is clear from this formula that we have the analytic continuation of $J(f, s | Y)$ to all values of s along with the functional equation:
$$J(f, s | Y) = |Y|^{-k/2} J(f^*, n/2 - s | Y^{-1}), \qquad (5.33)$$

using Exercises 21 and 22. Set $n = k$ and $Y = I$ to obtain the functional equation in part (5) of Theorem 2, thus completing the proof of that theorem. □

Exercise 23 (The Eigenvalue for Koecher's Zeta Function).

(a) Let the differential operator D_1 be defined by (5.32) and the polynomial $h(s)$ be defined by

$$D_1^*|X|^s = (-1)^k h(s)|X|^s, \quad \text{for } X \in \mathscr{P}_k, s \in \mathbb{C}.$$

Show that

$$h(s) = \prod_{j=0}^{k-1}\left(s - \frac{j}{2}\right).$$

(b) Use part (a) to show that if ∂_X is defined by (5.30) then

$$\partial_X |X|^s = h(s + (k-1)/2)|X|^{s-1}.$$

(c) Let D be the operator defined in Exercise 21 and let $\lambda_{D^*}(r(f,s))$ be the eigenvalue defined in Exercise 18. Suppose that f is the automorphic form that is identically one. Show that

$$\lambda_{D^*}(r(1,s)) = h(s)h(n/2 - s).$$

Hint.

$$(-1)^k \Gamma_k(0,\ldots,0,s)h(s) = \int_{\mathscr{P}_k} \exp\{-\text{Tr}(X)\} D_1^*|X|^s \, d\mu_k(X)$$

$$= \int_{\mathscr{P}_k} (D_1 \exp\{-\text{Tr}(X)\})|X|^s \, d\mu_k(X)$$

$$= (-1)^k \Gamma_k(0,\ldots,0,s+1).$$

Now we can consider the analytic continuation of the Eisenstein series defined in formula (5.3).

Corollary (Analytic Continuation and Functional Equation of the Eisenstein Series $E_{k,n-k}(f,s|Y)$). *Suppose that $f \in \mathscr{A}^0(GL(k,\mathbb{Z}),\lambda)$ is an automorphic form on the determinant one surface such that f is integrable over the fundamental domain $\mathscr{SM}_k$ and let f be an eigenfunction of the Hecke operators T_m for all positive integers m. Suppose also that $L_f(s)$ is the L function associated to f in part (5) of Theorem 2. Then the Eisenstein series $E_{k,n-k}(f,s|Y)$, $1 \le k < n$, defined in formula (5.3) of Section 4.5.1 has analytic continuation to all $s \in \mathbb{C}$ as a meromorphic function. Moreover it satisfies the functional equation below, using the notation of part (5) of Theorem 2:*

$$J(f,s|Y) = 2\pi^{-ks}\Gamma_k(r(f,s))L_{f^*}(2s)E_{k,n-k}(f^*,s|Y)$$
$$= |Y|^{-k/2} J(f^*, n/2 - s|Y^{-1}),$$

where $f^*(W) = f(W^{-1})$.

4.5. Automorphic Forms for $GL(n, \mathbb{Z})$ and Harmonic Analysis on $\mathscr{P}_n/GL(n, \mathbb{Z})$

PROOF. The only chore that remains is the proof of (5.28) when $1 \leq k < n$. To do this, we need to show that the *zeta function* defined by:

$$Z(f, s|Y) = \sum_{A \in \mathbb{Z}^{n \times k} \text{ rank } k/GL(k, \mathbb{Z})} |Y[A]|^{-s} f(Y[A]^0), \qquad \text{Re } s > n/2, \quad (5.34)$$

has the factorization

$$Z(f, s|Y) = L_f(2s) E(f, s|Y). \tag{5.35}$$

Use Lemma 1 in Section 4.5.1 to see that summing over rank k matrices $A \in \mathbb{Z}^{n \times k}$ modulo $GL(k, \mathbb{Z})$ is equivalent to summing $A = BC$ over

$$B \in \mathbb{Z}^{n \times k}, \qquad (B^*) \in GL(k, \mathbb{Z})/P(k, n - k),$$
$$C \in \mathbb{Z}^{k \times k}, \qquad \text{rank } k/GL(k, \mathbb{Z}).$$

Here $P(k, n - k)$ is the parabolic subgroup defined in formula (4.58) of Section 4.4.4. It follows that

$$Z(f, s|Y) = \sum_{B, C} |Y[BC]|^{-s} f(Y[BC]^0)$$
$$= \sum_B |Y[B]|^{-s} \sum_{r \geq 1} r^{-2s} \sum_{|C| = r} f((Y[B]^0)[C]^0),$$

which completes the proof of (5.35). The corollary follows from the proof of part (5) of Theorem 2. In particular, the functional equation of the Eisenstein series is formula (5.33). □

Remarks.

(1) The method of analytic continuation using differentiated theta functions is elegant, but it does not allow one to find the residues of the zeta function (5.34) at $s = n/2$, for example. For this, one must investigate the divergent integrals, as Siegel did in [1, Vol. III, pp. 328–333]. There is also an approach using Fourier expansions of automorphic forms (see Terras [8]) and we will consider this method in the next section (see (5.66)–(5.68) and Exercise 34 in Section 4.5.3).

(2) The gamma factors $\Gamma_k(r(f, s))$ appearing in the functional equations of the Dirichlet series $L_f(2s)$ resemble those in the functional equations of Hecke L-functions with grössencharacter (see Hecke [1, pp. 215–234, 249–287]). Of course, for proper congruence subgroups of $GL(2)$, this fact allowed Maass [3] to prove the existence of nonholomorphic cusp forms. For $GL(n, \mathbb{Z})$, one must generalize Theorem 5 to congruence subgroups and prove some kind of a converse theorem in order to obtain results similar to those of Maass. Jacquet, Piatetski-Shapiro, and Shalika [1] manage this in the language of representations of the adelic version of $GL(3)$ and thus prove the existence of an adelic automorphic representation of $GL(3)$ corresponding to L-functions for cubic number fields.

(3) There is a problem with the method of analytic continuation which was presented in this section since we don't know where $L_f(2s)$ vanishes. In the case of Epstein's zeta function the L-function was Riemann's zeta function.

For a general automorphic form f for $GL(n, \mathbb{Z})$ there is much less known about L_f than there is about Riemann's zeta function. The main reference producing a nonvanishing theorem for the adelic version of L-functions associated to automorphic forms for $GL(n)$ is the paper of Jacquet and Shalika [2].

(4) We should perhaps say something about the more general Eisenstein series defined by (5.6). Let us consider the case when $q = 2$ and the parabolic subgroup is maximal. When $n = n_1 + n_2$ write:

$$Y = a[v] \quad \text{with } a = \begin{pmatrix} a_1(Y) & 0 \\ 0 & a_2(Y) \end{pmatrix}, \; a_i(Y) \in \mathscr{P}_{n_i}, \; i = 1, 2,$$

$$\text{and} \quad v = \begin{pmatrix} I_{n_1} & X \\ 0 & I_{n_2} \end{pmatrix}, \; X \in \mathbb{R}^{n_1 \times n_2}. \tag{5.36}$$

Let $f_i \in \mathscr{A}(GL(n_i, \mathbb{Z}), \lambda_i)$, $i = 1, 2$, $Y \in \mathscr{P}_n$, $n = n_1 + n_2$, and define the *Eisenstein series*:

$$E_{n_1, n_2}(f_1, f_2 | Y) = \sum_{A \in GL(n, \mathbb{Z})/P(n_1, n_2)} f_1(a_1(Y[A])) f_2(a_2(Y[A])). \tag{5.37}$$

If $f_i(Y) = |Y|^{r_i} v_i(Y^\circ)$, with $v_i \in \mathscr{A}^\circ(GL(n_i, \mathbb{Z}), \eta_i)$, the series (5.37) will converge if v_i is integrable on $\mathscr{SM}_{n_i}$ and $\operatorname{Re} r_i$ is sufficiently large. For one can use an integral test based on the integral formula in Exercise 25 of Section 4.4.4. In order to obtain an analytic continuation similar to that given in the corollary to Theorem 2, one must relate the Eisenstein series (5.37) to a zeta function:

$$Z_{n_1, n_2}(f_1, f_2 | Y) = \sum_{A \in \mathbb{Z}^{n \times n} \text{ rank } n/P(n_1, n_2)} f_1(a_1(Y[A])) f_2(a_2(Y[A])). \tag{5.38}$$

Exercise 24. Suppose that the automorphic forms f_i in (5.37) can be written:

$$f_i(W) = |W|^{r_i} v_i(W^\circ), \quad \text{with } v_i \in \mathscr{A}^\circ(GL(n_i, \mathbb{Z}), \eta_i), \; r_i \in \mathbb{C},$$

with $\operatorname{Re} r_i$ sufficiently large for convergence of the Eisenstein series. Assume that the automorphic forms v_i are eigenfunctions of all the Hecke operators for $GL(n_i, \mathbb{Z})$. Let $L_{v_i}(s)$ be the L function associated to such an automorphic form as in part (5) of Theorem 2. Show that

$$Z(f_1, f_2 | Y) = E(f_1, f_2 | Y) L_{v_1}(2r_1) L_{v_2}(2r_2).$$

Hint. You need an analogue of Lemma 1 in Section 4.5.1 to be able to write a rank n matrix $A \in \mathbb{Z}^{n \times n}$ modulo $P(n_1, n_2)$ in the form $A = BC$ with

$$B \in GL(n, \mathbb{Z})/P(n_1, n_2),$$

$$C = \begin{pmatrix} C_1 & D \\ 0 & C_2 \end{pmatrix}, \quad C_i \in \mathbb{Z}^{n_i \times n_i} \text{ rank } n_i/GL(n_i, \mathbb{Z}).$$

(5) There are many generalizations of the zeta functions and Eisenstein series considered here. For example, Maass [6] deals with functions generalizing $Z(f, s | Y)$ by adding a spherical function $u: \mathbb{R}^{n \times k} \to \mathbb{C}$. See also Maass [7],

[8] and Christian [3]. In introducing these Maass zeta functions, Maass was motivated by the problem of studying the number of representations of a positive matrix $T \in \mathscr{P}_k$ in the form $T = S[G]$ for $G \in \mathbb{Z}^{n \times k}$, with $n > k$ and $S \in \mathscr{P}_n$ fixed. Maass wanted to study the zeta functions for $GL(n)$ with the additional variable coming from spherical functions for $\mathbb{R}^{n \times k}$ in order to employ a method analogous to that used by Hecke for similar problems in algebraic number fields. The last theorem in the paper of Maass [6] gives the analytic continuation and functional equations of these zeta functions.

The Eisenstein series (5.3) are eigenfunctions of the Hecke operators (5.19). The following formula gives an explicit expression for the eigenvalue.

Proposition 1 (Eigenvalues for the Action of Hecke Operators on Eisenstein Series). *Using the definitions (5.3) of the Eisenstein series $E(\varphi, s | Y)$ and (5.19) for the Hecke operator $T_k^{(n)} = T_k$, we have:*

$$T_k^{(n)} E(\varphi, s | Y) = u_k^{(n)} E(\varphi, s | Y),$$

where

$$u_k^{(n)} = k^{2ms/n} \sum_{t | k} d_{n-m}\left(\frac{k}{t}\right) t^{n-m-2s} u_t^{(m)}.$$

Here $T_t^{(m)} \varphi = u_t^{(m)} \varphi$ and

$$d_r(v) = \sum_{v = \prod_{j=1}^{r} d_j} d_1^{r-1} d_2^{r-2} \cdots d_{r-1},$$

where the d_j are all non-negative integers.

PROOF. Let us change our notation slightly from (5.19) and let $M_k^{(n)}$ denote the set of all $n \times n$ matrices of determinant k. Clearly we need representatives of $M_k^{(n)}/P(m, n - m)$. One can write $A \in M_k^{(n)}$ as $A = BC$, with

$$B = (B_1 *) \in \Gamma_n / P(m, n - m),$$

$$C = \begin{pmatrix} F & H \\ 0 & G \end{pmatrix} \in M_k^{(n)}, \quad F \in \mathbb{Z}^{m \times m}, \quad G \in \mathbb{Z}^{(n-m) \times (n-m)}.$$

Here $P(m, n - m)$ is the maximal parabolic subgroup defined in (4.58).

It follows that the sum $A \in M_k^{(n)}/P(m, n - m)$ is the same as the double sum over

$$(B_1 *) \in \Gamma_n / P(m, n - m) \quad \text{and} \quad F \in M_t^{(m)}/\Gamma_m,$$

$$G \in M_{k/t}^{(n-m)}/\Gamma_{n-m}, \quad H \bmod F,$$

for all divisors t of k. The notation "$H \bmod F$" denotes a complete set of representatives for the equivalence relation:

$$H \sim H' \Leftrightarrow H' = FU + H, \quad \text{for some } U \in \mathbb{Z}^{m \times (n-m)}.$$

The number of $H \bmod F$ is easily seen to be $|F|^{n-m} = t^{n-m}$. The number of $G \in M_{k/t}^{(n-m)}/\Gamma_{n-m}$ is $d_{n-m}(k/t)$, using the definition of $d_r(v)$ to be found in the statement of the proposition.

Putting all this together and setting $P = P(m, n-m)$, we obtain:

$$T_k^{(n)} E_{m,n-m}(\varphi, s | Y) = \sum_{B \in M_k^{(n)}/\Gamma_n} \sum_{\substack{A = (A_1*) \in \Gamma_n/P \\ A_1 \in \mathbb{Z}^{n \times m}}} \varphi(Y[BA_1]^0) | Y[k^{-1/n} BA_1]|^{-s}$$

$$= k^{2ms/n} \sum_{\substack{(A_1*) \in M_k^{(n)}/P \\ A_1 \in \mathbb{Z}^{n \times m}}} \varphi(Y[A_1]^0) | Y[A_1]|^{-s}$$

$$= k^{2ms/n} \sum_{\substack{(B_1*) \in \Gamma_n/P \\ B_1 \in \mathbb{Z}^{n \times m}}} \sum_{t | k} d_{n-m}(k/t) t^{n-m-2s} u_t$$

$$\times \varphi(Y[B_1]^0) | Y[B_1]|^{-s}.$$

This is easily seen to be equal to $u_k^{(n)} E_{m,n-m}(\varphi, s | Y)$. □

The next exercise shows that the Eisenstein series (5.37) always satisfy a trivial functional equation.

Exercise 25 (Trivial Functional Equation of the Eisenstein Series). Consider the Eisenstein series (5.37). Define $f^*(Y) = f(Y^{-1})$. Prove that

$$E_{k,n-k}(f_1, f_2 | Y) = E_{n-k,k}(f_2^*, f_1^* | Y^{-1}).$$

There is a similar argument relating Eisenstein series for the parabolic subgroup $P(n_1, \ldots, n_q)$ to those for the associated parabolic subgroup $P(n_{\sigma(1)}, \ldots, n_{\sigma(q)})$, for any permutation σ of q elements.

Hint. If Y is expressed in the form

$$Y = \begin{pmatrix} a_1 & 0 \\ 0 & a_2 \end{pmatrix} \begin{bmatrix} I_{n_1} & X \\ 0 & I_{n_2} \end{bmatrix},$$

then we have

$$Y^{-1} \begin{pmatrix} 0 & I \\ I & 0 \end{pmatrix} = \begin{pmatrix} a_2^{-1} & 0 \\ 0 & a_1^{-1} \end{pmatrix} \begin{bmatrix} I_{n_2} & {}^t X \\ 0 & I_{n_1} \end{bmatrix}.$$

Next note that if

$$\omega = \begin{pmatrix} 0 & I \\ I & 0 \end{pmatrix},$$

and we set $\gamma = {}^t \omega\, {}^t \tau^{-1} \omega$, then γ runs through $\Gamma_n/P(k, n-k)$ as fast as τ runs through $\Gamma_n/P(n-k, k)$.

The functional equation in Exercise 25 does not extend the domain of convergence. See Exercise 38 below for a similar result.

4.5.3. Fourier Expansions of Eisenstein Series

Next we plan to study Fourier expansions of automorphic forms for $GL(n, \mathbb{Z})$ using methods modeled on those of Siegel [1, Vol. II, pp. 97–137]. Before proceeding further, the reader should review the Fourier expansions of Maass wave forms for $SL(2, \mathbb{Z})$ which we found in Exercise 3 of Section 3.5 in Volume I. The main results in the present section are Fourier expansions of Eisenstein series for $GL(n)$ given in Theorems 3, 4, and 5 from the papers Terras [4] and Imai and Terras [1]. Similar results are obtained in Terras [3], Takhtadzhyan and Vinogradov [1], and Proskurin [1]. These results should also be compared with those of Bump [1] and we will do so later in this section.

Why look at Fourier expansions of automorphic forms for $GL(n)$? There are many reasons beyond simple curiosity. We will see in Section 4.5.4 that harmonic analysis on the fundamental domain $\mathscr{SP}_n/GL(n, \mathbb{Z})$ requires knowledge of the "constant term" in the Fourier expansion of Eisenstein series just as it did for $SL(2, \mathbb{Z})$ in Lemma 1 of Section 3.7 in Volume I. We will discuss the constant term in a rather simple-minded way. A more elegant theory of the constant term has been developed (see Arthur's talk in Borel and Casselman [1, Vol. I, pp. 253–274], Harish-Chandra [1], and Langlands [1]). Of course the constant term will not be a constant in our case. It was not even a constant in the case of Maass wave forms for $SL(2, \mathbb{Z})$.

There are also reasons for considering the "non-constant terms" in these Fourier expansions. For example, we saw in Section 3.6 of Volume I that, in the case of $SL(2, \mathbb{Z})$, a knowledge of the exact form of the non-constant terms was useful in our quest to understand Maass's extension of Hecke's correspondence between modular forms and Dirichlet series (Theorem 2 of Section 3.6, Vol. I). In particular, the Fourier expansion of the Eisenstein series for $SL(2, \mathbb{Z})$, given in Exercise 4 of Section 3.5, Vol. I, has had many applications in number theory. Many of these applications stem from the Kronecker limit formula (to be found in Exercise 6 of Section 3.5, Vol. I). Hecke [1, pp. 198–207] noticed that the Fourier expansion of Epstein's zeta function (see Proposition 2) gives a Kronecker limit formula for zeta functions of number fields. (See Vol. I, pp. 70–74 and Bump and Goldfeld [1].) However the analogue of $\eta(z)$ has yet to be completely understood for $GL(n)$, $n > 2$. Efrat [1] obtains an analogue of $|\eta(z)|$ which is a harmonic automorphic form for $SL(3, \mathbb{Z})$ of weight $\frac{1}{2}$ and considers the consequences for cubic number fields. Siegel [4] found that Hecke's result could be generalized to Hecke L-functions with grössencharacter. Takhtadzhyan and Vinogradov [1] have also announced applications of Fourier expansions of Eisenstein series for $GL(3)$ to the theory of divisor functions.

There has, in fact, been much work on Fourier expansions of automorphic forms for general discrete groups Γ acting on symmetric spaces $X = K\backslash G$ such that the fundamental domain X/Γ has "cusps." For example, Siegel considered $X = SU(n)\backslash Sp(n, \mathbb{R})$ and $\Gamma = Sp(n, \mathbb{Z})$, the Siegel modular group. In this case, X can be identified with the Siegel upper half plane which will be considered in Chapter 5. Siegel obtained the Fourier expansions of holomorphic Eisenstein series (see Siegel [1, Vol. II, pp. 97–137], Baily [1, pp. 228–240], and Chapter 5). Baily [1, p. 238] uses the rationality and bounded denominators of the Fourier coefficients in these expansions to show that the Satake compactification of X/Γ is defined over the field of rational numbers as an algebraic variety. Fourier expansions of non-holomorphic Eisenstein series for $Sp(n, \mathbb{Z})$ have been obtained by Maass [2, Section 18]. The arithmetic parts of both the holomorphic and non-holomorphic Eisenstein series are "singular series" or divisor-like functions (see the discussion in Exercise 4, Section 3.5, Vol. I). The non-arithmetic or analytic part in the holomorphic case is $\exp\{-\text{Tr}(NY)\}$, for $Y \in \mathcal{P}_n$, N a non-negative symmetric half-integral $n \times n$ matrix. Here "half-integral" means that $2n_{ij} \in \mathbb{Z}$, when $i \neq j$ and $n_{ii} \in \mathbb{Z}$. And N "non-negative" means that $N[x] \geq 0$ for all $x \in \mathbb{R}^n$. In the non-holomorphic case the non-arithmetic part is a matrix analogue of a confluent hypergeometric function of the sort which was studied in Section 4.2.2. We will obtain similar results for $GL(n)$.

It is also possible to obtain Fourier expansions of Eisenstein series for congruence subgroups of $SL(2, O_K)$, where O_K is the ring of integers of a number field. This will be discussed in the next chapter. Such Fourier expansions have been used by number theorists to study Gauss sums and elliptic curves for example (see Kubota [2], Heath-Brown and Patterson [1], Goldfeld, Hoffstein, and Patterson [1]).

Many authors take an adelic representation theoretic approach to the subject of Fourier expansions for $GL(n)$. See, for example, Jacquet, Piatetski-Shapiro, and Shalika [1]—a paper which makes use of Whittaker models for representations. Stark [3] indicates a way to bridge the gap between the classical and adelic points of view. See also Rhodes [1].

From our earlier study of the case of $SL(2)$, in view of work of Harish-Chandra and Langlands, we should be willing to believe that the spectral measure in the spectral resolution of the Laplacian on $L^2(\mathcal{SP}_n/GL(n, \mathbb{Z}))$ comes from the asymptotics and functional equations of the Eisenstein series. Knowing the asymptotic behavior of the Eisenstein series as the argument $Y \in \mathcal{P}_n$ approaches the boundary of the fundamental domain is the same as knowing the "constant term" in the Fourier expansion. Let us attempt to find a simple-minded way of obtaining this constant term in the region where the Dirichlet series defining $E_P(\varphi, s | Y)$ converges. That is we want to find a generalization of the method we had in mind for part (b) of Exercise 2 in Section 3.5 of Volume I.

As our first example, consider the Eisenstein series $E_{m, n-m}(\varphi, s | Y)$ defined in (5.3). Let $Y \in \mathcal{P}_n$ have partial Iwasawa decomposition:

$$Y = \begin{pmatrix} V & 0 \\ 0 & W \end{pmatrix} \begin{bmatrix} I & X \\ 0 & I \end{bmatrix}, \qquad V \in \mathcal{P}_m, \qquad W \in \mathcal{P}_{n-m}, \qquad X \in \mathbb{R}^{m \times (n-m)}.$$

Look at the Eisenstein series defined in (5.3) of Section 4.5.1 by:

$$E_{m, n-m}(\varphi, s | Y) = \sum_{A = (A_1 *) \in \Gamma_n/P} \varphi(Y[A_1]^0) |Y[A_1]|^{-s}, \qquad \text{for Re } s > n/2,$$

4.5. Automorphic Forms for $GL(n, \mathbb{Z})$ and Harmonic Analysis on $\mathscr{P}_n/GL(n, \mathbb{Z})$ 217

where $\Gamma_n = GL(n, \mathbb{Z})$ and $P = P(m, n - m)$. Here we assume that the automorphic form $\varphi \in \mathscr{A}^0(GL(m, \mathbb{Z}), \lambda)$ is bounded on the determinant one surface of the fundamental domain. Write

$$A_1 = \begin{pmatrix} B \\ C \end{pmatrix}, \quad B \in \mathbb{Z}^{m \times m}, \quad \text{so that } Y[A] = V[B + XC] + W[C].$$

If W approaches infinity in the sense that the diagonal entries in its Iwasawa decomposition all approach infinity, then it is not too hard to see that when $\operatorname{Re} s > n/2$, the term $|Y[A]|^{-s}$ must approach zero unless $C = 0$. It follows that $A \in P$ and so $A = I$. Thus we find that for fixed s with $\operatorname{Re} s > n/2$

$$E_{m, n-m}(\varphi, s | Y) \sim \varphi(V^0) |V|^{-s},$$

as W goes to infinity in the sense described above. When s is not in the region of convergence of the Dirichlet series, the functional equation can add in other terms to this asymptotic formula—a phenomenon that we saw already in the case of $SL(2, \mathbb{Z})$ in Section 3.5 of Volume I.

The preceding example is a little too simple-minded perhaps. So let us try to be a little more explicit about the approach to the boundary of the fundamental domain while attempting to consider the more general Eisenstein series defined by formula (5.37) in Section 4.5.2. We also want to relate all this to the theory of roots, parabolic subgroups, and Bruhat decompositions.

These considerations lead us to examine the possible ways of approaching the boundary of our fundamental domain. For concreteness, let us consider only the case of $GL(4)$. In this case we can write

$$Y = A[n], \quad \text{where } A = \begin{pmatrix} a_1 & 0 & 0 & 0 \\ 0 & a_2 & 0 & 0 \\ 0 & 0 & a_3 & 0 \\ 0 & 0 & 0 & a_4 \end{pmatrix} \quad \text{and} \tag{5.39}$$

$$n = \begin{pmatrix} 1 & * & * & * \\ 0 & 1 & * & * \\ 0 & 0 & 1 & * \\ 0 & 0 & 0 & 1 \end{pmatrix}.$$

The ways of approaching the boundary of $\mathscr{SP}_4/GL(4, \mathbb{Z})$ can be described by subsets of the set of quotients a_i/a_j, $i < j$, which are allowed to approach zero. These quotients are the multiplicative version of "roots" to be discussed in Chapter 5. See Figure 4.18 for a diagram of the various sets of roots giving ways of approaching the boundary, as well as the corresponding parabolic subgroups.

The general theory of parabolic subgroups says that they correspond to sets J of simple roots $\alpha_i(a) = a_i/a_{i+1}$ (see Borel's article in Borel and Mostow [1, pp. 1–19]) via:

$$\bigcap_{\alpha \in J} \ker \alpha = S_J, \quad P_J = Z(S_J)N, \tag{5.40}$$

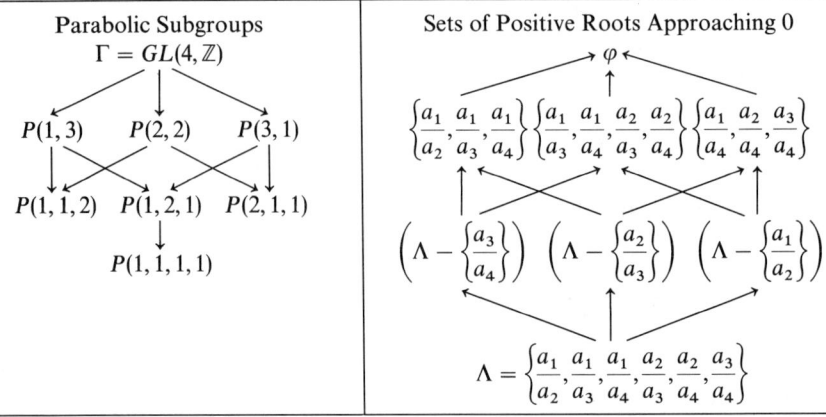

Figure 4.18. Parabolic subgroups of $GL(4)$ and ways to approach the boundary of $\mathscr{P}_4/GL(4,\mathbb{Z})$. Arrows indicate containment.

where N is the nilpotent or unipotent subgroup of upper triangular matrices with ones on the diagonal and $Z(S_J)$ denotes the centralizer of S_J. Let W be the Weyl group of all permutation matrices in K (cf. Exercise 29 below). Denote by w_i the Weyl group element:

$$w_i = \begin{pmatrix} 1 & & & & & & & \\ & \ddots & & & & & & \\ & & 1 & & & 0 & & \\ & & & 0 & 1 & & & \\ & & & -1 & 0 & & & \\ & & & & & 1 & & \\ & 0 & & & & & \ddots & \\ & & & & & & & 1 \end{pmatrix}, \quad (5.41)$$

with $i-1$ and 1 indicated,

which represents the permutation that interchanges i and $i+1$ leaving all else fixed. Then set

$W_J =$ the group generated by the w_i for $\alpha_i \in J$.

One can show that $P_J = BW_J B$, where B is the Borel or minimal parabolic subgroup $B = P(1,1,\ldots,1)$, using the notation (4.58). The *Bruhat decomposition* of $GL(n,\mathbb{R})$ corresponding to parabolic subgroups P_J and $P_{J'}$ is:

$$GL(n,\mathbb{R}) = \bigcup_{w \in W_J \backslash W/W_{J'}} P_J w P_{J'} \quad \text{(disjoint)}. \quad (5.42)$$

This result was discussed by Curtis [1]. We will consider only some special cases in what follows (see Exercises 28 and 35 as well as Lemma 5 below).

Now let's look at some simple examples of (5.40).

4.5. Automorphic Forms for $GL(n, \mathbb{Z})$ and Harmonic Analysis on $\mathscr{P}_n/GL(n, \mathbb{Z})$

Example 1. $P(3, 1)$ corresponds to the group generated by w_1, w_2 and the set
$$J = \{a_1/a_2, a_2/a_3\}$$
of simple roots.

The roots in Figure 4.17 which correspond to the parabolic subgroup $P(3, 1)$ are $\{a_1/a_4, a_2/a_4, a_3/a_4\}$. The complement of this set is the set formed from roots in J and products of roots in J.

Example 2. $P(2, 2)$ corresponds to the group generated by w_1 and w_3 and the set
$$J = \{a_1/a_2, a_3/a_4\}$$
of simple roots.

Now the roots in Figure 4.17 which correspond to $P(2, 2)$ form the set $\{a_1/a_3, a_1/a_4, a_2/a_3, a_2/a_4\}$ which is the complement of J.

Now we want to know how the Eisenstein series $E_{2,2}$ behaves as Y approaches the boundary in the direction corresponding to the parabolic subgroup $P(2, 2)$. We use the notation:

$$A_i = \begin{pmatrix} a_i & 0 \\ 0 & a_{i+1} \end{pmatrix} \begin{pmatrix} 1 & x_i \\ 0 & 1 \end{pmatrix}, \qquad i = 1, 3;$$

$$Y = I\left[\begin{pmatrix} A_1 & 0 \\ 0 & A_3 \end{pmatrix} \begin{pmatrix} I_2 & X_3 \\ 0 & I_2 \end{pmatrix}\right].$$

Regard again the Eisenstein series $E_{2,2}(f_1, f_2 | Y)$ in (5.37) with $f_i(W) = |W|^{r_i} v_i(W^0)$, $v_i \in \mathscr{A}^0(GL(n_i, \mathbb{Z}), \eta_i)$. It will be shown that for $\text{Re}(r_1 - r_2) > 2$, there is the following asymptotic formula:

$$E_{2,2}(f_1, f_2 | Y) \sim f_1(A_1) f_2(A_3), \qquad \text{as } A_1 A_3^{-1} \to 0. \tag{5.43}$$

Here "$A_1 A_3^{-1} \to 0$" means that $a_1/a_3, a_1/a_4, a_2/a_3, a_2/a_4$ all approach zero.

To prove (5.43), note that upon setting

$$a = \begin{pmatrix} A_1 & 0 \\ 0 & A_3 \end{pmatrix}, \qquad n' = \begin{pmatrix} I_2 & X_3 \\ 0 & I_2 \end{pmatrix},$$

$$\gamma = \begin{pmatrix} \gamma_{11} & \gamma_{12} \\ \gamma_{21} & \gamma_{22} \end{pmatrix}, \qquad \gamma_{ij} \in \mathbb{Z}^{2 \times 2},$$

then $Y = I[an]$ and

$$Y[\gamma] = I[an\gamma] = I[ana^{-1} a \gamma a^{-1} a]$$

$$= I\left[\begin{pmatrix} I_2 & A_1 X_3 A_3^{-1} \\ 0 & I_2 \end{pmatrix} \begin{pmatrix} A_1 \gamma_{11} A_1^{-1} & A_1 \gamma_{12} A_3^{-1} \\ A_3 \gamma_{21} A_1^{-1} & A_3 \gamma_{22} A_3^{-1} \end{pmatrix} a\right].$$

The first matrix in the last bracket approaches the identity as $A_1 A_3^{-1}$ approaches zero (see the discussion following for more details). In order for the 2nd matrix in brackets to remain finite (as well as the corresponding term in the Eisenstein series), the block entry γ_{21} *must vanish*. Thus γ must be in $P(2,2)$. So we must be looking at a term of the Eisenstein series which corresponds to the coset of the identity matrix. To see this in more detail, note that

$$A_1 X_3 A_3^{-1} = \begin{pmatrix} 1 & 0 \\ x_1 & 1 \end{pmatrix} \begin{pmatrix} a_1 & 0 \\ 0 & a_2 \end{pmatrix} X_4 \begin{pmatrix} a_3^{-1} & 0 \\ 0 & a_4^{-1} \end{pmatrix} \begin{pmatrix} 1 & 0 \\ -x_3 & 1 \end{pmatrix},$$

where

$$X_4 = \begin{pmatrix} u & v \\ y & z \end{pmatrix} = \begin{pmatrix} 1 & x_1 \\ 0 & 1 \end{pmatrix} X_3 \begin{pmatrix} 1 & -x_3 \\ 0 & 1 \end{pmatrix}.$$

It follows that

$$A_1 X_3 A_3^{-1} = \begin{pmatrix} 1 & 0 \\ x_1 & 1 \end{pmatrix} \begin{pmatrix} a_1 u a_3^{-1} & a_1 v a_4^{-1} \\ a_2 y a_3^{-1} & a_2 z a_4^{-1} \end{pmatrix} \begin{pmatrix} 1 & 0 \\ -x_3 & 1 \end{pmatrix}.$$

This certainly approaches zero as $a_1/a_3, a_1/a_4, a_2/a_3, a_2/a_4$ all approach zero. Similarly, each entry in $A_3 \gamma_{21} A_1^{-1}$ must approach infinity unless γ_{21} vanishes. Note that the corresponding term of the Eisenstein series involves

$$\left| I \begin{bmatrix} A_1(\gamma_{11} + X_3 \gamma_{21}) A_1^{-1} \\ A_3 \gamma_{21} A_1^{-1} \end{bmatrix} \right|^{r_2 - r_1}.$$

So we are looking at a term $|I[B] + I[C]|^s$, where $\operatorname{Re} s < 0$, and the eigenvalues of B remain bounded while those of C become infinite unless γ_{21} vanishes. Therefore this term goes to zero unless γ_{21} is zero and then $\gamma = I$. So we obtain (5.43).

To complete the discussion of $E_{2,2}$, one should note that if we let Y approach the boundary in a direction dictated by some other parabolic subgroup such as $P(3, 1)$, then we will see it approach zero, under the hypothesis that v_2 is a cusp form. For example, consider the behavior of $E_{2,2}$ as Y approaches the boundary in the direction corresponding to $P(3, 1)$; i.e., as $a_i/a_4 \to 0$, for $i = 1, 2, 3$. Note that we can let a_i/a_4 approach zero so that the quotients:

$$\frac{a_1}{a_3} = \frac{a_1/a_4}{a_3/a_4} \quad \text{and} \quad \frac{a_2}{a_3} = \frac{a_2/a_4}{a_3/a_4}$$

both approach zero. This puts us in the situation that we just encountered, except that, in addition, we are letting a_3/a_4 approach zero. Now if v_2 is a cusp form

$$v_2(A_3^0) \to 0 \quad \text{as} \quad a_3/a_4 \to 0,$$

because

$$A_3^0 = \begin{pmatrix} \sqrt{a_3/a_4} & 0 \\ 0 & \sqrt{a_4/a_3} \end{pmatrix} \begin{pmatrix} 1 & x_2 \\ 0 & 1 \end{pmatrix}.$$

4.5. Automorphic Forms for $GL(n, \mathbb{Z})$ and Harmonic Analysis on $\mathscr{P}_n/GL(n, \mathbb{Z})$

The usual discussions of the asymptotics of Eisenstein series involve the computation of integrals like the following integral which represents the zeroth Fourier coefficient of $E_{m,n-m}$ with respect to $P(m, n-m)$:

$$\int_{X \in (\mathbb{R}/\mathbb{Z})^{m \times (n-m)}} E_{m,n-m}\left(f \left| Y \begin{bmatrix} I & X \\ 0 & I \end{bmatrix}\right.\right) dX.$$

See, for example, Langlands [1]. The general result obtained by Langlands says that the constant term for E_P involves a sum over the Weyl group W_J associated to $P = P_J$ by (5.40). The terms in this sum are of the sort obtained in (5.43) multiplied by factors appearing in the functional equations of the Eisenstein series. In order to connect the two methods, one must understand the asymptotic behavior of the terms in the Fourier expansions of the Eisenstein series.

It would also be interesting to clarify the connection between the various notions of approach to the boundary of the fundamental domain and the compactifications of the fundamental domain that have been considered by authors such as Satake [3], Baily and Borel [1], Borel and Serre [1], and Ash, Mumford, Rapoport, and Tai [1]. See also Freitag [1]. To create a compactification, Satake adjoins to a fundamental domain $\mathscr{F}_n$ for $GL(n, \mathbb{Z})$ a union of lower rank fundamental domains $\mathscr{F}_m$ and defines approach of $Y \in \mathscr{F}_n$ to an element V^* of the lower rank fundamental domain $\mathscr{F}_m$ if

$$Y = \begin{pmatrix} V & 0 \\ 0 & W \end{pmatrix} \begin{bmatrix} I & X \\ 0 & I \end{bmatrix}, \quad V \in \mathscr{P}_m, \quad W \in \mathscr{P}_{n-m}, \quad X \in \mathbb{R}^{m \times (n-m)}, \quad (5.44)$$

and V approaches V^* while W approaches infinity in the sense that the diagonal part in the Iwasawa decomposition of W goes to infinity.

It would also be useful to clarify the connections with the truncations used by Arthur [1] in the continuous spectrum integrals of Eisenstein series appearing in the trace formula.

We will discuss several methods of obtaining Fourier expansions. The first method comes from Terras [4, 11] and goes back to the paper of Chowla and Selberg [1] which makes use of theta functions. The other methods are analogous to that used by Siegel [1, Vol. II, pp. 97–137] for $Sp(n, \mathbb{Z})$ and come from Imai and Terras [1].

First note that any automorphic form $f(Y)$, for $f \in \mathscr{A}(GL(n, \mathbb{Z}), \lambda)$, must be a periodic function of each entry of the matrix X when Y has the partial Iwasawa decomposition (5.44) above. Therefore $f(Y)$ must have a Fourier expansion:

$$f(Y) = \sum_{N \in \mathbb{Z}^{n \times (n-m)}} c_N(V, W) \exp\{2\pi i \operatorname{Tr}({}^t N X)\}. \quad (5.45)$$

Many authors have obtained such expansions. The case that $|Y| = 1$, $n = 2$, is considered in Exercise 3, Section 3.5, Volume I. Here we are mainly motivated by the work of Siegel [1, Vol. II, pp. 97–137] and Maass [2] for $Sp(n, \mathbb{Z})$. See also Freitag [1].

The terms $c_N(V, W)$ in the Fourier expansion (5.45) always have two parts. For Eisenstein series, one part is *arithmetic*—either a singular series (for an Eisenstein series defined as a sum over Γ_n/P) or a divisor-type function (for an Eisenstein series defined as a sum over rank m matrices in $\mathbb{Z}^{n \times m}/\Gamma_m$). When $n = 2$, the arithmetic part of the kth Fourier coefficient of the Eisenstein series is:

$$k^s \sum_{0 < t | k} t^{1-2s} = k^s \sigma_{1-2s}(k). \tag{5.46}$$

The singular series version of this is:

$$\sum_{\substack{c > 0, d \bmod c \\ (d,c)=1}} c^{-2s} \exp(2\pi i k d/c) = \sigma_{1-2s}(k)/\zeta(2s). \tag{5.47}$$

We found in Terras [3] (see Exercise 35) that the arithmetic part of the term corresponding to $N \in \mathbb{Z}^{m \times (n-m)}$ of rank m in $E_{m,n-m}(1,s|Y)$ is essentially the singular series:

$$\sum_{R \in (\mathbb{Q}/\mathbb{Z})^{m \times m}} v(R)^{-2s} \exp\{2\pi i \operatorname{Tr}({}^t R N)\}. \tag{5.48}$$

Here $v(R)$ is the product of the reduced denominators of the elementary divisors of R. Siegel [1, Vol. II, pp. 97–137] obtains an analogous result for holomorphic Eisenstein series for $Sp(n, \mathbb{Z})$. Maass [2, Section 18] does the non-holomorphic case for $Sp(n, \mathbb{Z})$. Lower rank terms are more complicated to describe. For example, the "most singular" term listed in Maass [2, p. 307] required quite a long computation. We will find here that the arithmetic part of Fourier expansions of $E(\varphi, s|Y)$ cannot be separated out so easily for general $\varphi \in \mathscr{A}^0(\Gamma_m, \eta)$.

The terms in the Fourier expansion (5.45) of an automorphic form $f \in \mathscr{A}(\Gamma_n, \lambda)$ will have a second part which is *analytic*—a matrix argument confluent hypergeometric function. For $GL(n, \mathbb{Z})$ one obtains analogues either of K-Bessel or Whittaker functions. We work mostly with K-Bessel functions because we are attempting to stay close to the Siegel-type Fourier expansions. We should caution the reader that most researchers use a slightly different formulation (see Jacquet [1], Bump [1], Proskurin [1]). We will discuss the connections between the K-Bessel function expansions and the Whittaker function expansions later in the section (see (5.63) below).

As we mentioned, the Fourier expansion (5.45) is analogous to that for Siegel modular forms. Koecher [1] proved that in the definition of holomorphic Siegel modular forms for $Sp(n, \mathbb{Z})$, one needs no hypothesis on the behavior of the form at infinity when $n \geq 2$ (see also Maass [2, pp. 185–187] and Chapter 5) to rule out terms $\exp\{-\operatorname{Tr}(NY)\}$, where N is not non-negative. It would be interesting to know whether a similar phenomenon occurs for $GL(n, \mathbb{Z})$—ruling out the analogues of I-Bessel functions in Fourier expansions (5.45) without explicitly assuming the third hypothesis in the definition (5.1) of automorphic form for $GL(n, \mathbb{Z})$.

I. Method of Chowla-Selberg

This method is a generalization of that of Chowla and Selberg [1]. In Proposition 2 we consider the method for the special case of Epstein's zeta function. Then in Theorem 3 we consider only the terms of maximal rank in Fourier expansions of Eisenstein series $E_{m,n-m}$ with respect to the parabolic subgroup $P(m, n - m)$.

Proposition 2 (Fourier Expansions of Epstein's Zeta Function). *Let $Z_{1,n-1}(1, s|Y)$ be Epstein's zeta function, defined for $\operatorname{Re} s > n/2$ and $Y \in \mathcal{P}_n$ by:*

$$Z_{1,n-1}(1,s|Y) = \frac{1}{2} \sum_{a \in \mathbb{Z}^n - 0} Y[a]^{-s}.$$

Here we use the notation (5.34) in Section 4.5.2 rather than (4.2) of Section 4.4.1. If Y has the partial Iwasawa decomposition (5.44), then we have the following Fourier expansion of the normalized Eisenstein series:

$$J_{1,n-1}(1,s|Y)$$
$$= 2\pi^{-s}\Gamma(s)Z_{1,n-1}(1,s|Y)$$
$$= J_{1,m-1}(1,s|V) + |V|^{-1/2} J_{1,n-m-1}(1, s - m/2|W)$$
$$+ 2|V|^{-1/2} \sum_{\substack{b \in \mathbb{Z}^m - 0 \\ c \in \mathbb{Z}^{n-m} - 0}} \exp\{2\pi i\, {}^tbXc\} \left(\frac{V^{-1}[b]}{W[c]}\right)^{(2s-m)/4}$$
$$\times K_{s-m/2}(2\pi\sqrt{V^{-1}[b]W[c]}).$$

Here $K_s(y)$ is the usual K-Bessel function defined by the formula in Exercise 1 on p. 136 of Volume I.

PROOF. (Compare Epstein [1], Terras [11].) We know that (as in part (a) of Exercise 7 on p. 60 of Volume I):

$$J_{1,n-1}(1,s|Y) = \int_{t>0} t^{s-1}\theta_1(Y,t)\,dt,$$

where

$$\theta_1(Y,t) = \sum_{a \in \mathbb{Z}^n - 0} \exp\{-\pi Y[a]t\}.$$

Write

$$a = \binom{b}{c}, \quad \text{with } b \in \mathbb{Z}^m.$$

Then $Y[a] = V[b + Xc] + W[c]$. We can split the sum defining θ_1 into two parts—that consisting of terms with $c = 0$ and the rest. The part of the sum with $c = 0$ gives $J_{1,m-1}(1,s|V)$. The rest of the terms have $c \neq 0$ and thus b is summed over all of $\mathbb{Z}^m$ and we can apply Poisson summation to obtain:

$$\sum_{b \in \mathbb{Z}^m} \exp\{-\pi V[b + Xc]t\} = |V|^{-1/2} t^{-m/2} \sum_{b \in \mathbb{Z}^m} \exp\{2\pi i\, {}^tbXc - \pi V^{-1}[b]t^{-1}\}.$$

Substitute this into our integral and obtain the result, since the $b = 0$ term gives

$$|V|^{-1/2} J_{1,n-m-1}(1, s - m/2 | W)$$

and the rest of the terms come from summing over non-zero b and c:

$$|V|^{-1/2} \exp\{2\pi i\, {}^tbXc\} \int_{t>0} t^{s-1-m/2} \exp\{-\pi(V^{-1}[b]t^{-1} + W[c]t)\}\, dt.$$

The integral is easily evaluated in terms of the ordinary K-Bessel function to complete the proof. $\square$

Exercise 26.

(a) Prove that if Y has the partial Iwasawa decomposition (5.44) with $v > 0$, $W \in \mathcal{P}_{n-1}$, $X \in \mathbb{R}^{1 \times (n-1)}$ in the case $m = 1$, and if $\operatorname{Re} s > n/2$, then

$$E_{1,n-1}(1,s|Y) \sim v^{-s},$$

as W approaches infinity in the sense that the diagonal elements in the Iwasawa decomposition of W all approach infinity.

(b) Use Proposition 2 to deduce the functional equation of Epstein's zeta function.

(c) Use Proposition 2 to find an analogue of the Kronecker limit formula from Exercise 6 on p. 210 of Volume I.

Hint. See Terras [11] and Efrat [1].

Theorem 3 (Highest Rank Terms in the Fourier Expansion of an Eisenstein Series). *Consider the Eisenstein series $E_{m,n-m}(\varphi, s|Y)$ defined in (5.3) of Section 4.5.1 in the special case that φ is Selberg's Eisenstein series; i.e., $\varphi(W) = E_{(m)}(r|W)$, $r \in \mathbb{C}^{n-1}$, as defined in (5.7) of Section 4.5.1. Suppose that $m < n/2$. Then when $N \in \mathbb{Z}^{m \times m}$ is non-singular, the Nth Fourier coefficient in the expansion (5.45) of the normalized Eisenstein series*

$$J(\varphi, s|Y) = 2\pi^{-ms} \Gamma_m(r(\varphi, s)) L_{\varphi^*}(2s) E_{m,n-m}(\varphi^*, s|Y)$$

as in (5.28) and Exercise 18 of Section 4.5.2 is:

$$2|V|^{-m/2} \sum_{\substack{D | N \\ D \in \mathbb{Z}^{m \times m}/\Gamma_m}} p_{-r}(I[D]) \sum_{a \in P_{(m)} \backslash \Gamma_m} K_m(-r, s - m/2 | \pi W[{}^tN\, {}^ta], \pi V^{-1}[a^{-1}]),$$

where "$D|N$" means that there is a matrix $C \in \mathbb{Z}^{(n-m) \times m}$ such that $N = D\,{}^tC$.

4.5. Automorphic Forms for $GL(n, \mathbb{Z})$ and Harmonic Analysis on $\mathscr{P}_n/GL(n, \mathbb{Z})$

This says that each elementary divisor in D divides the corresponding elementary divisor in N.

PROOF (Terras [4]). The proof will be similar to that of Proposition 2. Begin with formulas (5.27) and (5.28) for the normalized Eisenstein series:

$$J(f, s|Y) = \int_{H \in \mathscr{M}_m} \varphi(H^0)|H|^s \theta_m(Y, H) \, d\mu_m(H). \tag{5.49}$$

Since we have the partial Iwasawa decomposition (5.44), we find that the partial theta function can be written in the following way for $B \in \mathbb{Z}^{m \times m}$:

$$\theta_m(Y, H) = \sum_{\binom{B}{C} \in \mathbb{Z}^{n \times m},\, \text{rank } m} \exp\left\{-\pi \operatorname{Tr}\left(Y \begin{bmatrix} B \\ C \end{bmatrix} H\right)\right\}$$

$$= \sum_{\binom{B}{C} \in \mathbb{Z}^{n \times m},\, \text{rank } m} \exp\{-\pi \operatorname{Tr}(V[B + XC]H + W[C]H)\}.$$

The terms of maximal rank in the Fourier expansion of $E_{m, n-m}$ correspond to the matrices C of rank m in $\mathbb{Z}^{(n-m) \times m}$. (We shall not consider the other terms here.) For such terms B is summed over the full lattice $\mathbb{Z}^{m \times m}$ and we can use Poisson summation on B. This leads to the following formula for the terms with C of rank m in $\theta_m(Y, H)$:

$$|H|^{-m/2} |V|^{-m/2} \sum_{B \in \mathbb{Z}^{m \times m}} \exp\{2\pi i \operatorname{Tr}({}^t BXC) - \pi \operatorname{Tr}(V^{-1}[B]H^{-1})\}.$$

Substitute this into (5.49) to find that the terms with C of rank m are:

$$|V|^{-m/2} \sum_{\substack{B \in \mathbb{Z}^{m \times m}\, \text{rank } m/\Gamma_m \\ C \in \mathbb{Z}^{(n-m) \times m}\, \text{rank } m}} \exp\{2\pi i \operatorname{Tr}(C\,{}^t BX)\} I(\varphi, s - m/2 | \pi W[C], \pi V^{-1}[B]),$$

(5.50)

where for $F, G \in \mathscr{P}_m$, $s \in \mathbb{C}$:

$$I(\varphi, s | F, G) = \int_{H \in \mathscr{P}_m} |H|^s \varphi(H^0) \exp\{-\operatorname{Tr}(FH + GH^{-1})\} \, d\mu_m(H).$$

Now we need a lemma.

Lemma 4. *Suppose $A, B \in \mathscr{P}_m$, $r \in \mathbb{C}^m$, $\operatorname{Re} r_i > 1$. If $E_{(m)}(r|Y)$ is the Eisenstein series (5.7), we have the following expression for the integral above in terms of K-Bessel functions from (2.21) in Section 4.2.2:*

$$I(E_{(m)}(r|*), s | F, G) = \int_{H \in \mathscr{P}_m} E_{(m)}(r, H) |H|^s \exp\{-\operatorname{Tr}(FH + GH^{-1})\} \, d\mu_m(H)$$

$$= 2 \sum_{u \in P_{(m)} \backslash \Gamma_m} K_m(-r + (0, s) | F[{}^t u], G[u^{-1}]).$$

PROOF. Let N be the nilpotent subgroup of $GL(n, \mathbb{R})$ consisting of all upper triangular matrices with ones on the diagonal. We find that

$$2\int_{H\in\mathcal{P}_m} E_{(m)}(r|H)\exp\{-\mathrm{Tr}(FH+GH^{-1})\}\,d\mu_m(H)$$

$$=\int_{Y\in\mathcal{P}_m/\Gamma_m}\sum_{\tau\in\Gamma_m/\Gamma_m\cap N}p_{-r}(Y[\tau])\sum_{\gamma\in\Gamma_m}\exp\{-\mathrm{Tr}(FY[\gamma]+G(Y[\gamma])^{-1}\}\,d\mu_m(Y)$$

$$=\int_{Y\in\mathcal{P}_m/N} p_{-r}(Y)\int_{n\in N/N\cap\Gamma_m}\sum_{\gamma\in\Gamma_m}\exp\{-\mathrm{Tr}(FY[n\gamma]+G(Y[n\gamma])^{-1})\}\,dn\,d\mu_m(Y).$$

Here the factor of "2" comes from the order of the center of Γ_m. On the other hand,

$$K_m(-r|A,B)$$
$$=\int_{Y\in\mathcal{P}_m/N}p_{-r}(Y)\int_{n\in N}\exp\{-\mathrm{Tr}(AY[n]+B(Y[n])^{-1}\}\,dn\,d\mu_m(Y).$$

Thus it suffices to show the easily verified identity:

$$\int_{n\in N/N\cap\Gamma_m}\sum_{\gamma\in\Gamma_m}\exp\{-\mathrm{Tr}(FY[n\gamma]+G(Y[n\gamma])^{-1})\}\,dn$$
$$=\sum_{u\in\Gamma_m\cap N\backslash\Gamma_m}\int_{n\in N}\exp\{-\mathrm{Tr}(F[{}^tu]Y[n]+G[u^{-1}](Y[n])^{-1})\}\,dn.$$

This completes the proof of the lemma. □

Now we return to the proof of Theorem 3. From the lemma, we see that (5.50) is equal to

$$2|V|^{-m/2}\sum_{\substack{B\in\mathbb{Z}^{m\times m}\text{ rank }m/\Gamma_m \\ C\in\mathbb{Z}^{(n-m)\times m}\text{ rank }m}}\exp\{2\pi i\,\mathrm{Tr}(C^tBX)\}$$
$$\times\sum_{u\in P_{(m)}\backslash\Gamma_m}K_m\left(-r,s-\frac{m}{2}\bigg|\pi W[C^tu],\pi V^{-1}[Bu^{-1}]\right).$$

To finish this proof, we need to make use of another kind of Hecke operator to move the B around to be next to C in

$$I(\varphi,s|\pi W[C],\pi V^{-1}[B]).$$

These Hecke operators are associated to a matrix $N\in\mathbb{Z}^{m\times(n-m)}$ and defined for $f:\mathcal{P}_m\to\mathbb{C}$ by;

$$T_N f(Y)=\sum_{\substack{B\in\mathbb{Z}^{m\times m}/\Gamma_m \\ N=B^tC,\text{ for some }C\in\mathbb{Z}^{(n-m)\times m}}}f(Y[B]).$$

Then we have the following exercise which finishes the proof of Theorem 3.

Exercise 27.

(a) Show that for $N\in\mathbb{Z}^{m\times(n-m)}$ of rank m (assuming that $m>n-m$),

$$T_N E_{(m)}(r|Y)=a_N(r)E_{(m)}(r|Y),$$

4.5. Automorphic Forms for $GL(n, \mathbb{Z})$ and Harmonic Analysis on $\mathscr{P}_n/GL(n, \mathbb{Z})$

where

$$a_N(r) = \sum_{\substack{D|N \\ D \in \mathbb{Z}^{m \times m}/\Gamma_m}} p_{-r}(I[D]).$$

Here "$D|N$" means that there is a matrix $C \in \mathbb{Z}^{(n-m) \times m}$ such that $N = D\,{}^tC$.
(b) Use part (a) to finish the proof of Theorem 3. □

It would be interesting to use similar methods to find the lower rank Fourier coefficients and to deal with the case that φ is a cusp form rather than an Eisenstein series, but we shall not do that here.

Now let us consider other methods for finding Fourier expansions of Eisenstein series.

II. Methods of Imai and Terras [1] and Terras [3]

The prerequisite for Fourier expansions of Eisenstein series using the method of Siegel (see Baily [1, pp. 228–240], Maass [2, pp. 300–308], Siegel [1, Vol. II, pp. 97–137], and Terras [3]) is the Bruhat decomposition or some related matrix matrix decomposition.

Exercise 28 (The Bruhat Decomposition of $SL(n, \mathbb{Q})$ with Respect to $P(n-1, 1)$). Let $P(n-1, 1) = P \subset G = SL(n, \mathbb{Q})$ be defined as in (4.58) of Section 4.4.4 by:

$$P = \left\{ \begin{pmatrix} A & b \\ 0 & c \end{pmatrix} \in G \,\middle|\, A \in GL(n-1, \mathbb{Q}), b \in \mathbb{Q}^{n-1}, c \in \mathbb{Q} - 0 \right\}.$$

Show that we have the following disjoint union:

$$SL(n, \mathbb{Q}) = P \cup (P\sigma P),$$

where

$$\sigma = \begin{pmatrix} 0 & 0 & 1 \\ 0 & I_{n-2} & 0 \\ -1 & 0 & 0 \end{pmatrix}.$$

Hint. Note that a matrix

$$\begin{pmatrix} E & F \\ g & h \end{pmatrix} \in SL(n, \mathbb{Q}) \quad \text{with} \quad E \in \mathbb{Q}^{(n-1) \times (n-1)}, h \in \mathbb{Q},$$

lies in $P\sigma P$ if and only if the rank of g is one.

The result of Exercise 28 should be compared with the general Bruhat decomposition (5.42). Next let us consider another view of the Weyl group W of all permutation matrices in K. Define the following subgroups of $K = O(n)$:

M = the centralizer of A in K = $\{k \in K | kak^{-1} = a, \text{ for all } a \in A\}$,

M' = the normalizer of A in K = $\{k \in K | kak^{-1} \in A, \text{ for all } a \in A\}$.
(5.51)

Here A is the group of all positive diagonal $n \times n$ matrices, as usual. The *Weyl group* W of $GL(n)$ can then be defined to be

$$W = M'/M. \quad (5.52)$$

We can identify W with the group of permutations of n elements, as the following exercise shows.

Exercise 29.

(a) Show that M defined by (5.51) consists of all diagonal matrices with entries ± 1 (see the definition of the boundary of G/K in (1.19) of Section 4.1.3).
(b) Show that M' in (5.51) consists of all matrices such that each row or column has exactly one non-zero entry equal to ± 1.
(c) Show that the Weyl group of $GL(n, \mathbb{R})$ can be identified as the group of permutations of n elements.

Exercise 30. Show that formula (5.42) agrees with the decomposition in Exercise 28 when $n = 3$.

Hint. When $P = P(2, 1)$, what is J? You need to look at

$$\begin{pmatrix} b_1 & b_2 & b_3 \\ 0 & b_4 & b_5 \\ 0 & 0 & b_6 \end{pmatrix} \begin{pmatrix} g_1 & g_2 & g_3 \\ h_1 & h_2 & h_3 \\ 0 & 0 & j_3 \end{pmatrix}.$$

To put this matrix into $B = P(1, 1, 1)$, the Borel or minimal parabolic subgroup, requires the ability to interchange the rows of g's with the row's of h's. So J corresponds to the permutation matrix w_1 in the notation (5.41).

What is $W_J \backslash W / W_J$? Clearly representatives are the permutation matrices corresponding to the identity and the transposition (13).

Lemma 5 (The Bruhat Decomposition for the Minimal Parabolic or Borel Subgroup $B = P(1, \ldots, 1)$). *Let k denote any field. Then we have the disjoint union:*

$$GL(n, k) = \bigcup_{w \in W} BwB,$$

where W is the Weyl group of all permutation matrices.

PROOF. Note that given $g \in GL(n, k)$, there is an element $b \in B$ such that the first non-zero entry in each row of the matrix bg must occur in a different position for each row. Thus we can find an element $w \in W$ to put the rows of wbg in the correct order to form an element of B. □

4.5. Automorphic Forms for $GL(n, \mathbb{Z})$ and Harmonic Analysis on $\mathscr{P}_n/GL(n, \mathbb{Z})$

Exercise 31. Use Lemma 5 to obtain part of Lemma 2 of Section 4.3.3 which says that $G = SL(n, \mathbb{R})$ has the form:

$$G = (\bar{N}B) \cup (\text{something lower dimensional}),$$

where $\bar{N}$ denotes the subgroup of all lower triangular matrices with ones on the diagonal and B is the Borel or minimal parabolic subgroup of all upper triangular matrices in G.

Hint. Note that

$$\begin{pmatrix} 0 & & 1 \\ & \cdot\cdot\cdot & \\ 1 & & 0 \end{pmatrix} N \begin{pmatrix} 0 & & 1 \\ & \cdot\cdot\cdot & \\ 1 & & 0 \end{pmatrix} = \bar{N}.$$

Lemma 6 (Coset Representatives à la Bruhat). *Suppose that $P = P(n - 1, 1)$. The cosets in $SL(n, \mathbb{Z})/P$ can be represented by*

$$S_1^* \cup S_2^*,$$

where $S_1^* = \{I\}$ and

$$S_2^* = \left\{ \begin{pmatrix} {}^t A^{-1} & 0 \\ 0 & 1 \end{pmatrix} n_q \sigma p_q \;\middle|\; \begin{array}{l} A \in SL(n-1, \mathbb{Z})/P(1, n-2) \\ q = e/f, f \geq 1, (e, f) = 1 \\ e, f \in \mathbb{Z} \end{array} \right\},$$

with

$$n_q = \begin{pmatrix} 1 & 0 & q \\ 0 & I_{n-2} & 0 \\ 0 & 0 & 1 \end{pmatrix}, \quad \sigma = \begin{pmatrix} 0 & 0 & 1 \\ 0 & I_{n-2} & 0 \\ -1 & 0 & 0 \end{pmatrix}, \quad p_q = \begin{pmatrix} f & 0 & g \\ 0 & I_{n-2} & 0 \\ 0 & 0 & 1/f \end{pmatrix},$$

if $eg \equiv 1 \pmod{f}$, $0 \leq g < f$.

PROOF (Imai). The general idea is to use the method of Baily [1] and Terras [3]. This requires Exercises 28 and 32. We shall write $P_\mathbb{Q}$ when we wish to consider the parabolic subgroup in $SL(n, \mathbb{Q})$ and $P_\mathbb{Z}$ when we want the parabolic subgroup of $SL(n, \mathbb{Z})$. Define

$$T: P_\mathbb{Q} \to (SL(n, \mathbb{Z}) \cap P_\mathbb{Q} \sigma P_\mathbb{Q})/P_\mathbb{Z},$$

by $T(p) = p\sigma p' \pmod{P_\mathbb{Z}}$, for $p \in P_\mathbb{Q}$. Here p' is chosen in $P_\mathbb{Q}$ to put $p\sigma p'$ in $SL(n, \mathbb{Z})$. This is possible by Exercise 32.

Then matrix multiplication shows $\sigma p = p'\sigma$ is equivalent to:

$$p = \begin{pmatrix} a & 0 & 0 \\ c & D & e \\ 0 & 0 & g \end{pmatrix} \quad \text{with} \quad D \in \mathbb{Q}^{(n-2) \times (n-2)}.$$

So define

$$P_{\mathbb{Q}}^* = \left\{ p = \begin{pmatrix} a & 0 & 0 \\ c & D & 0 \\ 0 & 0 & g \end{pmatrix} \middle| p \in SL(n, \mathbb{Q}), D \in \mathbb{Q}^{(n-2) \times (n-2)} \right\}.$$

Finding the coset representatives for $SL(n, \mathbb{Z})/P_{\mathbb{Z}}$ is the same as reducing $p \in P_{\mathbb{Q}}$ modulo $P_{\mathbb{Q}}^*$. Representatives for $P_{\mathbb{Q}}/P_{\mathbb{Q}}^*$ are:

$$\begin{pmatrix} {}^tA^{-1} & {}^tA^{-1}c \\ 0 & 1 \end{pmatrix}, \quad A \in SL(n-1, \mathbb{Z})/P(1, n-1)_{\mathbb{Z}}, \quad c = \begin{pmatrix} q \\ 0 \end{pmatrix}, \quad q \in \mathbb{Q}.$$

The equality $p\sigma p' = p_1 \sigma p_1'$ with $p, p_1, p', p_1' \in P_{\mathbb{Q}}$ implies $p_1^{-1} p \in P_{\mathbb{Q}}^*$. Thus if

$$p = \begin{pmatrix} {}^tA^{-1} & 0 \\ 0 & 1 \end{pmatrix} n_q,$$

then $T(p) = p\sigma p'$ gives a complete set of representatives for

$$SL(n, \mathbb{Z}) \cap (P_{\mathbb{Q}} \sigma P_{\mathbb{Q}})/P_{\mathbb{Z}}.$$

Finally it must be proved that if $q = e/f$, $f > 1$, $(e, f) = 1$, then

$$p' = \begin{pmatrix} f & 0 & g \\ 0 & I_{n-2} & 0 \\ 0 & 0 & 1/f \end{pmatrix}, \quad \text{with } eg \equiv 1 \pmod{f}.$$

To see this, write

$${}^tA^{-1} = \begin{pmatrix} a & b \\ c & d \end{pmatrix}, \quad D \in \mathbb{Z}^{(n-2) \times (n-2)}.$$

Then

$$\begin{pmatrix} {}^tA^{-1} & 0 \\ 0 & 1 \end{pmatrix} \begin{pmatrix} 1 & 0 & q \\ 0 & I_{n-2} & 0 \\ 0 & 0 & 1 \end{pmatrix} \begin{pmatrix} 0 & 0 & 1 \\ 0 & I_{n-2} & 0 \\ -1 & 0 & 0 \end{pmatrix} p'$$

$$= \begin{pmatrix} -ae & b & a(-qg + 1/f) \\ -ce & d & c(-qg + 1/f) \\ -f & 0 & -g \end{pmatrix}.$$

Clearly the matrix on the right lies in $SL(n, \mathbb{Z})$, assuming that $q = e/f$ is as stated in the lemma. According to Exercise 28, the proof of Lemma 6 is now complete. □

Exercise 32. Show that $SL(n, \mathbb{Q}) = SL(n, \mathbb{Z}) B_{\mathbb{Q}}$, where $B_{\mathbb{Q}} = P(1, 1, \ldots, 1)_{\mathbb{Q}}$ is the minimal parabolic or Borel subgroup of upper triangular matrices in $SL(n, \mathbb{Q})$.

Hint. Use induction. First observe that for $A \in SL(n, \mathbb{Q})$ there is a matrix $U \in SL(n, \mathbb{Z})$ such that the first $n - 1$ elements in the last row of UA vanish.

4.5. Automorphic Forms for $GL(n, \mathbb{Z})$ and Harmonic Analysis on $\mathcal{P}_n/GL(n, \mathbb{Z})$

For $n-1$ homogeneous linear equations in n unknowns with rational coefficients have relatively prime integral solutions. Recall Lemma 2 of Section 4.4.2.

Imai used Lemma 6 to give the Fourier expansion of $E_{2,1}(f, s | Y)$ by a method like the third method of Exercise 4, Section 3.5, Vol. I for the Eisenstein series in the case of $GL(2, \mathbb{Z})$. Now we want to consider a decomposition that leads to an analogue of the second method in that same exercise from Volume I.

Exercise 33 (Coset Representatives sans* Bruhat). Show that the cosets of $\mathbb{Z}^{n \times (n-1)}$ rank $(n-1)/GL(n-1, \mathbb{Z})$ can be represented by $S_1 \cup S_2$, where

$$S_1 = \left\{ \begin{pmatrix} B \\ 0 \end{pmatrix} \middle| B \in \mathbb{Z}^{(n-1) \times (n-1)} \text{ rank } (n-1)/GL(n-1, \mathbb{Z}) \right\},$$

$$S_2 = \left\{ \begin{pmatrix} B \\ {}^t c \end{pmatrix} \middle| \begin{array}{l} B = HD, \ H \in GL(n-1, \mathbb{Z})/{}^t P, \\ d_j > 0, j = 2, \ldots, n, \\ d_{ij} \bmod d_j, j = 2, \ldots, n, \\ {}^t c = (c_1 0 \cdots 0), c_1 > 0 \end{array} \right. \quad D = \begin{pmatrix} d_1 & & 0 \\ & \ddots & \\ & d_{ij} & d_{n-1} \end{pmatrix} \right\}.$$

Here ${}^t P$ denotes the lower triangular subgroup of $GL(n-1, \mathbb{Z})$. The point of the inequalities on the lower triangular matrix D in S_2 is that d_1 is an arbitrary integer while $d_2, \ldots, d_n$ are all positive integers and the d_{ij} can be taken to lie between 0 and $d_j - 1$.

Hint. Write $A \in \mathbb{Z}^{n \times (n-1)}$ of rank $(n-1)$ as

$$A = \begin{pmatrix} B \\ {}^t c \end{pmatrix} \quad \text{with } B \text{ in } \mathbb{Z}^{(n-1) \times (n-1)} \text{ and } c \in \mathbb{Z}^{n-1}.$$

If $c = 0$, then A lies in the set of representatives S_1. Otherwise there is a matrix W in $GL(n-1, \mathbb{Z})$ such that

$$AW = \begin{pmatrix} b_1 & B_2 \\ c_1 & 0 \end{pmatrix} \quad \text{with } c_1 > 0, B_2 \in \mathbb{Z}^{(n-1) \times (n-2)}, b_1 \in \mathbb{Z}^{n-1}.$$

Moreover we can write:

$$\begin{pmatrix} b_1 & B_2 \\ c_1 & 0 \end{pmatrix} = \begin{pmatrix} b_1' & B_2' \\ c_1' & 0 \end{pmatrix} \begin{pmatrix} x & y \\ v & W \end{pmatrix} \quad \text{if and only if} \quad x = 1 \text{ and } y = 0 \in \mathbb{Z}^{n-1}.$$

So we need to take $A = \begin{pmatrix} B \\ {}^t c \end{pmatrix}$ modulo the subgroup of $GL(n-1, \mathbb{Z})$ of matrices of the form

$$\begin{pmatrix} 1 & 0 \\ v & W \end{pmatrix}.$$

---

* Without.

Thus W must be in $GL(n-2, \mathbb{Z})$. It must be shown that this puts A in S_2. Elementary divisor theory writes $B \in \mathbb{Z}^{(n-1) \times (n-1)}$ in the form $B = HD$ with H in $GL(n-1, \mathbb{Z})$ and

$$D = \begin{pmatrix} d_1 & 0 \\ d_{12} & D_2 \end{pmatrix},$$

with d_1 in $\mathbb{Z}$ and with a lower triangular, non-singular D_2 in $\mathbb{Z}^{(n-2) \times (n-2)}$. And we can reduce H modulo the lower triangular group ${}^tP$.

Next we use the preceding matrix decompositions to obtain some explicit Fourier expansions for Eisenstein series $E_{2,1}$ from Imai and Terras [1]. Let $\varphi \in \mathscr{A}^0(SL(2, \mathbb{Z}), \chi)$ have the Fourier expansion:

$$\varphi(U) = \alpha_0 k_{1,1}(1 - r|U, 0) + \alpha_0' k_{1,1}(r|U, 0) + \sum_{n \neq 0} \alpha_n k_{1,1}(r|U, n), \quad (5.53)$$

if

$$U = \begin{pmatrix} y & 0 \\ 0 & 1/y \end{pmatrix} \begin{bmatrix} 1 & 0 \\ x & 1 \end{bmatrix}, \quad y > 0, \quad x \in \mathbb{R},$$

and $k_{1,1}(r|U, a)$ denotes the k-Bessel function defined in formula (2.20) of Section 4.2.2. Formula (5.53) is just a restatement of Exercise 3 in Section 3.5 of Volume I, using the notation of Section 4.2.2.

Theorem 4 (Imai). *Let $\varphi_r \in \mathscr{A}^0(SL(2, \mathbb{Z}), r(r-1))$ be a cusp form having Fourier expansion (5.53) with Fourier coefficients α_k, $k \neq 0$, and $\alpha_0 = \alpha_0' = 0$. Then when*

$$Y = \begin{pmatrix} U & 0 \\ 0 & w \end{pmatrix} \begin{bmatrix} I_2 & x \\ 0 & 1 \end{bmatrix}, \quad U \in \mathscr{P}_2, \quad w = |U|^{-1}, \quad x \in \mathbb{R}^2,$$

we have the following Fourier expansion of the Eisenstein series $E_{2,1}(\varphi_r, s|Y)$ defined in (5.3)

$$E_{2,1}(\varphi_r, s|Y) = |U|^{-s}\varphi_r(U^0)$$
$$+ \sum \exp\{2\pi i \, {}^t x A m\} c_f(n) \alpha_k f^{1-2s-r} k_{2,1}(s - r/2, r|U(A, w), m),$$

where the sum runs over $k \in \mathbb{Z} - 0$, $n \in \mathbb{Z}$, $f \geq 1$, $A \in SL(2, \mathbb{Z})/P(1, 1)$,

$$m = \begin{pmatrix} n \\ -kf \end{pmatrix}, \quad U(A, w) = \begin{pmatrix} U[{}^tA^{-1}] & 0 \\ 0 & w \end{pmatrix},$$

and $P(1, 1)$ is the minimal parabolic subgroup of $SL(2, \mathbb{Z})$. Here $c_f(n)$ is Ramanujan's sum:

$$c_f(n) = \sum_{\substack{0 < e < f \\ (e, f) = 1}} \exp(2\pi i n e/f),$$

and $k_{2,1}$ is the Bessel function from (2.20) in Section 4.2.2.

4.5. Automorphic Forms for $GL(n, \mathbb{Z})$ and Harmonic Analysis on $\mathscr{P}_n/GL(n, \mathbb{Z})$

PROOF. Set

$$\tilde{Y} = Y \begin{bmatrix} {}^t A^{-1} & 0 \\ 0 & 1 \end{bmatrix} = \begin{pmatrix} \tilde{U} & 0 \\ 0 & \tilde{w} \end{pmatrix} \begin{bmatrix} I_2 & \tilde{x} \\ 0 & 1 \end{bmatrix}. \tag{5.54}$$

It is easily seen that

$$\tilde{U} = U[{}^tA^{-1}], \qquad \tilde{U}^{-1} = U^{-1}[A], \qquad \tilde{w} = W, \tag{5.55}$$

$$\tilde{x} = \begin{pmatrix} \tilde{x}_1 \\ \tilde{x}_2 \end{pmatrix} = {}^tAx = \begin{pmatrix} {}^ta_1 x \\ {}^ta_2 x \end{pmatrix}, \qquad \text{if } A = (a_1 a_2).$$

Using Lemma 6 with $n = 3$, we are led to compute $Y[(n_q \sigma p_q)_1]$, where the subscript "1" means that we must take the first two columns of the 3×3 matrix $n_q \sigma p_q$. Recall that $q = e/f$ and

$$(n_q \sigma p_q)_1 = \begin{pmatrix} -qf & 0 \\ 0 & 1 \\ -f & 0 \end{pmatrix}.$$

So we set

$$Y^{\#} = \tilde{Y}[(n_q \sigma p_q)_1] = \tilde{U}\begin{bmatrix} -f(q + \tilde{x}_1) & 0 \\ -f\tilde{x}_2 & 1 \end{bmatrix} + \begin{pmatrix} wf^2 & 0 \\ 0 & 0 \end{pmatrix}.$$

In order to use the Fourier expansion (5.53), we must set

$$Y^{\#} = |Y^{\#}|^{1/2} \begin{pmatrix} y & 0 \\ 0 & y^{1/2} \end{pmatrix} \begin{bmatrix} 1 & 0 \\ x & 1 \end{bmatrix} = \begin{pmatrix} * & * \\ x|Y^{\#}|^{1/2}/y & |Y^{\#}|^{1/2}/y \end{pmatrix} \tag{5.56}$$

and

$$\tilde{U} = \begin{pmatrix} t & 0 \\ 0 & v \end{pmatrix} \begin{bmatrix} 1 & 0 \\ p & 1 \end{bmatrix} = \begin{pmatrix} * & * \\ vp & v \end{pmatrix}. \tag{5.57}$$

It follows that

$$Y^{\#} = \begin{pmatrix} t & 0 \\ 0 & v \end{pmatrix} \begin{bmatrix} -f(q + \tilde{x}_1) & 0 \\ -f\{p(q + \tilde{x}_1) + \tilde{x}_2\} & 1 \end{bmatrix} + \begin{pmatrix} f^2 w & 0 \\ 0 & 0 \end{pmatrix}.$$

Putting all this together, we find that

$$|Y^{\#}| = vf^2 \{t(q + \tilde{x}_1)^2 + w\}, \qquad y = \sqrt{\frac{f^2}{v}} \sqrt{t(q + \tilde{x}_1)^2 + w},$$

$$x = -f\{p(q + \tilde{x}_1) + \tilde{x}_2\}.$$

By Lemma 6, q runs over all of the field of rational numbers. So we break this sum up into a sum over $q \in \mathbb{Q}/\mathbb{Z}$ and a sum over $n \in \mathbb{Z}$. Then use Poisson summation on the variable n to see that

$$E_{2,1}(\varphi_r, s | Y) = |U|^{-s} \varphi_r(U^0) + \sum_{A,k,q,n} \alpha_k T(s, r | A, k, q, n),$$

where the sum is over $A \in SL(2, \mathbb{Z})/P(1,1)$, $k \neq 0$, $q \in \mathbb{Q}/\mathbb{Z}$, $q = e/f$, $f \geq 1$, $(e,f) = 1$, $n \in \mathbb{Z}$, and

$$T = T(s, r | A, k, q, n)$$
$$= \int_{z \in \mathbb{R}} (vf^2\{t(z+q+\tilde{x}_1)^2 + w\})^{-s} \left(\frac{f^2}{v}\{t(z+q+\tilde{x}_1)^2 + w\}\right)^{(1-r)/2}$$
$$\times k_{1,1}\left(r \middle| I_2, k \frac{f}{\sqrt{v}} \sqrt{t(z+q+\tilde{x}_1)^2 + w}\right)$$
$$\times \exp(-2\pi i k f \{p(z+q+\tilde{x}_1) + \tilde{x}_2\} - 2\pi i n z) \, dz.$$

Next let $u = (t/w)^{1/2}(z + q + \tilde{x}_1)$ and use part (5) of Theorem 2 of Section 4.2.2 to obtain:

$$T = \exp\{2\pi i(nq + n\tilde{x}_1 - kf\tilde{x}_2)\} f^{-2s+1-r} v^{-s-(1-r)/2} t^{-1/2} w^{-s+(2-r)/2}$$
$$\times k_{2,1}\left(s - \frac{r}{2}, r \middle| I_3, \left(\sqrt{\frac{w}{t}}(kpf + n), -kf\sqrt{\frac{w}{v}}\right)\right).$$

Now the last argument of $k_{2,1}$ is the vector:

$$\sqrt{w}\begin{pmatrix} t^{-1/2} & 0 \\ 0 & v^{-1/2} \end{pmatrix}\begin{pmatrix} 1 & -p \\ 0 & 1 \end{pmatrix}\begin{pmatrix} n \\ -kf \end{pmatrix} = \sqrt{w}\, M \begin{pmatrix} n \\ -kf \end{pmatrix}, \quad \text{with}$$

$$M = \begin{pmatrix} t^{-1/2} & 0 \\ 0 & v^{-1/2} \end{pmatrix}\begin{pmatrix} 1 & -p \\ 0 & 1 \end{pmatrix}.$$

And ${}^t MM = \tilde{U}^{-1} = U^{-1}[A]$. Part (4) of Theorem 2 in Section 4.2.2 says that if ${}^t m = (n, -kf)$:

$$k_{2,1}\left(s - \frac{r}{2}, r \middle| I_3, \sqrt{w}\, Mm\right)$$
$$= p_{s-r/2, r-3/2}(U^{-1}[A]) k_{2,1}\left(s \middle| \begin{pmatrix} U[{}^t A^{-1}] & 0 \\ 0 & w \end{pmatrix}, m\right).$$

Next note that

$$v^{-s-(1-r)/2} w^{-s+(2-r)/2} t^{-1/2} = p_{s-r/2, r-3/2}(U^{-1}[A])^{-1}.$$

Thus the power functions cancel and we find that

$$T = \exp\{2\pi i(nq + n\tilde{x}_1 - kf\tilde{x}_2)\} f^{-2s+1-r} k_{2,1}\left(s - \frac{r}{2}, r \middle| \begin{pmatrix} U[{}^t A^{-1}] & 0 \\ 0 & w \end{pmatrix}, m\right).$$

This completes the proof of Theorem 4. $\square$

Next we want to use Exercise 33 to obtain an alternate Fourier expansion.

Theorem 5. *Let $\varphi_r \in \mathscr{A}^0(SL(2, \mathbb{Z}), r(r-1))$ be a cusp form having Fourier expansion (5.53) with Fourier coefficients α_k, $k \neq 0$, and $\alpha_0 = \alpha'_0 = 0$. Suppose*

4.5. Automorphic Forms for $GL(n,\mathbb{Z})$ and Harmonic Analysis on $\mathscr{P}_n/GL(n,\mathbb{Z})$

that

$$Y = \begin{pmatrix} U & 0 \\ 0 & w \end{pmatrix} \begin{bmatrix} I_2 & x \\ 0 & 1 \end{bmatrix}, \quad U \in \mathscr{P}_2, \quad w = |U|^{-1}, \quad x \in \mathbb{R}^2.$$

Then the Eisenstein series $E_{2,1}(\varphi_r, s | Y)$ defined by (5.3) has the following Fourier expansion with respect to the parabolic subgroup $P(2,1) \subset GL(3,\mathbb{Z})$:

$$L_{\varphi_r}(2s) E_{2,1}(\varphi_r, s | Y) = L_{\varphi_r}(2s) \varphi_r(U^0) |U|^{-s}$$

$$+ \sum \alpha_k c^{2-2s-r} d_2^{r-2s} \exp(2\pi i\, {}^t x Am) k_{2,1}\left(s - r/2, r \left| \begin{pmatrix} U[{}^tA^{-1}] & 0 \\ 0 & w \end{pmatrix}, m \right.\right),$$

where ${}^t m = c(d_1, k/d_2) \in \mathbb{Z}^2$ and the sum is over $A \in SL(2,\mathbb{Z})/P(1,1)$, $c > 0$, $d_1 \in \mathbb{Z}$, $0 < d_2 | k$, $k \neq 0$. The parabolic subgroup $P(1,1)$ of $SL(2,\mathbb{Z})$ consists of the upper triangular matrices of determinant one. The L function $L_{\varphi_r}(2s)$ is the one that is associated with φ_r by part (5) of Theorem 2 in Section 4.5.2 (see also (5.35)). Here $k_{2,1}$ is the Bessel function from (2.20) in Section 4.2.2.

PROOF. Everything goes as it did in Theorem 4, except that we use Exercise 33 rather than Lemma 6. Define $\tilde{Y}$ as in (5.54) and (5.55). Using Exercise 33 we must set

$$Y^\# = \tilde{Y} \begin{bmatrix} D \\ {}^t g \end{bmatrix}, \quad \text{where } {}^t g = (c \;\; 0),\, c > 0,\, D = \begin{pmatrix} d_1 & 0 \\ d_{12} & d_2 \end{pmatrix}, \qquad (5.58)$$

$$d_1 \in \mathbb{Z}, d_2 > 0, d_{12} \bmod d_2.$$

Suppose that $\tilde{U}$ is again given by (5.57). Then

$$Y^\# = \begin{pmatrix} t & 0 \\ 0 & v \end{pmatrix} \begin{bmatrix} d_1 + \tilde{x}_1 c & 0 \\ p(d_1 + \tilde{x}_1 c) + d_{12} + \tilde{x}_2 c & d_2 \end{bmatrix} + \begin{pmatrix} wc^2 & 0 \\ 0 & 0 \end{pmatrix}.$$

We compute $|Y^\#|$, x, y in (5.56) to be:

$$|Y^\#| = (vd_2^2)\{t(d_1 + \tilde{x}_1 c)^2 + wc^2\},$$

$$y = \frac{1}{\sqrt{vd_2}}\sqrt{t(d_1 + \tilde{x}_1 c)^2 + wc^2}, \qquad (5.59)$$

$$x = \frac{1}{d_2}\{p(d_1 + \tilde{x}_1 c) + d_{12} + \tilde{x}_2 c\}.$$

Since Exercise 33 says that the sum defining

$$Z(\varphi_r, s | Y) = L_{\varphi_r}(2s) E(\varphi_r, s | Y)$$

in (5.34) and (5.35) runs over all $d_1 \in \mathbb{Z}$, we can use Poisson summation to find that:

$$Z(\varphi_r, s | Y) = L_{\varphi_r}(2s) \varphi_r(U^0) |U|^{-s} + \sum \alpha_k T(s, r | A, c, D, k),$$

where the sum is over:

$$D = \begin{pmatrix} d_1 & 0 \\ d_{12} & d_2 \end{pmatrix}, \quad d_1 \in \mathbb{Z}, \quad d_2 > 0, \quad d_{12} \bmod d_2,$$

$A \in SL(2, \mathbb{Z})/P(1,1)$, $c > 0$, and $k \neq 0$. We define

$$T = T(s, r | A, c, D, k)$$

$$= \int_{z \in \mathbb{R}} \exp\left\{2\pi i \left(\frac{k}{d_2}[p(z + \tilde{x}_1 c) + d_{12} + \tilde{x}_2 c] - zd_1\right)\right\}$$

$$\times [vd_2^2\{t(z + \tilde{x}_1 c)^2 + wc^2\}]^{-s} \left(\frac{t(z + x_1 c)^2 + wc^2}{vd_2^2}\right)^{(1-r)/2}$$

$$\times k_{1,1}\left(r \bigg| I_2, \frac{k}{\sqrt{vd_2}} \sqrt{t(z + x_1 c)^2 + wc^2}\right) dz.$$

Now use the fact that

$$\sum_{0 \leq d_{12} < d_2} \exp\{2\pi i k d_{12}/d_2\} = \begin{cases} 0, & \text{if } d_2 \nmid k \\ d_2, & \text{if } d_2 | k \end{cases} \equiv \chi(d_2, k). \quad (5.60)$$

Therefore

$$T = \chi(d_2, k)(vd_2^2)^{-s-(1-r)/2} \exp\{2\pi i(p\tilde{x}_1 + \tilde{x}_2)kc/d_2\}$$

$$\times \int_{z \in \mathbb{R}} \exp\{-2\pi i z(d_1 - kp/d_2)\}(t(z + \tilde{x}_1 c)^2 + wc^2)^{-s+(1-r)/2}$$

$$\times k_{1,1}\left(r \bigg| I_2, k\sqrt{\frac{t(z + \tilde{x}_1 c)^2 + wc^2}{vd_2^2}}\right) dz.$$

As in the proof of Theorem 4, set $u = (wc^2/t)^{-1/2}(z + \tilde{x}_1 c)$ and use part (5) of Theorem 2 of Section 4.2.2 to obtain

$$T = \chi(d_2, k)d_2^{-2s+r-1}c^{-2s-r+3/2}\exp\{2\pi i c(d_1\tilde{x}_1 + \tilde{x}_2 k/d_2)\}$$

$$\times v^{-s-(1-r)/2}w^{1-s-r/2}t^{-1/2}$$

$$\times k_{2,1}\left(s - r/2, r \bigg| I_3, \sqrt{w}\left(\frac{c}{\sqrt{t}}\left(\frac{kp}{d_2} - d_1\right), \frac{kc}{\sqrt{vd_2}}\right)\right).$$

Set $M = \begin{pmatrix} t^{-1/2} & 0 \\ 0 & v^{-1/2} \end{pmatrix}\begin{pmatrix} 1 & -p \\ 0 & 1 \end{pmatrix}$. Then ${}^t MM = \tilde{U}^{-1}$ and if we set ${}^t m = c(d_1, k/d_2)$, part (4) of Theorem 2 of Section 4.2.2 says that:

$$T = \chi(d_2, k)d_2^{r-2s-1}c^{2-2s-r}\exp\{2\pi i {}^t xAm\}$$

$$\times k_{2,1}\left(s - r/2, r \bigg| \begin{pmatrix} U^{-1}[{}^t A^{-1}] & 0 \\ 0 & w \end{pmatrix}, m\right).$$

For, again the power functions of $\tilde{U}^{-1}$ cancel. This completes the proof of Theorem 5. □

4.5. Automorphic Forms for $GL(n, \mathbb{Z})$ and Harmonic Analysis on $\mathcal{P}_n/GL(n, \mathbb{Z})$

Next we consider the case that φ_r in $E_{2,1}(\varphi_r, s|Y)$ is itself an Eisenstein series. By Exercise 2 in Section 4.5.1, using the notation of (5.3) and (5.7) in that section, we know that:

$$E_{2,1}(E_r, s|Y) = E_{(3)}(r, s - r/2, 0|Y). \tag{5.61}$$

Instead of considering the Eisenstein series itself, we obtain the Fourier expansion of the zeta function $Z_{2,1}(\varphi_r, s|Y)$ of formula (5.34) in Section 4.5.2, where $\varphi_r = \pi^{-r}\Gamma(r)Z_{1,1}(1, r|Y)$ and $Z_{1,1}(1, r|Y)$ is Epstein's zeta function from Proposition 2. Of course, the Eisenstein series $E_{2,1}$ is related to the zeta function $Z_{2,1}$ by formula (5.35) in Section 4.5.2.

Theorem 6 (Fourier Expansion of Selberg's Eisenstein Series for $GL(3, \mathbb{Z})$). Suppose that $\varphi_r(W) = \pi^{-r}\Gamma(r)Z_{1,1}(1, r|W)$ where $Z_{1,1}$ denotes Epstein's zeta function for $GL(2, \mathbb{Z})$ from Proposition 2. If

$$Y = \begin{pmatrix} U & 0 \\ 0 & w \end{pmatrix}\begin{bmatrix} I_2 & x \\ 0 & 1 \end{bmatrix}, \quad U \in \mathcal{P}_2, \quad w > 0, \quad x \in \mathbb{R}^2,$$

the Fourier expansion of the normalized Eisenstein series $E_{2,1}(\varphi_r, s|Y)$ defined by (5.3) in Section 4.5.2 as a periodic function of $x \in \mathbb{R}^2$ is:

$$\pi^{-(s-r/2)}\Gamma(s - r/2)\pi^{-(s-(1-r)/2)}\Gamma(s - (1-r)/2)Z_{2,1}(\varphi_r, s|Y)$$

$$= c(s, r) + c((6 - 2s - 3r)/4, s - r/2) + c((3 + 3r - 2s)/4, s - (1-r)/2)$$

$$+ \sum_{\substack{k=0 \\ A,c,d_1 \neq 0, d_2}} \alpha_0' c^{2-2s-r} d_2^{r-2s} \exp\{2\pi i\, {}^t x A m\}$$

$$\times k_{2,1}\left(s - r/2, r \left| \begin{pmatrix} U[{}^tA^{-1}] & 0 \\ 0 & w \end{pmatrix}, \pi m\right.\right)$$

$$+ \sum_{\substack{k=0 \\ A,c,d_1 \neq 0, d_2}} \alpha_0 c^{1-2s+r} d_2^{1-r-2s} \exp\{2\pi i\, {}^t x A m\}$$

$$\times k_{2,1}\left(s - (1-r)/2, 1 - r \left| \begin{pmatrix} U[{}^tA^{-1}] & 0 \\ 0 & w \end{pmatrix}, \pi m\right.\right)$$

$$+ \sum_{\substack{k \neq 0 \\ A,c,d_1,d_2}} \alpha_k c^{2-2s-r} d_2^{r-2s} \exp\{2\pi i\, {}^t x A m\}$$

$$\times k_{2,1}\left(s - r/2, r \left| \begin{pmatrix} U[{}^tA^{-1}] & 0 \\ 0 & w \end{pmatrix}, \pi m\right.\right).$$

Here the zeta function $Z_{2,1}$ is defined by (5.34) in Section 4.5.2 and relates to the Eisenstein series $E_{2,1}$ via (5.35) in that section. And we define the following quantities:

$$\alpha_0 = \Lambda(s, r)/B(\tfrac{1}{2}, \tfrac{1}{2} - r), \qquad \alpha_0' = \Lambda(s, r)/B(\tfrac{1}{2}, r - \tfrac{1}{2}),$$

$$B(x, y) = \Gamma(x)\Gamma(y)/\Gamma(x+y), \qquad \text{the beta function,}$$

$$\alpha_k = \Lambda(s,r)\sigma_{1-2r}(k)/\zeta(2r),$$

$$\Lambda(s,r) = \pi^{-(s-r/2)}\Gamma(s-r/2)\pi^{-(s-(1-r)/2)}\Gamma(s-(1-r)/2),$$

$$c(s,r) = \Lambda(r)\Lambda(s-r/2)\Lambda(s-(1-r)/2)E_r(U^0)|U|^{-s},$$

$$\Lambda(r) = \pi^{-r}\Gamma(r)\zeta(2r),$$

$E_r(U^0) =$ *the Eisenstein series for* $GL(2,\mathbb{Z})$ *defined in formula* (3.66) *in Section 3.5 of Volume I.*

The three sums in the formula above are over $A \in SL(2,\mathbb{Z})/P(1,1)$, *where* $P(1,1)$ *is the subgroup of upper triangular matrices of determinant one,* $c > 0$, $d_1 \in \mathbb{Z}$ (*with* $d_1 \neq 0$ *in the first two sums*), $d_2 > 0$, $d_2|k$, $k \in \mathbb{Z}$ ($k \neq 0$ *in the third sum*). *And the vector* $m \in \mathbb{Z}^2$ *is defined by* ${}^t m = c(d_1, k/d_2)$. *Here* $k_{2,1}$ *denotes the Bessel function from formula* (2.20) *in Section 4.2.2.*

PROOF. The proof is the same as that of Theorem 5 except that α_0 and α'_0 are not zero. We need to use formula (5.53) with the Fourier coefficients given by Exercise 4 in Section 3.5 of Volume I. The constant term in the Fourier expansion of:

$$\pi^{-(s-r/2)}\Gamma(s-r/2)\pi^{-(s-(1-r)/2)}\Gamma(s-(1-r)/2)Z_{2,1}(\varphi_r,s|Y)$$

is:

$$\Lambda(s,r)|U|^{-s}E(r|U^0)L_{\varphi_r}(2s) + \alpha_0 k_{2,1}(s-(1-r)/2)|I_3,0)$$

$$\times \sum_{A,c,d_2} d_2^{-2s+1-r} c^{-2s+1+r} |U|^{3(1-r)/4+s/2-3/2} U^{-1}[a_1]^{-s+(1-r)/2}$$

$$+ \alpha'_0 k_{2,1}(s-r/2,r|I_3,0) \sum_{A,c,d_2>0} d_2^{r-2s} c^{2-2s-r} |U|^{3r/4+s/2-3/2} U^{-1}[a_1]^{-s+r/2}.$$

The computation of Harish-Chandra's c-function (i.e., the calculation after formula (3.35) in Section 4.3.3) shows that

$$k_{2,1}(s-r/2,r|I_3,0) = B(\tfrac{1}{2},r/2+s-1)B(\tfrac{1}{2},r-\tfrac{1}{2}). \tag{5.62}$$

In order to compute the L-function corresponding to the form φ_r, recall that Exercise 4 of Section 3.5, Vol. I, showed that the nth Fourier coefficient of φ_r is:

$$c_n = 4\pi^{-r}\Gamma(r)\zeta(2r)|n|^{r-1/2}\sigma_{1-2r}(n).$$

The theory of Hecke operators for $SL(2,\mathbb{Z})$, to be found in part (5) of Theorem 4 in Section 3.6 of Volume I, shows that if T_k denotes the kth Hecke operator for $SL(2,\mathbb{Z})$ and

$$T_k\varphi_r = u_k\varphi_r, \quad k \geq 1,$$

then (since our Hecke operators T_k and $k^{1/2}$ times those of Section 3.6, Vol. I):

$$u_n = \frac{c_n}{c_1}n^{1/2} = n^r\sigma_{1-2r}(n).$$

4.5. Automorphic Forms for $GL(n, \mathbb{Z})$ and Harmonic Analysis on $\mathcal{P}_n/GL(n, \mathbb{Z})$

Thus
$$L_{\varphi_r}(2s) = \sum_{n \geq 1} n^{-2s} u_n = \sum_{n \geq 1} \sum_{0 < d|n} n^{r-2s} d^{1-2r}$$
$$= \sum_{m \geq 1} \sum_{d \geq 1} (md)^{r-2s} d^{1-2r} = \zeta(2s-r)\zeta(2s+r-1).$$

This shows that the first part of the constant term is indeed $c(s, r)$. The third part of the constant term is:

$$\Lambda(s, 1-r) B\left(\frac{1}{2}, \frac{r}{2} + s - 1\right) B\left(\frac{1}{2}, r - \frac{1}{2}\right) \zeta(2s - r) \zeta\left(2s + \frac{3}{2} - r\right)$$
$$\times |U|^{-(3/2 - s/2 - 3r/4)} E(s - r/2 | U^\circ)/B\left(\frac{1}{2}, r - \frac{1}{2}\right)$$
$$= c((6 - 2s - 3r)/4, s - r/2).$$

So the second part of the constant term must be:
$$c((3 - 2s + 3r)/4, s - (1 - r)/2).$$

The rest of the proof of Theorem 6 proceeds as in Theorem 5. □

Exercise 34 (Remarks on the Constant Term). Let $E_r(U)$ be the Eisenstein series for $GL(2, \mathbb{Z})$ defined in formula (3.66) of Section 3.5, Volume I. Recall that by Exercise 1,
$$E_{2,1}(E_r, s|Y) = E_{(3)}(r, s - r/2, 0|Y),$$
using definitions (5.3) and (5.7) in Section 4.5.1. Consider Selberg's change of variables (5.14) in Section 4.5.1:
$$r = z_2 - z_1 + \tfrac{1}{2}, \qquad s - r/2 = z_3 - z_2 + \tfrac{1}{2}.$$

Show that the three parts of the constant term in the Fourier expansion of $E_{2,1}(E_r, s|Y)$ in Theorem 6 correspond to the permutations (1), (23), and (13) of the z-variables. This is a special case of a very general phenomenon described by Langlands [1].

Exercise 35 (Another Bruhat Decomposition and Its Consequences).

(a) Suppose that $n \geq 2m$ and $P = P(m, n - m)$. Show that we have the following disjoint union (Bruhat decomposition):
$$SL(n, \mathbb{Q}) = \bigcup_{r=0}^{m} P_\mathbb{Q} \sigma_r P_\mathbb{Q}, \quad \text{with } \sigma_r = \begin{pmatrix} 0 & 0 & I_r & 0 \\ 0 & I_{m-r} & 0 & 0 \\ -I_r & 0 & 0 & 0 \\ 0 & 0 & 0 & I_{n-m-r} \end{pmatrix}.$$

(b) Obtain a complete set of representatives for $(SL(n, \mathbb{Z}) \cap (P_\mathbb{Q} \sigma_r P_\mathbb{Q}))/P_\mathbb{Z}$ of the form $p\sigma_r p'$ where σ_r is as in part (a),

$$p = \begin{pmatrix} {}^tA^{-1} & 0 \\ 0 & B \end{pmatrix} \begin{pmatrix} I_r & 0 & U & 0 \\ 0 & I_{m-r} & 0 & 0 \\ 0 & 0 & I_r & 0 \\ 0 & 0 & 0 & I_{n-m-r} \end{pmatrix},$$

$U \in \mathbb{Q}^{r \times r}$, $A \in SL(m, \mathbb{Z})/P(r, m-r)$, $B \in SL(n-m, \mathbb{Z})/P(r, n-m-r)$. The element $p' \in P_\mathbb{Q}$ is fixed, once p is.

(c) Use part (b) to obtain the Fourier expansion of the Eisenstein series $E_{m,n-m}(1, s|Y)$ with respect to $P(m, n-m)$. Here the automorphic form f in $\mathscr{A}^0(GL(m, \mathbb{Z}), \lambda)$ is chosen to be identically one in formula (5.3) of Section 4.5.1.

Hint (Terras [3]). If

$$Y = \begin{pmatrix} V & 0 \\ 0 & W \end{pmatrix} \begin{bmatrix} I & Q \\ 0 & I \end{bmatrix}, \quad V \in \mathscr{P}_m, \quad W \in \mathscr{P}_{n-m}, \quad Q \in \mathbb{R}^{m \times (n-m)},$$

the Fourier coefficient corresponding to $N \in \mathbb{Z}^{m \times (n-m)}$ of rank r, $0 < r \leq m$, involves the Bessel function K_r or $k_{r,r}$ from §4.2.2, plus the *singular series*

$$\sigma(C, s) = \sum_{R \in \mathbb{Q}^{r \times r}/\mathbb{Z}^{r \times r}} v(R)^{-2s} \exp(2\pi i \operatorname{Tr}({}^tRC)),$$

where $v(R)$ denotes the product of the reduced denominators of the elementary divisors of R. This is an analogue of the singular series appearing in the Fourier coefficients of Eisenstein series for the Siegel modular group (see Siegel [1, Vol. II, pp. 97–137]).

Let us now compare the Fourier expansions (5.45) with those obtained by Bump [1]. For example, in the case of a cusp form for $GL(3, \mathbb{Z})$ (as defined in the beginning of Section 4.5.1), Bump obtains the equivalent of the expansion:

$$f(Y) = \sum_{A \in SL(2, \mathbb{Z})/P(1,1)} \sum_{r_1, r_2 \geq 1} a_{r_1 r_2} W\left(s \middle| Y \begin{bmatrix} A & 0 \\ 0 & 1 \end{bmatrix}, r\right), \tag{5.63}$$

where $W(s|Y, r)$ denotes the Whittaker function defined at the end of Section 4.2.2. This expansion is obtained by starting with our usual sort of expansion (5.45) for the parabolic subgroup $P(2, 1)$:

$$f(Y) = \sum_{r \in \mathbb{Z}^2} f_r(Y), \quad \text{where } f_r(Y) = \int_{x \in \mathbb{R}^2} f\left(Y \begin{bmatrix} I_2 & x \\ 0 & 1 \end{bmatrix}\right) \exp(-2\pi i\, {}^trx)\, dx.$$

This means that

$$f_r\left(Y \begin{bmatrix} I_2 & x \\ 0 & 1 \end{bmatrix}\right) = \exp(2\pi i\, {}^trx) f_r(Y), \quad \text{for all } x \in \mathbb{R}^2.$$

Then note that for $A \in SL(2, \mathbb{Z})$ and $r \in \mathbb{R}^2$, we have:

$$f_{{}^tAr}\left(Y \begin{bmatrix} A & 0 \\ 0 & 1 \end{bmatrix}\right) = f_r(Y). \tag{5.64}$$

4.5. Automorphic Forms for $GL(n, \mathbb{Z})$ and Harmonic Analysis on $\mathcal{P}_n/GL(n, \mathbb{Z})$ 241

If $r \in \mathbb{Z}^2$, then there is a matrix $A \in SL(2, \mathbb{Z})$ such that ${}^t\!Ar = {}^t(0, m)$ for some positive integer m. Furthermore, if $a \in \mathbb{Z}$, one has

$$f_r\left(Y\begin{bmatrix} 1 & a & 0 \\ 0 & 1 & 0 \\ 0 & 0 & 1 \end{bmatrix}\right) = f_{{}^t(r_1, r_2 - r_1 a)}(Y). \tag{5.65}$$

This shows that when $r_1 = 0$, the coefficient f_r must be invariant under

$$Y \mapsto Y\begin{bmatrix} 1 & a & 0 \\ 0 & 1 & 0 \\ 0 & 0 & 1 \end{bmatrix}, \quad \text{for } a \in \mathbb{Z}.$$

Thus one can find the Fourier expansion with respect to the x_{12} variable in N, the group of upper triangular matrices with ones on the diagonal. That takes us from Bessel to Whittaker functions. It is the multiplicity one theorem of Shalika [1] which says that the resulting functions must be multiples of Whittaker functions. A summary of work on modular forms for $GL(3)$ in the language of Bump [1] can be found in Friedberg [2].

Next let us consider another application of these Fourier expansions—an application to the analytic continuation of Eisenstein series $E_{m,n-m}(f, s | Y)$ for automorphic forms $f \in \mathcal{A}^0(GL(m, \mathbb{Z}), \lambda)$. See Jacquet and Shalika [1] for a similar adelic argument. According to formulas (5.27) and (5.28) in Section 4.5.2, we have:

$$J(f, s | Y) = 2\pi^{-ms}\Gamma_m(r(f, s))L_{f*}(2s)E_{m,n-m}(f^*, s | Y)$$

$$= \int_{X \in \mathcal{M}_m, |X| \geq 1} \theta_m(Y, X) f(X^0) |X|^s \, d\mu_m$$

$$+ \int_{X \in \mathcal{M}_m, |X| \geq 1} \theta_m(Y, X^{-1}) f^*(X^0) |X|^{-s} \, d\mu_m.$$

Therefore, by Exercise 17 of Section 4.5.2:

$$J(f, s | Y) = \int_{X \in \mathcal{M}_m, |X| \geq 1} \theta_m(Y, X) f(X^0) |X|^s \, d\mu_m$$

$$+ |Y|^{-m/2} \int_{X \in \mathcal{M}_m, |X| \geq 1} \theta_m(Y^{-1}, X) f^*(X^0) |X|^{n/2-s} \, d\mu_m \tag{5.66}$$

$$+ \sum_{k=0}^{m-1} I_k(f, s | Y),$$

where

$$I_k(f, s | Y)$$

$$= \int_{X \in \mathcal{M}_m, |X| \geq 1} \left(|Y|^{-m/2} |X|^{n/2-s} f^*(X^0) \theta_k(Y^{-1}, X) - |X|^s f(X^0) \theta_k(Y, X) \right) d\mu_m. \tag{5.67}$$

The term $I_0(f, s | Y)$ is no problem:

$$I_0(f, s | Y) = \begin{cases} 0, & \text{if } f \text{ is orthogonal to the constants,} \\ |Y|^{-m/2}\left(\dfrac{n}{2} - s\right)^{-1} - s^{-1}, & \text{if } f \text{ is identically one.} \end{cases}$$

(5.68)

To study I_k, for $0 < k < m$, we need the following exercise.

Exercise 36. (Study of the Integrals Occurring in the Analytic Continuation of Eisenstein Series by the Method of Theta Functions) (Terras [8]).

(a) Show that we can express every rank k matrix $A \in \mathbb{Z}^{n \times m}$, for $1 \leq k \leq m - 1$, uniquely in the form:

$$A = B^t C, \qquad B \in \mathbb{Z}^{n \times k}, \qquad \text{rank } k,$$
$$(C \ *) \in GL(m, \mathbb{Z}) / P(k, m - k), \qquad C \in \mathbb{Z}^{m \times k}.$$

(b) Obtain the Jacobian of the following change of variables for $Y \in \mathcal{P}_n$:

$$t^{-1/n} Y = \begin{pmatrix} u^{-1}T & 0 \\ 0 & uV \end{pmatrix} \begin{bmatrix} I & Q \\ 0 & I \end{bmatrix}, \qquad u > 0, \qquad T \in \mathcal{SP}_p,$$

$$V \in \mathcal{SP}_q, \qquad Q \in \mathbb{R}^{p \times q}, \qquad n = p + q.$$

Answer. $d\mu_n(Y) = (2pq/n)u^{-pq-1}t^{-1}\, dt\, du\, dT\, dV$. Compare Exercise 19 of Section 4.4.3.

(c) Rewrite the integral $I_k(f, s | Y)$ in (5.67) using parts (a) and (b).

(d) Now suppose that $f \in \mathcal{A}^0(GL(m, \mathbb{Z}), \lambda)$ has the Fourier expansion:

$$f(W) = \sum_{N \in \mathbb{Z}^{k \times (m-k)}} A_{N,f}(u^{-1}T, uV) \exp\{2\pi i\, \mathrm{Tr}({}^t N Q)\},$$

if W is expressed as in part (b). Show that

$m/(2k(m-k))I_k(f, s | Y)$

$$= \int_{T, V, t, u} \left(A_{0, f^*}(u^{-1}T, uV) t^{n/2-s} |Y|^{-m/2} \sum_{B \in \mathbb{Z}^{n \times k} \text{ rank } k} \exp\{-\pi\, \mathrm{Tr}(Y^{-1}[B]t^{1/m}u^{-1}T)\} \right.$$

$$\left. - A_{0, f}(u^{-1}T, uV) t^{-s} \sum_{B \in \mathbb{Z}^{n \times k} \text{ rank } k} \exp\{-\pi\, \mathrm{Tr}(Y[B]t^{-1/m}u^{-1}T)\} \right) u^{-\tau} t^{-1}\, du\, dt\, dT\, dV,$$

where $\tau = k(m - k) + 1$ and the integral is over $T \in \mathcal{SM}_k$, $V \in \mathcal{SM}_{m-k}$, $t \geq 1$, $u \geq 0$.

(e) What hypotheses on f are necessary to justify the preceding arguments? Be careful. It was just this sort of divergent integral problem that led to

4.5. Automorphic Forms for $GL(n, \mathbb{Z})$ and Harmonic Analysis on $\mathscr{P}_n/GL(n, \mathbb{Z})$

gaps in many papers, as we mentioned in the remarks on part (5) of Theorem 2 in Section 4.5.2.

(f) In the special case $n = 3$, $m = 2$, $f(X) = E_r(X)$, $X \in \mathscr{SP}_2$, the Eisenstein series from formula (3.66) of Section 3.5, Volume I and Exercise 2 of Section 4.5.1, let

$$E_{2,1}(E_r, s|Y) = E_{(3)}(r, s - r/2, 0|Y)$$

be continued as above and show that then the integrals I_k are:

$$I_1(E_r, s|Y) = |Y|^{-1} \frac{\Lambda_3(Y^{-1}, 1-r)}{s - 1 - r/2} - \frac{\Lambda_3(Y, 1-r)}{s + (r-1)/2}$$
$$+ \frac{c(r)|Y|^{-1}\Lambda_3(Y^{-1}, r)}{s + (r-3)/2} - \frac{c(r)\Lambda_3(Y, r)}{s - r/2},$$

where $\Lambda_3(Y, r) = \Lambda(r) E_r(Y)$, $c(r) = \Lambda(1-r)/\Lambda(r)$, $\Lambda(r) = \pi^{-r}\Gamma(r)\zeta(2r)$.

(g) What happens when $f(X) \in \mathscr{A}^0(GL(m, \mathbb{Z}), \lambda)$ is a cusp form?

This ends our discussion of Fourier expansions of automorphic forms for $GL(n, \mathbb{Z})$. Many questions remain. In particular, we have certainly not obtained the most general sorts of Fourier expansions. One would also like to build up a Hecke correspondence by making use of the Fourier expansions. In this regard, note that we already have the Mellin transform of the K-Bessel functions from Exercise 18 in Section 4.2.2. However it is not useful to attempt to do the Mellin transforms of the k-Bessel functions in the following exercise.

Exercise 37 (Mellin Transforms That Diverge).

(a) Show that the following integral diverges in general

$$\int_{t>0} t^r k_{1,1}\left(s\left|\begin{pmatrix} t & 0 \\ 0 & 1 \end{pmatrix}, n\right.\right) dt.$$

(b) Do the same for

$$\int_{U \in \mathscr{M}_2} |U|^{-r} f(U^0) k_{2,1}\left(s\left|\begin{pmatrix} U & 0 \\ 0 & w \end{pmatrix}, n\right.\right) \exp(-\operatorname{Tr}(U)) dU,$$

when $r \in \mathbb{C}$, $f \in \mathscr{A}^0(GL(2, \mathbb{Z}), \lambda)$.

See Bump [1] and Bump and Friedberg [2] for information on Mellin transforms of Whittaker functions.

The next exercise should be compared with Exercise 25 in Section 4.5.2.

Exercise 38 (Another Functional Equation for Eisenstein Series).

(a) Use part (4) of Proposition 1 of Section 4.2.1 to show that Selberg's Eisenstein series $E_{(n)}(s|Y)$ defined by (5.7) of Section 4.5.1 satisfies the functional equation:

$$E_{(n)}(s|Y^{-1}) = E_{(n)}(s^*|Y), \quad s^* = \left(s_{n-1}, \ldots, s_1, -\sum_{j=1}^{n} s_j\right), \text{ for } s \in \mathbb{C}^n.$$

(b) Apply Exercise 10 of Section 4.5.1 to the Eisenstein series $E_{k,n-k}(1, s_k|Y)$, in the notation (5.3), to show that this Eisenstein series is essentially a specialization of $E_{(n)}(s|Y)$ arrived at by setting all but one variable equal to zero; more specifically,

$$E_{k,n-k}(1, s|Y) = E_{(n)}(s|Y)|_{s_j=0, \text{ for all } j \neq k}.$$

(c) Show that the Eisenstein series $E_{k,n-k}(1, s|Y)$ satisfies the functional equation:

$$E_{k,n-k}(1, s|Y^{-1}) = |Y|^s E_{n-k,k}(1, s|Y).$$

In particular, this means that if $Z_{k,n-k}(1, s|Y)$ denotes Koecher's zeta function, using the notation (5.34) of Section 4.5.2 rather than (4.2) of Section 4.4.1, then the following equality holds:

$$Z_{2,1}(1, s|Y) = \zeta(2s-1)\zeta(2s)|Y|^{-s}E_{1,2}(1, s|Y^{-1})$$
$$= \zeta(2s-1)|Y|^{-s}Z_{1,2}(1, s|Y^{-1}).$$

Thus, in this special case, Koecher's zeta function is just a product of Riemann and Epstein zeta functions. Use this result to check the formula obtained in part (f) of Exercise 36.

Hint. (c) You will need to take the limit of the quantity

$$r(r-1)J_{2,1}(E_r, s|Y)$$

in (5.66) and Exercise 36 as r approaches zero. In particular, this leads to the formula:

$$\lim_{r \to 0} r(r-1)\Lambda(r)I_1(E_r, s|Y) = \frac{|Y|^{-1}J_{1,2}(1,1|Y)}{s-1} - \frac{J_{1,2}(1,1|Y)}{s-1/2}$$
$$+ \frac{\Lambda(1)|Y|^{-1}}{s-3/2} - \frac{\Lambda(1)}{s}.$$

Thus Exercise 36 gives the analytic continuation of Koecher's zeta function and agrees with formula (3.16) in Koecher [1].

The last exercise in this section shows that one must be careful in obtaining the analytic continuation of Koecher's zeta function.

Exercise 39 (Double Poles). Consider the special case of Koecher's zeta function $Z_{m,0}(1, s|Y)$, in the notation of (5.34) in Section 4.5.2—a case in which the zeta function factors into a product of Riemann zeta functions as in (4.3) of Section 4.4.1. Then form the normalized function from (5.28) of Section 4.5.2:

4.5. Automorphic Forms for $GL(n, \mathbb{Z})$ and Harmonic Analysis on $\mathscr{P}_n/GL(n,\mathbb{Z})$

$$J_{m,0}(1,s|Y) = 2\pi^{-ms}\Gamma_m(s)Z_{m,0}(1,s|Y)$$

$$= 2|Y|^{-s}\prod_{j=0}^{m-1}\Lambda(s-j/2), \quad \text{where}$$

$$\Lambda(s) = \pi^{-s}\Gamma(s)\zeta(2s).$$

Show that $J_{m,0}(1, s|Y)$ has simple poles at $s = 0, m/2$ and has double poles at $s = \frac{1}{2}, 1, \frac{3}{2}, \ldots, (m-2)/2, (m-1)/2$.

This exercise demonstrates that Koecher [1] formula (3.16) is incorrect, in general, as this formula says that only simple poles occur. Another reference on the location of poles of Eisenstein series is Feit [1].

Other references on L-functions and Eisenstein series are: Böcherer [1], Duke [1], and Garrett [1].

We close this section with some remarks on cusp forms. Recall that in Section 4.5.1 we defined an automorphic form $f \in \mathscr{A}^0(GL(n,\mathbb{Z}),\lambda)$ to be a *cusp form* if for every $k = 1, \ldots, n-1$, the following integral vanishes for all $Y \in \mathscr{P}_n$:

$$a_0^k(Y) = \int_{X \in (\mathbb{R}/\mathbb{Z})^{k \times (n-k)}} f\left(Y \begin{bmatrix} I & X \\ 0 & I \end{bmatrix}\right) dX. \tag{5.69}$$

Thus a cusp form has a zero constant term for each one of the Fourier expansions with respect to maximal parabolic subgroups $P(k, n-k)$, $1 \le k \le n-1$.

A reader familiar with the definition of cusp form for the Siegel modular group might ask whether our definition implies that all of the Fourier coefficients $a_N(U, V)$ vanish for $N \in \mathbb{Z}^{k \times (n-k)}$ not of maximal rank in (5.45).

Another question raised by Siegel's approach to the definition of cusp form is that of determining whether the following operation leads from a modular form for $GL(n, \mathbb{Z})$ to one for $GL(n-1, \mathbb{Z})$:

$$(f|\Phi)(V) = \lim_{w \to \infty} f\begin{pmatrix} V & 0 \\ 0 & w \end{pmatrix}, \quad \text{for } V \in \mathscr{P}_{n-1}, w > 0. \tag{5.70}$$

This is an analogue of Siegel's Φ-operator. Siegel would define f to be cuspidal if $f|\Phi = 0$ (see Chapter 5, Maass [2, pp. 187–198] or Freitag [1]). Is this equivalent to (5.69)?

Answers to the questions of the preceding paragraphs require more information on the Fourier expansions than we have at our disposal. So we leave them open at this time.

4.5.4. Remarks on Harmonic Analysis on the Fundamental Domain

Recall that the Euclidean Poisson summation formula for a Schwartz function $f: \mathbb{R}^n/\mathbb{Z}^n \to \mathbb{C}$ says:

$$f(x+a) = \sum_{a \in \mathbb{Z}^n} \hat{f}(a) \exp(2\pi i\,{}^t ax),$$

where $\hat{f}$ denotes the Fourier transform on $\mathbb{R}^n$ defined by

$$\hat{f}(a) = \int_{y \in \mathbb{R}^n} f(y) \exp(-2\pi i\,{}^t ay)\,dy.$$

We have seen in Section 1.4 of Volume I that there are many applications. For example, setting $f(x) = \exp(-Y[x])$ for $Y \in \mathcal{P}_n$, we obtain the transformation formula for the theta function. This transformation formula allows one to prove the analytic continuation and functional equation of Epstein's zeta function. It also gives information about the asymptotics of the fundamental solution of the heat equation on $\mathbb{R}^n/\mathbb{Z}^n$.

In Theorem 3 of Section 3.7 of Volume I, we found an analogue of Poisson summation for functions on the fundamental domain for $SL(2, \mathbb{Z})$ in the Poincaré upper half plane. Now we want to examine a generalization of this result to $GL(n, \mathbb{Z})$. The result we seek to prove follows easily from the generalization of the Roelcke-Selberg spectral decomposition of the Laplacian to $GL(n, \mathbb{Z})$, a result stated by Arthur [1]. For a proof of this highly non-trivial theorem, one needs the discussion of Langlands [1] or Harish-Chandra [1], or Osborne and Warner [1]. Here we suppose for simplicity that f lies in $C_c^\infty(\mathcal{SP}_n/GL(n,\mathbb{Z}))$ and that f is $O(n)$-invariant. This implies that the Helgason-Fourier transform $\hat{f}$ defined in (3.1) of Section 4.3.1 is independent of the rotation variable $k \in O(n)$. Thus we can write $\hat{f}$ as a function of $s \in \mathbb{C}^n$ alone:

$$\hat{f}(s) = \int_{Y \in \mathcal{P}_n} f(Y)\overline{p_s(Y)}\,d\mu_n(Y). \tag{5.71}$$

Poisson summation for $\Gamma_n = GL(n, \mathbb{Z})$ says:

$$\sum_{\gamma \in \Gamma_n/\pm I} f(I[a\gamma b^{-1}]) = \sum_{m \geq 0} \hat{f}(s_m) w_m(I[a]) \overline{w_m(I[b])}$$

$$+ \sum_P \sum_{v \in A(P)} \kappa_P \int_{\substack{r \in \mathbb{C}^{|P|-1} \\ \operatorname{Re} r = \text{constant}}} \hat{f}(s_P(v,r)) E_P(v,r|I[a]) \overline{E_P(v,r|I[b])}\,dr. \tag{5.72}$$

Here the sum over P runs over all non-associated parabolic subgroups $P = P(n_1, \ldots, n_q)$ as in (4.58) and we write $|P| = q$. We say that two parabolic subgroups $P(n_1, \ldots, n_q)$ and $P(m_1, \ldots, m_q)$ are *associated* if the m_j are a permutation of the n_i. The sum over $m \geq 0$ corresponds to a sum over the discrete spectrum of the G-invariant differential operators in $D(\mathcal{SP}_n)$ on $L^2(\mathcal{SP}_n/\Gamma_n)$, $\Gamma_n = GL(n, \mathbb{Z})$. We will use the notation:

$$A(GL(n)) = \{w_m | m \geq 0\} \tag{5.73}$$

to represent an orthogonal basis of $L^2(\mathcal{SP}_n/\Gamma_n)$ consisting of eigenfunctions of $D(\mathcal{SP}_n)$. We will let w_0 be the constant function. This may be viewed as a residue of an Eisenstein series. There will also be cusp forms as defined in (5.69) of the preceding section. Thus $A(GL(n))$ has both "cuspidal" and

4.5. Automorphic Forms for $GL(n, \mathbb{Z})$ and Harmonic Analysis on $\mathscr{P}_n/GL(n, \mathbb{Z})$

"residual" parts. It is somewhat confusing to refer to these parts as the "cuspidal spectrum" and the "residual spectrum" since the term "residual spectrum" has a different meaning in spectral theory. See, for example Reed and Simon [1, pp. 188, 194] where it is proved that self-adjoint operators have *no* residual spectrum in the sense of functional analysis. So let us just say that the spectrum has "cuspidal" and "non-cuspidal" components. The "non-cuspidal" part has proved to be problematical for higher rank versions of the trace formula.

To continue with the definitions of the terms in (5.72), now suppose that we are given the parabolic subgroup $P = P(n_1, \ldots, n_q)$ and write the corresponding partial Iwasawa decomposition of $Y \in \mathscr{SP}_n$ as:

$$Y = \begin{pmatrix} a_1 & & 0 \\ & \ddots & \\ 0 & & a_q \end{pmatrix} \begin{bmatrix} I_{n_1} & & * \\ & \ddots & \\ 0 & & I_{n_q} \end{bmatrix}, \quad a_j \in \mathscr{SP}_{n_j}. \tag{5.74}$$

Here

$$\prod_{j=1}^{q} |a_j| = 1, \quad \text{assuming } |Y| = 1.$$

And if $n_j > 1$, we have a non-empty discrete spectrum $A(GL(n_j))$ as in (5.73). We define

$$A(P(n_1, \ldots, n_q)) = \{v = (v_1, \ldots, v_q) | v_j \in A(GL(n_j))\}, \tag{5.75}$$

where if $n_j = 1$, we write $A(GL(1)) = \{1\}$.

As in (5.6) of Section 4.5.1, the *Eisenstein series* in the continuous spectrum are given by:

$$E_P(v, r | Y) = \sum_{\gamma \in \Gamma_n/P} \varphi(Y[\gamma]), \tag{5.76}$$

for

$$\varphi(Y) = \prod_{j=1}^{q} v(a_j^0) |a_j|^{-r_j}, \quad r \in \mathbb{C},$$

with $v = (v_1, \ldots, v_q) \in A(P)$, Re r_j suitably restricted for convergence of (5.76). We are using the notation:

$$a_j^0 = |a_j|^{-1/n_j} a_j \in \mathscr{SP}_{n_j}.$$

Actually, since the determinant of Y is 1, we have only $q - 1$ independent r-variables. So we just integrate over the first $q - 1$ of them. Also, we must continue the Eisenstein series outside of the region of convergence of (5.76) to reach the spectrum of the invariant differential operators. Since, $E_P(v, r | Y)$ is an eigenfunction of all the G-invariant differential operators on $\mathscr{SP}_n$, it determines, as in Proposition 4 of Section 4.2.3, a vector of powers $s_P(v, r) = s \in \mathbb{C}^n$ to put in the Helgason-Fourier transform $\hat{f}(s(r))$ in (5.71). That is, E_P and p_s have the same eigenvalues for the G-invariant differential operators L:

$LE_P(v,r|Y) = \lambda_L E_P(v,r|Y)$ and $Lp_s = \lambda_L p_s$ for all $L \in D(\mathscr{SP}_n)$,   (5.77)

where $p_s(Y)$ denotes the power function defined in (2.1) of Section 4.2.1. Finally, the κ_P in formula (5.72) denotes a positive constant.

The proof of Poisson's summation formula comes from the spectral decomposition of the G-invariant differential operators on $L^2(\mathscr{SP}_n/GL(n,\mathbb{Z}))$ plus Selberg's basic lemma that eigenfunctions of G-invariant differential operators are eigenfunctions of G-invariant integral operators (Proposition 4 of Section 4.2.3). We spell out the details later in the section.

One of the first applications of this formula is to study an analogue of the circle problem on the asymptotics of the number of lattice points in a circle in $\mathbb{R}^2$ of radius x as x approaches infinity. In our case the statistical function is

$$\#\{\gamma \in GL(n,\mathbb{Z}) | \text{Tr}({}^t\gamma\gamma) \leq x\}, \qquad (5.78)$$

and one should obtain the asymptotic character of this function as x approaches infinity. In order to imitate the discussion for $n = 2$ that appears in pages 266–268 of Volume I, it is necessary to study the Helgason transform of the function $\exp(-p\,\text{Tr}(Y))$ for $p > 0$ and Y in $\mathscr{SP}_n$. When $n = 2$, this is a K-Bessel function (though it looks like a gamma function). We have not yet carried out the details for $GL(n)$. Results like (4.2) have been obtained by many authors, going back to Delsarte in the 1940's for cocompact subgroups. Results for general compact K/Γ are obtained by Bartels [1]. The case of $G = SL(2,\mathbb{C})$, coming from imaginary quadratic fields is considered by Mennicke [1].

Mennicke has also proposed that the Poisson summation formula can be used to study properties of the cusp forms themselves. It would be interesting to carry out this investigation for $SL(3,\mathbb{Z})$ or Γ having compact fundamental domain in $\mathscr{SP}_n$. As far as I know, no one has done much in this direction for $SL(3,\mathbb{Z})$. Of course, there are problems in using (5.72), since one has little knowledge about the discrete spectrum of $SL(3,\mathbb{Z})$. It would be nice to be able to turn the tables on the formula and extract such knowledge anyway.

We should also mention that the spectral decomposition of the $GL(n,\mathbb{R})$-invariant differential operators on $\mathscr{SP}_n/GL(n,\mathbb{Z})$ can be used to obtain a converse theorem for an analogue of Hecke's correspondence that takes place between Siegel modular forms and Dirichlet series in several variables. This is worked out for $Sp(2,\mathbb{Z})$ by Kaori Imai [1], extending work of Koecher [1] and Maass [2]. We will say a little more about the subject of Hecke's correspondence for Siegel modular forms in Chapter 5. Weissauer [1] obtains a very general converse theorem for congruence subgroups of $Sp(n,\mathbb{Z})$.

After this brief introduction, let us attempt to understand the discrete spectrum better. To do this we need to develop an analogue of Theorem 2 in Section 3.7 of Volume I. So we define:

$L_0^2(\mathscr{SP}_n/GL(n,\mathbb{Z}))$

$= \{h \in L^2(\mathscr{SP}_n/GL(n,\mathbb{Z})) | a_0^k(Y) = 0, \text{for almost all } Y \in \mathscr{SP}_n, 1 \leq k \leq n-1\},$
(5.79)

4.5. Automorphic Forms for $GL(n, \mathbb{Z})$ and Harmonic Analysis on $\mathscr{P}_n/GL(n, \mathbb{Z})$

where $a_0^k(v)$ is defined by (5.69) in the previous section as the 0th Fourier coefficient of h with respect to $P(k, n - k)$. Gelfand and Piatetski-Shapiro have proved the $GL(n)$-analogue of Theorem 2 of Section 3.7, Volume I. Let us give a sketch of Godement's proof of this result (see Borel and Mostow [1, pp. 225–234]). The argument is very close to the one we gave in Volume I, but it is complicated by the non-abelian nature of the nilpotent subgroup $N \subset G = SL(n, \mathbb{R})$, where N consists of upper triangular matrices with ones on the diagonal.

We want to prove the compactness on $L_0^2(\mathscr{P}_n/SL(n, \mathbb{Z}))$ of the convolution operators C_h defined below for $h: \mathscr{P}_n \to \mathbb{C}$ infinitely differentiable with compact support. As usual, we identify functions on $\mathscr{P}_n$ with functions on $G = SL(n, \mathbb{R})$ via $f(I[g]) = f(g)$ for all $g \in G$ and write the convolution operator as in (1.24) of Section 4.1.3:

$$C_h f(a) = \int_{b \in G = SL(n, \mathbb{R})} h(ab^{-1}) f(b) \, db. \tag{5.80}$$

Theorem 7 (Gelfand and Piatetski-Shapiro). *The convolution operator C_h defined in (5.80) gives a compact operator on the space of square integrable functions on the fundamental domain with vanishing constant terms; i.e., on $L_0^2(\mathscr{P}_n/GL(n, \mathbb{Z}))$ defined in (5.79).*

PROOF. This discussion follows a similar path to that laid out in the proof of Theorem 2 in Section 3.7, Volume I. Here we will consider only the case that $n = 3$.

Just as for $n = 2$, it suffices, by the Theorem of Arzelà and Ascoli (see Kolmogorov and Fomin [1]) to show that:

$$|C_h f(a)| \leq k \|f\|_2, \quad \text{for some positive constant } k,$$

where

$$\|f\|_2^2 = \int_{\mathscr{P}_3/\Gamma_3} |f(W)|^2 \, dW, \quad \Gamma_3 = GL(3, \mathbb{Z}).$$

We assume that $h(b) = h(b^{-1})$ so that:

$$C_h f(x) = \int_{\mathscr{P}_3} f(W) h(W[x^{-1}]) \, dW$$

$$= \int_{\mathscr{P}_3/N_{\mathbb{Z}}} f(W) \sum_{n \in N_{\mathbb{Z}}} h(W[nx^{-1}]) \, dW,$$

where $N_{\mathbb{Z}}$ denotes the integral upper triangular 3×3 matrices with ones on the diagonal.

Now apply ordinary Poisson summation as stated at the beginning of this section to rewrite the sum over $N_{\mathbb{Z}}$. You find that this sum is equal to:

$$\sum_{R \in N_{\mathbb{Z}}} \hat{h}(R), \quad \text{where } \hat{h}(R) = \int_{N_{\mathbb{R}}} h(W[nx^{-1}]) \exp\left(2\pi i \sum_{1 \leq j < k \leq 3} n_{jk} r_{jk}\right) dn.$$

Here $N_\mathbb{R}$ denotes the real 3×3 upper triangular matrices with ones on the diagonal.

We may assume that h lies in a Dirac delta sequence, in particular, that h is $K = O(3)$ bi-invariant. Write the Iwasawa decomposition of $x \in G$ as $x = kav$, where $k \in K = O(3)$, $v \in N_\mathbb{R}$, and a is positive diagonal with jth diagonal entry a_j. We want to bound $C_h f(x)$ as a_i/a_{i+1} approaches zero for $i = 1, 2$. Let

$$W = b[t] = \begin{pmatrix} b_1 & 0 & 0 \\ 0 & b_2 & 0 \\ 0 & 0 & b_3 \end{pmatrix} \begin{bmatrix} 1 & t_{12} & t_{13} \\ 0 & 1 & t_{23} \\ 0 & 0 & 1 \end{bmatrix}, \quad \text{for } b_j > 0.$$

Then

$$\hat{h}(R) = \int_{n \in N_\mathbb{R}} h(b[tnv^{-1}a^{-1}]) \exp\left(2\pi i \sum_{j<k} n_{jk} r_{jk}\right) dn.$$

Next change variables via $m = tnv^{-1}$, $n = t^{-1}mv$. Recall that

$$t = \begin{pmatrix} 1 & t_{12} & t_{13} \\ 0 & 1 & t_{23} \\ 0 & 0 & 1 \end{pmatrix} \text{ implies } t^{-1} = \begin{pmatrix} 1 & -t_{12} & t_{12}t_{23} - t_{13} \\ 0 & 1 & -t_{23} \\ 0 & 0 & 1 \end{pmatrix}$$

and

$$\begin{pmatrix} 1 & m_{12} & m_{13} \\ 0 & 1 & m_{23} \\ 0 & 0 & 1 \end{pmatrix} \begin{pmatrix} 1 & v_{12} & v_{13} \\ 0 & 1 & v_{23} \\ 0 & 0 & 1 \end{pmatrix} = \begin{pmatrix} 1 & m_{12} + v_{12} & v_{13} + m_{12}v_{23} + m_{13} \\ 0 & 1 & m_{23} + v_{23} \\ 0 & 0 & 1 \end{pmatrix}.$$

From this, we see that, if $n = t^{-1}mv$ and we use the same notation for entries of n as for entries of t, then we obtain:

$$n_{12} = -t_{12} + m_{12} + v_{12},$$

$$n_{23} = -t_{23} + m_{23} + v_{23},$$

$$n_{13} = v_{13} + m_{12}v_{23} + m_{13} - t_{12}(m_{23} + v_{23}) + t_{12}t_{23} - t_{13}.$$

It follows that, with the n_{ij} as above:

$$\hat{h}(R) = \int_{m \in N_\mathbb{R}} h(b[ma^{-1}]) \exp\{2\pi i(r_{12}n_{12} + r_{13}n_{13} + r_{23}n_{23})\} dm.$$

Then we must change variables via $n = ama^{-1}$, with Jacobian:

$$\alpha(a) = \prod_{i<j} a_j/a_i.$$

This gives:

$$\hat{h}(R) = \alpha(a) \int_{n \in N_\mathbb{R}} h(b[a^{-1}n]) \exp\{2\pi i T(n)\} dn,$$

where

4.5. Automorphic Forms for $GL(n, \mathbb{Z})$ and Harmonic Analysis on $\mathcal{P}_n/GL(n, \mathbb{Z})$

$$T(n) = r_{12}\left(\frac{a_2}{a_1}n_{12} - t_{12} + v_{12}\right) + r_{23}\left(\frac{a_3}{a_2}n_{23} - t_{23} + v_{23}\right)$$

$$+ r_{13}\left\{v_{13} + \frac{a_2}{a_1}n_{12}v_{23} + \frac{a_3}{a_1}n_{13} - t_{12}\left(\frac{a_3}{a_2}n_{23} + v_{23}\right) + t_{12}t_{23} - t_{13}\right\}.$$

It will be easier if we approximate h by functions such that:

$$h(b[t]) = h_0(b)h_{12}(t_{12})h_{13}(t_{13})h_{23}(t_{23}).$$

Then we have:

$$\hat{h}(R) = \alpha(a)h_0(b[a^{-1}])\exp\{2\pi i L(r, v, t)\}\hat{h}_{12}\left(\frac{a_2}{a_1}(r_{12} + r_{13}v_{23})\right)$$

$$\times \hat{h}_{13}\left(\frac{a_3}{a_1}r_{13}\right)\hat{h}_{23}\left(\frac{a_3}{a_2}(r_{23} - r_{13}t_{12})\right),$$

where

$$L(r, v, t) = r_{12}(v_{12} - t_{12}) + r_{13}(v_{13} - t_{12}v_{23} + t_{12}t_{23} - t_{13}) + r_{23}(v_{23} - t_{23}).$$

If $R = (r_{12}, r_{13}, r_{23}) = 0$, then the fact that f is a cusp form says that the integral over $t_{ij} \in [0, 1]$ is zero. Thus we need only consider $R \neq 0$.

Since h is infinitely differentiable with compact support (i.e., $h \in C_c^\infty$), we can apply differential operators to h and stay in C_c^∞. But then we can obtain the following bound:

$$\sum_{R \neq 0} \hat{h}(R) \leq \alpha(a)h_0(b[a^{-1}])Z(W(a, v, t), s),$$

where $Z(W, s)$ is Epstein's zeta function from Section 1.4 of Volume I and its first argument is the positive matrix:

$$W(a, v, t) = I\begin{bmatrix} a_2/a_1 & v_{23}a_2/a_1 & 0 \\ 0 & a_3/a_1 & 0 \\ 0 & -t_{12}a_3/a_2 & a_3/a_2 \end{bmatrix}.$$

Here s is any sufficiently large integer. Epstein's zeta function can in turn be bounded by

$$\max(\delta_1^{-s}, \delta_2^{-s}, \delta_3^{-s}) \sum_{R \in \mathbb{Z}^3 - 0} Y(v, t)[R]^{-s},$$

where $\delta_1 = a_2/a_1$, $\delta_2 = a_3/a_1$, $\delta_3 = a_3/a_2$, and

$$Y(v, t) = I\begin{bmatrix} 1 & v_{23} & 0 \\ 0 & 1 & 0 \\ 0 & -t_{12} & 1 \end{bmatrix}.$$

Since v and t can be assumed to be bounded, our only problem is to deal with $h_0(b[a^{-1}])$ for a_i/a_{i+1} near zero. But, because h_0 has compact support, $h_0(b[a^{-1}]) \neq 0$ implies

$$c_1 < \frac{b_i}{b_{i+1}} \frac{a_{i+1}}{a_i} < c_2.$$

If we assume that the quotient a_i/a_{i+1} is sufficiently near zero, then

$$b_i/b_{i+1} < c,$$

for as small a constant c as we like.

Thus we have bounded $|C_h f(x)|$ by a constant multiplied by the product of some powers of a_i/a_{i+1} and the following integral, if $x = kav$, as above:

$$\int_{|t_{ij}| \leq 1/2} \int_{b_i \leq cb_{i+1}} f(b[t]) b_1 \, db_1 \, db_2 \, dt_{12} \, dt_{13} \, dt_{23}.$$

To complete the proof of the bound on $C_h f(x)$, one must note that when $c = 1$, the domain of integration in the integral above must lie in the fundamental domain $\mathscr{S}\mathscr{F}_3$ of Section 4.4.3. For this, see Exercise 17 of Section 4.4.3. Therefore the desired inequality follows from the Cauchy-Schwartz inequality. □

This is the end of our discussion of the discreteness of the spectrum of the cusp forms for $SL(n, \mathbb{Z})$. Of course, we have not proved that $L_0^2(\mathscr{S}\mathscr{P}_n/GL(n, \mathbb{Z}))$ has anything non-trivial in it. So we don't really know that cusp forms exist. I seem to remember that Dorothy (Wallace) Andreoli and Isaac Efrat have told me that they have generalizations of Exercise 6 of Section 3.7 in Volume I which would give the existence of some analogue of the odd cusp forms.

There is another question one might raise at this point, as Kaori Imai pointed out to me. Our discussion of the discreteness of the spectrum of the cusp forms for $SL(3, \mathbb{Z})$ used the vanishing of the integral defined by:

$$\iiint_{t_{ij} \in [0, 1]} f(Y[t]) \, dt_{12} \, dt_{13} \, dt_{23}, \quad t = \begin{pmatrix} 1 & t_{12} & t_{13} \\ 0 & 1 & t_{23} \\ 0 & 0 & 1 \end{pmatrix}. \quad (5.81)$$

The definition of cusp form given in (5.69) requires the vanishing of both of the integrals below:

$$\iint_{t_{ij} \in [0, 1]} f(Y[t]) \, dt_{12} \, dt_{13}, \quad t = \begin{pmatrix} 1 & t_{12} & t_{13} \\ 0 & 1 & 0 \\ 0 & 0 & 1 \end{pmatrix},$$

$$\iint_{t_{ij} \in [0, 1]} f(Y[t]) \, dt_{13} \, dt_{23}, \quad t = \begin{pmatrix} 1 & 0 & t_{13} \\ 0 & 1 & t_{23} \\ 0 & 0 & 1 \end{pmatrix}. \quad (5.82)$$

Note that (5.82) implies (5.80) but the converse is not clear.

Next let us say a little more about the proof of (5.72). Note that we saw in Section 4.5.3 for a special case that if P and Q are non-associated parabolic

4.5. Automorphic Forms for $GL(n, \mathbb{Z})$ and Harmonic Analysis on $\mathscr{P}_n/GL(n, \mathbb{Z})$

subgroups of $GL(n)$, then:

$$E_P(v, r | Y) \sim \prod_{j=1}^{q} v_j(a_j^0) |a_j|^{-r_j}, \quad \text{as } Y \to \partial \mathscr{S}\mathscr{F}_n,$$

in the direction of P. Here $\partial \mathscr{S}\mathscr{F}_n$ denotes the boundary of the fundamental domain $\mathscr{S}\mathscr{F}_n$ for $\mathscr{S}\mathscr{P}_n/GL(n, \mathbb{Z})$. However, because P and Q are non-associated parabolic subgroups and if v is a cusp form:

$$E_Q(v, r | Y) \sim 0,$$

as Y approaches the boundary of the fundamental domain in the direction corresponding to Q. One expects that this implies that E_P and E_Q are orthogonal provided that P and Q are non-associated and at least one of the Eisenstein series has a cusp form floating around in it. An argument similar to that of Exercise 36, Section 4.5.3 should work here provided that one has proved the vanishing of the 0th Fourier coefficient of E_Q with respect to the parabolic subgroup P.

Of course, if P and Q are associated parabolic subgroups, then one can use an argument similar to that given in the proof of Proposition 1 in Section 4.5.3 to see that E_P and E_Q are really the same function after a trivial change of variables.

Thus it appears likely that a generalization of the principle of asymptotics and functional equations from Section 4.3 should lead to the spectral resolution of the G-invariant differential operators on $L^2(\mathscr{S}\mathscr{P}_n/GL(n, \mathbb{Z}))$. One could attempt to imitate the approach given for Theorem 1 of Section 3.7 in Volume I for $SL(2, \mathbb{Z})$, using incomplete theta series attached to the non-associated parabolic subgroups. We will not do this here. We would, for example, need to see that someone orthogonal to all the Eisenstein series E_P for all P must be in L_0^2 and thus in the span of the discrete spectrum $A(GL(n))$. The ultimate result would be that any $g \in L^2(\mathscr{S}\mathscr{P}_n/GL(n, \mathbb{Z}))$ has the *spectral decomposition*:

$$g(Y) = \sum_{w \in A(G)} (g, w) w(Y)$$
$$+ \sum_P \sum_{v \in A(P)} \kappa_P \int_{\substack{r \in \mathbb{C}^{q-1} \\ \operatorname{Re} r_j = \text{constant}}} (g, E_P) E_P(v, r | Y) \, dr, \quad (5.83)$$

where (f, g) is the inner product on $L^2(\mathscr{S}\mathscr{P}_n/GL(n, \mathbb{Z}))$ with respect to the G-invariant measure. We use the notation of (5.72)–(5.77) above. See Langlands [1], Harish-Chandra [1], Osborne and Warner [1], and Arthur's lecture in Borel and Casselman [1, Vol. I, pp. 253–274].

In order to derive the Poisson summation formula (5.72) from (5.83), one must use Selberg's basic lemma which says that eigenfunctions of G-invariant differential operators are eigenfunctions of G-invariant integral operators (see Proposition 4 in Section 4.2.3). This shows that if we write $f(I[a]) = f(a)$ for all $a \in G$ and

$$g(a) = \sum_{\gamma \in \Gamma_n/\pm I} f(a\gamma b^{-1}), \quad \text{with } f \text{ as in (5.72)},$$

then by Proposition 4 of Section 4.2.1, since we are assuming that $f(a) = f(a^{-1})$, we have:

$$(g, E_P) = \int_{a \in G/\Gamma_n} \sum_{\gamma \in \Gamma_n/\pm I} f(a\gamma b^{-1}) \overline{E_P(I[a])} \, da$$

$$= \int_G f(ab^{-1}) \overline{E_P(I[a])} \, da$$

$$= \int_G f(ba^{-1}) \overline{E_P(I[a])} \, da = C_f(\bar{E}_P)(b) = \hat{f}(s_P) \overline{E_P(I[b])}.$$

Here $s_P = s_P(v, r)$ is chosen as in (5.77). A similar argument applies to the discrete spectrum terms in (5.72). This completes our very sketchy discussion of Poisson summation for $GL(n, \mathbb{Z})$.

The main application of the Poisson summation formula (5.72) is to the trace formula. If we are looking at a discrete subgroup Γ of $GL(n, \mathbb{R})$ acting on $\mathscr{SP}_n$ so as to have a *compact* fundamental domain, then the trace formula is easily obtained, since (5.72) then involves no continuous or residual spectrum. See Mostow [2] for an example of such a group Γ. In what follows we give a sketchy discussion of the Selberg trace formula, imitating what we said in the last part of Section 3.7 in Volume I.

Stage 1.

$$\int_{K \backslash G/\Gamma} \sum_{\gamma \in \Gamma/\pm I} f(b\gamma b^{-1}) \, dg = \sum_{m \geq 0} \hat{f}(s_m). \tag{5.84}$$

Formula (5.84) is obtained by integrating (5.72) and using the orthonormality of the various elements of the spectrum of the G-invariant differential operators on $\mathscr{SP}_n/GL(n, \mathbb{Z})$.

If the fundamental domain for Γ is not compact, divergent integrals arise on the left in (5.84) and must be cancelled against integrals from components of the continuous spectrum, as in the case of $SL(2, \mathbb{Z})$ in Section 3.7 of Volume I. This is reminiscent of renormalization methods in quantum field theory. There are even problems showing that one can take the trace. Some references are Arthur [1–2], A.B. Venkov [1], Wallace (Andreoli) [1–6], and Warner's paper in the volume of Hejhal, Sarnak, and Terras [1, pp. 529–534].

Stage 2. Let $\{\gamma\} = \{a\gamma a^{-1} | a \in \Gamma/\pm I\}$ = the *conjugacy class* of γ in $\Gamma/\pm I$. And let Γ_γ be the *centralizer* of γ in Γ. Then it is easily shown, as in (3.129) of Section 3.7, Volume I, that

$$\sum_{m \geq 0} \hat{f}(s_m) = \sum_{\{\gamma\}} c_f(\gamma), \tag{5.85}$$

where the sum on the right is over all distinct conjugacy classes $\{\gamma\}$ in $\Gamma/\pm I$, and the *orbital integral* is:

$$c_f(\gamma) = \int_{K \backslash G/\Gamma_\gamma} f(b\gamma b^{-1}) \, db. \tag{5.86}$$

4.5. Automorphic Forms for $GL(n,\mathbb{Z})$ and Harmonic Analysis on $\mathscr{P}_n/GL(n,\mathbb{Z})$

Stage 3. Now one needs to evaluate orbital integrals. If γ is in the *center* of G, this is easy, since it is clear that:

$$c_f(\gamma) = f(\gamma)\mathrm{Vol}(K\backslash G/\Gamma), \qquad \text{for } \gamma \text{ in the center of } \Gamma.$$

This can be evaluated in terms of $\hat{f}(s)$ using the inversion formula from Theorem 1 in Section 4.3.1.

Next suppose that γ is *hyperbolic*; i.e., γ is conjugate in $G = SL(n,\mathbb{R})$ to a matrix with n distinct real eigenvalues d_i none of which equal one. Let us assume for simplicity that $n = 3$. In this case, one finds that the orbital integral for γ can be replaced by one in which the argument of f has the following form:

$$\begin{pmatrix} y_1 & 0 & 0 \\ 0 & y_2 & 0 \\ 0 & 0 & y_3 \end{pmatrix} \begin{pmatrix} 1 & x_1 & x_3 \\ 0 & 1 & x_2 \\ 0 & 0 & 1 \end{pmatrix} \begin{pmatrix} d_1 & 0 & 0 \\ 0 & d_2 & 0 \\ 0 & 0 & d_3 \end{pmatrix} \begin{pmatrix} 1 & x_1 & x_3 \\ 0 & 1 & x_2 \\ 0 & 0 & 1 \end{pmatrix}^{-1} \begin{pmatrix} 1/y_1 & 0 & 0 \\ 0 & 1/y_2 & 0 \\ 0 & 0 & 1/y_3 \end{pmatrix}$$

$$= \begin{pmatrix} d_1 & x_1 y_1(d_2 - d_1)/y_2 & x_3 y_1(d_3 - d_1)/y_3 + x_1 x_2 y_1(d_1 - d_2)/y_3 \\ 0 & d_2 & x_2 y_2(d_3 - d_2)/y_3 \\ 0 & 0 & d_3 \end{pmatrix}.$$

Now look at the fundamental domain $K\backslash G/\Gamma_\gamma$ for the centralizer of γ. We can assume that we have conjugated everything so that Γ_γ consists of diagonal matrices with diagonal entries c_j, $j = 1, 2, 3$. The c_j must be units in a totally real cubic number field. The elements of $K\backslash G$ are represented by matrices of the form:

$$\begin{pmatrix} y_1 & 0 & 0 \\ 0 & y_2 & 0 \\ 0 & 0 & y_3 \end{pmatrix} \begin{pmatrix} 1 & x_1 & x_3 \\ 0 & 1 & x_2 \\ 0 & 0 & 1 \end{pmatrix}, \qquad y_3 = \frac{1}{y_1 y_2}.$$

Then γ acts on g by multiplying the y_j by c_j. So we can take the y_j in a fundamental domain for this action. From number theory (Dirichlet's theorem on units in algebraic number fields which was mentioned in Section 1.4 of Volume I), one knows that there is a compact fundamental domain for the action of the c's on the y's. The volume of this compact fundamental domain is measured by what number theorists call a regulator. See pages 67–72 of Section 1.4 of Volume I for the definition and properties of the regulator.

Now make a change of variables in the orbital integral via:

$$u_1 = x_1 y_1(d_2 - d_1)/y_2, \qquad u_2 = x_2 y_2(d_3 - d_2)/y_3,$$
$$u_3 = x_3 y_1(d_3 - d_1)/y_3 + x_1 x_2 y_1(d_1 - d_2)/y_3.$$

You obtain

$$H(d) \int_{\substack{y \bmod c \\ u \in \mathbb{R}^3}} f \begin{pmatrix} d_1 & u_1 & u_3 \\ 0 & d_2 & u_2 \\ 0 & 0 & d_3 \end{pmatrix} y_1^{-3} y_2^{-2} \, dy \, du,$$

where

$$H(d) = \prod_{i<j} |d_i - d_j|^{-1}.$$

Thus we end up with the Harish transform of f over N as in the hint to Exercise 3 in Section 4.3.1. The Helgason-Fourier transform (5.71) is a composition of Mellin transforms over the y-variables and the Harish-transform over the x-variables. See Wallace (Andreoli) [3] for more details on this calculation.

Since we have seen that units in cubic number fields appear in these formulas, one expects to generalize Sarnak's result (Theorem 6 of Section 3.7 in Volume I) to cubic and higher degree totally real fields once the full trace formula for $SL(n, \mathbb{Z})$ is worked out in excruciating detail. We should also note that there will be other sorts of orbital integrals from elliptic and mixed terms.

Stage 4. In this stage one must deal with the problems which arise when the fundamental domain for Γ is not compact. The orbital integrals for parabolic $\gamma \in \Gamma$ (i.e., γ some of whose eigenvalues are 1) require truncation, as do the inner products of continuous spectrum terms. The most explicit formulas in this direction can be found in A.B. Venkov [1] and Wallace (Andreoli) [1–6]). One is somewhat disturbed to see that there are infinite sums of divergent things and that moreover some of the terms from the Maass-Selberg relations are not explicitly computed. The Maass-Selberg relations are just the Green's identity for $(f, \Delta g)_T$ where the subscript means that the inner product is taken over a truncated fundamental domain. In the higher rank case, the Fourier expansion is much messier, as are the constant terms and thus this result is complicated to state. We won't do it here. Arthur [1] has developed a formalism for these results involving truncation operators and Arthur polynomials (cf. Osborne's first paper in the volume edited by Hejhal, Sarnak, and Terras [1]). There are still questions about the traceability of the integral operators as we mentioned above.

It seems likely that one will have to use the trace formula for $SL(n-1, \mathbb{Z})$ in order to prove that the cancellation occurs and we get a trace formula for $SL(n, \mathbb{Z})$. Even when $n = 3$, it appears that complicated calculations will be needed, if one is to develop the result in a way analogous to that used in the case $n = 2$ in Section 3.7 of Volume I. Many very large sheets of paper are needed seemingly.

Of course, we should mention that various cases of the trace formula have already been worked out; for example in the case of discrete or discontinuous group actions on the quaternionic upper half space $SL(2, \mathbb{C})/SU(2)$, or actions on products of upper half planes, or actions on Siegel's upper half space $Sp(n, \mathbb{R})/SU(n)$. See Chapter 5 for more information on such examples. References for this and related topics are: Arakawa [2], Arthur [1–3], Christian [4], Efrat [2], Eie [1], Elstrodt, Grunewald, and Mennicke [1–3], Gangolli and Warner [1,2], Godement [6], Hashimoto [2], Hejhal, Sarnak, and Terras [1, pp. 253–276], Langlands [1–7], Mennicke [1–2],

4.5. Automorphic Forms for $GL(n, \mathbb{Z})$ and Harmonic Analysis on $\mathscr{P}_n/GL(n, \mathbb{Z})$

Morita [1], Müller [1], Petra Ploch [1], Sarnak [1], Selberg [1, 3], Shimizu [1], Tanigawa [1], A.B. Venkov [1], Marie-France Vignéras [3–5], Dorothy Wallace-(Andreoli) [1–6], Warner [2], and Zograf [1]. Langlands [6] has obtained information on dimensions of spaces of automorphic forms by noting that since it is not too hard to compute some of the terms in the trace formula, e.g., the identity and hyperbolic terms, then one can make use of a good vanishing result for the remaining terms or a way of showing them asymptotically negligible. See Langlands talk in Borel and Mostow [1, pp. 251–257] and Warner's discussion of the Selberg principle [1, Vol. II, p. 370].

Adelic versions of the trace formula have also been obtained. See Flicker [1]. And there is a trace formula for $N_\mathbb{R}/N_\mathbb{Z}$, where N is the nilpotent group of 3×3 upper triangular matrices with ones on the diagonal (see Osborne's 2nd paper in Hejhal, Sarnak, and Terras [1, pp. 375–385]). It would be natural to use this result in our derivation of the trace formula for $SL(3, \mathbb{Z})$, but, so far as I know, no one has done this. There are also twisted trace formulas which are needed for base change and the work of Langlands [3] on Artin's conjecture.

We have not posed any official exercises in this section. Let us just close Chapter 4 by saying that if you want something to work on, you could try to translate anything in Section 3.7 in Volume I over to $SL(3, \mathbb{Z})$. Many have slaved on this project. But there is still much to do.

CHAPTER V

The General Noncompact Symmetric Space

"These things will become clear to you," said the old man gently, "at least," he added with a slight doubt in his voice, "clearer than they are at the moment."

> From *The Hitchhiker's Guide to the Galaxy*, by Douglas Adams, Pocket Books, NY, 1981. Reprinted by permission of The Crown Publishing Group.

5.1. Geometry and Analysis on G/K

The first four chapters of this tome considered various examples and applications of symmetric spaces X, along with harmonic analysis on X and X/Γ for discrete groups Γ of isometries of X. Here we consider some aspects of analysis on a general noncompact symmetric space $X = G/K$. Our discussion will be very sketchy. The main goal is to lay the groundwork for extension of the results of the preceding chapters to other symmetric spaces which are of interest for applications; in particular, the Siegel upper half plane $\mathcal{H}_n$ (which can be identified with $Sp(n, \mathbb{R})/U(n)$) and hyperbolic three space $\mathcal{H}^c$ (which can be viewed as $SL(2, \mathbb{C})/SU(2)$). We will also be interested in the fundamental domains $\mathcal{H}_n/Sp(n, \mathbb{Z})$ for the Siegel modular group as well as the fundamental domain $\mathcal{H}^c/SL(2, \mathbb{Z}[i])$ for the Picard group. It is possible to generalize just about everything we did in the earlier chapters for such examples; e.g., the Selberg trace formula. And our main motivations for doing so come from number theory. Because it is time consuming and sometimes not so enlightening to do each of these examples separately, we have decided to present some results on the general symmetric space. Those interested in number theoretic applications may find this equally tedious and attempt to

5.1. Geometry and Analysis on G/K

jump to the next section. But I think it is useful to know what a general Iwasawa decomposition is, for example, in order to find the right coordinates to use in solving a given problem on the symmetric space. Of course, others will say that the discussion which follows is neither sufficiently general, detailed, or rigorous. We refer those characters to the texts of other authors which are listed below.

Some topics in physics that lead one to study these other symmetric spaces are: quantum statistical mechanics and quantum field theory (see Hurt [1]), particle physics (see Wybourne [1, Ch. 21]), coherent states (see Monastyrsky and Perelomov [1], Perelomov [1], and Hurt [1]), boson fields (see Shale [1] and Cartier's article in Borel and Mostow [1, pp. 361–386]), solitons (see Dubrovin, Matveev, and Novikov [1], McKean and Trubowitz [1], Lonngren and Scott [1], and Novikov [1]), rotating tops (see Sonya Kovalevsky [1], Linda Keen [1], Pelageya Kochina [1], and Cooke [1]), and string theory (see Polyakov [1]).

Many branches of number theory steer one into these realms; e.g., the theory of quadratic forms (see Siegel [1–5]), algebraic number theory (see Siegel [4] and Hecke [1]). The study of the ring $\mathbb{Z}[i]$ of Gaussian integers and similar rings for various algebraic number fields leads one to think that anything one can do for $\mathbb{Z}$ should be generalizable to $\mathbb{Z}[i]$. In particular, we will see that the theory of Maass wave forms for $SL(2, \mathbb{Z})$ has an analogue for $SL(2, \mathbb{Z}[i])$. This leads to some interesting formulas for the Dedekind zeta function of $\mathbb{Q}(i)$, among other things. See the discussion in Section 5.2 which follows. There is also an analogue of Selberg's trace formula (see Sarnak [1]).

Finally electrical engineering has many applications of these symmetric spaces as well (see Blankenship [1] and Helton [1–3]). We saw in Section 3.1 of Volume I that 2-port microwave circuits lead to quantities in $SL(2, \mathbb{R})$. Similarly more complicated circuits lead to higher rank Lie groups.

References for this section include: Baily [1], Barut and Rączka [1], Broecker and tom Dieck [1], Chevalley [1], Choquet-Bruhat et al. [1], Dieudonné [1], Gangolli [1–4], Harish-Chandra [1–2], Helgason [1–10], Hermann [1–2], Hua [1], Loos [1], Maass [2], Piatetski-Shapiro [2], Sagle and Walde [1], Séminaire Cartan [1], Siegel [1–5], Varadarajan [1–3], Wallach [1–3], Warner [1], and Wybourne [1].

We will assume that the reader has had a decent course in multivariable calculus. Our favorite books for this are Lang [1, 2]. The notions of differential, tangent space, matrix exponential, Taylor's formula are all covered there. You may also need to refer to a book like that of Sagle and Walde [1] for more details on various arguments. The true story of everything is found in Helgason's big green books. Varadarajan [1] is also useful, for example, as a source for all the details of the root space calculations.

Élie Cartan obtained the basic theory of symmetric spaces between 1914 and 1927. Then, beginning in the 1950's, Harish-Chandra, Helgason, and others developed harmonic analysis and representation theory on these spaces and their Lie groups of isometries.

A *symmetric space* M is a Riemannian manifold (as in the discussion at the beginning of Chapter 2, Volume I) such that at each point $P \in M$ there is a geodesic-reversing isometry

$$s_P: M \to M;$$

i.e., s_P preserves the Riemannian metric.

Our first goal is to produce a multitude of examples of symmetric spaces. We always start with a *Lie group* G; i.e., a real analytic manifold which is also a group such that the mapping

$$G \times G \to G$$
$$(g, h) \mapsto gh^{-1}$$

is analytic. We will only consider Lie groups of real or complex matrices here.

The *Lie algebra* $\mathfrak{g}$ of a Lie group G is the tangent space to G at the identity, once it has been provided with an additional operation called "the Lie bracket." It is traditional that the Lie algebra is written as the lower case German letter corresponding to the upper case Latin letter which is the group.

The Lie bracket operation is defined by identifying the Lie algebra

$$\mathfrak{g} = T_e(G) = \text{the tangent space to } G \text{ at the identity } e \in G,$$

with the space of left-invariant vector fields on G. These vector fields are first order differential operators on G (with real analytic coefficients) which commute with left translation. This identification is achieved by making use of the left translation $L_g(x) = gx$, for $x, g \in G$. If $\tilde{X}$ is a left-invariant vector field, then

$$\tilde{X}_g = dL_g(X), \qquad \text{for } g \in G, X \in \mathfrak{g}.$$

Here dL_g denotes the differential of left multiplication on G. Now we define the Lie bracket of two left invariant vector fields $\tilde{X}, \tilde{Y}$ by;

$$[\tilde{X}, \tilde{Y}] = \tilde{X}\tilde{Y} - \tilde{Y}\tilde{X},$$

which is also a left invariant vector field; i.e., the bracket of two first order differential operators is actually a first order and not a second order differential operator. Write $[X, Y]$ for the corresponding bracket of elements X, Y in the Lie algebra $\mathfrak{g}$. What makes $\mathfrak{g}$ a *Lie algebra*? The answer is that the bracket can be shown to have the following defining properties of such an algebra:

1. $[X, Y]$ is a bilinear map of $\mathfrak{g} \times \mathfrak{g}$ into $\mathfrak{g}$;
2. $[X, Y] = -[Y, X]$;
3. $[X, [Y, Z]] + [Y, [Z, X]] + [Z, [X, Y]] = 0$ (Jacobi's identity).

Then one defines subalgebra, ideal, homomorphism, isomorphism, etc., for Lie algebras in the usual way (see the references). For example, an ideal $\mathfrak{a} \subset \mathfrak{g}$ is a vector subspace $\mathfrak{a}$ of $\mathfrak{g}$ such that $[\mathfrak{a}, \mathfrak{g}] \subset \mathfrak{a}$.

5.1. Geometry and Analysis on G/K

Exercise 1. Prove that if $G = GL(n, \mathbb{R})$, then the corresponding Lie algebra can be identified with the vector space $\mathbb{R}^{n \times n}$ of all $n \times n$ real matrices with bracket defined by

$$[A, B] = AB - BA, \quad \text{for } A, B \in \mathbb{R}^{n \times n}.$$

Here AB denotes the usual matrix product. Thus $\mathfrak{gl}(n, \mathbb{R})$ is identified with $\mathbb{R}^{n \times n}$.

Hint. See Dieudonné [1, Vol. VI, pp. 145–146] or Sagle and Walde [1, pp. 117–118]. The vector space $\mathbb{R}^{n \times n}$ can certainly be identified with the tangent space to $GL(n, \mathbb{R})$, since $GL(n, \mathbb{R})$ is an open subset of $\mathbb{R}^{n \times n}$. In fact, using the matrix exponential, we can make the identification as follows. Let $A \in \mathbb{R}^{n \times n}$, $f: GL(n, \mathbb{R}) \to \mathbb{C}$, and $g \in GL(n, \mathbb{R})$. Then

$$(\tilde{A}f)(g) = \frac{d}{dt} f(g \exp tA)|_{t=0}.$$

One has for $A, B \in \mathbb{R}^{n \times n}$ and $g \in GL(n, \mathbb{R})$:

$$(\tilde{A}\tilde{B}f)(g) = \frac{d}{dt}\frac{d}{ds} f(g \exp tA \exp sB)|_{t=s=0}.$$

Use the chain rule to see that at $g = e$ this is $f_e''(A, B) + f_e'(AB)$. If you interchange A and B and then subtract, the second order terms cancel and you get $f_e'(AB - BA)$. This shows that the identification of $\mathfrak{g}$ with $\mathbb{R}^{n \times n}$ does preserve brackets.

There is a representation of any (real) Lie algebra $\mathfrak{g}$ in $\mathfrak{gl}(n, \mathbb{R})$, where $n = \dim_{\mathbb{R}} \mathfrak{g}$. This representation is called the *adjoint representation* defined as follows, thinking of $\mathfrak{gl}(n, \mathbb{R})$ as the space of linear transformations of $\mathfrak{g}$ into itself:

$$\begin{aligned}\text{ad}: \mathfrak{g} \to \mathfrak{gl}(n, \mathbb{R}), \quad n = \dim_{\mathbb{R}} \mathfrak{g}, \\ (\text{ad } X)Y = [X, Y], \quad \text{for } X, Y \in \mathfrak{g}.\end{aligned} \quad (1.1)$$

Exercise 2. Show that the adjoint representation defined by (1.1) above does indeed preserve brackets; i.e., $[\text{ad } X, \text{ad } Y] = \text{ad}[X, Y]$.

The *Killing form* of a Lie algebra $\mathfrak{g}$ is defined to be the bilinear form:

$$B_{\mathfrak{g}} = B(X, Y) = \text{Tr}(\text{ad } X \text{ ad } Y), \quad \text{for } X, Y \in \mathfrak{g}. \quad (1.2)$$

A Lie algebra is called *semisimple* if the Killing form B is nondegenerate; i.e., $B(X, Y) = 0$ for all $Y \in \mathfrak{g}$ implies $X = 0$. A Lie algebra $\mathfrak{g}$ is *simple* if it is semisimple, and if, in addition, it has no ideals but $\{0\}$ and itself.

Example ($GL(n, \mathbb{R})$). Since matrices of the form aI, $a \in \mathbb{R}$, commute with $n \times n$ matrices, it is clear that $\text{ad}(aI) = 0$ and thus that $B(aI, Y) = 0$ for all $Y \in \mathbb{R}^{n \times n}$.

Thus $\mathfrak{gl}(n, \mathbb{R}) \cong \mathbb{R}^{n \times n}$ is not semisimple. It will be useful to compute the Killing form for $\mathfrak{gl}(n, \mathbb{R})$. One can do this as follows. Let E_{ij} be the $n \times n$ matrix with i, j entry equal to one and the rest zero. Let H be the diagonal matrix with ith diagonal entry h_i. Then $\operatorname{ad}(H)E_{ij} = (h_i - h_j)E_{ij}$. Therefore

$$B(H, H) = \operatorname{Tr}(\operatorname{ad} H \operatorname{ad} H) = \sum_{i,j=1}^{n} (h_i - h_j)^2 = 2n \operatorname{Tr}(H^2) - 2(\operatorname{Tr} H)^2.$$

Note that it suffices to compute the Killing form on diagonal matrices. For the map $X \mapsto gXg^{-1}$, with $g \in GL(n, \mathbb{R})$ and $X \in \mathfrak{gl}(n, \mathbb{R})$, leaves the Killing form invariant. And matrices conjugate to a diagonal matrix are dense in $\mathfrak{gl}(n, \mathbb{R})$.

Exercise 3.

(a) Show that if σ is an automorphism of $\mathfrak{g}$, then

$$B(X, Y) = B(\sigma X, \sigma Y), \quad \text{for all } X, Y \in \mathfrak{g}.$$

(b) Show that

$$B(X, [Y, Z]) = B(Y, [Z, X]) = B(Z, [X, Y]) \quad \text{for all } X, Y, Z \in \mathfrak{g}.$$

There is an analogue of the matrix exponential for any Lie group G. It is, appropriately enough, called the *exponential map* and it maps the Lie algebra into the Lie group such that if $X \in \mathfrak{g}$, $g \in G$, and $f: G \to \mathbb{C}$ is infinitely differentiable, then

$$\tilde{X}_g f = \frac{d}{dt} f(g \exp tX)|_{t=0}. \tag{1.3}$$

For matrix groups the matrix exponential is the Lie group exponential map. For general Lie groups, the existence of $\exp: \mathfrak{g} \to G$ comes from standard results in ordinary differential equations. Let us list a few properties of exp. The curve

$$\mathbb{R} \to G$$
$$t \mapsto \exp(tX), \quad \text{for } X \in \mathfrak{g},$$

is a *one-parameter subgroup* of G; i.e., $\exp(0) = e$, the identity in G, and

$$\exp(tX)\exp(sX) = \exp(t + s)X, \quad \text{for all real numbers } s \text{ and } t. \tag{1.4}$$

And *Taylor's formula* for G says that:

$$f(g \exp X) = \sum_{n \geq 0} \frac{1}{n!} (\tilde{X}^n f)(g), \quad \text{for } g \in G, X \in \mathfrak{g}, \tag{1.5}$$

where f is a real analytic function on G. *The exponential map allows one to relate multiplication on the Lie group with bracket on the Lie algebra via:*

5.1. Geometry and Analysis on G/K

$$\exp tX \exp tY = \exp\{t(X + Y) + \tfrac{1}{2}t^2[X, Y] + O(t^3)\}, \quad (1.6)$$

for $X, Y \in \mathfrak{g}$, and $t \in \mathbb{R}$. It is possible to continue the expansion inside the braces in (1.6) and the result is called the Campbell-Hausdorff formula.

Exercise 4. Prove formula (1.6).

Hint. First consider the case of $GL(n, \mathbb{R})$. The same sort of proof works in general using Taylor's formula (1.5).

It is possible to compute the *differential of exp* and obtain:

$$(d\exp)_X Y = (dL_{\exp X})_e \left(\frac{1 - e^{-\operatorname{ad} X}}{\operatorname{ad} X} \right)(Y), \quad \text{for } X, Y \in \mathfrak{g}. \quad (1.7)$$

Formula (1.7) implies, in particular that the mapping from X to $\exp X$ is a diffeomorphism from an open neighborhood of 0 in $\mathfrak{g}$ onto an open neighborhood of the identity e in G.

Let us prove (1.7) in the case of matrix exp. First note that

$$\lim_{t \to 0} (e^{X+tY} - e^X)/t = \lim_{t \to 0} \sum_{n \geq 0} \frac{1}{n!t} \{(X + tY)^n - X^n\}$$

$$= \sum_{n \geq 0} \frac{1}{(n+1)!} \{X^n Y + X^{n-1} YX + \cdots + YX^n\}.$$

Beware that $XY \neq YX$, in general, so that you cannot blindly use the binomial theorem. However, it is possible to be clever (although that is unworthy of a Vulcan), since right and left multiplication by X do commute as operators. Define $R_X Y = YX$ and $L_X Y = XY$. Observe that $\operatorname{ad} X = L_X - R_X$. The three operators R_X, L_X, and $\operatorname{ad} X$ will commute. Thus we can apply the binomial theorem to obtain:

$$R_X^m = (L_X - \operatorname{ad} X)^m = \sum_{k=0}^{m} \binom{m}{k} L_X^{m-k} (-\operatorname{ad} X)^k.$$

This allows us to write:

$$X^n Y + X^{n-1} YX + \cdots + YX^n = \sum_{i=0}^{n} X^i \sum_{k=0}^{n-i} \binom{n-i}{k} X^{n-i-k} (-\operatorname{ad} X)^k Y$$

$$= \sum_{k=0}^{n} \sum_{i=0}^{n-k} \binom{n-i}{k} X^{n-k} (-\operatorname{ad} X)^k Y,$$

upon reversing sums. It is an *exercise* in the properties of binomial coefficients to show that

$$\sum_{i=0}^{n-k} \binom{n-i}{k} = \binom{n+1}{k+1}. \quad (1.8)$$

Therefore

$$(d\exp)_X Y = \sum_{n\geq 0} \frac{1}{(n+1)!} \sum_{k=0}^{n} \binom{n+1}{k+1} X^{n-k}(-\operatorname{ad} X)^k Y$$

$$= \sum_{k\geq 0} \sum_{n\geq k} \frac{1}{(k+1)!(n-k)!} X^{n-k}(-\operatorname{ad} X)^k Y$$

$$= \sum_{k\geq 0} \sum_{r\geq 0} \frac{1}{(k+1)!r!} X^r(-\operatorname{ad} X)^k Y$$

$$= e^X \sum_{k\geq 0} \frac{1}{(k+1)!}(-\operatorname{ad} X)^k Y.$$

This completes the proof of (1.7) in the case of the matrix exponential. The general result is proved in a similar way (see Helgason [1]).

Exercise 5. Prove formula (1.8) for a general Lie group.

One of the most important tools in Lie group theory is the *dictionary* that allows one to translate between Lie groups and Lie algebras. We list a few results from the dictionary. For the proofs, see references such as Helgason's big green books, Sagle and Walde [1], or Varadarajan [1].

For each Lie group G with Lie algebra $\mathfrak{g}$ and for each Lie subalgebra $\mathfrak{h}$ of $\mathfrak{g}$, there is a unique connected Lie subgroup H of G with Lie algebra $\mathfrak{h}$. However, H may not have the induced topology; e.g., consider the densely wound line in the torus:

$$\mathbb{R} \to (\mathbb{R}/\mathbb{Z})^2 = \mathbb{T}^2$$
$$t \mapsto (e^{it}, e^{iat}), \qquad \text{when } a \in \mathbb{R} \text{ is irrational.}$$

If $f: G_1 \to G_2$ is a Lie group homomorphism of connected Lie groups, then the differential $(df)_e: \mathfrak{g}_1 \to \mathfrak{g}_2$ is a Lie algebra homomorphism. Moreover, we have the following relations between images and kernels:

$$\text{Lie algebra } (f(G_1)) = (df)_e \mathfrak{g}_1,$$

$$\text{Lie algebra } (\ker f) = \ker(df)_e.$$

If $\lambda: \mathfrak{g}_1 \to \mathfrak{g}_2$ is a Lie algebra homomorphism and G_1, G_2 are connected Lie groups with Lie algebras $\mathfrak{g}_1, \mathfrak{g}_2$, respectively, and if, in addition, G_1 is simply connected, then there exists a unique Lie group homomorphism $f: G_1 \to G_2$ such that $(df)_e = \lambda$.

The hypothesis that G_1 be simply connected cannot be removed in the preceding result. For example, $\mathbb{R}/\mathbb{Z}$ and $\mathbb{R}$ have the same Lie algebra. But the identity mapping of $\mathbb{R}$ onto itself cannot be the differential of a Lie group homomorphism from $\mathbb{R}/\mathbb{Z}$ to $\mathbb{R}$.

5.1. Geometry and Analysis on G/K

Exercise 6.

(a) Show that the exponential map $\mathfrak{g} \to G$ need not be onto.
(b) Show that exp: $\mathfrak{gl}(n, \mathbb{C}) \to GL(n, \mathbb{C})$ is onto.

Hints.
(a) Take $G = SL(2, \mathbb{R})$ and consider
$$A = \begin{pmatrix} r & 0 \\ 0 & 1/r \end{pmatrix} \quad \text{for } r < -1.$$
If $A = \exp(X)$, consider the eigenvalues of X.
(b) Use the Jordan canonical form.

The final dictionary result that we list here concerns a closed subgroup H of a Lie group G. Then H must have the induced topology and
$$\text{Lie algebra } (H) = \mathfrak{h} = \{X \in \mathfrak{g} | \exp(tX) \in H \text{ for all } t \in \mathbb{R}\}. \tag{1.9}$$
Formula (1.9) provides a quick way to compute Lie algebras. For example, since $\det(e^X) = e^{\text{Tr } X}$, for matrices X, it follows that the Lie algebra $\mathfrak{sl}(n, \mathbb{R})$ consists of all $n \times n$ real matrices of trace zero. Here we use the notation that the Lie algebra of a group G is in lower case German letters so that $\mathfrak{sl}(n, \mathbb{R})$ is the Lie algebra of $SL(n, \mathbb{R})$. One can show that the Killing form of $\mathfrak{sl}(n, \mathbb{R})$ is:
$$B_{\mathfrak{sl}(n, \mathbb{R})}(X, Y) = 2n \, \text{Tr}(XY), \quad \text{for } X, Y \in \mathfrak{sl}(n, \mathbb{R}).$$
Therefore $\mathfrak{sl}(n, \mathbb{R})$ is a semisimple Lie algebra. In fact, it is actually a simple Lie algebra.

Exercise 7.

(a) Verify the comments made in the last paragraph.
(b) Find the Lie algebra of the symplectic group:
$$Sp(n, \mathbb{R}) = \{g \in \mathbb{R}^{2n \times 2n} | \, {}^tg J_n g = J_n\}, \quad \text{for } J_n = \begin{pmatrix} 0 & I_n \\ -I_n & 0 \end{pmatrix}.$$
(c) Find the Lie algebra of the Lorentz-type group
$$O(p, q) = \{g \in \mathbb{R}^{n \times n} | \, {}^t g I_{p,q} g = I_{p,q}\},$$
where $n = p + q$, and
$$I_{p,q} = \begin{pmatrix} I_p & 0 \\ 0 & -I_q \end{pmatrix}.$$

Hint. The answer to part (a) is given in formula (1.10).

There is an analogue of the adjoint representation on the group level. To obtain it, proceed as follows. If $g \in G$, define $\text{Int}(g)x = gxg^{-1}$, for all $x \in G$.

Then define $\mathrm{Ad}(g) = (d\,\mathrm{Int}(g))_e$, where e is the identity of G. Then we have a commutative diagram:

$$\begin{array}{ccc} \mathfrak{g} & \xrightarrow{\mathrm{Ad}(g)} & \mathfrak{g} \\ \exp \downarrow & & \downarrow \exp \\ G & \xrightarrow{\mathrm{Int}(g)} & G. \end{array}$$

It can be proved that $(d\,\mathrm{Ad})_e X = \mathrm{ad}\,X$, for all $X \in \mathfrak{g}$. Thus we have another commutative diagram:

$$\begin{array}{ccc} \mathfrak{g} & \xrightarrow{\mathrm{ad}} & \mathfrak{gl}(\mathfrak{g}) \\ \exp \downarrow & & \downarrow \text{matrix exp} \\ G & \xrightarrow{\mathrm{Ad}} & GL(\mathfrak{g}). \end{array}$$

If G is a matrix group already, then $\mathrm{Ad}(g) = \mathrm{Int}(g)$, since $\mathrm{Int}(g)x$ is a linear function of x. If $\mathfrak{g}$ is semisimple, then the kernel of ad is $\{0\}$ and the kernel of Ad is the center of G, which must then be discrete.

Exercise 8. Prove that $(d\,\mathrm{Ad})_e X = \mathrm{ad}\,X$, for all $X \in \mathfrak{g}$.

It is possible to classify all the simple Lie algebras over the complex numbers. Except for a finite number of exceptional Lie algebras, the *simple Lie algebras over* $\mathbb{C}$ are in the following list (with J_n as in Exercise 7 above):

$$\mathfrak{a}_n = \mathfrak{sl}(n+1, \mathbb{C}), \qquad n \geq 1;$$

$$\mathfrak{b}_n = \mathfrak{so}(2n+1, \mathbb{C}) = \{X \in \mathbb{C}^{(2n+1)\times(2n+1)} \mid {}^tX = -X\}, \qquad n \geq 2;$$

$$\mathfrak{c}_n = \mathfrak{sp}(n, \mathbb{C}) = \{X \in \mathbb{C}^{(2n)\times(2n)} \mid {}^tXJ_n + J_nX = 0\} \qquad (1.10)$$

$$= \left\{ \begin{pmatrix} A & B \\ C & -{}^tA \end{pmatrix} \middle| A, B, C \in \mathbb{R}^{n\times n}, B, C \text{ symmetric} \right\}, \qquad n \geq 3;$$

$$\mathfrak{d}_n = \mathfrak{so}(2n, \mathbb{C}) = \{X \in \mathbb{C}^{(2n)\times(2n)} \mid {}^tX = -X\}, \qquad n \geq 4.$$

The indices n are restricted because in low dimensions some strange things happen; e.g., $\mathfrak{d}_1$ and $\mathfrak{d}_2$ are not simple, since $\mathfrak{d}_1$ is abelian and $\mathfrak{d}_2 \cong \mathfrak{a}_1 \oplus \mathfrak{a}_1$. Also $\mathfrak{a}_1 \cong \mathfrak{b}_1 \cong \mathfrak{c}_1, \mathfrak{b}_2 \cong \mathfrak{c}_2, \mathfrak{d}_3 \cong \mathfrak{a}_3$. These things can be proved using Dynkin diagrams. You can find the details in Varadarajan [1]. The Lie groups corresponding to the Lie algebras in this list are $SL(n, \mathbb{C})$, the special linear group of $n \times n$ complex matrices of determinant one, $SO(n, \mathbb{C})$, the special orthogonal group of $n \times n$ complex matrices g of determinant one such that ${}^tgg = I$, and $Sp(n, \mathbb{C})$, the complex symplectic group of $(2n) \times (2n)$ matrices g with the property that ${}^tgJ_ng = J_n$, for J_n as in Exercise 7 above.

Cartan's classification of symmetric spaces makes use of the preceding classification of complex simple Lie algebras. It also uses the surprising, but

5.1. Geometry and Analysis on G/K

simple, observation that *the group I(M) of isometries of a symmetric space M acts transitively on M*. To see this fact, it helps to recall the *Hopf-Rinow Theorem* in differential geometry (see Helgason [2, p. 56]) which says that if M is a Riemannian manifold, then the following are equivalent:

(i) M is a complete metric space;
(ii) each maximal geodesic $\gamma(t)$ in M can be extended to all $t \in \mathbb{R}$;
(iii) each bounded closed subset of M is compact.

If M is a complete Riemannian manifold, then any two points P, Q in M can be joined by a geodesic whose length is the metric space distance between P and Q. To see that a symmetric space M must be complete, note that if a point P lies on the geodesic γ of M and s_P denotes the geodesic-reversing isometry at P, then $s_P\gamma$ is an extension of γ. Thus each maximal geodesic of a symmetric space must have domain the set of all real numbers. Then to see that the group $I(M)$ of isometries of M acts transitively on M, note that if P and Q are in M, then the geodesic-reversing isometry at the midpoint of the geodesic connecting P to Q will exchange P and Q.

It is possible to prove that $I(M)$ is a Lie group such that the connected component of the identity in $I(M)$ still acts transitively on M (see Helgason [2, Ch. 4]). *Let G be the connected component of the identity in I(M).*

Now fix a point O to be called the *origin* of the symmetric space M. And let K denote the subgroup of G consisting of elements which fix O. Then K is compact and we can identify M with G/K. Suppose next that s_O denotes the geodesic-reversing isometry at the origin. Then consider the map:

$$\sigma: G \to G$$

$$g \mapsto s_O g s_O.$$

Note that σ is an involutive automorphism of G (i.e., σ is an automorphism in the sense of Lie groups such that σ^2 is the identity). Moreover, setting

$$K_\sigma = \{g \in G | \sigma g = g\}$$

and

$$(K_s)_O = \text{the connected component of the identity in } K_\sigma,$$

we have

$$(K_\sigma)_O \subset K \subset K_\sigma.$$

This means that K and K_σ have the same Lie algebra.

Now consider the consequences of the preceding remarks about symmetric spaces and Lie groups of isometries on the Lie algebras of these groups, using the dictionary relating Lie groups and Lie algebras. One sees that:

$$(d\sigma)_e: \mathfrak{g} \to \mathfrak{g}$$

is an involutive Lie algebra automorphism which fixes $\mathfrak{k}$, the Lie algebra of K. Moreover, the eigenspace decomposition of $(d\sigma)_e$ on $\mathfrak{g}$ is:

$$\mathfrak{g} = \mathfrak{k} \oplus \mathfrak{p},$$

where $\mathfrak{k} = \{X \in \mathfrak{g} | (d\sigma)_e X = X\}$ and $\mathfrak{p} = \{X \in \mathfrak{g} | (d\sigma)_e X = -X\}$; that is, $\mathfrak{k}$ is the space of eigenvectors corresponding to the eigenvalue $+1$ while $\mathfrak{p}$ consists of eigenvectors corresponding to the eigenvalue -1.

If $\pi: G/K \to M$ is the natural identification, then $(d\pi)_0$ maps $\mathfrak{k}$ to $\{0\}$ and identifies $\mathfrak{p}$ with the tangent space $T_O(M)$.

To proceed further with the classification of symmetric spaces, one must reduce to semisimple Lie algebras, using the following result of E. Cartan (see Helgason [2, Ch. 5]).

Symmetric Space Decomposition. Suppose that M is a simply connected symmetric space. Then M is a product:

$$M = M_o \times M_c \times M_n,$$

where M_o is of *Euclidean* type, M_c is *compact with semisimple Lie group* of isometries, and M_n is *non-compact with semisimple Lie group* of isometries having Lie algebra with a Cartan decomposition described below.

We say that a semisimple Lie algebra is of *compact type* if its Killing form is negative definite (see Helgason [2, p. 122]).

A *Cartan decomposition* of a noncompact semisimple Lie algebra $\mathfrak{g}$ is a vector space direct sum decomposition $\mathfrak{g} = \mathfrak{k} \oplus \mathfrak{p}$ such that the Killing form of $\mathfrak{g}$ is negative definite on $\mathfrak{k}$ and positive definite on $\mathfrak{p}$. Also the mapping $\theta: \mathfrak{g} \to \mathfrak{g}$ with $\theta(X + Y) = X - Y$, for $X \in \mathfrak{k}$ and $Y \in \mathfrak{p}$, must be an automorphism of $\mathfrak{g}$. We call θ the *Cartan involution*.

Example. Consider the simple Lie algebra $\mathfrak{sl}(n, \mathbb{R})$ of all trace zero $n \times n$ real matrices. Set

$$\mathfrak{so}(n) = \{\text{skew-symmetric } n \times n \text{ real matrices of trace } 0\}$$

and

$$\mathfrak{p}_n = \{\text{symmetric } n \times n \text{ real matrices of trace } 0\}.$$

Clearly we have the direct sum decomposition:

$$\mathfrak{sl}(n, \mathbb{R}) = \mathfrak{so}(n) \oplus \mathfrak{p}_n,$$

with Cartan involution $\theta(X) = -{}^t X$. The Killing form on $\mathfrak{sl}(n, \mathbb{R})$ is $B(X, Y) = 2n \operatorname{Tr}(XY)$, and it is easy to see that this is negative definite on $\mathfrak{so}(n)$ and positive definite on $\mathfrak{p}_n$. It is also easy to check that the Cartan involution preserves the Lie bracket in $\mathfrak{sl}(n, \mathbb{R})$, which is $[X, Y] = XY - YX$.

Exercise 9. Prove all the claims made in the preceding example.

There is a mirror image of the *Cartan decomposition on the Lie group level*:

$$G = KP,$$

5.1. Geometry and Analysis on G/K

where $P = \exp \mathfrak{p}$. For the example above, we have

$$SL(n, \mathbb{R}) = SO(n)\mathscr{SP}_n, \qquad (1.11)$$

where, as in Chapter 4, $\mathscr{SP}_n$ denotes the positive $n \times n$ real matrices of determinant one. The proof of (1.11) is easy (see Exercise 5 of Section 4.1.1).

We have not given more than a rough sketch of the preceding arguments on classification of symmetric spaces because we are more interested in studying the examples. Thus we will give more attention to the question: *How does one obtain symmetric spaces out of Cartan decompositions of semisimple Lie algebras?* Suppose that $\mathfrak{g} = \mathfrak{k} \oplus \mathfrak{p}$ is a Cartan decomposition of the semisimple Lie algebra $\mathfrak{g}$ with Cartan involution θ. Let G be a connected real semisimple Lie group with Lie algebra $\mathfrak{g}$, and let K be a Lie subgroup of G having Lie algebra $\mathfrak{k}$. Then G/K has a unique analytic manifold structure such that the mapping of $\mathfrak{p}$ into G/K defined by sending X to $(\exp X)K$ is a diffeomorphism. If $\mathfrak{g}$ is of noncompact type, it can be proved that K is closed, connected, and equal to the fixed point set of an involutive automorphism $t: G \to G$ such that $(dt)_e$ is the Cartan involution θ. Such a map t clearly exists if G is simply connected (making use of the dictionary between group and algebra). But one does not really have to assume that G is simply connected in the non-compact case. Moreover K is compact if and only if the center of G is finite and then K is a *maximal* compact subgroup of G. For proofs of these results, see Helgason [2, Ch. 6].

To make G/K a symmetric space, we use the Killing form B of $\mathfrak{g}$. Let $\pi: G \to G/K$ be defined by $\pi(g) = gK$. Define the Riemannian metric Q on G/K by translating the Killing form on the space $\mathfrak{p}$:

$$Q_{gK}((d\pi)_g \tilde{X}_g, (d\pi)_g \tilde{Y}_g), \qquad \text{for all } X, Y \in \mathfrak{p}. \qquad (1.12)$$

Here $\tilde{X}$ denotes the left invariant vector field corresponding to $X \in \mathfrak{p}$. The metric Q is well defined because the Killing form is invariant under $\operatorname{Ad} k$, for $k \in K$. It is clear that the metric is positive from the definition of the Cartan decomposition. And it is easily seen that the metric is G-invariant.

The *geodesic-reversing isometry* s_O at the origin O, which is the coset K in G/K, is obtained from the involutive automorphism $t: G \to G$ as follows:

$$s_O: G/K \to G/K$$

$$gK \mapsto t(g)K.$$

Translate by elements of G to obtain the geodesic-reversing isometries at other points of G/K.

Example. The Riemannian structure on $SL(n, \mathbb{R})/SO(n)$ obtained from (1.12) above is just the same as that defined in Chapter 4. To see this, first note that one has an identification:

$$SL(n, \mathbb{R})/SO(n) \to \mathscr{SP}_n$$

$$g\, SO(n) \mapsto g\,{}^tg.$$

The action of $g \in SL(n, \mathbb{R})$ on $Y \in \mathscr{SP}_n$ is given by $a_g(Y) = Y[{}^t g]$. The differential is $(da_g)_I = a_g$ since $a_g(Y)$ is a linear function of Y. So we find that if $Y = g{}^t g$, for $g \in SL(n, \mathbb{R})$, and if u, v are in $T_Y(\mathscr{SP}_n)$, the tangent space to $\mathscr{SP}_n$ at the point Y, then:

$$Q_Y(u, v) = 2n \operatorname{Tr}((da_g)_I^{-1} u \cdot (da_g)_I^{-1} v) = 2n \operatorname{Tr}(g^{-1} u {}^t g^{-1} \cdot g^{-1} v {}^t g^{-1})$$
$$= 2n \operatorname{Tr}(Y^{-1} u Y^{-1} v).$$

This is exactly the Riemannian structure of Chapter 4.

Before considering more examples, let us record a few more general facts. Suppose again that we have a Cartan decomposition $\mathfrak{g} = \mathfrak{k} \oplus \mathfrak{p}$ of a semisimple noncompact Lie algebra over the real numbers. Assume that the Lie group $\tilde{G}$ is the universal covering group of G. Then there is a unique involutive automorphism $\tilde{t}: \tilde{G} \to \tilde{G}$ such that the differential $(d\tilde{t})_e$ is θ, the Cartan involution. It can be proved that the center $\tilde{Z}$ of $\tilde{G}$ is contained in $\tilde{K}$, where $\tilde{K}$ is the analytic subgroup of $\tilde{G}$ with Lie algebra $\mathfrak{k}$ (see Helgason [2, p. 216]). Now G is a quotient $\tilde{G}/N$ for some $N \subset \tilde{Z}$. Thus $\tilde{t}$ induces an involutive automorphism of G. Setting $K = \tilde{K}/N$, we have:

$$G/K \cong (\tilde{G}/N)/(\tilde{K}/N) \cong \tilde{G}/\tilde{K}.$$

So the symmetric space G/K is independent of the choice of Lie group G with Lie algebra $\mathfrak{g}$. So we may assume that G is simply connected whenever we need this. Furthermore, it can be proved that all the K's are conjugate (see Helgason [2, p. 256]). Note, however, that the K's need not be semisimple.

The preceding arguments fail for symmetric spaces of compact type. For example, the center of G need not lie in K; e.g., consider $G = SU(n)$, $K = SO(n)$. Also K need not be connected; e.g., $SO(3)/K = P^2$, the real projective plane, with K the subgroup of $SO(3)$ leaving a line through the origin invariant. Finally, the Cartan involution need not correspond to an automorphism of G in the compact case.

Another difference between compact and noncompact symmetric spaces is that the noncompact ones are topologically (though not geometrically) identifiable with the Euclidean space $\mathfrak{p}$. However, the compact symmetric spaces are not topologically trivial (see Greub, Halperin, and Vanstone [1]). This fact makes the compact and non-compact symmetric spaces very different. However, there is a duality between the two types, as we shall see.

Now we intend to manufacture many examples of symmetric spaces by exploring the connection between real forms of complex simple Lie algebras and Cartan decompositions of real Lie algebras.

A *real form* $\mathfrak{g}$ of a complex simple Lie algebra $\mathfrak{g}^c$ is defined by the equality of the complexification of $\mathfrak{g}$ with $\mathfrak{g}^c$; i.e.,

$$\mathfrak{g}^c = \mathfrak{g} \otimes_\mathbb{R} \mathbb{C} = \mathfrak{g} \oplus i\mathfrak{g}.$$

It is possible to list the real forms $\mathfrak{g}$ of $\mathfrak{g}^c$ by listing the conjugations of $\mathfrak{g}^c$. By a *conjugation* of $\mathfrak{g}^c$, we mean a mapping $C: \mathfrak{g}^c \to \mathfrak{g}^c$ which is conjugate linear,

5.1. Geometry and Analysis on G/K

bracket preserving, and such that C^2 is the identity. See Helgason [2, Chs. 3, 10] or Loos [1, Vol. II, Ch. VII].

It can also be shown that any complex semisimple Lie algebra $\mathfrak{g}^c$ has a compact real form $\mathfrak{u}$; i.e., the Killing form of $\mathfrak{u}$ is negative definite. Then to make a *list of symmetric spaces of noncompact type coming from complex simple Lie algebras* $\mathfrak{g}^c$, one must follow through the following plan of action.

Plan for Construction of Noncompact Symmetric Spaces of Type III

I. List the conjugations (involutive automorphisms) of $\mathfrak{g}^c$. The fixed points will be real forms of $\mathfrak{g}^c$. One of these real forms $\mathfrak{u}$ will be compact.

II. For the noncompact real forms $\mathfrak{g}$ of $\mathfrak{g}^c$, the Cartan decomposition is:
$$\mathfrak{g} = (\mathfrak{g} \cap \mathfrak{u}) \oplus (\mathfrak{g} \cap i\mathfrak{u}).$$

Note that the Killing form of $\mathfrak{g}$ has the correct behavior on the decomposition since $\mathfrak{u}$ is compact; i.e., the Killing form is negative definite on $\mathfrak{u}$. Furthermore, if τ is the conjugation of $\mathfrak{g}^c$ corresponding to the compact real form $\mathfrak{u}$, then the restriction of τ to $\mathfrak{g}$ is θ, the Cartan involution corresponding to this Cartan decomposition.

III. Form the symmetric space G/K by taking Lie groups $G \supset K$ with Lie algebras $\mathfrak{g}$, $\mathfrak{k}$, respectively. Here $\mathfrak{k} = \mathfrak{g} \cap \mathfrak{u}$.

Type a Examples

I. *Real Forms* of $\mathfrak{sl}(n, \mathbb{C})$.

1. $\mathfrak{sl}(n, \mathbb{R})$ = *normal real form* = fixed points of the conjugation $\tau(X) = \bar{X}$.
2. $\mathfrak{su}(n, \mathbb{R})$ = *compact real form* = fixed points of the conjugation $\tau(X) = -{}^t\bar{X}$.
3. $\mathfrak{su}(p, q)$ = fixed points of the conjugation $\tau(X) = -I_{p,q}{}^t\bar{X}I_{p,q}$, where
$$I_{p,q} = \begin{pmatrix} I_p & 0 \\ 0 & -I_q \end{pmatrix}, \qquad n = p + q.$$
4. $\mathfrak{su}^*(2m)$ = fixed points of $\tau(X) = J_m \bar{X} J_m^{-1}$, where
$$J_m = \begin{pmatrix} 0 & I_m \\ -I_m & 0 \end{pmatrix}, \qquad n = 2m \text{ (for even } n\text{)}.$$

II. *Cartan Decompositions* of Noncompact Real Forms of $\mathfrak{sl}(n, \mathbb{C})$.

1. $\mathfrak{sl}(n, \mathbb{R}) = \mathfrak{so}(n) \oplus \mathfrak{p}_n$, where
$$\mathfrak{so}(n) = \{X \in \mathfrak{sl}(n, \mathbb{R}) | {}^t X = -X\},$$
$$\mathfrak{p}_n = \{X \in \mathfrak{sl}(n, \mathbb{R}) | {}^t X = X\}.$$
The Cartan involution is $\theta(X) = -{}^t X$, $X \in \mathfrak{sl}(n, \mathbb{R})$.

2. $\mathfrak{su}(p,q) = \mathfrak{k} \oplus \mathfrak{p}$, where

$$\mathfrak{k} = \left\{ \begin{pmatrix} A & 0 \\ 0 & B \end{pmatrix} \middle| A \in \mathfrak{u}(p), B \in \mathfrak{u}(q), \text{Tr}(A+B) = 0 \right\},$$

$$\mathfrak{p} = \left\{ \begin{pmatrix} 0 & Z \\ {}^t\bar{Z} & 0 \end{pmatrix} \middle| Z \in \mathbb{C}^{p \times q} \right\}.$$

The Cartan involution is $\theta(X) = I_{p,q} X I_{p,q}$, $X \in \mathfrak{su}(p,q)$.

3. $\mathfrak{su}^*(2m) = \mathfrak{k} \oplus \mathfrak{p}$, where

$$\mathfrak{k} = \mathfrak{sp}(m, \mathbb{C}) \cap \mathfrak{u}(2m) = \mathfrak{sp}(m) \text{ (by definition)},$$

$$\mathfrak{p} = \mathfrak{su}^*(2m) \cap (i\mathfrak{u}(2m)).$$

The Cartan involution is $\theta(X) = -J_m {}^t X J_m^{-1}$.

III. *The Noncompact Symmetric Spaces* Corresponding to the Noncompact Real Forms.

1. $SL(n, \mathbb{R})/SO(n)$.
2. $SU(p,q)/S(U_p \times U_q)$, where $n = p + q$ and

$$SU(p,q) = \{g \in SL(n, \mathbb{C}) | {}^t\bar{g} I_{p,q} g = I_{p,q}\},$$

$$U(p) = \{g \in \mathbb{C}^{p \times p} | {}^t\bar{g} g = I_p\} = \text{the unitary group},$$

$$S(U_p \times U_q) = \left\{ g \in SL(n, \mathbb{C}) \middle| g = \begin{pmatrix} A & 0 \\ 0 & B \end{pmatrix}, A \in U(p), B \in U(q) \right\}.$$

3. $SU^*(2n)/Sp(n)$, where

$$SU^*(2n) = \{g \in SL(2n, \mathbb{C}) | g = J_n \bar{g} J_n^{-1}\},$$

$$Sp(n) = Sp(n, \mathbb{C}) \cap U(2n),$$

$$Sp(n, \mathbb{C}) = \{g \in \mathbb{C}^{2n \times 2n} | {}^t g J_n g = J_n\} = \text{the complex symplectic group}.$$

IV. The Corresponding *Compact Symmetric Spaces*.

4. $SU(n)/SO(n)$.
5. $SU(p+q)/S(U_p \times U_q)$.
6. $SU(2n)/Sp(n)$.

Type c Examples

I. *Real Forms of* $\mathfrak{sp}(n, \mathbb{C})$.

1. $\mathfrak{sp}(n, \mathbb{R}) = $ *normal real form* $=$ fixed points of the conjugation $\tau(X) = \bar{X}$.
2. $\mathfrak{sp}(n) = $ *compact real form* $=$ fixed points of the conjugation $\tau(X) = -{}^t\bar{X}$.
   Note that $\mathfrak{sp}(n) = \mathfrak{sp}(n, \mathbb{C}) \cap \mathfrak{u}(2n)$.

5.1. Geometry and Analysis on G/K

3. $\mathfrak{sp}(p,q)$ = fixed points of the conjugation $\tau(X) = -K_{p,q} {}^t\bar{X} K_{p,q}$, where

$$K_{p,q} = \begin{pmatrix} I_p & 0 & 0 & 0 \\ 0 & -I_q & 0 & 0 \\ 0 & 0 & I_p & 0 \\ 0 & 0 & 0 & -I_q \end{pmatrix}, \quad p+q=n.$$

II. *Cartan Decompositions of Noncompact Real Forms* of $\mathfrak{sp}(n,\mathbb{C})$.

1. $\mathfrak{sp}(n,\mathbb{R}) = \mathfrak{k} \oplus \mathfrak{p}$, where

$$\mathfrak{k} = \left\{ \begin{pmatrix} A & B \\ -B & A \end{pmatrix} \middle| A, B \in \mathbb{R}^{n \times n}, B = {}^tB, A = -{}^tA \right\} \cong \mathfrak{u}(n),$$

$$\mathfrak{p} = \{X \in \mathfrak{sp}(n,\mathbb{R}) | X = {}^tX\}.$$

To see that $\mathfrak{k} \cong \mathfrak{u}(n)$, map

$$\begin{pmatrix} A & B \\ -B & A \end{pmatrix} \in \mathfrak{k} \text{ to } A + iB \in \mathfrak{u}(n).$$

The Cartan involution is $\theta(X) = -{}^tX$.

2. $\mathfrak{sp}(p,q) = \mathfrak{k} \oplus \mathfrak{p}$, where

$$\mathfrak{k} = \left\{ \begin{pmatrix} X_{11} & 0 & X_{13} & 0 \\ 0 & X_{22} & 0 & X_{24} \\ -X_{13} & 0 & X_{11} & 0 \\ 0 & -X_{24} & 0 & X_{22} \end{pmatrix} \middle| \begin{array}{l} X_{11} \in \mathfrak{u}(p), X_{22} \in \mathfrak{u}(q) \\ X_{13} \in \mathbb{C}^{p \times p}, {}^tX_{13} = X_{13}, \\ X_{24} \in \mathbb{C}^{q \times q}, {}^tX_{24} = X_{24} \end{array} \right\}$$

$\cong \mathfrak{sp}(p) \times \mathfrak{sp}(q)$.

The Cartan involution is $\theta(X) = K_{p,q} X K_{p,q}$.

III. The Corresponding *Noncompact Symmetric Spaces*.

1. $Sp(n,\mathbb{R})/U(n)$.

Here $G = Sp(n,\mathbb{R})$ is the symplectic group defined in Exercise 7 while $U(n)$ is really the subgroup $K = G \cap O(2n)$ which is isomorphic to the unitary group

$$U(n) = \{g \in \mathbb{C}^{n \times n} | {}^t\bar{g}g = I\},$$

by part (b) of Lemma 1 below.

There are two equivalent but rather different ways to view this symmetric space. The first is as the space $\mathscr{P}_n^*$ of positive symplectic $2n \times 2n$ real matrices.

The second version of $Sp(n,\mathbb{R})/U(n)$ is the Siegel upper half space $\mathscr{H}_n$ defined by:

$$\mathscr{H}_n = \{Z \in \mathbb{C}^{n \times n} | {}^tZ = Z, \operatorname{Im} Z \in \mathscr{P}_n\}.$$

This example is the most important one for the rest of this book. We will discuss the various identifications of $Sp(n,\mathbb{R})/U(n)$ below.

2. $Sp(p,q)/Sp(p) \times Sp(q)$.

Here $Sp(p,q) = \{g \in SL(p+q, \mathbb{C}) | {}^t g K_{p,q} \bar{g} = K_{p,q}\}$, where

$$K_{p,q} = \begin{pmatrix} -I_p & & & 0 \\ & I_q & & \\ & & -I_p & \\ 0 & & & I_q \end{pmatrix}.$$

Exercise 10. Check the computations for the type A and C noncompact symmetric space examples above.

This is just about all the examples of symmetric spaces that we shall discuss. Table 5.1 lists some others. We will also be interested in the symmetric space $SL(2, \mathbb{C})/SU(2)$, which can be identified with the quaternionic upper half space or hyperbolic 3-space. It is considered at the end of this section. It is the symmetric space of a complex Lie group considered as a real group. We do not discuss compact symmetric spaces here, except to note that there is a duality between symmetric spaces U/K' of the compact type and symmetric spaces G/K of the noncompact type. This duality is obtained on the Lie algebra level by writing

$$\mathfrak{g} = \mathfrak{k} \oplus \mathfrak{p}$$

and

$$\mathfrak{u} = \mathfrak{k} \oplus i\mathfrak{p},$$

where $\mathfrak{u}$ is a compact real form of the complexification of $\mathfrak{g}$. See Helgason [2, Ch. 5] for more details. Helgason [2, p. 321] gives a global duality result for bounded symmetric domains (which will be defined below) allowing them to be viewed as open submanifolds of a compact Hermitian space. This is the Borel embedding theorem (see Borel [3, 4]). Such results can be applied to compute dimensions of spaces of automorphic forms via the Hirzebruch-Riemann-Roch theorem (see Hirzebruch [2, pp. 162–165]). Healy [1] provides an example of the implications of this duality for harmonic analysis on $SU(2)$ and hyperbolic 3-space.

Example of the Duality Between Compact and Noncompact Symmetric Spaces

This example shows that hyperbolic geometry is dual to spherical geometry. We begin with the two Cartan decompositions:

$$\mathfrak{sl}(2, \mathbb{R}) = \mathfrak{so}(2) \oplus \mathfrak{p}_2, \quad \mathfrak{su}(2) = \mathfrak{so}(2) \oplus i\mathfrak{p}_2,$$

where

$$\mathfrak{p}_2 = \{X \in \mathbb{R}^{2 \times 2} | {}^t X = X, \operatorname{Tr} X = 0\}.$$

5.1. Geometry and Analysis on G/K

The symmetric space $SL(2, \mathbb{R})/SO(2)$ can be viewed as the hyperbolic upper half plane of Chapter 3, Volume I, while $SU(2)/SO(2)$ can be viewed as the sphere S^2 in $\mathbb{R}^3$, which is the symmetric space considered in Chapter 2 of Volume I.

The Siegel upper half plane $\mathscr{H}_n$ is a *Hermitian symmetric space*; i.e., a symmetric space with a complex structure, invariant under each geodesic-reversing symmetry. It turns out that the Hermitian symmetric spaces of the compact or noncompact type have non-semisimple K (see Helgason [2, p. 281] or Loos [1, Vol. II, p. 161]). Such is indeed the case for $G = Sp(n, \mathbb{R})$, $K = U(n)$. It also turns out that the Hermitian symmetric spaces of noncompact type are the *bounded symmetric domains* D in complex n-space. Here *symmetric* means that for every $z \in D$ there is a biholomorphic involutive map on D having z as an isolated fixed point (see Helgason [2, pp. 311–322] or Loos [1, Vol. II, p. 164]). Koecher [3] found a way of constructing all the Hermitian symmetric spaces from Jordan algebras. We shall see in Exercise 27 that the Siegel upper half plane is identifiable with a bounded symmetric domain, namely the *generalized unit disc*:

$$\mathscr{D}_n = \{W \in \mathbb{C}^{n \times n} | {}^t W = W, I - \overline{W}W \in \mathscr{P}_n\}.$$

The identification map is the generalized Cayley transform:

$$\mathscr{H}_n \to \mathscr{D}_n$$
$$Z \mapsto (Z - iI)(Z + iI)^{-1}.$$

Cartan proved in 1935 that there are only 6 types of irreducible homogeneous bounded symmetric domains (see Helgason [3, p. 518]). It is possible to generalize many results from analysis and number theory to these classical domains (see Hua [1], Piatetski-Shapiro [2], and Siegel [5], [1, Vol. II, pp. 274–369]).

We could also have differentiated between the three types of symmetric spaces M according to their *sectional curvature*. The sectional curvature is defined as $-g(R(u,v)u, v)$, where g is the Riemannian metric for M, R is the curvature tensor, and u, v are orthonormal tangent vectors in $T_P(M)$, the tangent space to M at a point P. For a symmetric space, the curvature tensor at the origin is:

$$R_0(X, Y)Z = -[[X, Y], Z], \quad \text{for } X, Y, Z \in \mathfrak{p}$$

(see Helgason [2, p. 180]).

Then one has the *classification of types of symmetric spaces M by sectional curvature* (see Helgason [2, p. 205]):

M is of noncompact type $\Leftrightarrow$ the sectional curvature of M is ≤ 0;

M is of compact type $\Leftrightarrow$ the sectional curvature of M is ≥ 0;

M is of Euclidean type $\Leftrightarrow$ the sectional curvature of M is $= 0$.

Table 5.1. Irreducible Riemannian Symmetric Spaces of Types I and III for the Non-exceptional Groups

	Noncompact	Compact
AI	$SL(n, \mathbb{R})/SO(n)$	$SU(n)/SO(n)$
AII	$SU^*(2n)/Sp(n)$	$SU(2n)/Sp(n)$
AIII	$SU(p, q)/S(U_p \times U_q)$	$SU(p + q)/S(U_p \times U_q)$
BDI	$SO_o(p, q)/SO(p) \times SO(q)$	$SO(p + q)/SO(p) \times SO(q)$
DIII	$SO^*(2n)/U(n)$	$SO(2n)/U(n)$
CI	$Sp(n, \mathbb{R})/U(n)$	$Sp(n)/U(n)$
CII	$Sp(p, q)/Sp(p) \times Sp(q)$	$Sp(p + q)/Sp(p) \times Sp(q)$

It is possible to prove the *conjugacy of all maximal compact subgroups of noncompact semisimple real Lie groups G* using Cartan's fixed point theorem, which says that if a compact group K_1 acts on a simply connected Riemannian manifold of negative curvature such as G/K, there must be a fixed point. And $x^{-1}K_1 x \subset K$ means xK is fixed. See Helgason [2, p. 75] for more details on Cartan's theorem.

The grand finale of the classification theory is the listing of the four types of irreducible symmetric spaces given in Helgason [2, Ch. 9] and [3, pp. 515–518]. Here *irreducible* means that the corresponding Lie group is simple. The irreducible symmetric spaces of types I and III which come from non-exceptional Lie groups are in Table 5.1.

THE FOUR TYPES OF IRREDUCIBLE SYMMETRIC SPACES ARE:

I. G/K, where G is a compact connected simple real Lie group and K is the subgroup of points fixed by an involutive automorphism of G.

II. G is a compact, connected simple Lie group provided with a left and right invariant Riemannian structure unique up to constant factor.

III. G/K where G is a connected non-compact simple real Lie group and K is the subgroup of points fixed by an involutive automorphism of G (a maximal compact subgroup).

IV. G/U, where G is a connected Lie group whose Lie algebra is a simple Lie algebra over $\mathbb{C}$ viewed as a real Lie algebra, and U is a maximal compact subgroup of G.

Here $SO^*(2n) = \{g \in SO(2n, \mathbb{C}) | {}^t\bar{g} J_n g = J_n\}$ and $SO_o(p, q)$ is the identity component of $SO(p, q) = \{g \in SL(n, \mathbb{R}) | {}^t g I_{p,q} g = I_{p,q}\}$, where $n = p + q$ and $I_{p,q}$ is defined in part (3) of the list of real forms of $\mathfrak{sl}(n, \mathbb{C})$.

From now on, our emphasis will be upon the symmetric space $Sp(n, \mathbb{R})/U(n)$. Our first task is to study the various realizations of this space. We begin with the realization as the space of *positive symplectic matrices*:

$$Sp(n, \mathbb{R})/U(n) \cong \mathscr{P}_n^* = \{Y \in Sp(n, \mathbb{R}) | Y \in \mathscr{P}_{2n}\}. \qquad (1.13)$$

The proof of (1.13) involves the global or group level Cartan decomposition.

5.1. Geometry and Analysis on G/K

Let $\mathfrak{g}$ be a noncompact semisimple (real) Lie algebra with Cartan decomposition $\mathfrak{g} = \mathfrak{k} \oplus \mathfrak{p}$. Suppose that K and G are the corresponding connected Lie groups with Lie algebras $\mathfrak{k}$ and $\mathfrak{g}$, respectively. Then we have the *global Cartan decomposition*:

$$G = KP, \qquad P = \exp \mathfrak{p},$$

and G is diffeomorphic to $K \times \mathfrak{p}$. Note that G and K are Lie groups but P is not. The main idea of the proof of this result is to use the Adjoint representation of G to deduce the Cartan decomposition of G from that for $GL(n, \mathbb{R})$, which is:

$$GL(n, \mathbb{R}) = O(n) \cdot \mathscr{P}_n \qquad (1.14)$$

(see Exercise 5 of Section 4.1.1). Proofs of the general Cartan decomposition can be found in Helgason [2, p. 215] or Loos [1, Vol. I, p. 156]. We shall only consider the special case of interest.

Lemma 1 (The Cartan Decomposition for the Symplectic Group).

(a) *The Cartan decomposition for $G = Sp(n, \mathbb{R})$ comes from the Cartan decomposition (1.14) for $GL(2n, \mathbb{R})$ by taking intersections; i.e.,*

$$G = Sp(n, \mathbb{R}) = K \cdot \mathscr{P}_n^*, \qquad \text{with } \mathscr{P}_n^* = \mathscr{P}_{2n} \cap G \text{ and } K = O(2n) \cap G.$$

(b) *The maximal compact subgroup K of G given in part (a) can be identified with the unitary group $U(n) = \{g \in \mathbb{C}^{n \times n} | {}^t\bar{g}g = I\}$. It follows also that the symmetric space $G/K = Sp(n, \mathbb{R})/U(n)$ can be identified with $\mathscr{P}_n^*$.*

PROOF.

(a) See Helgason [2, p. 345] and [3, p. 450].

Observe that (1.14) says that $g \in G$ can be written as $g = up$ with $u \in O(2n)$ and $p \in \mathscr{P}_{2n}$. We need to show that both u and p lie in $Sp(n, \mathbb{R})$. To see this, note that $p^2 = {}^tgg$. Moreover, $g \in G$ implies that ${}^tg^{-1}$ and ${}^tg$ both also lie in G (a situation really brought about by the existence of an involution of G with differential the Cartan involution of $\mathfrak{g}$). Thus $p^2 \in G$.

Now we must show that $p^2 \in G$ implies that p lies in G. To do this, note that G is a pseudoalgebraic group, meaning that there is a finite set of polynomials

$$f_j \in \mathbb{C}[X_1, \ldots, X_{4n^2}]$$

such that a matrix g lies in G if and only if g is a root of all the f_j. Now there is a rotation matrix $k \in O(2n)$ such that

$$k^{-1}p^2k = \begin{pmatrix} e^{h_1} & & 0 \\ & \ddots & \\ 0 & & e^{h_{2n}} \end{pmatrix}.$$

And $k^{-1}Gk$ is also a pseudoalgebraic group. Thus the diagonal matrices

$$k^{-1}p^2k = \begin{pmatrix} e^{rh_1} & & 0 \\ & \ddots & \\ 0 & & e^{rh_{2n}} \end{pmatrix}$$

satisfy a certain set of polynomial equations for any integer r. But if an exponential polynomial

$$F(t) = \sum_{j=1}^{B} c_j \exp(b_j t)$$

vanishes for all integers t, then it must vanish for all real numbers t as well. Thus, in particular, p must lie in the group G, as will all elements

$$p_t = \exp(tX), \quad \text{for } t \in \mathbb{R}, \quad \text{if } p^2 = \exp(2X).$$

But then $p \in G$ implies that $u = gp^{-1} \in G$. This completes the proof of part (a) except to show the uniqueness of the expression $g = up$ and the fact that G is diffeomorphic to $K \times \mathfrak{p}$. We leave these proofs as an exercise.

(b) To see that K is isomorphic to $U(n)$, proceed as follows. We have $K = G \cap O(2n)$. Thus if $J = J_n$ is as defined in Exercise 7, then

$$M = \begin{pmatrix} A & B \\ C & D \end{pmatrix} \in K \Leftrightarrow JM = MJ \text{ and } {}^tMJM = J$$

$$\Leftrightarrow C = -B, \quad D = A, \quad {}^tAB = {}^tBA \text{ and } {}^tAA + {}^tBB = I.$$

The last statement is equivalent to saying that $A + iB \in U(n)$. Thus the identification of K and $U(n)$ on the group level is the same as that on the Lie algebra level which was discussed when we listed the Cartan decompositions corresponding to noncompact real forms of $\mathfrak{sp}(n, \mathbb{C})$. In fact,

$$\sigma\begin{pmatrix} A & B \\ -B & A \end{pmatrix} = A + iB$$

defines a mapping which preserves matrix multiplication as well as addition. The map σ identifies K with $U(n)$. A good reference for these things is Séminaire Cartan [1, Exp. 3]. The proof of Lemma 1 is now complete. □

Exercise 11. Fill in all the details in the proof of Lemma 1.

Note that most calculations are far easier on the Lie algebra level than on the group level. For an example of the difference between the algebra and the group, note that it is clear that $\mathfrak{sp}(n, \mathbb{R})$ is contained in $\mathfrak{sl}(2n, \mathbb{R})$, but it is not obvious that $Sp(n, \mathbb{R})$ is contained in $SL(2n, \mathbb{R})$, though it is true.

Exercise 12. Prove the last statement.

Hint. Show that $Sp(n, \mathbb{R})$ is connected. See Chevelley [1, p. 36] for the useful result which says that H and G/H connected implies G connected, where H is a closed subgroup of the topological group G.

5.1. Geometry and Analysis on G/K

Next we seek to generalize the *Iwasawa decomposition* from Exercise 12 in Section 4.1.2:

$$G = GL(n, \mathbb{R}) = KAN, \qquad (1.15)$$

where K is the compact group $O(n)$, A is the abelian group of positive diagonal matrices in G, and N is the nilpotent group of upper triangular matrices in G with ones on the diagonal. In order to obtain such an Iwasawa decomposition for any noncompact semisimple (real) Lie group G, one must discuss the *root space decomposition* of the Lie algebra of G. We do not give a detailed discussion of root spaces, except for several examples. The details for the general case can be found in Helgason [2, 3] or Loos [1].

Some definitions are needed to discuss the root space decomposition of the Lie algebra $\mathfrak{g}$ of G. Define $\mathfrak{a}$ to be a *maximal abelian subspace of* $\mathfrak{p}$. Here $\mathfrak{p}$ comes from the Cartan decomposition of $\mathfrak{g}$. Then for any *real linear functional (root)* $\alpha: \mathfrak{a} \to \mathbb{R}$, define the *root space*:

$$\mathfrak{g}_\alpha = \{X \in \mathfrak{g} | (\operatorname{ad} H)X = \alpha(X)X, \text{ for all } H \in \mathfrak{a}\}.$$

If $\mathfrak{g}_\alpha \neq \{0\}$, then we say that the linear functional α is a *restricted root*. Let Λ denote the set of all *non-zero restricted roots*. When we are considering a normal real form such as $\mathfrak{sl}(n, \mathbb{R})$, the restricted roots are restrictions of roots of the complexification of $\mathfrak{g}$.

Next set $\mathfrak{m}$ equal to the *centralizer* of $\mathfrak{a}$ in $\mathfrak{k}$, where $\mathfrak{k}$ comes from the Cartan decomposition of $\mathfrak{g}$; i.e.,

$$\mathfrak{m} = \{X \in \mathfrak{k} | [X, \mathfrak{a}] = 0\}.$$

In fact, $\mathfrak{m}$ will always be zero for normal or split real forms such as $\mathfrak{sp}(n, \mathbb{R})$.

Finally, *the root space decomposition of the real noncompact semisimple Lie algebra* $\mathfrak{g}$ *is*:

$$\mathfrak{g} = \mathfrak{a} \oplus \mathfrak{m} \oplus \sum_{\alpha \in \Lambda} \mathfrak{g}_\alpha.$$

To prove the validity of this decomposition, consider the positive definite bilinear form F on $\mathfrak{g}$ defined as follows, using the Killing form B and the Cartan involution θ of $\mathfrak{g}$:

$$F(X, Y) = B(X, \theta Y), \qquad \text{for } X, Y \in \mathfrak{g}.$$

If $X \in \mathfrak{p}$, then ad X is symmetric with respect to F and thus is a diagonalizable linear transformation of $\mathfrak{g}$. Therefore the commuting family of all the ad X for $X \in \mathfrak{a}$ is simultaneously diagonalizable with real eigenvalues. It remains to show that the eigenspace corresponding to the zero functional is:

$$\mathfrak{g}_0 = (\mathfrak{g}_0 \cap \mathfrak{k}) \oplus (\mathfrak{g}_0 \cap \mathfrak{p}) = \mathfrak{m} \oplus \mathfrak{a}.$$

This comes from the definitions.

Note that if $\mathfrak{g}$ is $\mathfrak{sl}(n, \mathbb{R})$ or $\mathfrak{sp}(n, \mathbb{R})$, then $\mathfrak{m} = \{0\}$ and restricted roots are the same as the roots of the complexifications $\mathfrak{sl}(n, \mathbb{C})$ and $\mathfrak{sp}(n, \mathbb{C})$ restricted to the normal real form.

One can define the set of *positive restricted roots* Λ^+ as a subset of Λ such that Λ is the disjoint union of Λ^+ and $-\Lambda^+$. We will soon see how to find such sets of positive roots in our favorite cases.

We need to use the positive roots to construct a certain nilpotent Lie subalgebra $\mathfrak{n}$ of $\mathfrak{g}$. By definition, a Lie algebra $\mathfrak{n}$ is said to be *nilpotent* if the lower central series $\mathfrak{n}^k$ defined by

$$\mathfrak{n}^0 = \mathfrak{n}, \qquad \mathfrak{n}^1 = [\mathfrak{n}, \mathfrak{n}], \qquad \mathfrak{n}^{k+1} = [\mathfrak{n}, \mathfrak{n}^k]$$

terminates; i.e., $\mathfrak{n}^k = \{0\}$, for some k.

Suppose that Λ^+ denotes the chosen set of positive roots of $\mathfrak{g}$. Define the *nilpotent Lie subalgebra* $\mathfrak{n}$ *of* $\mathfrak{g}$ by:

$$\mathfrak{n} = \sum_{\alpha \in \Lambda^+} \mathfrak{g}_\alpha.$$

We can also define the *opposite nilpotent subalgebra* $\bar{\mathfrak{n}}$ of $\mathfrak{g}$ by:

$$\bar{\mathfrak{n}} = \sum_{\alpha \in \Lambda^+} \mathfrak{g}_{-\alpha}.$$

To prove that $\mathfrak{n}$ is nilpotent, it suffices to know the following simple facts about roots.

Simple Facts about Roots

Λ^+ is a finite set;

$[\mathfrak{g}_\alpha, \mathfrak{g}_\beta] \subset \mathfrak{g}_{\alpha+\beta}$;

$\alpha, \beta \in \Lambda^+$ implies that $\alpha + \beta$ is either a positive root or not a root at all.

Furthermore, if θ denotes the Cartan involution of $\mathfrak{g} = \mathfrak{k} \oplus \mathfrak{p}$, then θ interchanges the nilpotent algebra $\mathfrak{n}$ and its opposite $\bar{\mathfrak{n}}$; i.e., $\theta \mathfrak{n} = \bar{\mathfrak{n}}$. To see this, note that $\theta(X) = -X$ for all X in $\mathfrak{a} \subset \mathfrak{p}$ and θ preserves the Lie bracket.

From the preceding considerations, it is easy to obtain the *Iwasawa decomposition of the noncompact real semisimple Lie algebra*:

$$\mathfrak{g} = \mathfrak{k} \oplus \mathfrak{a} \oplus \mathfrak{n}.$$

For clearly, one has $\mathfrak{g} = \bar{\mathfrak{n}} \oplus \mathfrak{m} \oplus \mathfrak{a} \oplus \mathfrak{n}$. And $\mathfrak{k} \oplus \mathfrak{a} \oplus \mathfrak{n}$ is a direct sum, since

$$X + H + Y = 0, \qquad \text{for } X \in \mathfrak{k}, H \in \mathfrak{a}, Y \in \mathfrak{n},$$

implies that

$$0 = \theta(X + H + Y) = X - H + \theta(Y).$$

Subtract the two equations to see that $2H + Y - \theta(Y) = 0$. This implies that $H = 0$ by the fact that the root space decomposition is a direct sum. Thus $Y = \theta(Y) = 0$ and $X = 0$.

To complete the proof of the Lie algebra Iwasawa decomposition, we need

5.1. Geometry and Analysis on G/K

only show that the dimensions are correct. It suffices to look at the following mapping:

$$\mathfrak{m} \oplus \bar{\mathfrak{n}} \xrightarrow{1-1, \text{onto}} \mathfrak{k}$$

$$X + Y \longmapsto X + Y + \theta(Y), \quad \text{for } X \in \mathfrak{m}, Y \in \bar{\mathfrak{n}}.$$

Next we want to consider three examples: $\mathfrak{sl}(n, \mathbb{R})$, $\mathfrak{sp}(n, \mathbb{R})$, and $\mathfrak{su}(3, 1)$. The first two examples are *split* or *normal*, so that $\mathfrak{m} = \{0\}$, the restricted roots are restrictions of complex roots of the complexification, and all the roots spaces are one dimensional real vector spaces.

Three Examples of Iwasawa Decompositions of Real Semisimple Lie Algebras

Example 1 ($\mathfrak{sl}(n, \mathbb{R})$).
Recall that the Cartan decomposition is $\mathfrak{sl}(n, R) = \mathfrak{k} \oplus \mathfrak{p}_n$, where

$$\mathfrak{k} = \mathfrak{so}(n) = \{X \in \mathbb{R}^{n \times n} | {}^t X = -X\},$$

$$\mathfrak{p}_n = \{X \in \mathbb{R}^{n \times n} | {}^t X = X, \operatorname{Tr} X = 0\}.$$

One can show that a maximal abelian subspace of $\mathfrak{p}_n$ is:

$$\mathfrak{a} = \{H \in \mathbb{R}^{n \times n} | H \text{ is diagonal of trace } 0\}.$$

Next let E_{ij} for $1 \leq i, j \leq n$ denote the matrix with 1 in the i, j place and 0's elsewhere. Then set $e_i(H) = h_i$ if H is a diagonal matrix with h_i as its ith diagonal entry. Let $\alpha_{ij} = e_i - e_j$. Then $[H, E_{ij}] = \alpha_{ij}(H) E_{ij}$ and we find the root space decomposition involves the

$$\mathfrak{g}_{\alpha_{ij}} = \mathbb{R} E_{ij} \quad \text{with} \quad \Lambda^+ = \{\alpha_{ij} | 1 \leq i < j \leq n\}.$$

Thus $\mathfrak{n} = \sum_{1 \leq i < j \leq n} \mathbb{R} E_{ij} =$ *the upper triangular real $n \times n$ matrices with 0 on the diagonal*.

Thus the Lie algebra analogue of the Iwasawa decomposition of $SL(n, \mathbb{R})$ coming from (1.15) says: $\mathfrak{sl}(n, \mathbb{R}) = \mathfrak{so}(n) \oplus \mathfrak{a} \oplus \mathfrak{n}$, with $\mathfrak{so}(n)$ denoting the skew symmetric $n \times n$ real matrices, $\mathfrak{a}$ equal to the $n \times n$ real diagonal trace zero matrices, and $\mathfrak{n}$ equal to the upper triangular $n \times n$ real matrices with zeros on the diagonal.

Example 2 ($\mathfrak{sp}(n, \mathbb{R})$).
Recall that $\mathfrak{sp}(n, \mathbb{R})$ consists of matrices $\begin{pmatrix} A & B \\ C & -{}^t A \end{pmatrix} = (A, B, C)$, with $A, B, C \in \mathbb{R}^{n \times n}$ and B, C symmetric. We found the Cartan decomposition had:

$$\mathfrak{k} = \{(A, B, -B) | B \text{ symmetric}, A \text{ skew-symmetric}\},$$

$$\mathfrak{p} = \{(A, B, B) | A, B \text{ symmetric}\}.$$

A calculation shows that a maximal abelian subspace of $\mathfrak{p}$ is:

$$\mathfrak{a} = \{(H,0,0) | H \text{ is real } n \times n \text{ diagonal}\}.$$

Suppose that the E_{ij}, $1 \leq i < j \leq n$, are as in Example 1. Set $G_{pq} = E_{pq} + E_{qp}$, for $1 \leq p \leq q \leq n$. Then, if we abuse notation and write $H = (H,0,0)$, we have

$$[H,(E_{ij},0,0)] = (e_i - e_j)(H)(E_{ij},0,0),$$

$$[H,(0,G_{pq},0)] = (e_p + e_q)(H)(0,G_{pq},0),$$

$$[H,(0,0,G_{pq})] = -(e_p + e_q)(H)(0,0,G_{pq}).$$

It follows that $\Lambda^+ = \{e_i - e_j | 1 \leq i < j \leq n\} \cup \{e_p + e_q | 1 \leq p \leq q \leq n\}$. Thus

$$\mathfrak{n} = \sum_{1 \leq i < j \leq n} \mathbb{R}(E_{ij},0,0) + \sum_{1 \leq p \leq q \leq n} \mathbb{R}(0,G_{pq},0)$$

$$= \{(A,B,0) | A \text{ upper triangular, } 0 \text{ on diagonal, } B \text{ symmetric}\}.$$

So the *Iwasawa decomposition* is:

$$\mathfrak{sp}(n,\mathbb{R}) = \mathfrak{k} \oplus \mathfrak{a} \oplus \mathfrak{n},$$

where

$$\mathfrak{k} = \{(A,B,-B) | B \text{ symmetric}, A \text{ skew-symmetric}\},$$

$$\mathfrak{a} = \{(H,0,0) | H \text{ diagonal}\},$$

$$\mathfrak{n} = \{(A,B,0) | A \text{ upper triangular with } 0 \text{ on the diagonal}, B \text{ symmetric}\}.$$

Here we use the notation $(A,B,C) = \begin{pmatrix} A & B \\ C & -{}^tA \end{pmatrix}$.

Example 3 ($\mathfrak{su}(3,1)$).

First recall that

$$\mathfrak{su}(3,1) = \{X \in \mathfrak{sl}(4,\mathbb{C}) | -I_{3,1} {}^t\bar{X} I_{3,1} = X\}$$

where

$$I_{3,1} = \begin{pmatrix} 1 & & & 0 \\ & 1 & & \\ & & 1 & \\ 0 & & & -1 \end{pmatrix}.$$

The corresponding Lie group is $SU(3,1) = \{g \in SL(4,\mathbb{C}) | {}^t\bar{g} I_{3,1} g = I_{3,1}\}$.

One sees easily that

$$\mathfrak{su}(3,1) = \{(A,b,c) | A \in \mathfrak{u}(3), b \in \mathbb{R}, c \in \mathbb{C}^{3 \times 1}, \text{Tr } A + ib = 0\},$$

where

$$(A,b,c) = \begin{pmatrix} A & c \\ {}^t\bar{c} & ib \end{pmatrix}.$$

5.1. Geometry and Analysis on G/K

We saw that the Cartan decomposition of $\mathfrak{su}(3,1)$ involves

$$\mathfrak{k} = \{(A, b, 0) | A \in \mathfrak{u}(3), b \in \mathbb{R}, \operatorname{Tr} A + ib = 0\},$$
$$\mathfrak{p} = \{(0, 0, c) | c \in \mathbb{C}^{3 \times 1}\}.$$

A maximal abelian subspace of $\mathfrak{p}$ is $\mathfrak{a} = \mathbb{R}(0, 0, e_1)$ where $e_1 = {}^t(1, 0, 0)$. Note that $(0, 0, ie_1)$ does not commute with $(0, 0, e_1)$. You need to multiply matrices to check these things (see Exercise 13 below). Similarly you can show that:

$$\mathfrak{m} = \left\{ \begin{pmatrix} ib & 0 & 0 & 0 \\ 0 & & & 0 \\ 0 & & U & 0 \\ 0 & 0 & 0 & ib \end{pmatrix} \middle| b \in \mathbb{R}, U \in \mathfrak{u}(2), \operatorname{Tr} U + 2ib = 0 \right\}.$$

To prove this, one must show that the matrices of $\mathfrak{m}$ centralize $\mathfrak{a}$ and that nothing else in $\mathfrak{k}$ does the same trick. Once again, this is checked by multiplying matrices. Thus we have come upon an example of a non-zero and rather fat $\mathfrak{m}$. The root space decomposition of $\mathfrak{su}(3,1)$ is rather complicated. We find roots λ such that 2λ is also a root. Such things cannot happen for complex semisimple Lie algebras. And one finds root spaces $\mathfrak{g}_\lambda$ of dimension greater than one over $\mathbb{R}$.

The positive roots of $\mathfrak{su}(3,1)$ are $\Lambda^+ = \{\lambda, 2\lambda\}$, where $\lambda((0, 0, e_1)) = 1$. And the root space decomposition of $\mathfrak{su}(3,1)$ is:

$$\mathfrak{su}(3,1) = \mathfrak{a} \oplus \mathfrak{m} \oplus \mathfrak{g}_\lambda \oplus \mathfrak{g}_{-\lambda} \oplus \mathfrak{g}_{2\lambda} \oplus \mathfrak{g}_{-2\lambda}$$

$$\dim_\mathbb{R} \mathfrak{a} = 1, \qquad \dim_\mathbb{R} \mathfrak{m} = 4, \qquad \dim_\mathbb{R} \mathfrak{g}_\lambda = \dim_\mathbb{R} \mathfrak{g}_{-\lambda} = 4,$$
$$\dim_\mathbb{R} \mathfrak{g}_{2\lambda} = \dim_\mathbb{R} \mathfrak{g}_{-2\lambda} = 1.$$

To see this note that

$$\left[\begin{pmatrix} 0 & 0 & 0 & 1 \\ 0 & 0 & 0 & 0 \\ 0 & 0 & 0 & 0 \\ 1 & 0 & 0 & 0 \end{pmatrix}, \begin{pmatrix} 0 & a & b & 0 \\ -\bar{a} & 0 & 0 & c \\ -\bar{b} & 0 & 0 & d \\ 0 & \bar{c} & \bar{d} & 0 \end{pmatrix} \right] = \begin{pmatrix} 0 & \bar{c} & \bar{d} & 0 \\ -c & 0 & 0 & \bar{a} \\ -d & 0 & 0 & \bar{b} \\ 0 & a & b & 0 \end{pmatrix}$$

$$= k \begin{pmatrix} 0 & a & b & 0 \\ -\bar{a} & 0 & 0 & c \\ -\bar{b} & 0 & 0 & d \\ 0 & \bar{c} & \bar{d} & 0 \end{pmatrix}$$

implies that $k = \pm 1$ and that $ka = \bar{c}, kb = \bar{d}$. Thus if $k = 1$, we find that

$$\mathfrak{g}_\lambda = \left\{ \begin{pmatrix} 0 & a & b & 0 \\ -\bar{a} & 0 & 0 & \bar{a} \\ -\bar{b} & 0 & 0 & \bar{b} \\ 0 & a & b & 0 \end{pmatrix} \middle| (a, b) \in \mathbb{C}^2 \right\} \text{ which is 4-dimensional over } \mathbb{R}.$$

If $k = -1$ then we find that

$$\mathfrak{g}_{-\lambda} = \left\{ \begin{pmatrix} 0 & a & b & 0 \\ -\bar{a} & 0 & 0 & -\bar{a} \\ -\bar{b} & 0 & 0 & -\bar{b} \\ 0 & -a & -b & 0 \end{pmatrix} \middle| (a,b) \in \mathbb{C}^2 \right\}$$

which is again 4-dimensional over $\mathbb{R}$.

Recalling what it means to be in $\mathfrak{su}(3,1)$, we see that it remains to deal with

$$\mathfrak{g}_{2\lambda} = \mathbb{R}\begin{pmatrix} i & 0 & 0 & -i \\ 0 & 0 & 0 & 0 \\ 0 & 0 & 0 & 0 \\ i & 0 & 0 & -i \end{pmatrix} \quad \text{and} \quad \mathfrak{g}_{-2\lambda} = \mathbb{R}\begin{pmatrix} i & 0 & 0 & i \\ 0 & 0 & 0 & 0 \\ 0 & 0 & 0 & 0 \\ -i & 0 & 0 & -i \end{pmatrix}.$$

The nilpotent Lie algebra $\mathfrak{n}$ is then:

$$\mathfrak{n} = \mathfrak{g}_\lambda \oplus \mathfrak{g}_{2\lambda} = \left\{ \begin{pmatrix} ic & a & b & -ic \\ -\bar{a} & 0 & 0 & \bar{a} \\ -\bar{b} & 0 & 0 & \bar{b} \\ ic & a & b & -ic \end{pmatrix} \middle| a, b \in \mathbb{C}, c \in \mathbb{R} \right\}.$$

Exercise 13.

(a) Check the calculations in the preceding three examples.
(b) Perform the analogous calculation to that of part (a) in the case of the Lorentz algebra $\mathfrak{so}(3,1)$.

Our next goal is to understand *the group level Iwasawa decomposition of a noncompact semisimple connected real Lie group G*:

$$G = KAN,$$

where K, A, N are connected Lie subgroups of G with Lie algebras $\mathfrak{k}$, $\mathfrak{a}$, $\mathfrak{n}$, respectively. G is actually diffeomorphic to the product $K \times A \times N$. The exponential maps $\mathfrak{k}$ onto the compact group K. And exp is a diffeomorphism which maps $\mathfrak{a}$ onto the abelian group A while taking addition to multiplication. The exponential is a diffeomorphism of $\mathfrak{n}$ onto the nilpotent group N. Recall that the exponential does not in general map $\mathfrak{g}$ onto G, nor is exp $\mathfrak{a}$ *diffeomorphism* in general (see Exercise 6). For a proof that the exponential map is onto for abelian, nilpotent, and compact Lie groups, see Helgason [2, pp. 229, 56–58, 188–189].

In our discussion of the group level Iwasawa decomposition, we shall only consider the special case of the symplectic group. The proof of the global Iwasawa decomposition in the general case uses the Adjoint representation (see Helgason [2, 3] or Loos [1]).

5.1. Geometry and Analysis on G/K

Recall the following definition:

$$Sp(n, \mathbb{R}) = \text{the symplectic group} = \{g \in SL(2n, \mathbb{R}) | {}^t g J_n g = J_n\},$$

where

$$J_n = \begin{pmatrix} 0 & I_n \\ -I_n & 0 \end{pmatrix}.$$

It follows that

$$Sp(n, \mathbb{R}) = \left\{ \begin{pmatrix} A & B \\ C & D \end{pmatrix} \middle| {}^t AC = {}^t CA, {}^t BD = {}^t DB, {}^t AD - {}^t CB = I_n \right\}. \quad (1.16)$$

And *Lie subgroups* of $Sp(n, \mathbb{R})$ which correspond to the Lie subalgebras $\mathfrak{k}, \mathfrak{a}, \mathfrak{n}$ in the Iwasawa decomposition $\mathfrak{g} = \mathfrak{k} \oplus \mathfrak{a} \oplus \mathfrak{n}$ are:

$$K_n^* = \left\{ \begin{pmatrix} A & B \\ -B & A \end{pmatrix} \middle| A + iB \in U(n) \right\},$$

$$A_n^* = \left\{ \begin{pmatrix} H & 0 \\ 0 & H^{-1} \end{pmatrix} \middle| H \text{ positive diagonal} \right\}, \quad (1.17)$$

$$N_n^* = \left\{ \begin{pmatrix} A & B \\ 0 & {}^t A^{-1} \end{pmatrix} \middle| A \text{ upper triangular, 1 on diagonal}; A\, {}^t B = B\, {}^t A \right\}.$$

We have seen that the *symmetric space* associated to $Sp(n, \mathbb{R})$ is:

$$Sp(n, \mathbb{R})/K_n^* \cong \mathscr{P}_n^* = \mathscr{P}_{2n} \cap Sp(n, \mathbb{R}) \quad (1.18)$$

(see Lemma 1). Now we wish to find (along with the Iwasawa decomposition) another realization of this symmetric space—*Siegel's upper half plane*:

$$\mathscr{H}_n = \{ Z \in \mathbb{C}^{n \times n} | {}^t Z = Z, \operatorname{Im} Z \in \mathscr{P}_n \}. \quad (1.19)$$

Observe that we can define the following *actions of* $G = Sp(n, \mathbb{R})$ on the three versions of the symmetric space:

action of $g \in G$ on G/K is $a_g(xK) = gxK$ for $x \in G$;

action of $g \in G$ on $\mathscr{P}_n^*$ is $b_g(Y) = Y[g] = {}^t g Y g$ for $Y \in \mathscr{P}_n^*$;

action of $g \in G$ on $\mathscr{H}_n$ is $c_g(Z) = (AZ + B)(CZ + D)^{-1}$ (1.20)

$$\text{if } g = \begin{pmatrix} A & B \\ C & D \end{pmatrix}, Z \in \mathscr{H}_n.$$

Exercise 14.

(a) Prove formula (1.16).
(b) Check that $c_{hg}(Z) = c_h(c_g(Z))$ in formula (1.20).

The following lemma will allow us to identify all the versions of the symmetric space associated to the symplectic group. To see this, study the following diagram of mappings:

$$Sp(n, \mathbb{R})/K_n^* \to \mathscr{P}_n^* \to \mathscr{H}_n \tag{1.21}$$

$$g \mapsto g\,{}^t g = S \mapsto X + iY.$$

Here X, Y come from the *partial Iwasawa decomposition* of $S \in \mathscr{P}_n^*$:

$$S = \begin{pmatrix} Y & 0 \\ 0 & Y^{-1} \end{pmatrix} \begin{bmatrix} I & 0 \\ X & I \end{bmatrix}, \quad \text{for } X = {}^t X \in \mathbb{R}^{n \times n}, Y \in \mathscr{P}_n. \tag{1.22}$$

Lemma 2 below gives the existence and uniqueness of this decomposition for every positive symplectic matrix. There is an equivalent Iwasawa decomposition obtained by applying matrix inverse to formula (1.22):

$$S = \begin{pmatrix} Y^{-1} & 0 \\ 0 & Y \end{pmatrix} \begin{bmatrix} I & -X \\ 0 & I \end{bmatrix}, \quad \text{for } X = {}^t X \in \mathbb{R}^{n \times n}, Y \in \mathscr{P}_n. \tag{1.22}'$$

Exercise 15.

(a) Show that the composition of the two maps in (1.21) takes gK_n^* with

$$g = \begin{pmatrix} A & B \\ C & D \end{pmatrix}$$

to $(Ai + B)(Ci + D)^{-1}$ in $\mathscr{H}_n$.

(b) Show also that the maps in (1.21) preserve the group actions in (1.20). More precisely, define $i_1({}^tgK_n) = I[g]$ for $g \in G$ and define $i_2(S) = X + iY$ for S with partial Iwasawa decomposition (1.22). Prove that

$$i_1 \circ a_g = b_{t_g} \circ i_1 \quad \text{and} \quad i_2 \circ b_{t_g} = c_g \circ i_2.$$

(c) Suppose that $Z^* = c_g(Z) = (AZ + B)(CZ + D)^{-1}$ for $Z \in \mathscr{H}_n$ and g as in part (a). Show that the imaginary part of Z^* is $Y^* = Y\{(CZ + D)^{-1}\}$, where Y is the imaginary part of Z and $Y\{W\} = {}^t\overline{W}YW$. Then show that $Z^* \in \mathscr{H}_n$.

(d) Show that the Jacobian $|\partial Z^*/\partial Z| = |CZ + D|^{-n-1}$.

Hint. See Maass [2, p. 33].
(a) Note that $Z = (Ai + B)(Ci + D)^{-1}$
$= (Ai + B)(-{}^tCi + {}^tD)(-{}^tCi + {}^tD)^{-1}(Ci + D)^{-1}$.
(c) Note that $c_g(W) - c_g(Z) = (W{}^tC + {}^tD)^{-1}(W - Z)(CZ + D)^{-1}$. To find Y^*, let $W = \bar{Z}$.

Lemma 2 (Iwasawa Decomposition for the Symplectic Group). *Here we use the notation* (1.16)–(1.22)'.

(a) *Every positive symplectic matrix has the partial Iwasawa decomposition given in* (1.22) *or* (1.22)'. *Thus the mappings in* (1.21) *are identifications of the three differentiable manifolds.*

5.1. Geometry and Analysis on G/K

(b) *The Iwasawa decomposition of $G = Sp(n, \mathbb{R})$ says that*
$$G = K_n^* A_n^* N_n^*, \qquad \text{with } K_n^*, A_n^*, N_n^* \text{ as in (1.17).}$$

PROOF.

(a) We know from (1.15) that we can write $S \in \mathcal{P}_n^*$ as:
$$S = \begin{pmatrix} A & 0 \\ 0 & B \end{pmatrix} \begin{bmatrix} I & X \\ 0 & I \end{bmatrix} \qquad \text{with } A, B \in \mathcal{P}_n, X \in \mathbb{R}^{n \times n}.$$

Since S is symplectic, it follows that for J_n as defined in Exercise 7, we have $SJ_n S = J_n$. Thus $J_n S J_n = -S^{-1}$.

The only way for $J_n S J_n$ to be equal to $-S^{-1}$, when S has the given partial Iwasawa decomposition is that
$$A = B^{-1} \quad \text{and} \quad X = {}^t X.$$

This is easily seen using again the fact that $J_n^2 = -I$. For
$$J_n S J_n = J_n \begin{pmatrix} I & 0 \\ {}^t X & I \end{pmatrix} J_n J_n \begin{pmatrix} A & 0 \\ 0 & B \end{pmatrix} J_n J_n \begin{pmatrix} I & X \\ 0 & I \end{pmatrix} J_n = -\begin{pmatrix} B & 0 \\ 0 & A \end{pmatrix} \begin{bmatrix} I & 0 \\ -X & I \end{bmatrix}.$$

(b) Use part (a). This allows one to write $S \in \mathcal{P}_n^*$ as
$$S = \begin{pmatrix} A & 0 \\ 0 & A^{-1} \end{pmatrix} \begin{bmatrix} I & B \\ 0 & I \end{bmatrix}, \qquad \text{for } A \in \mathcal{P}_n, B = {}^t B \in \mathbb{R}^{n \times n}.$$

Then express A as $A = H[Q]$ with H positive diagonal and Q upper triangular with 1's on the diagonal. This is possible by the Iwasawa decomposition for $GL(n, \mathbb{R})$. Thus
$$Y = \begin{pmatrix} H & 0 \\ 0 & H^{-1} \end{pmatrix} \begin{bmatrix} Q & QB \\ 0 & {}^t Q^{-1} \end{bmatrix},$$

which is the full Iwasawa decomposition of $Y \in \mathcal{P}_n^*$. This translates to the Iwasawa decomposition for an element of the symplectic group using the Cartan decomposition (Lemma 1), completing the proof of Lemma 2. □

Exercise 16. Show that the Killing form on $\mathfrak{g} = \mathfrak{sp}(n, \mathbb{R})$ is
$$B(X, Y) = 4(n + 1) \operatorname{Tr}(XY).$$

Hint. Use the root space decomposition of $\mathfrak{g}$.

The Riemannian metric on $Sp(n, \mathbb{R})/K_n^* \cong \mathcal{P}_n^*$ is given by:
$$Q_Y(u, v) = \operatorname{Tr}(Y^{-1} u Y^{-1} v),$$

for $Y \in \mathcal{P}_n^*$, $u, v \in T_Y(\mathcal{P}_n^*)$ = the tangent space to $\mathcal{P}_n^*$ at Y (see (1.12) and the analogous result for $SL(n, \mathbb{R})$). Here we have dropped the constant in the Killing form of Exercise 16. Using the notation of formula (1.11) in Section

4.1.2, if $dY = (dy_{ij}) \in T_Y(\mathscr{P}_n^*)$, $Y \in \mathscr{P}_n^*$, then the *arc length* on $\mathscr{P}_n^*$ is

$$ds^2 = \text{Tr}(Y^{-1} dY Y^{-1} dY). \tag{1.23}$$

We want to show that the geodesics for this metric come from matrix exp and thus that $\mathscr{P}_n^*$ is a totally geodesic submanifold of $\mathscr{P}_{2n}$. We can use partial Iwasawa coordinates from Lemma 2 for this purpose. Now $W \in \mathscr{P}_n^*$ has partial Iwasawa decomposition

$$W = \begin{pmatrix} V & 0 \\ 0 & V^{-1} \end{pmatrix} \begin{bmatrix} I & X \\ 0 & I \end{bmatrix}, \qquad V \in \mathscr{P}_n, \quad X = {}^t X \in \mathbb{R}^n. \tag{1.24}$$

Just as in Exercise 14 of Section 4.1.2, we obtain the following formula for the arc length on $\mathscr{P}_n^*$ in partial Iwasawa coordinates (1.24):

$$ds^2 = \text{Tr}((V^{-1} dV)^2 + (V d(V^{-1}))^2 + 2 V^{-1} {}^t dX \, V^{-1} dX). \tag{1.25}$$

Exercise 17. Prove formula (1.25). Then note that $d(V^{-1}) = -V^{-1} dV V^{-1}$ and use this to show that the arc length on $\mathscr{P}_n^*$ can be expressed as follows using partial Iwasawa coordinates (1.24):

$$ds^2 = 2 \text{Tr}((V^{-1} dV)^2 + (V^{-1} dX)^2).$$

Show that the action of $G = Sp(n, \mathbb{R})$ on $\mathscr{P}_n^*$ leaves the arc length invariant.

Using Exercise 17 we find that the *arc length on the Siegel upper half plane* $\mathscr{H}_n$ is:

$$ds^2 = 2 \text{Tr}(V^{-1} dZ \, V^{-1} d\bar{Z}), \quad \text{if } Z = U + iV \in \mathscr{H}_n. \tag{1.26}$$

This is indeed the arc length considered by Siegel [1, Vol. II, p. 276].

Before proceeding to the study of geodesics in $\mathscr{P}_n^*$ or $\mathscr{H}_n$, we need to consider the analogue of polar coordinates in these spaces.

Lemma 3 (The Polar Decomposition of a Noncompact Semisimple Real Lie Group). *Let G be a noncompact real semisimple Lie group with connected Lie subgroups K and A, as in the Cartan and Iwasawa decompositions. Then G has the polar decomposition*:

$$G = KAK.$$

PROOF. First we show that if the Lie algebra $\mathfrak{g}$ of G has Cartan decomposition $\mathfrak{g} = \mathfrak{k} \oplus \mathfrak{p}$ and Iwasawa decomposition $\mathfrak{g} = \mathfrak{k} \oplus \mathfrak{a} \oplus \mathfrak{n}$, then

$$\mathfrak{p} = \text{Ad}(K)\mathfrak{a}, \tag{1.27}$$

where the Adjoint representation Ad is defined at the beginning of this section. To prove (1.27), choose H in $\mathfrak{a}$ so that its centralizer in $\mathfrak{p}$ is $\mathfrak{a}$; i.e., take $H \in \mathfrak{a}$ such that $\alpha(H) \neq 0$ for all roots $\alpha \in \Lambda$. Set K^* equal to $\text{Ad}_G K$ and suppose that X is in $\mathfrak{p}$. Now there is an element k_0 in K^* such that

$$B(H, \text{Ad}(k_0)X) = \text{Min}\{B(H, \text{Ad}(k)X) | k \in K^*\}.$$

5.1. Geometry and Analysis on G/K

Suppose that $T \in \mathfrak{k}$. Then the derivative at $t = 0$ of the following function of the real variable t must be 0 by the first derivative test:

$$B(H, \mathrm{Ad}(\exp tT)\,\mathrm{Ad}(k_0)X).$$

This implies, using the fact that the derivative of Ad is ad:

$$B(H, (\mathrm{ad}\, T)(\mathrm{Ad}(k_0)X)) = 0 \quad \text{for all } T \text{ in } \mathfrak{k}.$$

Thus (by Exercise 3)

$$B(T, [H, \mathrm{Ad}(k_0)X]) = 0 \quad \text{for all } T \text{ in } \mathfrak{k}.$$

Since $[\mathfrak{p}, \mathfrak{p}] \subset \mathfrak{k}$, and B is negative definite on $\mathfrak{k}$, it follows that $[H, \mathrm{Ad}(k_0)X] = 0$ which says that $\mathrm{Ad}(k_0)X \in \mathfrak{a}$, by the definition of H. The proof of Lemma 3 is completed by observing that (1.27) implies

$$\exp \mathfrak{p} = \exp(\mathrm{Ad}(K))\mathfrak{a} = \mathrm{Int}(K)(\exp \mathfrak{a}) = \bigcup_{k \in K} kAk^{-1}.$$

Lemma 3 follows from this equality and Lemma 1. □

Next we consider some examples.

Examples of The Polar Decomposition

Example 1 ($GL(n, \mathbb{R}) = O(n)A_n O(n)$, Where A_n Consists of All Positive Diagonal Matrices). This is equivalent (via the Cartan decomposition) to saying that for any positive matrix Y in $\mathscr{P}_n$, there is an orthogonal matrix k in $O(n)$ and a positive diagonal matrix a in A_n such that $Y = k^{-1}ak$. Thus the polar decomposition is just the *spectral theorem* for positive definite matrices, as we noted already in formula (1.22) of Section 4.1.3.

The next question is: *How unique are the a and k in the polar decomposition of Y in $\mathscr{P}_n$?* We saw in Exercise 24 of Section 4.1.3 that these coordinates give a $(2^n n!)$-fold covering of $\mathscr{P}_n$, since the entries of a are unique up to the action of the Weyl group of permutations of the diagonal entries and the matrices in $O(n)$ that commute with all the diagonal matrices must themselves be diagonal with entries ± 1.

Example 2 (Euler Angle Decomposition of the Compact Group $SO(3)$). A reference is Hermann [1, pp. 30–39]. Set $G = SO(3)$,

$$k = \begin{pmatrix} SO(2) & 0 \\ 0 & 1 \end{pmatrix}.$$

The Cartan decomposition of $\mathfrak{g} = \mathfrak{k} \oplus \mathfrak{p}$ is:

$$\mathfrak{so}(3) = \begin{pmatrix} \mathfrak{so}(2) & 0 \\ 0 & 0 \end{pmatrix} \oplus \left\{ \begin{pmatrix} 0 & c \\ -{}^t c & 0 \end{pmatrix} \,\bigg|\, c \in \mathbb{R}^2 \right\}.$$

Then we can take the maximal abelian subspace of $\mathfrak{p}$ to be

$$\mathfrak{a} = \mathbb{R}\begin{pmatrix} 0 & 0 & 1 \\ 0 & 0 & 0 \\ -1 & 0 & 0 \end{pmatrix}.$$

And

$$\exp\left\{t\begin{pmatrix} 0 & 0 & 1 \\ 0 & 0 & 0 \\ -1 & 0 & 0 \end{pmatrix}\right\} = \begin{pmatrix} \cos t & 0 & \sin t \\ 0 & 1 & 0 \\ -\sin t & 0 & \cos t \end{pmatrix}.$$

which is easily seen by writing out the series for the matrix exponential. Thus the *Euler angle decomposition* of g in $SO(3)$ is:

$$\begin{pmatrix} \cos u & \sin u & 0 \\ -\sin u & \cos u & 0 \\ 0 & 0 & 1 \end{pmatrix}\begin{pmatrix} \cos t & 0 & \sin t \\ 0 & 1 & 0 \\ -\sin t & 0 & \cos t \end{pmatrix}\begin{pmatrix} \cos v & \sin v & 0 \\ -\sin v & \cos v & 0 \\ 0 & 0 & 1 \end{pmatrix}.$$

The three Euler angles are u, t, v. Thus any rotation in 3-space is a product of three rotations about two axes.

Example 3 (The Dual Noncompact Group to $SO(3)$ Is the Lorentz Group $SO(2, 1)$). The group $SO(2, 1)$ again has an Euler angle decomposition that is well known to physicists. You need two angular variables and one real variable. One finds that the maximal abelian subalgebra $\mathfrak{a}$ of $\mathfrak{p}$ is:

$$\mathfrak{a} = \mathbb{R}\begin{pmatrix} 0 & 0 & 1 \\ 0 & 0 & 0 \\ 1 & 0 & 0 \end{pmatrix}.$$

Then

$$\exp\left\{t\begin{pmatrix} 0 & 0 & 1 \\ 0 & 0 & 0 \\ 1 & 0 & 0 \end{pmatrix}\right\} = \begin{pmatrix} \cosh t & 0 & \sinh t \\ 0 & 1 & 0 \\ \sinh t & 0 & \cosh t \end{pmatrix}.$$

For $SO(3, 1)$, these matrices are called "Lorentz boosts" (see Misner, Thorne, and Wheeler [1, p. 67]). The A-part of this group does not get wound up like the A-part of the compact group in Example 3.

Example 4 (Euler Angles for $U(3, 1)$). The physicist Wigner [1] considers this example, for which KAK is:

$$\begin{pmatrix} A & 0 \\ 0 & u \end{pmatrix}\begin{pmatrix} \cosh t & 0 & 0 & \sinh t \\ 0 & 0 & 0 & 0 \\ 0 & 0 & 0 & 0 \\ \sinh t & 0 & 0 & \cosh t \end{pmatrix}\begin{pmatrix} B & 0 \\ 0 & v \end{pmatrix},$$

for A, B in $U(3)$ and u, b in $i\mathbb{R}$.

5.1. Geometry and Analysis on G/K

Example 5 (SU(2) Has the Euler Angles: $(0 \leq \theta \leq \pi, 0 \leq \varphi \leq 2\pi, 0 \leq \psi \leq 4\pi)$).

$$\begin{pmatrix} \exp(i\varphi/2) & 0 \\ 0 & \exp(-i\varphi/2) \end{pmatrix} \begin{pmatrix} \cos(\theta/2) & \sin(\theta/2) \\ -\sin(\theta/2) & \cos(\theta/2) \end{pmatrix} \begin{pmatrix} \exp(i\psi/2) & 0 \\ 0 & \exp(-i\psi/2) \end{pmatrix}.$$

Exercise 18. Fill in the details of the derivations of polar decompositions in Examples 2–5. How unique are these decompositions?

Example 6 (The Symplectic Group $Sp(n, \mathbb{R})$). The polar decomposition of $Sp(n, \mathbb{R})$ says:

$$Sp(n, \mathbb{R}) = K_n^* A_n^* K_n^*,$$

where

$$K_n^* = O(2n) \cap Sp(n, \mathbb{R}) \cong U(n),$$

$$A_n^* = A_{2n} \cap Sp(n, \mathbb{R}) = \{\text{positive diagonal symplectic matrices}\}.$$

How unique is this decomposition? This time it is not legal to permute all the $2n$ entries of the diagonal matrix in A_n^* because the matrix has to remain symplectic. The matrix looks therefore like

$$\begin{pmatrix} H & 0 \\ 0 & H^{-1} \end{pmatrix} \quad \text{with} \quad H = \begin{pmatrix} a_1 & & 0 \\ & \ddots & \\ 0 & & a_n \end{pmatrix}, \quad a_j \text{ positive.}$$

Certainly it is legal to permute all the a_j. One can also send a_j to a_j^{-1}. The group generated by such transformations of the elements of A_n^* is the *Weyl group* of $Sp(n, \mathbb{R})$, which has order $n!2^n$. If we define A_n^{*+} to be the set of diagonal matrices of the form:

$$\begin{pmatrix} H & 0 \\ 0 & H^{-1} \end{pmatrix} \quad \text{with} \quad H = \begin{pmatrix} a_1 & & 0 \\ & \ddots & \\ 0 & & a_n \end{pmatrix}$$

such that $1 \leq a_1 \leq a_2 \leq \cdots \leq a_n$, then the polar decomposition

$$\mathscr{P}_n^* = A_n^{*+}[K_n^*]$$

is *unique*, up to the action of M_n^* = the centralizer of A_n^* in K_n^*, which has order 2^n.

In order to discuss the uniqueness of the general polar decomposition, one needs to discuss the Weyl group for a general semisimple noncompact real Lie group. However, let us postpone this until we have obtained the geodesics in the symmetric space for the symplectic group.

Theorem 1 (Geodesics in the Symmetric Space of the Symplectic Group).

(a) *A geodesic segment in $\mathscr{P}_n^*$ of the form $T(t)$, for $0 \leq t \leq 1$, with $T(0) = I$ and $T(1) = Y \in \mathscr{P}_n^*$ has the expression:*

$$T(t) = \exp\{tB[U]\}, \quad \text{for } 0 \le t \le 1,$$

provided that Y has polar decomposition from Lemma 3 with

$$Y = \exp B[U], \quad \text{for } U \in O(n) \text{ and}$$

$$B = \begin{pmatrix} H & 0 \\ 0 & -H \end{pmatrix} \quad \text{with} \quad H = \begin{pmatrix} h_1 & & 0 \\ & \ddots & \\ 0 & & h_n \end{pmatrix}, \quad h_j \in \mathbb{R}, \quad 1 \le j \le n.$$

The length of the geodesic segment is:

$$\left(2 \sum_{j=1}^n h_j^2\right)^{1/2}.$$

(b) *Consider the geodesic through* Z_0 *and* Z_1 *in* $\mathcal{H}_n$. *Set*

$$\rho(Z_1, Z_0) = (Z_1 - Z_0)(\bar{Z}_1 - Z_0)^{-1}(\bar{Z}_1 - \bar{Z}_0)(Z_1 - \bar{Z}_0)^{-1}.$$

A given pair of points Z_0, Z_1 *in* $\mathcal{H}_n$ *can be transformed by the same matrix* $M \in Sp(n, \mathbb{R})$ *into another pair of points* W_0, W_1 *in* $\mathcal{H}_n$ *if and only if the matrices* $\rho(Z_0, Z_1)$ *and* $\rho(W_0, W_1)$ *have the same eigenvalues.*

If $r_1, \ldots, r_n$ *are the eigenvalues of the matrix* $\rho(Z_0, Z_1)$, *then the symplectic distance between* Z_1 *and* Z_0 *is:*

$$s(Z_0, Z_1) = \sqrt{2} \left(\sum_{j=1}^n \log^2 \frac{1 + \sqrt{r_j}}{1 - \sqrt{r_j}} \right)^{1/2}.$$

PROOF. See Maass [2, p. 39].

(a) The proof proceeds exactly as in the proof of Theorem 1 of Section 4.1.2. In the partial Iwasawa decomposition (1.24) of $T(t)$, we only decrease the arc length by taking X to be identically zero. Then by Exercise 17,

$$ds^2 = 2 \operatorname{Tr}((V^{-1} dV)^2),$$

and we know from Theorem 1 of Section 4.1.2 that this arc length is minimized by taking V to be diagonal. The rest of part (a) is immediate.

(b) Note that if $W_j = (AZ_j + B)(CZ_j + D)^{-1}$, for $j = 0, 1$,

$$\begin{pmatrix} A & B \\ C & D \end{pmatrix} \in Sp(n, \mathbb{R}),$$

then

$$\rho(W_1, W_0) = (Z_1{}^tC + {}^tD)^{-1} \rho(Z_1, Z_0)(Z_1{}^tC + {}^tD). \qquad (*)$$

Using part (a), we need only observe that with H as in part (a), we have

$$\rho(iH, iI) = (H - I)^2(H + I)^{-2}.$$

The eigenvalues of $\rho(iH, iI)$ are $r_j = (h_j - 1)^2(h_j + 1)^{-2}$. Thus

5.1. Geometry and Analysis on G/K

$$h_j = \frac{1 + \sqrt{r_j}}{1 - \sqrt{r_j}}.$$

This completes the proof of Theorem 1. □

It is possible to generalize Theorem 1 to all noncompact real symmetric spaces.

Exercise 19. Prove formula (∗) which was used in the proof of part (b) of Theorem 1.

Hint. First show that

$$W_1 - W_0 = (Z_1\,{}^tC + {}^tD)^{-1}(Z_1 - Z_0)(CZ_0 + D)^{-1}.$$

Theorem 2. *Suppose that G is a connected noncompact real semisimple Lie group.*

(a) *A geodesic in G/K which passes through gK has the form*:

$$\gamma_X(t) = g \exp(tX) K, \qquad \text{for some } X \in \mathfrak{p}, \qquad \text{with } t \in \mathbb{R}.$$

Here the Cartan decomposition of the Lie algebra $\mathfrak{g}$ of G is $\mathfrak{g} = \mathfrak{k} \oplus \mathfrak{p}$, and K is a connected Lie subgroup of G with Lie algebra $\mathfrak{k}$.

(b) *Geodesics of G/K have the form $\gamma_X(t)$, for all $t \in \mathbb{R}$, using the notation of part (a). This means that G/K is a complete Riemannian manifold. Moreover, any two points of G/K can be joined by a geodesic segment of length equal to the Riemannian distance between the points.*

(c) *A geodesic through the origin K in G/K has the form*

$$\gamma(t) = k \exp(tX) K \qquad \text{for some } k \in K, X \in \mathfrak{a}, \qquad \text{with } t \in \mathbb{R}.$$

Here the Iwasawa decomposition of $\mathfrak{g}$ is $\mathfrak{g} = \mathfrak{k} \oplus \mathfrak{a} \oplus \mathfrak{n}$.

PROOF.

(a) Let $g_0 K$ and $g_1 K$ be two cosets in G/K. Apply the transformation

$$a_{g_0^{-1}} \text{ from (1.20)}$$

to transform these cosets to K and $(g_0^{-1}g_1)K$. Now we have the polar decomposition (Lemma 3): $g_0^{-1}g_1 = k_1 a k_2$, with $k_i \in K$ and $a \in A$. So the transformation $a_{k_1^{-1}}$ sends these cosets to K and aK. Thus we have reduced the proof of the case that $\gamma(t)$ is a geodesic with $\gamma(0) = K$ and $\gamma(1) = aK$, with $a \in A$. Write

$$\gamma(t) = a(t)n(t)K, \qquad \text{using the Iwasawa decomposition.}$$

We want to show that $n(t) = e$, the identity in G. Then we would be reduced to the known result that straight lines in the Euclidean space $\mathfrak{a}$ are the geodesics.

The Riemannian structure on G/K comes from the Killing form B of formula (1.12). If $\pi: G \to G/K$ with $\pi(g) = gK$ and $\gamma(t) = w(t)K$, with $w(t) \in G$, $w(t) = a(t)n(t)$, then

$$Q_{\gamma(t)}(\gamma'(t), \gamma'(t)) = B([(d\pi)_{w(t)}(dL_{w(t)})_e]^{-1}\gamma'(t), \text{same}),$$

with $L_g(x) = gx$. Since $\gamma = \pi \circ w$, we have

$$Q_{\gamma(t)}(\gamma'(t), \gamma'(t)) = B(dL_{w(t)}^{-1} w'(t), dL_{w(t)}^{-1} w'(t)).$$

Now, we can calculate the *differential of multiplication* $w(t) = a(t)n(t)$ as:

$$w'(t) = (dL)_{w(t)}\{\operatorname{Ad}(n(t))^{-1} dL_{a(t)}^{-1}(a'(t)) + dL_{n(t)}^{-1}(n'(t))\}. \quad (1.28)$$

(See Exercise 20.) Thus

$$\begin{aligned}
Q_{\gamma(t)}(\gamma'(t), \gamma'(t)) &= B(\operatorname{Ad}(n(t))^{-1} dL_{a(t)}^{-1}(a'(t)) + dL_{n(t)}^{-1}(n'(t)), \text{same}) \\
&= B(dL_{a(t)}^{-1}(a'(t)) + \operatorname{Ad}(n(t)) dL_{n(t)}^{-1}(n'(t)), \text{same}) \\
&= Q_{a(t)}(a'(t), a'(t)) + Q_{n(t)}(n'(t), n'(t)),
\end{aligned}$$

since

$$0 = 2B(dL_{a(t)}^{-1}(a'(t)), \operatorname{Ad}(n(t)) dL_{n(t)}^{-1}(n'(t))),$$

because the first argument of the Killing form lies in $\mathfrak{a}$ and the second lies in $\mathfrak{n}$ (see Exercise 21). Thus the distance is only made smaller by setting $n(t) = e$. Thus we are reduced to the computation of the geodesics in the space $\mathfrak{a}$. The Killing form gives a metric on $\mathfrak{a}$ which is equivalent to the usual Euclidean metric. So the geodesics in $\mathfrak{a}$ are straight lines and the geodesics in $A = \exp \mathfrak{a}$ through e are of the form $\exp(tX)$, $t \in \mathbb{R}$, for some $X \in \mathfrak{a}$. This completes the proof of part (a) of Theorem 2.
(b) This is proved in Helgason [2, p. 56] using part (a).
(c) This follows from part (1) and the polar decomposition (Lemma 3). □

Exercise 20 (The Differential of Multiplication). Suppose that the Lie algebra $\mathfrak{g}$ can be decomposed into a direct sum of subalgebras

$$\mathfrak{g} = \mathfrak{m} \oplus \mathfrak{h}$$

and let $G \supset M, H$ be the corresponding connected Lie subgroups. If the map $\alpha: M \times H \to G$ is defined by $\alpha(m, h) = mh$, show that the differential is:

$$(d\alpha)_{(m,h)}(dL_m X, dL_h Y) = (dL_{mh})(\operatorname{Ad}(h^{-1})X + Y), \quad \text{for } X \in \mathfrak{m}, Y \in \mathfrak{h}.$$

Hint. Define $L_m \times L_h: M \times H \to M \times H$ by $(L_m \times L_h)(x, y) = (mx, hy)$, for $x \in M, y \in H$. Then

$$\alpha \circ (L_m \times L_h) = L_{mh} \circ \alpha \circ (\operatorname{Int}(h^{-1}) \times I).$$

And you can use formula (1.6) relating multiplication on G and Lie bracket on $\mathfrak{g}$ to show that:

5.1. Geometry and Analysis on G/K

$$(d\alpha)_{(e,e)}(X, Y)f = \frac{df}{dt}(\exp tX \exp tY)\Big|_{t=0} = (X + Y)f.$$

Exercise 21. Suppose that $\mathfrak{g}$ has the Iwasawa decomposition $\mathfrak{g} = \mathfrak{k} \oplus \mathfrak{a} \oplus \mathfrak{n}$ and Cartan involution θ. Consider the form $F(X, Y) = -B(X, \theta Y)$ for $X, Y \in \mathfrak{g}$. Then F is a positive definite bilinear form on $\mathfrak{g}$. Show that there is an orthonormal basis of $\mathfrak{g}$ with respect to F such that:

$X \in \mathfrak{k} \Rightarrow \text{ad } X$ is skew symmetric;

$X \in \mathfrak{a} \Rightarrow \text{ad } X$ is diagonal;

$X \in \mathfrak{n} \Rightarrow \text{ad } X$ is upper triangular with 0 on the diagonal.

Hint. (See Wallach [2, p. 166] or Helgason [2, p. 223].) You need to take an ordered set of positive roots: $\alpha_1, \alpha_2, \ldots, \alpha_m$. Then form an orthonormal basis of $\mathfrak{g}$ by taking orthonormal bases of

$$\mathfrak{g}_{\alpha_m}, \ldots, \mathfrak{g}_{\alpha_1}, \quad \mathfrak{a} \oplus \mathfrak{m}, \quad \theta(\mathfrak{g}_{\alpha_1}), \ldots, \theta(\mathfrak{g}_{\alpha_m}).$$

You need to use properties of the roots such as the fact that: $[\mathfrak{g}_\alpha, \mathfrak{g}_\beta] \subset \mathfrak{g}_{\alpha+\beta}$.

Next we consider the Weyl group of the symmetric space. See (5.52) of Section 4.5.3 for the definition in the case of $GL(n, \mathbb{R})$. As usual, suppose that G is a noncompact real semisimple Lie group with the standard definitions of K, $\mathfrak{a}$, etc. Define the following subgroups of K:

$$M = \text{the } \textit{centralizer} \text{ of } \mathfrak{a} \text{ in } K = \{k \in K \,|\, \text{Ad}(k)|_\mathfrak{a} = \text{identity}\},$$
$$M' = \text{the } \textit{normalizer} \text{ of } \mathfrak{a} \text{ in } K = \{k \in K \,|\, \text{Ad}(k)\mathfrak{a} \subset \mathfrak{a}\}. \quad (1.29)$$

Both M and M' are closed subgroups of K. The *Weyl group* of G/K is defined to be $W = M'/M$. Note that W is independent of the choice of $\mathfrak{a}$, by the conjugacy of all maximal abelian subspaces of $\mathfrak{p}$ (see the proof of Lemma 3). These definitions can also be made in the case that G is compact (see Helgason [2, p. 244]).

Theorem 3 (The Weyl Group).

(1) *The Weyl group is a finite group contained in the orthogonal group in $GL(\mathfrak{a})$ with respect to the inner product on $\mathfrak{a}$ defined by the Killing form of $\mathfrak{g}$.*
(2) *The Weyl group permutes the restricted roots. Define a* Weyl chamber *to be a connected component of $(\mathfrak{a} - \alpha^{-1}(0))$, for $\alpha \in \Lambda$. Note that $\alpha^{-1}(0)$ is a hyperplane in $\mathfrak{a}$. The Weyl group also permutes the Weyl chambers. And the action of the Weyl group on the Weyl chambers is simply transitive.*
(3) *For $\lambda \in \Lambda$, define $s_\lambda: \mathfrak{a} \to \mathfrak{a}$ by*

$$s_\lambda(H) = H - 2(\lambda(H)/\lambda(H_\lambda))H_\lambda, \quad \text{where } H_\lambda \in \mathfrak{a}$$

is defined by

$$B(H, H_\lambda) = \lambda(H), \quad \text{for all } H \in \mathfrak{a}.$$

Then s_λ is the reflection in the hyperplane $\lambda^{-1}(0)$. The Weyl group is generated by these reflections s_λ, for $\lambda \in \Lambda$.

PROOF. We shall only prove part (1). For the other parts of the theorem, see Helgason [2, Ch. 7] or Wallach [2, pp. 77, 168]. By the Lie group/Lie algebra dictionary,

the Lie algebra of $M = \text{Lie}(M) = \mathfrak{m} = \{X \in \mathfrak{k} \,|\, \text{ad}\, X|_\mathfrak{a} = 0\}$.

If we can show that M' has the same Lie algebra as M, then it will follow that the quotient M'/M is both discrete and compact (thus finite). Suppose that T is in the Lie algebra of M'. Then write out the root space decomposition of T:

$$T = Y + \sum_{\lambda \in \Lambda} X_\lambda, \quad \text{for } Y \in \mathfrak{m}, X_\lambda \in \mathfrak{g}_\lambda.$$

It follows that for all $H \in \mathfrak{a}$

$$[H, T] = \sum \lambda(H) X_\lambda \in \mathfrak{a} \text{ implies that } [H, T] = 0.$$

Here we have used the fact that the sum in the root space decomposition is direct.

To see that the group M'/M permutes the restricted roots is easy. To see that the reflections s_λ come from some $\text{Ad}(k)$, $k \in K$, is harder. To see that the s_λ generate the Weyl group is harder yet. Note that we cannot claim that the s_λ, with λ from a system of simple roots, generate the Weyl group. A system of simple roots has the property that any root is a linear combination of simple roots with integer coefficients that are either all positive or all negative (with $r = \dim \mathfrak{a}$ elements). Such simple root systems *do* give generators of the Weyl group in the case of *complex* semisimple Lie algebras. However, real Lie algebras are somewhat different, as we will see in the following examples. □

Exercise 22.

(a) Why is it reasonable to call a Lie algebra "semisimple" if the Killing form is nondegenerate? What is the connection with the standard notion that an algebraic object is semisimple if it is a direct sum of simple objects?
(b) Why do we call a semisimple Lie algebra "compact" if its Killing form is negative definite? What is the connection with compact Lie groups? Can we drop the hypothesis that the Lie algebra be semisimple?
(c) Recall that we said a Lie algebra is "nilpotent" if all sufficiently long brackets must vanish. What is the connection with the usual idea of a nilpotent linear transformation (such as ad X)?

Hints.
(a) See Helgason [2, pp. 121–122].
(b) See Helgason [2, pp. 120–122]. Think about $\mathbb{R}$ and $\mathbb{R}/\mathbb{Z}$.
(c) See Helgason [2, pp. 135–137].

Examples of Weyl Groups

Example 1 ($GL(n, \mathbb{R})$). Since Ad = Int for matrix groups, we have:

$$M = \{k \in O(n) | kXk^{-1} = X, \text{ for any diagonal matrix } X\},$$
$$M' = \{k \in O(n) | kXk^{-1} \text{ is diagonal, for any diagonal matrix } X\}.$$

It follows that

$$M = \{\text{diagonal matrices with entries } +1 \text{ or } -1\},$$
$$M' = \{\text{matrices with each row or column having exactly one non-zero entry of } \pm 1\}.$$

Thus the Weyl group of $GL(n, \mathbb{R})$ is the group of all permutations of n objects as we also saw in Exercise 29 of Section 4.5.3.

Example 2 ($Sp(n, \mathbb{R})$). Here

$$K = \left\{ \begin{pmatrix} A & B \\ -B & A \end{pmatrix} \bigg| A + iB \in U(n) \right\},$$

$$\mathfrak{a} = \left\{ \begin{pmatrix} H & 0 \\ 0 & -H \end{pmatrix} \bigg| H \;\; n \times n \text{ real diagonal} \right\},$$

$$M = \left\{ \begin{pmatrix} A & 0 \\ 0 & A \end{pmatrix} \bigg| A \text{ diagonal } n \times n \text{ entries } \pm 1 \right\},$$

$$M' = \left\{ \begin{pmatrix} A & B \\ -B & A \end{pmatrix} \bigg| A + B \text{ is in the } M' \text{ for } GL(n, \mathbb{R}) \right\}.$$

It follows that the Weyl group $W = M'/M$ contains all permutations of entries of H in

$$\begin{pmatrix} H & 0 \\ 0 & -H \end{pmatrix} \text{ in } \mathfrak{a},$$

as well as all possible changes of sign. So it has 2^n times $n!$ elements. For example, let $n = 3$ and

$$A = \begin{pmatrix} 0 & 1 & 0 \\ 1 & 0 & 0 \\ 0 & 0 & 0 \end{pmatrix}, \;\; B = \begin{pmatrix} 0 & 0 & 0 \\ 0 & 0 & 0 \\ 0 & 0 & 1 \end{pmatrix}, \;\; k = \begin{pmatrix} A & B \\ -B & A \end{pmatrix}, \;\; H = \begin{pmatrix} h_1 & & 0 \\ & h_2 & \\ 0 & & h_3 \end{pmatrix},$$

$$a = \begin{pmatrix} H & 0 \\ 0 & -H \end{pmatrix}, \;\; \text{then} \;\; \text{Ad}(k)a = \begin{pmatrix} H' & 0 \\ 0 & -H' \end{pmatrix},$$

where

$$H' = \begin{pmatrix} h_2 & 0 & 0 \\ 0 & h_1 & 0 \\ 0 & 0 & -h_3 \end{pmatrix}.$$

Exercise 23. Check the results stated in Example 2 above for $Sp(n, \mathbb{R})$.

Example 3 (SU(2, 1)). For this example,

$$K = \left\{ \begin{pmatrix} U & 0 \\ 0 & t \end{pmatrix} \middle| U \in U(2), t = (\det U)^{-1} \right\},$$

$$\mathfrak{a} = \mathbb{R} \begin{pmatrix} 0 & 0 & 1 \\ 0 & 0 & 0 \\ 1 & 0 & 0 \end{pmatrix},$$

$$M = \left\{ k = \begin{pmatrix} e^{i\alpha} & 0 & 0 \\ 0 & e^{i\beta} & 0 \\ 0 & 0 & e^{i\alpha} \end{pmatrix} \middle| \det k = 1 \right\},$$

$$M' = \left\{ k = \begin{pmatrix} e^{i\alpha} & 0 & 0 \\ 0 & e^{i\beta} & 0 \\ 0 & 0 & \pm e^{i\alpha} \end{pmatrix} \middle| \det k = 1 \right\}.$$

So the Weyl group has only 2 elements. The entries of the diagonal matrices in M' are supposed to be of complex norm 1.

Exercise 24. Verify the results stated in Example 3 for SU(2, 1).

Now that we have described the Weyl group, it is possible to discuss the *non-uniqueness of the polar decomposition*. The precise result is that if we set $\mathfrak{a}' = \{H \in \mathfrak{a} \mid \lambda(H) \neq 0, \text{ for all } \lambda \in \Lambda\}$ and $A' = \exp \mathfrak{a}'$, and define the map $f: (K/M) \times A' \to G/K$ by $f(kM, a) = (ka)K$, then the map f is $\#W$ to 1, regular, and onto an open submanifold of G/K whose complement in G/K has lower dimension (see Helgason [2, p. 381] or [3, p. 402] or Wallach [2]).

One should also consider the relation between the structure theory for $\mathfrak{g}$ a noncompact semisimple real Lie algebra and that for the complexification $\mathfrak{g}^c = \mathfrak{g} \otimes_\mathbb{R} \mathbb{C}$. And the same question could be asked for the compact real form of $\mathfrak{g}^c$. As an example, consider $SU(2, 1)$ again. The maximal abelian subalgebra $\mathfrak{h}$ of $\mathfrak{su}(2, 1)$ containing $\mathfrak{a}$ is:

$$\mathfrak{h} = \left\{ \begin{pmatrix} a & 0 & b \\ 0 & c & 0 \\ b & 0 & a \end{pmatrix} \middle| a, c \in i\mathbb{R}, b \in \mathbb{R} \right\}.$$

Clearly the complexification of $\mathfrak{h}$ is a Cartan subalgebra of the complexification of $\mathfrak{g}$. This shows that much is missing from the complexification of $\mathfrak{a}$. One can show that the restricted roots are really restrictions of roots of the complexified Lie algebra (see Helgason [2, Ch. 6]). Again, some roots from the complexification may be missing in the real version of the Lie algebra.

Our next topic is integral formulas for noncompact semisimple real Lie groups. First perhaps we should discuss the Haar measures in G, A, N, and

5.1. Geometry and Analysis on G/K

K. See our earlier comments on this subject in Chapters 2 and 4. More details about Haar measures can be found in Helgason [2, Chapter 10]. Because Haar measure is unique up to a positive scalar multiple, we can define the *modular function* $\delta\colon G \to \mathbb{R}^+$ by the formula (assuming $dg = $ left Haar measure):

$$\int f(gs^{-1})\,dg = \delta(s) \int f(g)\,dg.$$

For the left hand side of the equality is a left G-invariant integral for fixed s. Thus it must be a positive constant times the Haar integral of f. It follows easily that δ is continuous, $\delta(st) = \delta(s)\delta(t)$, and $d(gs) = \delta(s)\,dg$. Thus the modular function relates right and left Haar measure. By definition, a *unimodular group* has $\delta = 1$ identically. Furthermore, it is easy to see that $d(g^{-1}) = \delta(g^{-1})\,dg$. If G is a Lie group, one also has $d(s^{-1}gs) = d(gs) = \delta(s)\,dg$. Thus $\det(\operatorname{Ad}(s^{-1})) = \delta(s)$, for all s in G.

We prove that compact, semisimple and nilpotent Lie groups are all unimodular. Suppose first that K is compact. Then δ maps K onto a compact subgroup of $\mathbb{R}^+$ which must contain only one element, since otherwise powers would approach 0 or infinity. Suppose next that G is semisimple. Then $\operatorname{Ad}(s)$ leaves the Killing form invariant for s in G. But the Killing form of a semisimple group is nondegenerate and thus equivalent to

$$I_{p,q} = \begin{pmatrix} I_p & 0 \\ 0 & -I_q \end{pmatrix}, \qquad \text{for some } p, q.$$

If ${}^t g I_{p,q} g = I_{p,q}$, the determinant of g must have absolute value 1. Finally suppose that N is nilpotent and connected. Then $\det(\operatorname{Ad}(n)) = \exp(\operatorname{Tr}(\operatorname{ad}(\log n))) = 1$, for $n \in N$, since $\operatorname{ad}(\log n)$ is a nilpotent linear transformation.

Proposition 1 (The Integral Formula for the Iwasawa Decomposition). *Define $J(a) = \prod_{0 < \lambda \in \Lambda^+} \exp(\lambda(\log a))$ for $a \in A$. Then*

$$\int_A \int_N \int_K f(ank)\,da\,dn\,dk = \int_G f(g)\,dg,$$

where all the measures are left-invariant (and thus right-invariant) Haar measures on G, A, N, K. However, changing the order gives:

$$\int_K \int_A \int_N f(kan) J(a)\,dk\,da\,dn = \int_G f(g)\,dg.$$

PROOF. In order to compute the Jacobian of the Iwasawa decomposition, we proceed as in Exercise 20 of Section 4.1.3. Thus we need the differential of $\operatorname{Int}(a)n = ana^{-1}$, for $n \in N$ and $a \in A$. We know that the differential of $\operatorname{Int}(a)$ is $\operatorname{Ad}(a)$, by definition. Thus if $a = \exp H$ for $H \in \mathfrak{a}$, we find that:

$$\det(\operatorname{Ad}(a)) = \det(\exp(\operatorname{ad} H)) = \exp(\operatorname{Tr}(\operatorname{ad} H))$$

$$= \exp\left(\sum_{\lambda \in \Lambda^+} \lambda(H)\right) = \prod_{\lambda \in \Lambda^+} \exp \lambda(H),$$

which is simply $J(a)$, as defined in the proposition, since $H = \log a$. Here we have used the fact that:

$$\mathfrak{n} = \sum_{\lambda \in \Lambda^+} \mathfrak{g}_\lambda, \qquad \mathfrak{g}_\lambda = \{X \in \mathfrak{g} \mid \operatorname{ad} H(X) = \lambda(H)X, \text{ for all } H \in \mathfrak{a}\}.$$

Note that $\operatorname{Int}(a): N \to N$ for any $a \in A$.

The rest of the argument is really the same as that of Exercise 20 in Section 4.1.3, but we shall repeat it for completeness. First observe that the left Haar measures on G and K can be normalized so that if $d\bar{g}$ denotes the G-invariant measure on the symmetric space G/K, then the following equality prevails:

$$\int_G f(g)\, dg = \int_{\bar{g} = gK \in G/K} \int_{k \in K} f(gk)\, dk\, d\bar{g}.$$

Now G/K can be identified with AN. Thus we need only show

$$\int_A \int_N f(an)\, da\, dn$$

gives a left AN-invariant integral on AN. Let $a_1 \in A$ and $n_1 \in N$. Then we have

$$\int_A \int_N f(a_1 n_1 an)\, da\, dn = \int_A \int_N f(a_1 a n_2 n)\, da\, dn, \qquad \text{if } n_2 = a^{-1} n_1 a.$$

Since both da and dn are left invariant, the last integral is just

$$\int_A \int_N f(an)\, da\, dn.$$

Now we are ready to prove the 2nd version of the integral formula for the Iwasawa decomposition. Using the differential of $\operatorname{Int}(a)$ and the first integral formula, we get:

$$\int_G f(g)\, dg = \int_N \int_A \int_K f(nak) J(a)^{-1}\, dn\, da\, dk.$$

Now replace $f(g)$ by $f(g^{-1})$. This will reverse orders on the right-hand side and produce

$$\int_N \int_A \int_K f(k^{-1} a^{-1} n^{-1}) J(a)^{-1}\, dn\, da\, dk.$$

Finally the fact that G, N, A, K are all unimodular leads to the 2nd integral formula for the Iwasawa decomposition. □

5.1. Geometry and Analysis on G/K

Examples.
(1) $G = GL(n, \mathbb{R})$.
$$J(a) = \prod_{1 \leq i < j \leq n} a_i/a_j = \prod_{i=1}^{n} a_i^{n-2i+1}.$$

(2) $G = Sp(n, \mathbb{R})$.
$$J(a) = \prod_{1 \leq i < j \leq n} a_i/a_j \prod_{1 \leq i \leq j \leq n} a_i a_j = \prod_{i=1}^{n} a_i^{2(n+1-i)}.$$

In order to be more precise, we need to fix the invariant volumes on the symmetric spaces.

Invariant Volume Elements on the Symmetric Spaces of $GL(n, \mathbb{R})$ and $Sp(n, \mathbb{R})$

(1) We found in formula (1.16) of Section 4.1.3 that the $GL(n, \mathbb{R})$-invariant volume element on $\mathscr{P}_n$ is:
$$d\mu_n = |Y|^{-(n+1)/2} \prod_{1 \leq i \leq j \leq n} dy_{ij}, \quad \text{if } Y = (y_{ij}) \in \mathscr{P}_n.$$

(2) Next we want to find the invariant volume element on the Siegel upper half space $\mathscr{H}_n$. The argument following (1.16) of Section 4.1.3 can be imitated to show that the $Sp(n, \mathbb{R})$-invariant volume on $\mathscr{H}_n$ is:
$$d\mu_n^*(Z) = |Y|^{-(n+1)} \prod_{1 \leq i \leq j \leq n} dx_{ij} \, dy_{ij}, \quad \text{if } Z = X + iY \in \mathscr{H}_n,$$

with
$$X = (x_{ij}) \quad \text{and} \quad Y = (y_{ij}).$$

Let us prove this. As in the case of $GL(n, \mathbb{R})$, it suffices to find the Jacobian of the action of a diagonal symplectic matrix on $\mathscr{H}_n$. So observe that the image of $X + iY \in \mathscr{H}_n$ under the matrix

$$\begin{pmatrix} a & 0 \\ 0 & a^{-1} \end{pmatrix} \text{ in } Sp(n, \mathbb{R}), \quad a = \begin{pmatrix} a_1 & & 0 \\ & \ddots & \\ 0 & & a_n \end{pmatrix},$$

is $aZa = Z[a] = X[a] + iY[a]$, according to (1.20). Thus the Jacobian of the transformation is:
$$\prod_{1 \leq i \leq j \leq n} a_i a_j \prod_{1 \leq i \leq j \leq n} a_i a_j = |a|^{2(n+1)}.$$

This shows that the measure $d\mu_n^*$ is $Sp(n, \mathbb{R})$-invariant.

Proposition 2 (The Integral Formula for Polar Coordinates). *Suppose that the root space $\mathfrak{g}_\lambda$ has dimension m_λ for any restricted root λ. Then there is a positive* constant c such that if dg denotes the Haar measure on G, then

$$\int_G f(g)\,dg = c \int_K \int_A \int_K f(k_1 a k_2) D(a)\,dk_1\,da\,dk_2,$$

where

$$D(a) = \prod_{0 < \lambda \in \Lambda} |\sinh[\lambda(\log a)]|^{m_\lambda}, \text{ for } a \in A.$$

PROOF. See Helgason [2, Ch. 10]. The main step is the following lemma. Then one uses the fact that $\mathfrak{p} = \mathrm{Ad}(K)\mathfrak{a}$ from Lemma 3.

Lemma 4 (The Integral Formula for Exp Restricted to $\mathfrak{p}$ in the Cartan Decomposition). *There is a positive constant c such that:*

$$\int_{G/K} f(x)\,dx = c \int_{\mathfrak{p}} f(\exp Y) J(Y)\,dY,$$

where

$$J(X) = \det\left(\frac{\sinh \mathrm{ad}\,X}{\mathrm{ad}\,X}\right), \quad \text{for } X \in \mathfrak{p}.$$

PROOF. First recall our calculation of the differential of exp in formula (1.7) or see Helgason [2, p. 95] for the general result. Observe that if X, $Y \in \mathfrak{p}$, then

$$(d\exp)_X Y = (dL_{\exp X})_e \circ \left\{ \sum_{n \geq 0} \frac{1}{(2n+1)!}(-\mathrm{ad}\,X)^{2n} \right\} Y.$$

For $[\mathfrak{k},\mathfrak{p}] \subset \mathfrak{p}$ and $[\mathfrak{p},\mathfrak{p}] \subset \mathfrak{k}$ follow from the properties of the Cartan involution. Therefore

$$(\mathrm{ad}\,X)^{2n+1} Y \in \mathfrak{k} \quad \text{and} \quad (\mathrm{ad}\,X)^{2n} Y \in \mathfrak{p}.$$

Since $\mathfrak{k} \cap \mathfrak{p} = 0$, we have the vanishing of the sum of the odd powers of $\mathrm{ad}\,X$ in the series expression for the differential of exp.

We can write $X \in \mathfrak{p}$ in the form

$$X = Y + \sum_{\lambda \in \Lambda^+} (X_\lambda - \theta X_\lambda), \quad Y \in \mathfrak{a},$$

for X_λ in a basis for $\mathfrak{g}_\lambda$. Note also that $H \in \mathfrak{a}$, $X_\lambda \in \mathfrak{g}_\lambda$ implies that

$$(\mathrm{ad}\,H)(X_\lambda - \theta X_\lambda) = \lambda(H)(X_\lambda + \theta X_\lambda), \quad \text{if } X_\lambda \in \mathfrak{g}_\lambda.$$

It follows that

$$(\mathrm{ad}\,H)^2 (X_\lambda - \theta X_\lambda) = (\lambda(H))^2 (X_\lambda - \theta X_\lambda).$$

Thus the differential at H has determinant

$$|(d\exp|_{\mathfrak{p}})_H| = |e^H| \prod_{0 < \lambda \in \Lambda} (\sinh \lambda(H))/\lambda(H),$$

5.1. Geometry and Analysis on G/K

proving Lemma 4. This shows, in particular, that exp is a diffeomorphism when restricted to $\mathfrak{p}$. □

Now we return to the proof of Proposition 2. Observe first that $X \in \mathfrak{k}$ has the representation:

$$X = Y + \sum_{\lambda \in \Lambda^+} (X_\lambda + \theta X_\lambda), \quad \text{with } Y \in \mathfrak{m}, X_\lambda \in \mathfrak{g}_\lambda,$$

where $\mathfrak{m}$ and $\mathfrak{g}_\lambda$ are from the root space decomposition of $\mathfrak{g}$ (see Helgason [2, p. 224]).

Define $f: K \times A \to P$ by $f(k, a) = kak^{-1} = p$. Suppose that $Y \in \mathfrak{k}$, $H \in \mathfrak{a}$. Then

$$(df)_{(k,a)}(Y, H) = \lim_{t \to 0} \frac{1}{t} \{f(ke^{tY}, ae^{tH}) - f(k, a)\}$$

$$= \lim_{t \to 0} \frac{1}{t} \{ke^{tY} ae^{tH} e^{-tY} k^{-1} - kak^{-1}\}$$

$$= k \left\{ \lim_{t \to 0} \frac{1}{t} (e^{tY} ae^{tH} e^{-tY} - a) \right\} k^{-1}.$$

Now suppose that $a = \exp(H_0)$ for $H_0 \in \mathfrak{a}$. Then the object inside the last limit is:

$$e^{tY} e^{H_0 + tH} e^{-tY} - e^{H_0} = \exp(H_0 + tH + t[Y, H_0 + tH] + o(t^2)) - \exp H_0$$

$$= \exp(H_0 + t(H + [Y, H_0]) + o(t^2)) - \exp H_0.$$

Use the chain rule to evaluate the derivative of the preceding quantity with respect to t at $t = 0$ and obtain:

$$(d\exp)_{H_0}(H + [Y, H_0]).$$

Take a basis of $\mathfrak{p}$ coming from $\mathfrak{a}$ and vectors $X_\lambda - \theta X_\lambda$ and a basis of $\mathfrak{k}$ coming from $\mathfrak{m}$ and vectors $X_\lambda + \theta X_\lambda$ with X_λ in the root spaces $\mathfrak{g}_\lambda$. One sees that for $Y = X_\lambda + \theta X_\lambda$, the preceding is:

$$(d\exp)_{H_0}(H - \lambda(H_0)(X_\lambda - \theta X_\lambda)).$$

Now use the formula for the differential of exp along with the fact that the odd powers of $(\operatorname{ad} H_0)$ vanish once again. This yields:

$$(d\exp)_{H_0}(H + [X_\lambda + \theta X_\lambda, H_0])$$

$$= e^{H_0} \left\{ H - \lambda(H_0) \sum_{n \geq 0} \frac{1}{(2n+1)!} \lambda(H_0)^{2n}(X_\lambda - \theta X_\lambda) \right\}$$

$$= e^{H_0} \{H - \sinh \lambda(H_0)(X_\lambda - \theta X_\lambda)\}.$$

This completes our discussion of Proposition 2. □

If G is a real noncompact semisimple Lie group, K a compact subgroup coming from the Cartan decomposition of G, the *boundary* of G/K can be defined as K/M. The group M was defined in (1.29). And we can identify this boundary with G/B, if B is the Borel and minimal parabolic subgroup $B = MAN$, as in (1.19) of Section 4.1.3. This boundary can also be viewed as the set of Weyl chambers in $\mathfrak{a}$ (from the Iwasawa decomposition of the Lie algebra of G). Furstenberg [2] and Moore [1] show that G/B is a "maximal boundary" in a certain probabilistic sense.

Example ($G = SL(n, \mathbb{R})$). Here $B = MAN$ consists of all upper triangular matrices of determinant one. We can identify G/B as the *flag manifold*:

$$F_n = \{(V_1, \ldots, V_{n-1}) \mid V_i \text{ is a vector subspace of } \mathbb{R}^n, \dim_\mathbb{R} V_i = i, V_i \subset V_{i+1}\}.$$

The action of $g \in G$ on F_n is $g(V_1, \ldots, V_{n-1}) = (gV_1, \ldots, gV_{n-1})$. This action is easily seen to be transitive. To calculate the stability group of a point, let $e_i \in \mathbb{R}^n$ denote the column vector with ith coordinate one and the rest zero. Then set $V_i^0 = \mathbb{R}e_1 \oplus \cdots \oplus \mathbb{R}e_i$. Then g fixes V_i^0, for all i, means $g \in B$.

Exercise 25. Show that the Jacobian of the action of $g \in G$ on the boundary $G/MAN \cong K/M$ is given by the following integral formula:

$$\int_{K/M} f(\bar{k})\, d\bar{k} = \int_{\bar{k} = kM \in K/M} f(g(\bar{k})) J(a(gk))^{-1}\, d\bar{k},$$

where $a(g)$ is the A-part of the KAN-Iwasawa decomposition of g. And $J(a)$ is the Jacobian of the Iwasawa decomposition in Proposition 1. Here $d\bar{k}$ is any K-invariant measure on K/M.

It is also possible to show (using the Bruhat decomposition of G described in Section 4.5.3 for $GL(n)$) that if $\bar{N}$ denotes the opposite nilpotent subgroup corresponding to the Lie subalgebra $\bar{\mathfrak{n}}$,

$$\bar{\mathfrak{n}} = \sum_{0 > \alpha \in \Lambda} \mathfrak{g}_\alpha,$$

then $\bar{N}MAN$ is an open subset of G with lower dimensional complement. See Lemma 2 of Section 4.3.3 for a proof of this result when $G = SL(n, \mathbb{R})$. This allows us to identify K/M with $\bar{N}$ as far as integration is concerned. In Section 4.3, we applied such a result to obtain the asymptotics of spherical functions. For more information on boundaries and compactifications of symmetric spaces, see Gérardin [1], Helgason [1–7], Koranyi [1], and the references mentioned when we defined the boundary of $\mathscr{P}_n$.

This concludes our discussion of the basic integral formulas for symmetric spaces. Next let us consider differential operators on the symmetric space G/K when G is a noncompact real semisimple Lie group.

Let φ be a diffeomorphism of a manifold M. We say that a differential operator D on M is *invariant* under φ if D commutes with φ; i.e., if $D(f \circ \varphi) =$

5.1. Geometry and Analysis on G/K

$(Df) \circ \varphi$, for all infinitely differentiable functions f on M. For each $g \in G$, we have a diffeomorphism a_g of the symmetric space G/K defined by $a_g(xK) = (gx)K$, for $g, x \in G$. Define $D(G/K)$ to be the set of all differential operators on G/K which are a_g-invariant for all $g \in G$. So $D(G/K)$ is the *algebra of invariant differential operators on G/K*. The Laplacian will, of course, be such an operator. In general, however, there will be invariant differential operators on G/K which are not polynomials in the Laplacian, just as for $\mathscr{P}_n$ (see Theorem 2 of Section 4.1.4). The following theorem is proved in Helgason [2, p. 432]).

Theorem 4 (Harish-Chandra and Chevalley). *Suppose that G is a noncompact real semisimple Lie group with $\dim_\mathbb{R} \mathfrak{a} = r = \text{rank of } G/K$. Then the algebra $D(G/K)$ of all invariant differential operators on G/K is a commutative algebra. In fact, it is a polynomial ring with r algebraically independent generators.*

There is a close relation between $D(G/K)$ and $D(G) =$ the left-invariant differential operators on G or the universal enveloping algebra of G (see Helgason [2, Ch. 10]).

Question. Can one relate the invariant differential operators $D(G/K)$ for the following chain of inclusions of totally geodesic submanifolds?

$$\mathscr{P}_n \to \mathscr{H}_n \subset \mathscr{P}_{2n}^* \subset \mathscr{P}_{2n},$$
$$Y \mapsto iY.$$

We know, for example, that the arc length on $\mathscr{H}_n$ is given by:

$$ds^2 = 2\operatorname{Tr}((Y^{-1}\,dY)^2 + (Y^{-1}\,dX)^2), \qquad \text{for } Z = X + iY \in \mathscr{H}_n.$$

Therefore the Laplacian on $\mathscr{H}_n$ must be a sum of the Laplacian on $\mathscr{P}_n$ plus a term involving only differentiation with respect to X-variables. If follows that for functions of the Y-variable alone, the Laplacian on $\mathscr{H}_n$ coincides with that on $\mathscr{P}_n$, disregarding constants.

Let G be a noncompact real semisimple Lie group, as usual. A function $u: G/K \to \mathbb{C}$ is called *harmonic* if $Du = 0$ for any operator $D \in D(G/K)$ such that D annihilates constants. This definition was made by Godement [5]. Furstenberg [2] shows that, in fact, a bounded solution of $\Delta u = 0$ on G/K is automatically harmonic. Other references for harmonic functions on symmetric spaces are Helgason [1] and Koranyi [1].

Theorem 5 (Godement) (Mean Value Theorem). *Suppose $u: G/K \to \mathbb{C}$ is infinitely differentiable. Then u is harmonic if and only if*

$$\int_{k \in K} u(gkhK)\,dk = u(gK), \qquad \text{for all } g, h \in G.$$

PROOF. (Helgason [1, pp. 42–43]).
⇒ Let u be harmonic and

$$F(h) = \int_K u(gkhK)\, dk.$$

We want to show that $F(h) = F(e) = u(gK)$, $e =$ identity of G. Since F satisfies an elliptic partial differential equation with analytic coefficients, it follows by a theorem of Bernstein that F is analytic (see John [1, pp. 57, 142]). Now it suffices to show that

$$(DF)(e) = 0,$$

for every left invariant differential operator D on the Lie group G such that D annihilates the constants. To show that DF vanishes at the identity, we must merely relate differential operators in $D(G/K)$ with those in $D(G)$. This is done in detail in Helgason [2, Ch. 10]. We merely sketch the process. Let us use the following notation for a diffeomorphism φ of G:

$$D^\varphi f = D(f \circ \varphi) \circ \varphi^{-1}, \qquad \text{if } D \in D(G).$$

Then for $D \in D(G)$ write

$$D^\# f = \int_K D^{R_k}\, dk, \qquad \text{if } R_k x = xk \text{ for } x \in G.$$

Now it can be show that $D^\#$ is a differential operator which is invariant under all the R_k, $k \in K$ and thus gives rise to an operator $\tilde{D}$ in $D(G/K)$. And we find that by hypothesis:

$$(DF)(e) = (D^\# F)(e) = \int_K (\tilde{D}u)(gkK)\, dk = 0.$$

⇐ Assume that u has the mean value property stated in the theorem and that $D \in D(G/K)$ annihilates constants. As usual, set $a_g(hK) = ghK$, for g, h in G. Thus

$$\int_{k \in K} u(a_{gk}(x))\, dk = u(gK), \qquad \text{if } x \in G/K.$$

Apply D to both sides of this equation considered as functions of $x \in G/K$ to obtain

$$\int_{k \in K} (Du)(a_{gk}(x))\, dk = 0,$$

since D and a_g commute. Take x to be the coset K in G/K to see that $Du = 0$. □

Theorem 6 (Furstenberg) (Poisson Integral Formula). *Suppose $u: G/K \to \mathbb{C}$ is a bounded harmonic function. Then there is a bounded measurable function*

5.1. Geometry and Analysis on G/K

$\hat{u}: K/M \to \mathbb{C}$ such that

$$u(gK) = \int_{\bar{k} \in K/M} \hat{u}(g(\bar{k})) \, d\bar{k}, \qquad \text{for all } g \in G. \tag{1.30}$$

Here $d\bar{k}$ is the unique K-invariant measure on the boundary K/M such that

$$\int_{K/M} d\bar{k} = 1.$$

And conversely, given $\hat{u}$, as above, the function u on G/K defined by (1.30) is harmonic. We can rewrite formula (1.30) as:

$$u(x) = \int_{\bar{k} \in K/M} \hat{u}(\bar{k}) P(x, \bar{k}) \, d\bar{k},$$

where

$$P(gK, kM) = \text{Poisson's kernel} = d(g^{-1}(\bar{k}))/d\bar{k} = J^{-1}(a(g^{-1}k)).$$

Here J denotes the Jacobian of the Iwasawa decomposition from Proposition 1 and $a(g)$ is the A-part of the KAN Iwasawa decomposition of $g \in G$.

SKETCH OF THE PROOF. (See Helgason [1, pp. 42–52].)
$\Rightarrow$ We need to know that the Borel subgroup $B = MAN$ has the following *fixed point property*. Suppose that B acts continuously on a locally convex topological vector space by linear transformations leaving a nonempty compact convex set invariant. Then B has a fixed point in the convex set. Assuming this result, suppose $u: G/K \to \mathbb{C}$ is bounded and harmonic. Define the set

$$Q_u = \left\{ w \in L^\infty(G) \;\middle|\; \begin{array}{l} \|w\|_\infty = \text{l.u.b.}\{|w(h)| \,|\, h \in G\} \le \|u\|_\infty \\ u(gK) = \int_K w(gkh) \, dk, \quad \text{for all } g, h \in G \end{array} \right\}.$$

By Godement's Mean Value Theorem, $u \circ \pi = \tilde{u} \in Q_u$, where $\pi: G \to G/K$ is defined by $\pi(g) = gk$.

Suppose that MAN leaves u_1 in Q_u fixed. Set $\hat{u}(gMAN) = u_1(g)$ for all $g \in G$. Then $\hat{u}$ has the required property.

$\Leftarrow$ If $\hat{u}$ is as described in the theorem, then u defined by (1.30) is easily shown to have the mean value property. □

Exercise 26. Prove this last statement; i.e., the $\Leftarrow$ of Theorem 6.

Another standard result in potential theory generalizes as follows.

Theorem 7. *Suppose that F is continuous on the boundary of G/K and set*

$$u(gK) = \int_{K/M} P(gK, \bar{k}) F(\bar{k}) \, d\bar{k}, \qquad \text{for } g \in G \qquad (P(x, b) = \text{Poisson's kernel}).$$

Then u has boundary values given by F; i.e.,

$$\lim_{t \to \infty} u((k \exp tH)K) = F(kM), \quad \text{for } k \in K, H \in \mathfrak{a}^+,$$

where $\mathfrak{a}^+$ is a fixed Weyl chamber in $\mathfrak{a}$ (from the Iwasawa decomposition).

PROOF (Helgason [1, pp. 47–48]). First one must identify the boundary K/M with $\bar{N}$ the Lie subgroup of G corresponding to the Lie subalgebra

$$\bar{\mathfrak{n}} = \sum_{\alpha \in \Lambda^-} \mathfrak{g}_\alpha,$$

where the Weyl chamber is

$$\mathfrak{a}^+ = \{H \in \mathfrak{a} \mid \alpha(H) < 0 \text{ if } \alpha \in \Lambda^-\}.$$

Set $a_t = \exp(tH)$, for $t \in \mathbb{R}$. Write $k(g) =$ the K-part in the KAN Iwasawa decomposition of $g \in G$ and obtain:

$$\int_{K/M} F(a_t(\bar{k})) \, d\bar{k} = \int_{\bar{N}} F(k(\text{Int}(a_t)\bar{n})M) \frac{d\bar{k}}{d\bar{n}} \, d\bar{n},$$

since $a_t \bar{n} MAN = a_t \bar{n} a_t^{-1} MAN$. Set $\bar{n} = \exp \sum_{\alpha < 0} X_\alpha$, for $X_\alpha \in \mathfrak{g}_\alpha$. Then

$$\text{Int}(\exp tH)\bar{n} = \exp\left(\text{Ad}(\exp tH) \sum_{\alpha < 0} X_\alpha\right) = \exp\left(e^{\text{ad } tH} \sum_{\alpha < 0} X_\alpha\right)$$

$$= \exp\left(\sum_{\alpha < 0} e^{t\alpha(H)}\right) X_\alpha \to 0, \quad \text{as } t \to \infty,$$

because $\alpha(H) < 0$ if $H \in \mathfrak{a}^+$. $\square$

It is now possible to discuss various types of special functions on the symmetric space $K \backslash G$ of a noncompact real semisimple Lie group G. We shall view G as acting on the right in order to remain close to the notation that we used in Section 4.2. The basic eigenfunction of the invariant differential operators on $K \backslash G$ is the *power function $p(Kg)$* defined as follows. Let $\lambda : \mathfrak{a} \to \mathbb{C}$ be a linear functional over $\mathbb{R}$. For $g \in G$, with Iwasawa decomposition $g = kan$, write $H(g) = \log a \in \mathfrak{a}$. Then define

$$p_\lambda(Kg) = \exp(\lambda(H(g))). \tag{1.31}$$

The power function is indeed an eigenfunction for all the G-invariant differential operators $D \in D(K \backslash G)$. The proof is the same as that for Proposition 1 of Section 4.2.1. We know that if $t = a_1 n_1$, for $a_1 \in A$, $n_1 \in N$, we have

$$p_\lambda((Kx)t) = p_\lambda(Kx) p_\lambda(Kt),$$

since $x = kan$ implies that $xa_1 = kana_1 = kaa_1(a_1^{-1}na_1)$ with $a_1^{-1}na_1 \in N$. Then, if $D \in D(K \backslash G)$, $t \in AN$, and $Ky = (Kx)t$,

$$Dp_\lambda(Ky) = (Dp_\lambda)(Kxt) = (Dp_\lambda(Kx))p_\lambda(Kt).$$

5.1. Geometry and Analysis on G/K

Set $x = e =$ the identity, to complete the proof that the power function is indeed an eigenfunction for all the invariant differential operators on $K\backslash G$.

Define a *spherical function* of $K\backslash G$ to be a function $f: K\backslash G \to \mathbb{C}$ such that $f(K) = 1$ which is a K-invariant eigenfunction for all the invariant differential operators in $D(K\backslash G)$.

Spherical functions can be built up out of power functions as in Theorem 3 of Section 4.2.3. The following theorem is proved in Helgason [2, Ch. 10]. In fact, the proof that we gave in Section 4.2.3 generalizes. It is also possible to extend the rest of Theorem 3 of Section 4.2.3 to $K\backslash G$.

Theorem 8 (Harish-Chandra). *A spherical function has the form*

$$h_\lambda(Kg) = \int_K p_\lambda(Kgk)\, dk,$$

where $\lambda = i\mu - \rho$, $\rho = \frac{1}{2}\sum_{\alpha>0} \alpha$. *Moreover spherical functions* $h_{i\mu-\rho}$ *are invariant under the Weyl group acting on the μ-variable. Here* $\mu \in \mathfrak{a}^* =$ *the dual vector space of* $\mathfrak{a}$.

Harish-Chandra [2] obtained the *asymptotics of the spherical function*:

$$h_{i\mu-\rho}(\exp H) \sim e^{-\rho(H)} \sum_{s\in W} c(s\mu) e^{is\mu(H)}, \quad \text{as } H \to \infty, \quad H \in \mathfrak{a}^+, \tag{1.32}$$

$$c(\mu) = \int_{\bar{N}} \exp\{(-i\mu - \rho)(H(\bar{n}))\}\, d\bar{n}.$$

Gindikin and Karpelevic [1] obtained the explicit formula for the c-function:

$$c(\mu) = I(i\mu)/I(\rho),$$

$$I(v) = \prod_{\alpha>0} B\left(\frac{1}{2}m_\alpha, \frac{1}{4}m_{\alpha/2} + \frac{(v,\alpha)}{(\alpha,\alpha)}\right), \quad v \in \mathfrak{a}^*, \tag{1.33}$$

where B is the beta function, $m_\alpha = \dim_{\mathbb{R}} \mathfrak{g}_\alpha$, $\rho = \frac{1}{2}\sum_{\alpha>0}\alpha$. Here (v,α) denotes the inner product on the dual space $\mathfrak{a}^*$ induced by the Killing form of $\mathfrak{g}$ restricted to $\mathfrak{a}$ (a form which is automatically positive definite).

The asymptotics and functional equations of h_λ are sufficient to study the *Helgason-Fourier transform* of $f: K\backslash G \to \mathbb{C}$ defined by:

$$\mathcal{H}f(\lambda, \bar{k}) = \int_{x\in K\backslash G} f(x)\overline{p_\lambda(xk)}\, dx. \tag{1.34}$$

Here $\lambda \in (\mathfrak{a}^*)^c$, the complexification of the dual vector space to $\mathfrak{a}$, $\bar{k} = kM \in K/M$, xg, for $x \in K\backslash G$ and $g \in G$, denotes the right action given by $(Kh)g = K(hg)$, and dx denotes the G-invariant volume on the symmetric space $K\backslash G$.

The *inversion formula* for this transform is due to Harish-Chandra and Helgason (see Helgason [1, 4, 5, 7, 10]).

310 V. The General Noncompact Symmetric Space

$$f(x) = \int_{\mu \in \mathfrak{a}^*} \int_{B=K/M} \mathscr{H}f(i\mu + \rho, \bar{k}) p_{i\mu+\rho}(xk) |c(\mu)|^{-2} d\mu \, d\bar{k}, \quad (1.35)$$

with a suitable normalization of the Euclidean measure on the real vector space $\mathfrak{a}^*$, which is the dual space to $\mathfrak{a}$. The proof of (1.35) is analogous to that of Theorem 1 in Section 4.3. Helgason [4, Ch. IV] gives a detailed account of the transform for K bi-invariant functions on G. Information on the history of the subject can be found in the same place.

Example 1 ($G = Sp(n, \mathbb{R})$). Recall our identifications of $K \backslash G$ and $\mathscr{H}_n$ with $\mathscr{P}_n^*$ via:

$$W = \begin{pmatrix} Y^{-1} & 0 \\ 0 & Y \end{pmatrix} \begin{bmatrix} I & -X \\ 0 & I \end{bmatrix}, \quad {}^t X = X \in \mathbb{R}^{n \times n}, \quad Y \in \mathscr{P}_n.$$

Thus, the power function is:

$$p_s(W) = \prod_{j=1}^n |Y_j|^{s_j}, \quad \text{for } W, Y \text{ as above, } s \in \mathbb{C}^n.$$

Thus the power function on $\mathscr{P}_n^*$ restricts to the power function on $\mathscr{P}_n$ which was defined in Equation (2.1) of Section 4.2.1.

The analogue of the *gamma function* for $\mathscr{P}_n^*$ is the Helgason transform of $\exp(-\text{Tr}(W))$:

$$\int_{\mathscr{P}_n^*} \exp(-\text{Tr}(W)) p_s(W) d\mu_n^*(W)$$

$$= \int_{Y \in \mathscr{P}_n} \int_{\substack{X \in \mathbb{R}^{n \times n} \\ X = {}^t X}} \exp\{-\text{Tr}(Y + Y^{-1} + Y^{-1}[X])\}$$

$$\times p_s(Y) |Y|^{-(n+1)/2} d\mu_n(Y) dX$$

$$= \pi^{n(n+1)/4} K_n(s \# | I, I), \quad s\# = s - (0, \ldots, 0, \tfrac{1}{2}),$$

where K_n denotes the K-Bessel function for $\mathscr{P}_n$ defined by formula (2.21) of Section 4.2.2. We saw the case $n = 1$ of this result in formula (3.108) in Section 3.7 of Volume I. The present formula should allow one to generalize (3.109) of Section 3.7, Vol. I, to $Sp(n, \mathbb{Z})$ using the spectral resolution of the G-invariant differential operators on $L^2(\mathscr{H}_n/Sp(n, \mathbb{Z}))$.

Example 2 (The Heat Equation on $K \backslash G$) (Gangolli [1, pp. 108–109]). We want to find $u(Kx, t)$ such that

$$\Delta u = u_t, \quad \text{where} \quad \Delta = \text{the Laplacian for } K \backslash G,$$

$$u(Kx, 0) = f(Kx), \quad \text{for some given } K\text{-invariant function on } K \backslash G.$$

Now, it can be shown that

5.1. Geometry and Analysis on G/K

$$\Delta p_{i\mu-\rho} = -\{(\mu,\mu) + (\rho,\rho)\}p_{i\mu-\rho}.$$

Thus the same sort of argument that worked in §4.3.4 shows that

$$g_t(Kx) = \int_{\mu \in \mathfrak{a}^*} \exp(-\{(\mu,\mu) + (\rho,\rho)\}t) h_{i\mu-\rho}(Kx)|c(\mu)|^{-2}\,d\mu$$

and

$$u(t, Kx) = g_t * f, \quad \text{where the convolution is over } G.$$

Gangolli [loc. cit.] shows that $g_t(Kx)$ has the standard properties of the fundamental solution of the heat equation, just as we saw for $\mathscr{P}_n$ in Exercise 8 of Section 4.3.4.

Helgason [1, pp. 67–68] solves the wave equation on a symmetric space using the Radon transform on $K \backslash G$. He also discusses Huyghen's principle for a symmetric space.

Exercise 27 (The Cayley Transform). Show that $W = (Z - iI)(Z + iI)^{-1}$ maps $Z \in \mathscr{H}_n$ into W in the generalized unit disc defined by

$$\mathscr{D}_n = \{W \in \mathbb{C}^{n \times n} | {}^t W = W, I - W\overline{W} \in \mathscr{P}_n\}.$$

This mapping allows us to view the symmetric space of the symplectic group as a bounded symmetric domain.

It is shown by Helgason that eigenfunctions of $D(G/K)$ can be expressed as a Poisson integral over the boundary of the symmetric space. See our discussion after Exercise 22 of Section 4.1.3.

An Example of a Symmetric Space of Type IV. The Quaternionic Upper Half 3-Space

References for this example include Belinfante and Kolman [1], Bougerol [2], Elstrodt, Grunewald, and Mennicke [1–3], Jauch [1], Kubota [3–5], Maass [2, Ch. 1], Mennicke [1–2], Sarnak [1], and Vignéras [2].

First we need a brief review of *quaternions*. The latter, denoted $\mathbb{H}$ for Hamilton, form a division ring or non-commutative field:

$$\mathbb{H} = \mathbb{R} \oplus \mathbb{R}i \oplus \mathbb{R}j \oplus \mathbb{R}k,$$

where $ij = k = -ji, jk = i = -kj, ki = j = -ik, i^2 = j^2 = k^2 = -1$. The *norm* of a quaternion $q = a + bi + cj + dk$, with a, b, c, d real is

$$\|q\| = \sqrt{a^2 + b^2 + c^2 + d^2} = \sqrt{qq^c},$$

with the *conjugate* $q^c = a - bi - cj - dk$. All goes very much as with the complex numbers except that things do not commute.

It is possible to represent quaternions by complex 2 × 2 matrices via:

$$1 \mapsto \begin{pmatrix} 1 & 0 \\ 0 & 1 \end{pmatrix}$$

$$i \mapsto \begin{pmatrix} 0 & -i \\ -i & 0 \end{pmatrix} = -i\sigma_1$$

$$j \mapsto \begin{pmatrix} 0 & 1 \\ -1 & 0 \end{pmatrix} = -i\sigma_2$$

$$k \mapsto \begin{pmatrix} i & 0 \\ 0 & -i \end{pmatrix} = -i\sigma_3.$$

These matrices are $-i$ times the Pauli matrices $\sigma_1, \sigma_2, \sigma_3$ from quantum mechanics.

One can view $SU(2)$ as the unit quaternions via such an identification. Thus $SU(2)$ is simply connected. Call the preceding map from quaternions to matrices f. Then we claim $SU(2) \cong \{f(q) \mid \|q\| = 1\}$. (The group $SU(2)$ can be mapped onto $SO(3, \mathbb{R})$ via a homomorphism of fundamental importance in the Dirac theory of electron spin. The map is given by taking Q in $SU(2)$ to $A = (a_{ij})$ in $\mathbb{R}^{3 \times 3}$ via $a_{ij} = \text{Tr}(Q\sigma_i {}^t\bar{Q}\sigma_j)/2$—but this is not relevant at the moment. The map is onto with kernel the center of $SU(2)$).

After this brief discussion of quaternions, we can give various descriptions of a symmetric space that has been of interest to number theorists and physicists. The space is

$$SL(2, \mathbb{C})/SU(2).$$

It fits into type IV of Cartan's classification of symmetric spaces. We can identify this space as the space of positive Hermitian matrices of determinant one:

$$\mathscr{SP}_2^c = \{Y \in \mathbb{C}^{2 \times 2} \mid Y = {}^t\bar{Y},\ Y \text{ positive},\ |Y| = 1\}.$$

The identification is:

$$SL(2, \mathbb{C})/SU(2) \to \mathscr{SP}_2^c,$$

$$gSU(2) \mapsto g\,{}^t\bar{g}.$$

Here, we define a positive Hermitian matrix Y to be a Hermitian matrix $Y \in \mathbb{C}^{n \times n}$ so that $Y\{x\} = {}^t\bar{x}Yx > 0$ for all $x \in \mathbb{C}^n - 0$. These matrices are quite analogous to ordinary positive matrices. We could rewrite Chapters 3 and 4 in the Hermitian case, if we had the time.

By generalizing the Iwasawa decomposition, one sees that the coset representatives $g \in SL(2, \mathbb{C})$ for $SL(2, \mathbb{C})/SU(2)$ can be chosen of the form:

$$g = \begin{pmatrix} \sqrt{t} & z/\sqrt{t} \\ 0 & 1/\sqrt{t} \end{pmatrix}, \quad z \in \mathbb{C}, \quad t > 0, \quad z = x + iy. \quad (1.36)$$

This allows us to identify $SL(2, \mathbb{C})/SU(2)$ with *the quaternionic upper half plane*:

5.1. Geometry and Analysis on G/K

$$\mathcal{H}^c = \{z + kt = x + iy + kt \mid x, y \in \mathbb{R}, t > 0\}; \tag{1.37}$$

thus the elements of $\mathcal{H}^c$ are quaternions with j-coordinate equal to zero. The mapping from $SL(2, \mathbb{C})/SU(2)$ to $\mathcal{H}^c$ sends $gSU(2)$ with g given by (1.36) to $z + kt$.

The action of a matrix

$$g = \begin{pmatrix} a & b \\ c & d \end{pmatrix} \text{ in } SL(2, \mathbb{C})$$

on an element q of the quaternionic upper half plane is:

$$g(q) = (aq + b)(cq + d)^{-1} = q^* \quad \text{with} \quad t^* = t \|cq + d\|^{-2}.$$

Recall that it is all right to divide by quaternions, but it is not all right to interchange the order of multiplication.

The action of $g \in SL(2, \mathbb{C})$ on $Y \in \mathcal{SP}_2^c$ is

$$Y \mapsto Y\{g\} = {}^t\bar{g} Y g.$$

Using this action we identify our symmetric space as $SU(2) \backslash SL(2, \mathbb{C})$; i.e., we consider left rather than right cosets.

Exercise 28. Check that the three group actions are preserved in our identifications of $SL(2, \mathbb{C})/SU(2)$ with $\mathcal{H}^c$ and $\mathcal{SP}_2^c$.

Exercise 29. Show that the invariant arc length, volume element, and Laplacian on $\mathcal{H}^c$ are given by:

$$ds^2 = (dx^2 + dy^2 + dt^2)t^{-2};$$
$$d\mu = t^{-3} dx\, dy\, dt;$$
$$\Delta = t^2(\partial^2/\partial x^2 + \partial^2/\partial y^2 + \partial^2/\partial t^2) - t\partial/\partial t.$$

Exercise 30.

(a) Show that a spherical function on $\mathcal{SP}_2^c$ has the form

$$h_\lambda(Y) = \frac{2 \sin(\lambda r/2)}{\lambda \sinh r}, \quad \text{if } Y = a_r[k],$$

$$a_r = \begin{pmatrix} \exp(r/2) & 0 \\ 0 & \exp(-r/2) \end{pmatrix}, \quad k \in SU(2) = K.$$

Here r is the geodesic radial coordinate in the polar coordinate decomposition of Y in $\mathcal{SP}_2^c$.

(b) Show that if f is in $L^1(\mathcal{SP}_2^c/K)$, then

$$\int_{\mathcal{SP}_2^c} f(Y)\, d\mu = \int_{\mathbb{R}} f(a_r) \sinh^2 r\, dr.$$

(c) Use part (b) to show that the Helgason-Fourier transform for K-invariant functions on $\mathscr{SP}_2^c/K$ has the form:

$$\hat{f}(\lambda) = \frac{2}{\lambda i} \int_{\mathbb{R}} \exp(i\lambda t/2) f(a_t) \sinh t \, dt.$$

(d) Use the inversion formula for the ordinary Fourier transform from Section 1.2 of Volume I to show that the spectral measure for Fourier inversion on $\mathscr{SP}_2^c$ is:

$$\frac{|\lambda|^2}{16} d\lambda, \quad \text{where } d\lambda = \text{Lebesgue measure on } \mathbb{R}.$$

(e) Find the fundamental solution for the heat equation on $\mathscr{SP}_2^c$.

Hint. See Bougerol [2], Karpelevich, Tutubalin, and Shur [1], or Burridge and Papanicolaou [1].

Exercise 30 shows that harmonic analysis on $SL(2, \mathbb{C})$ is simpler than that on $SL(2, \mathbb{R})$. This is an example of a general phenomenon (see Helgason [7, p. 31] for the generalization of part (a) of Exercise 30).

This completes our brief sketch of the theory of harmonic analysis on general symmetric spaces. There are many applications, other than those mentioned so far. For example, Resnikoff [4] considers the consequences of using the geometries of the spaces $(\mathbb{R}^+)^3$ or $(\mathbb{R}^+ \times SL(2, \mathbb{R})/SO(2))$ as models for color perception. An experiment is posed for distinguishing which geometry gives a more accurate model. Another reference for the general theory is Wawrzyńczyk [1].

5.2. Geometry and Analysis on $\Gamma \backslash G/K$

Say what you know, do what you must, and whatever will be, will be.

Sonya Kovalevsky's maxim from her paper [1] quoted in Kochina [1, p. 168]

In this section we will give a very brief sketch of the theory of automorphic forms for certain subgroups Γ of G *acting discontinuously* on the symmetric space $X = G/K$. This means that for each $x \in X$, the set of images of x under Γ has no limit point in X. We will concentrate on two specific discontinuous groups: $GL(n, O_K)$, where O_K is the ring of integers in an algebraic number field K,* and the *Siegel modular group* $Sp(n, \mathbb{Z})$. Here $GL(n, O_K)$ is the *modular*

---

* Hopefully the beleaguered reader will not be too confused by our use of K for the maximal compact subgroup of G as well as an algebraic number field.

5.2. Geometry and Analysis on $\Gamma \backslash G/K$

group over an algebraic number field which consists of $n \times n$ matrices γ such that both γ and γ^{-1} have entries in O_K. See Section 1.4 for the necessary definitions from algebraic number theory. And $Sp(n, \mathbb{Z})$ consists of all symplectic $2n \times 2n$ integral matrices.

There are many reasons to study such discontinuous groups. Of course knowledge of $GL(n, O_K)$ and related groups leads to greater understanding of the arithmetic of K itself. For example, there are applications to explicit class field theory, distribution of Gauss sums, values of Dedekind zeta functions and L-functions, asymptotics of units, elliptic curves. References for these subjects include: Hecke [1, pp. 21–114], Siegel [1], [4], Borel and Casselman [1], Borel and Mostow [1], Elstrodt, Grunewald, and Mennicke [1–3], Gelbart [2–3], Goldfeld, Hoffstein and Patterson [1], Heath-Brown and Patterson [1], Jacquet and Langlands [1], Kubota [2–5], Langlands [3], Mennicke [1–2], Saito [1], Sarnak [1], Séminaire Cartan [1], Shimura [2], Shintani [2], Tunnell [1–2], and Weil [4].

The Siegel modular group $Sp(n, \mathbb{Z})$ and kindred groups appear in many diverse areas of physics, often via the connections with abelian integrals and Riemann theta functions which arise in many theories from boson fields to solitons. See Cartier's article in Borel and Mostow [1, pp. 361–386], Cooke [1], Dubrovin, Matveev, and Novikov [1], Keen [1], Kochina [1], Kovalevsky [1], Lion and Vergne [1], Lonngren and Scott [1], McKean and Trubowitz [1], Monastyrsky and Perelomov [1], Mumford [1], Novikov [1], Perelomov [1], Shale [1], and Wallach [3].

Theta functions also play a major role in the analytic theory of quadratic forms. See Siegel [1, Vol. I, pp. 326–405, 410–443, 469–548] and Weil [2, Vol. 2, pp. 1–157].

Here we seek to outline the foundations of a building which would ultimately encompass the generalization to these new discontinuous groups Γ of Sections 3.3–3.7 of Volume I. Our achievements will be pitiful compared with what is required. In particular, we will not touch on extensions of Section 3.7 of Volume I; i.e., the analogues of the noneuclidean Poisson summation formula and the Selberg trace formula. Such results have already found various arithmetic and geometric applications, e.g., in computing dimensions of spaces of holomorphic automorphic forms. There are also results on units in number fields over imaginary quadratic fields and elliptic curves over imaginary quadratic fields. References for such work include: Christian [4], Eie [1], Efrat [2], Elstrodt, Grunewald, and Mennicke [1–3], Hashimoto's article in Hejhal, Sarnak, and Terras [1, pp. 253–276], Hashimoto [2], Langlands [1–7], Mennicke [1–2], Morita [1], Müller [1], Petra Ploch [1], Sarnak [1], Tanigawa [1], Marie-France Vignéras [3–5], Yamazaki [1], and Zograf [1].

General references for this section include: Baily [1], Hel Braun [1–2], Christian [1], Freitag [1], Gelfand, Graev, and Piatetski-Shapiro [1], Hecke [1], Hirzebruch and Van der Geer [1], Maass [2], Mennicke [1], Séminaire Cartan [1], Shimura [2], Siegel [1, 4, 5], and Weil [2–4].

Our first topic is fundamental domains $K \backslash G/\Gamma$, for our favorite examples.

Example 1 (The Picard Group). Let K be the number field $\mathbb{Q}(i)$ with ring of integers
$$O_K = \mathbb{Z}[i] = \{x + iy | x, y \in \mathbb{Z}\}.$$
Here $i = \sqrt{-1}$. The *Picard group* is defined to be
$$\Gamma = SL(2, O_K) = \left\{ \gamma = \begin{pmatrix} a & b \\ c & d \end{pmatrix} \middle| a, b, c, d \in O_K, \det \gamma = 1 \right\}.$$
And $SL(2, O_K)$ acts discontinuously on the quaternionic upper half plane $\mathcal{H}^c$ defined by formula (1.37) in Section 5.1. An equivalent version of this action from formula (1.36) in Section 5.1 gives the action of $\gamma \in SL(2, O_K)$ on a positive determinant one Hermitian matrix $Y \in \mathcal{SP}_2^c$ via:

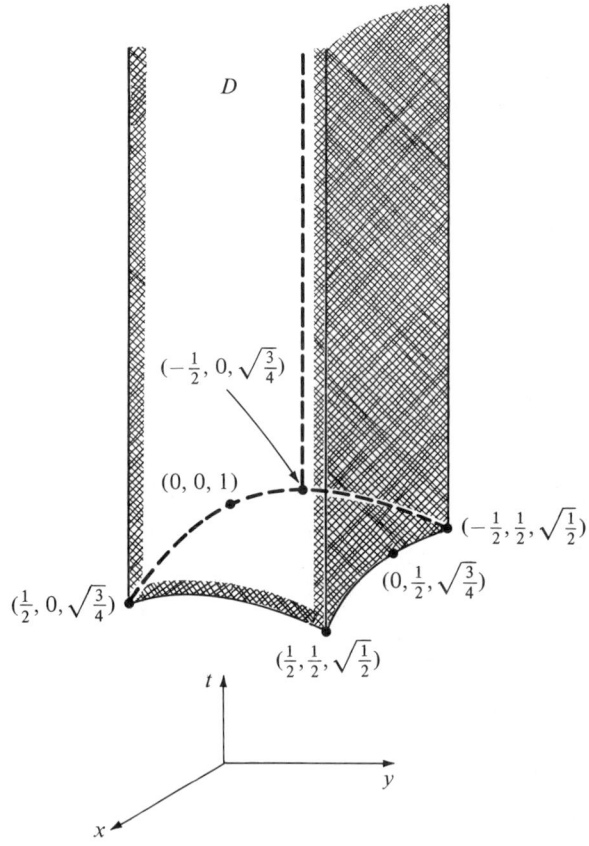

Figure 5.1. A fundamental domain for $SL(2, \mathbb{Z}[i])$ in the quaternionic upper half plane; $D = \{x + iy + kt | |x| \leq 1/2, 0 \leq y \leq 1/2, x^2 + y^2 + t^2 \geq 1, t > 0\}$.

5.2. Geometry and Analysis on $\Gamma \backslash G/K$

Figure 5.2. Tessellation of the quaternionic upper half plane from $SL(2, \mathbb{Z}[i])$ in stereo. It may help to put a division between the two halves of this figure and those that follow, in order to produce the 3D effect. (Drawn by the UCSD VAX computer and Mark Eggert.)

$$Y \mapsto Y\{\gamma\} = {}^t\bar{\gamma} Y \gamma.$$

A fundamental domain for this action was determined by Picard [1] and is pictured in Figure 5.1. A tessellation of hyperbolic 3-space $\mathcal{H}^c$ obtained by transforming this fundamental domain by elements of $\Gamma = SL(2, O_K)$ is shown in Figures 5.2–5.6. Figures 5.7 and 5.8 show Cayley transforms of this tessellation which are inside of the unit sphere in 3-space. The figures are shown in stereo. If you stare at the two versions of the picture, one for each eye, you should be able to see a 3D tessellation. All of the tessellations were drawn by Mark Eggert using one of UCSD's VAX computers.

As for $SL(2, \mathbb{Z})$ (see Exercise 1 of Section 3.3, Volume I), the sides of the fundamental domain are mapped to each other by generators of Γ, which are in this case:

$$\begin{pmatrix} 1 & 1 \\ 0 & 1 \end{pmatrix}, \quad \begin{pmatrix} 1 & i \\ 0 & 1 \end{pmatrix}, \quad \begin{pmatrix} i & 0 \\ 0 & -i \end{pmatrix}, \quad \begin{pmatrix} 0 & -1 \\ 1 & 0 \end{pmatrix}.$$

Given any imaginary quadratic number field $K = Q(\sqrt{D})$, of discriminant $D > 0$, one can consider $SL(2, O_K)$, O_K = the algebraic integers in K,

$$SL(2, O_K) = \left\{ \gamma = \begin{pmatrix} a & b \\ c & d \end{pmatrix} \middle| a, b, c, d \in O_K, \det \gamma = 1 \right\}.$$

This investigation was begun by Bianchi [1]. Humbert [2] showed that the

318 V. The General Noncompact Symmetric Space

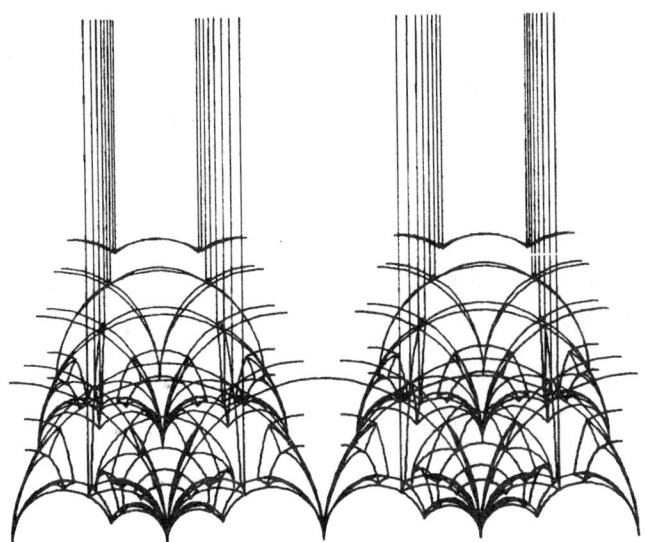

Figure 5.3. Tessellation of the quaternionic upper half plane from $SL(2, \mathbb{Z}[i])$ in stereo. (Drawn by the UCSD VAX computer and Mark Eggert.)

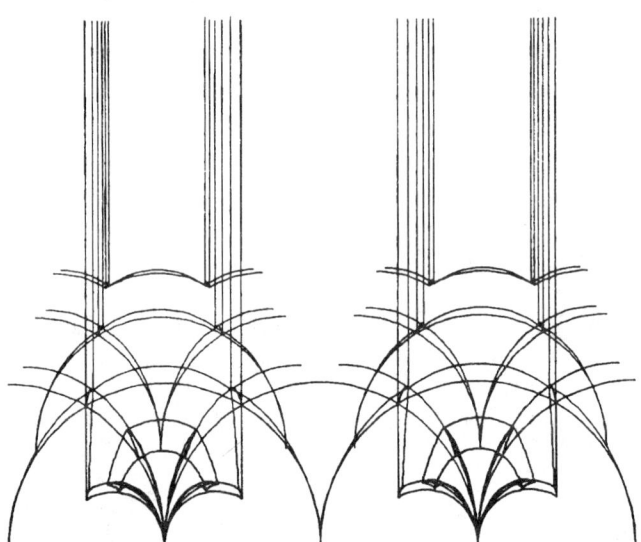

Figure 5.4. Tessellation of the quaternionic upper half plane from $SL(2, \mathbb{Z}[i])$ in stereo. (Drawn by the UCSD VAX computer and Mark Eggert.)

5.2. Geometry and Analysis on $\Gamma \backslash G/K$

Figure 5.5. Tessellation of the quaternionic upper half plane from $SL(2, \mathbb{Z}[i])$ in stereo. (Drawn by the UCSD VAX computer and Mark Eggert.)

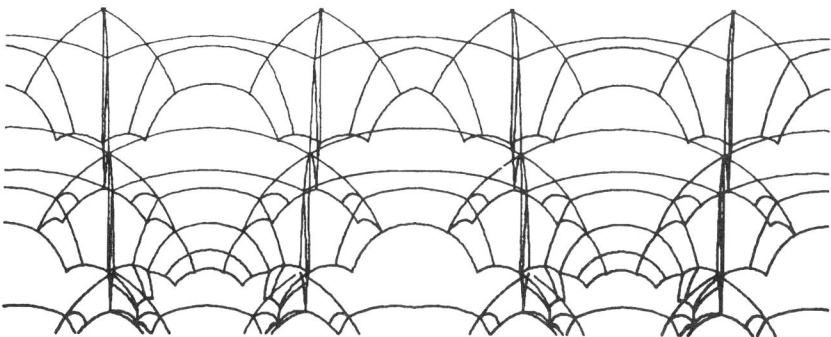

Figure 5.6. Tessellation of the quaternionic upper half plane from $SL(2, \mathbb{Z}[i])$ in stereo. (Drawn by the UCSD VAX computer and Mark Eggert.)

volume of the fundamental domain $SL(2, O_K) \backslash \mathcal{H}^c$ is

$$\frac{|D|^{3/2} \zeta_K(2)}{4\pi^2}, \qquad \zeta_K(s) = \text{the Dedekind zeta function of } K$$

(see Section 1.4 of Volume I). The geometry of the fundamental domain is thus closely associated with the arithmetic of the number field K. In particular, the number of cusps of the fundamental domain is equal to the class number of K, which was defined in Section 1.4 of Vol. I. We will demonstrate this fact in Proposition 1.

Siegel gives two methods to prove formulas for the volume of fundamental domains of this sort. See Siegel [1, Vol. I, pp. 464-465, Vol. II, pp. 330-331, and Vol. III, pp. 39-46, 328-333]. One of Siegel's methods is the one we used in Theorem 2 of Section 4.4.4 to find the volume of the fundamental domain for $GL(n, \mathbb{Z})$ via Siegel's integral formula. The other method involves finding

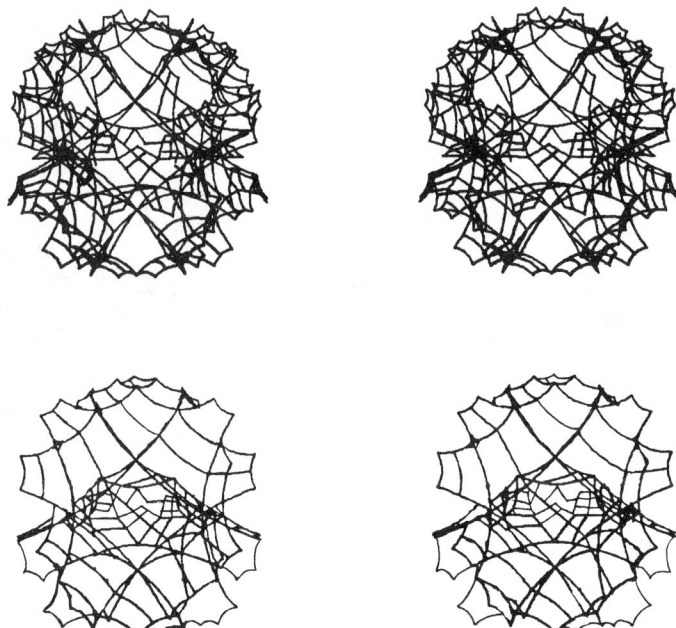

Figure 5.7. Stereo tessellation of the unit sphere obtained by mapping the preceding tessellations of the quaternionic upper half plane into the unit sphere by a Cayley transform. The mapping for this figure and Figure 5.8 is $(q - k)(-kq + 1)^{-1}$. (Drawn by the UCSD VAX computer and Mark Eggert.)

the residues of Eisenstein series like (4.3) in Section 4.4.1, using the method of theta functions.

References for fundamental domains in quaternionic 3-space include Ahlfors [1], Elstrodt, Grunewald, and Mennicke [1–3], Humbert [1–2], Kubota [2–5], Mennicke [1–2], Milnor [2], Sarnak [1], and Stark [1].

The next example to be considered is the analogue of Example 1 for real quadratic fields K.

Example 2 (The Hilbert Modular Group). Suppose that K is a real quadratic number field and $K = \mathbb{Q}(\sqrt{D})$ has positive discriminant D; e.g., $K = \mathbb{Q}(\sqrt{2})$. Such a field has, as mentioned in Section 1.4 of Volume I, two conjugations mapping K into the field of real numbers and denoted $x^{(1)}$, $x^{(2)}$, for $x \in K$. If $x = a + b\sqrt{d} \in K$, with $a, b \in \mathbb{Q}$, then $x^{(1)} = x$ and $x^{(2)} = a - b\sqrt{D}$. Form the group

$$\Gamma = SL(2, O_K) = \left\{ \gamma = \begin{pmatrix} a & b \\ c & d \end{pmatrix} \bigg| a, b, c, d \in O_K, \det \gamma = 1 \right\}.$$

5.2. Geometry and Analysis on $\Gamma \backslash G/K$

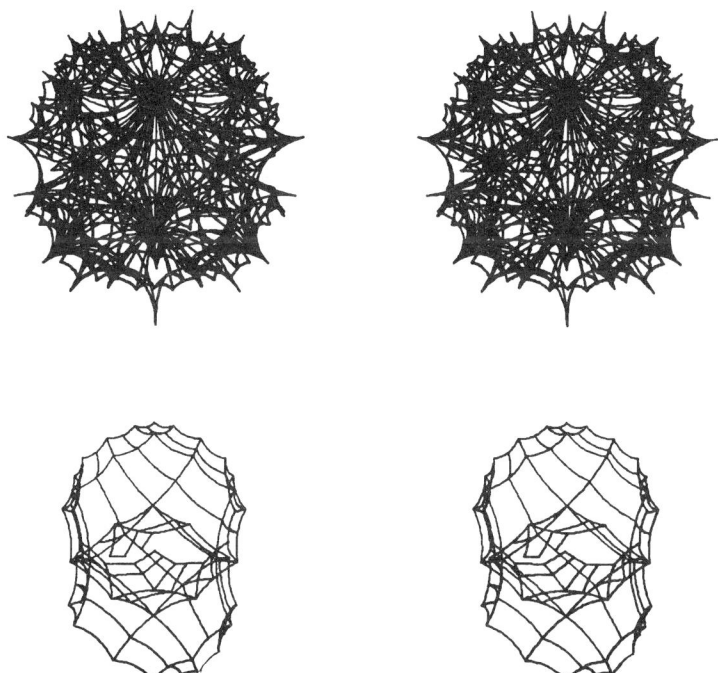

Figure 5.8. Stereo tessellation of the unit sphere obtained by mapping the preceding tessellations of the quaternionic upper half plane into the unit sphere by a Cayley transform. (Drawn by the UCSD VAX computer and Mark Eggert.)

This is called the Hilbert modular group for the field K. Then Γ acts discontinuously on the product $\mathcal{H}^2$ of two ordinary upper half planes via

$$\gamma(z^{(1)}, z^{(2)}) = (\gamma^{(1)} z^{(1)}, \gamma^{(2)} z^{(2)}), \qquad \text{for } z^{(j)} \in \mathcal{H},$$

and $\gamma^{(j)}$ denoting the matrix each of whose entries is obtained by taking the (j)-conjugate of the corresponding entry of $\gamma \in \Gamma$, $j = 1, 2$. Here $\gamma^{(j)}$ acts on $z^{(j)}$ by fractional linear transformation as in Chapter 3.

Exercise 1. Show that the action of $\Gamma = SL(2, O_K)$, for a real quadratic field K, on the ordinary upper half plane $\mathcal{H}$ via $z \mapsto \gamma z$, for $z \in \mathcal{H}$, is not discontinuous.

Hint. Make use of the units in K; that is $x \in O_K$ such that $x^{-1} \in O_K$.

It is an easy matter to generalize to $SL(2, O_K)$ where K is any totally real algebraic number field K; i.e., a number field K such that every conjugation is real-valued. Then $SL(2, O_K)$ acts discontinuously on $\mathcal{H}^m$, where m is the degree of K over $\mathbb{Q}$.

The fundamental domains $\Gamma\backslash\mathcal{H}^m$ have been rather intensively studied. They are $2m$-dimensional and complicated by the existence of units of infinite order in O_K. The formula for the volume of the fundamental domain is

$$2(-2\pi)^m \zeta_K(-1) = 2\pi^{-m} D_K^{3/2} \zeta_K(2),$$

where D_K is the absolute value of the discriminant of K and $\zeta_K(s)$ is Dedekind's zeta function of K. See Klingen [1] for a proof of a much more general result.

References for these results include: Blumenthal [1], H. Cohn [1], Giraud [1], Gundlach [1], Hammond [1], Hirzebruch [1], Hirzebruch and Van der Geer [1], Hirzebruch and Zagier [1–2], Humbert [1], Klingen [1–3], Maass [9, 10], Resnikoff [1], Shimizu [1], Shimura [2], Siegel [1, 4], Thomas and Vasquez [1], and Weisser [1]. See our earlier comments on volumes of fundamental domains for the Picard type groups.

Example 3 (The General Linear Group over Any Number Field). Suppose that K is any number field, O_K its ring of integers, and m its degree over $\mathbb{Q}$. Then, as in Section 1.4 of Volume I, K must be equal to $\mathbb{Q}(a)$ for some complex number a with minimal polynomial $f(x)$ and K is isomorphic to the quotient $\mathbb{Q}[x]/((f(x))$. So

$$K \otimes_\mathbb{Q} \mathbb{R} \cong \mathbb{R}[x]/(f(x)) \cong \sum_{j=1}^{r_1+r_2} {}^\oplus E_j, \quad (2.1)$$

$$E_j \cong \begin{cases} \mathbb{R}, & j = 1, \ldots, r_1; \\ \mathbb{C}, & j = r_1 + 1, \ldots, r_1 + r_2. \end{cases}$$

Therefore we have m conjugations sending K into E_j by mapping x to $x^{(j)}$, for $j = 1, \ldots, r_1$, and mapping x to $x^{(j)}$ or $\overline{x^{(j)}}$, for $j = r_1 + 1, \ldots, r_1 + r_2$.

What is a positive matrix over a number field? We are actually seeking the infinite prime part of an adelic symmetric space (see Cassels and Fröhlich [1], Gelbart [3], Gelfand, Graev, and Piatetski-Shapiro [1], and Weil [2–4]).

There is a long history of looking only at the infinite prime part of the symmetric space. Some references are Hecke [1, pp. 21–55], Humbert [1–2], Klingen [1–3], Ramanathan [1–2], Siegel [4], Weil [3], Weyl [1, Vol. IV, pp. 232–264]. Much of our discussion here was inspired by working with John Hunter who considered number-theoretic applications of analogues of Siegel's integral formula (Proposition 2 of Section 4.4.4) for $SL(n, O_K)$, K imaginary quadratic. I regret that John's death prevents publication of his thesis work (see Hunter [1]).

Definition. A *positive quadratic form Y over the number field K* is defined to be a vector

$$Y = (Y^{(1)}, \ldots, Y^{(r_1+r_2)}),$$

with $Y^{(j)} \in \mathscr{P}_n$, for $j = 1, \ldots, r_1$ and $Y^{(j)} \in \mathscr{P}_n^c$, $j = r_1 + 1, \ldots, r_1 + r_2$. Here $\mathscr{P}_n$ is the symmetric space of positive real $n \times n$ matrices studied in Chapter 4,

5.2. Geometry and Analysis on $\Gamma \backslash G/K$

while $\mathscr{P}_n^c$ is the symmetric space of positive $n \times n$ Hermitian complex matrices; i.e.,

$$\mathscr{P}_n^c = \{Y \in \mathbb{C}^{n \times n} | {}^t\bar{Y} = Y, Y \text{ positive}\} \cong U(n) \backslash GL(n, \mathbb{C}).$$

A complex Hermitian matrix Y is called *positive* if $Y\{x\} = {}^t\bar{x}Yx > 0$ for every column vector $x \in \mathbb{C}^n - 0$. Set $\mathscr{P}_n^K =$ the space of positive quadratic forms over K. Clearly this symmetric space will generalize the two preceding examples, if we restrict to the *determinant one subspace* $\mathscr{SP}_n^K = \{Y \in \mathscr{P}_n^K | |Y^{(j)}| = 1, j = 1, \ldots, r_1 + r_2\}$.

Set

$$\Gamma = GL(n, O_K) = \{\gamma \in O_K^{n \times n} | \gamma^{-1} \in O_K^{n \times n}\}.$$

Exercise 2. Show that $GL(n, O_K)$ consists of all matrices in $O_K^{n \times n}$ whose determinant is a unit in O_K.

The *action of the general linear group* $GL(n, O_K)$ on $Y \in \mathscr{P}_n^K$ is given by:

$$Y \mapsto Y\{A\}, \quad \text{with } (Y\{A\})^{(j)} = (Y^{(j)})\{A^{(j)}\}, j = 1, \ldots, r_1 + r_2.$$

Here $A^{(j)}$ denotes the matrix all of whose entries are the jth conjugate of the corresponding entries in A.

Fundamental domains for $\mathscr{P}_n^K/GL(n, O_K)$ were discussed by Humbert [1], who generalized many of the results that we presented in Section 4.4 for the case that K is the field of rational numbers. Siegel [1] obtains analogues of many of the results of Section 4.4 in various places. For example, Siegel [1, Vol. I, p. 475] gives an analogue of Lemma 2 of Section 4.4.1. And Siegel [loc. cit., pp. 464–465] obtains a formula for the volume of the fundamental domain, in a paper which was to be corrected later ([loc. cit., Vol. III, pp. 328–333]).

We choose not to rewrite all of Section 4.4 in this case. Instead we take up some aspects of the theory when $\Gamma = SL(2, O_K)$. In particular, we discuss a result of Maass [10] correcting an error of Blumenthal [1]. This error also appears in Hecke's first paper (see Hecke [1, pp. 21–55 and the notes at the end of the volume]). We want to show that the cusps of the fundamental domain $\mathscr{SP}_2^K/SL(2, O_K)$ are in one-to-one correspondence with the ideal classes of K. The cusps are the points of the fundamental domain which are equivalent to infinity under the action of $SL(2, K)$. Thus they are elements of $\hat{K} = K \cup \{\infty\}$.

Proposition 1 (The Cusp-Ideal Class Correspondence). *The cusps of the fundamental domain for* $\mathscr{SP}_2^K/SL(2, O_K)$, *$K$ any number field, are in one-to-one correspondence with the ideal class group I_K of K.*

PROOF. See Siegel [4, p. 242]. Let h denote the class number of K. Choose fixed integral ideals $\mathfrak{a}_1, \ldots, \mathfrak{a}_h$ representing the ideal class group I_K. We want to show that the elements of $\hat{K} = K \cup \{\infty\}$ are divided into h equivalence classes by the action of

$$\gamma = \begin{pmatrix} a & b \\ c & d \end{pmatrix} \in SL(2, O_K) \text{ on } x \in \hat{K} \quad \text{defined by} \quad \gamma(x) = \frac{ax+b}{cx+d}, \quad \gamma(\infty) = a/c.$$

Suppose that $x = p/s$ with $p, s \in O_K$. Here we write $\infty = 1/0$. Then define $f(x)$ to be the integral ideal (p, s) which is generated by p and s. Note that the ideal class of $f(x)$ is well-defined. For if $x = p_1/s_1 = p_2/s_2$, then $\mathfrak{a}_2 = k\mathfrak{a}_1$, for $k = p_2/p_1$, where $\mathfrak{a}_i = f(p_i/s_i)$.

So there is an induced map $\bar{f}: K/SL(2, O_K) \to I_K$, since if $\gamma \in SL(2, O_K)$ and

$$\gamma = \begin{pmatrix} a & b \\ c & d \end{pmatrix}, \quad \text{then} \quad f(\gamma(p/s)) = (ap + bs, cp + ds) \subset (p, s).$$

The reverse inclusion must hold as well because the determinant of γ is one.

The map $\bar{f}$ is onto since every ideal in O_K has at most two generators (see Pollard [1]).

In order to show that $\bar{f}$ is one-to-one, you will probably first think of the following argument. Suppose that $f(p_1/s_1) = kf(p_2/s_2)$ for some $k \in K$. If $k = \omega/\tau$, for $\omega, \tau \in O_K$, then we see that

$$\tau(ap_1 + bs_1) = \omega p_2, \quad \text{and} \quad \tau(cp_1 + ds_1) = \omega s_2,$$

for some

$$\gamma = \begin{pmatrix} a & b \\ c & d \end{pmatrix} \in GL(2, O_K).$$

It follows that

$$\frac{ap_1 + bs_1}{cp_1 + ds_1} = \frac{p_2}{s_2}.$$

This says that p_1/s_1 and p_2/s_2 are indeed equivalent modulo $GL(2, O_K)$. But unfortunately we need to know that they are equivalent modulo $SL(2, O_K)$. The difference between special and general linear groups over O_K can be rather large thanks to the presence of lots of units.

Since our boat seems to have stopped moving, we take a different tack. This time we follow Siegel's argument. Suppose that $\mathfrak{a}^{-1}$ denotes the inverse ideal to $\mathfrak{a}$. Then $\mathfrak{a}\mathfrak{a}^{-1} = O_K = (1)$ and thus

$$p_1 v_1 - s_1 u_1 = 1 \quad \text{and} \quad p_2 v_2 - s_2 u_2 = 1 \quad \text{for some } u_i, v_i \in \mathfrak{a}_i^{-1}.$$

Set

$$A_i = \begin{pmatrix} p_i & u_i \\ s_i & v_i \end{pmatrix}, i = 1, 2,$$

and note that although A_i only has entries in K, the product $A_2 A_1^{-1}$ is actually in $SL(2, O_K)$. For we have assumed that $\bar{f}(p_i/s_i)$ both equal $\mathfrak{a}_1$, say. Thus if you compute the first entry of $A_2 A_1^{-1}$, for example, you find it is $p_2 v_1 - u_2 s_1$, which is in the ideal $\mathfrak{a}_2 \mathfrak{a}_1^{-1} + \mathfrak{a}_1 \mathfrak{a}_2^{-1} = O_K$, since $\mathfrak{a}_1 = \mathfrak{a}_2$. It is important that we have chosen a fixed set of representatives of our ideal classes. But note here that $f(p_1/s_1) = kf(p_2/s_2)$, for $k \in K$, implies that we can assume $k = 1$ by replacing p_2/s_2 by $(kp_2)/(ks_2)$. To see that $A_2 A_1^{-1}$ does indeed take p_1/s_1 to p_2/s_2, note that A_i maps ∞ to p_i/s_i (acting by fractional linear transformation).

This completes the proof of Proposition 1. □

Lemma 1. *The stabilizer in $SL(2, O_K)$ of a cusp $x_i = p_i/s_i$ for the fundamental domain $\mathscr{SP}_2^K/SL(2, O_K)$ is defined by:*

$$\Gamma_{x_i} = \{\gamma \in SL(2, O_K) | \gamma x_i = x_i\}.$$

Suppose that $x_i = A_i \infty$ and that

$$A_i = \begin{pmatrix} p_i & u_i \\ s_i & v_i \end{pmatrix} \in SL(2, K), \quad \text{with } \mathfrak{a}_i = (p_i, s_i),\ u_i, v_i \in \mathfrak{a}_i^{-1},$$

for $i = 1, 2, \ldots, h$. Here the ideals $\mathfrak{a}_1, \ldots, \mathfrak{a}_h$ represent the ideal class group I_K. And $\mathfrak{a}^{-1}$ is the inverse ideal to $\mathfrak{a}$. Let U_K denote the group of units in O_K. Then the stabilizer of a cusp has the form:

$$\Gamma_{x_i} = \left\{ A_i \begin{pmatrix} w & z \\ 0 & w^{-1} \end{pmatrix} A_i^{-1} \,\middle|\, z \in \mathfrak{a}_i^{-2}, w \in U_K \right\}.$$

PROOF. See Exercise 3 below. The result is clearly true for the infinite cusp. □

Exercise 3. Prove the formula for the stabilizer of the infinite cusp in Lemma 1 above. Then deduce the result for an arbitrary cusp.

Hint. You can find the details in Siegel [4, p. 245].

If γ stabilizes x_i, then $A_i^{-1} \gamma A_i$ stabilizes infinity and therefore

$$A_i^{-1} \gamma A_i = \begin{pmatrix} w & z \\ 0 & w^{-1} \end{pmatrix}.$$

To see that z must lie in $\mathfrak{a}_i^{-2}$, just multiply out the matrices. And note that

$$(ap_i + bs_i)s_i = (cp_i + ds_i)p_i \quad \text{if} \quad \gamma = \begin{pmatrix} a & b \\ c & d \end{pmatrix}.$$

But then it follows that w must be a unit, since, after division by $\mathfrak{a}_i^2$, we see that:

$$\frac{(ap_i + bs_i)}{\mathfrak{a}_i} = \frac{(p_i)}{\mathfrak{a}_i}, \quad \frac{(cp_i + ds_i)}{\mathfrak{a}_i} = \frac{(s_i)}{\mathfrak{a}_i}.$$

You also have to multiply out the following matrices:

$$\begin{pmatrix} p_i & u_i \\ s_i & v_i \end{pmatrix} \begin{pmatrix} w & z \\ 0 & w^{-1} \end{pmatrix} = \begin{pmatrix} a & b \\ c & d \end{pmatrix} \begin{pmatrix} p_i & u_i \\ s_i & v_i \end{pmatrix},$$

to see that w is a unit.

Example 4 (The Siegel Modular Group). The Siegel modular group $Sp(n, \mathbb{Z})$ is the group of all symplectic matrices with integer entries. It acts discontinuously on Siegel's upper half plane $\mathscr{H}_n$ (or on the space of positive symplectic matrices $\mathscr{P}_n^*$) considered in Section 5.1.

Siegel [1, Vol. II, pp. 300–301] shows that a fundamental domain for $Sp(n, \mathbb{Z}) \backslash \mathscr{H}_n$ can be obtained by generalizing the method of perpendicular bisectors from Exercise 7 of Section 3.3 of Volume I. That is, taking $\Gamma = Sp(n, \mathbb{Z})$, we consider the domain:

$$\mathscr{D}_n = \{Z \in \mathscr{H}_n | d(Z, W) \le d(Z, \gamma W) \text{ for all } \gamma \in \Gamma\}.$$

Here $W \in \mathscr{H}_n$ is chosen to be a point not fixed by Γ.

As we quoted at the end of Section 4.4.3, Siegel [1, Vol. II, p. 309] said: "The application of the general method [stated above] ... would lead to a rather complicated shape of the frontier [boundary] of F." So Siegel goes on to consider another fundamental domain for $Sp(n, \mathbb{Z}) \backslash \mathscr{H}_n$. Before we can say what this new domain is, we need to make a definition. Call $C, D \in \mathbb{Z}^{n \times n}$ a *coprime symmetric pair* if $C^t D = D^t C$ and the matrices C and D relatively prime in the sense that: if for any matrix $G \in \mathbb{Q}^{n \times n}$, the matrices GC and GD are both integral, then G must be integral; i.e., in $\mathbb{Z}^{n \times n}$.

Exercise 4. Show that $C, D \in \mathbb{Z}^{n \times n}$ are a coprime symmetric pair if and only if $(C \ D)$ can be completed to a matrix

$$\begin{pmatrix} A & B \\ C & D \end{pmatrix} \text{ in } Sp(n, \mathbb{Z}).$$

Now we can obtain Siegel's fundamental domain $\mathscr{F}_n$ for $Sp(n, \mathbb{Z}) \backslash \mathscr{H}_n$, as in Siegel [1, Vol. II, p. 108], by finding an analogue of the highest point method from Exercise 1 of Section 3.3 of Vol. I. For this, we need to recall from Exercise 15 of Section 5.1 that if $W = (AZ + B)(CZ + D)^{-1}$, then the imaginary part of W is $Y\{(CZ + D)^{-1}\}$ if Y is the imaginary part of Z and $Y\{A\} = {}^t\bar{A} Y A$. We will take the *height* of $Z = X + iY \in \mathscr{H}_n$ to be the determinant $|Y|$. So we see that $|W| = |Y| \|CZ + D\|^{-2}$. This concept of height leads to the following construction for a fundamental domain by a highest point method.

A *fundamental domain* $\mathscr{F}_n$ for $Sp(n, \mathbb{Z}) \backslash \mathscr{H}_n$ can be taken to be the set of $Z \in \mathscr{H}_n$ such that the following 3 statements hold (if we ignore boundary identifications):

1. $\|CZ + D\| \ge 1$ for all coprime symmetric pairs C, D with $C \ne 0$;
2. $Y = \text{Im } Z \in \mathscr{M}_n = $ Minkowski's fundamental domain for $\mathscr{P}_n/GL(n, \mathbb{Z})$;
3. $X = \text{Re } Z$, $X = (x_{ij})$ with $|x_{ij}| \le 1/2$, $1 \le i, j \le n$.

5.2. Geometry and Analysis on $\Gamma\backslash G/K$

Again there is a certain relation between the fundamental domain and matrices in $Sp(n, \mathbb{Z})$:

1. $\begin{pmatrix} A & B \\ C & D \end{pmatrix} \in Sp(n, \mathbb{Z})$ with $C \neq 0$;

2. $\begin{pmatrix} U & 0 \\ 0 & {}^tU^{-1} \end{pmatrix}$, $U \in GL(n, \mathbb{Z})$;

3. $\begin{pmatrix} I & N \\ 0 & I \end{pmatrix}$, $N = {}^tN \in \mathbb{Z}^{n \times n}$.

In fact, Maass [2, §11] shows that $Sp(n, \mathbb{Z})$ can be generated by matrices of the form

$$\begin{pmatrix} I & N \\ 0 & I \end{pmatrix}, \quad N = {}^tN \in \mathbb{Z}^{n \times n} \quad \text{and} \quad J = \begin{pmatrix} 0 & -I \\ I & 0 \end{pmatrix}.$$

The fundamental domain $\mathscr{F}_n$ can be shown to be closed, connected, and bounded by finitely many algebraic hypersurfaces. Compactifications have been studied and their singularities resolved (see Ash, Mumford, Rapoport, and Tai [1], Chai [1], and Namikawa [1]). Gottschling [1] found the explicit list of inequalities defining the fundamental domain $\mathscr{F}_2$ for $Sp(2, \mathbb{Z})$. As far as I know, no one has written down the explicit inequalities for $\mathscr{F}_n$, when n is larger than 2. Other references for related facts are Christian [1], Freitag [1], Maass [2], Séminaire H. Cartan [1], and Siegel [1, 5].

The symplectic volume of the fundamental domain $Sp(n, \mathbb{Z}) \backslash \mathscr{H}_n$ was computed by Siegel [1, Vol. II, p. 279] to be:

$$2 \prod_{k=1}^{n} \Lambda(k), \quad \text{if } \Lambda(s) = \pi^{-s} \Gamma(s) \zeta(2s).$$

Setting $V_n = \text{Vol}(\mathscr{F}_n)$, it follows that

$$V_1 = \pi/3, \quad V_2 = \pi^3/270, \quad V_3 = \pi^6/127575, \quad V_4 = \pi^{10}/200930625.$$

Klingen [1] generalized this result to the *Hilbert-Siegel modular group* which is defined for any totally real algebraic number field K by:

$$Sp(n, O_K) = \{\gamma \in O_K^{n \times n} | {}^t\gamma J \gamma = J\}, \quad J = \begin{pmatrix} 0 & -I_n \\ I_n & 0 \end{pmatrix}.$$

It is possible to connect the volume of the fundamental domain for $Sp(n, \mathbb{Z})$ with the Euler characteristic of the fundamental domain via the Gauss-Bonnet theorem (see Siegel [1, Vol. II, p. 277, 331], Harder [1] and Klingen [1]). Harder's result is very general. However, the symmetric spaces involved do not include those like $\mathscr{SP}_n$ for $n > 2$ or the quaternionic upper half plane.

Mathematicians have studied much more general arithmetic groups and their fundamental domains, including that for $GL(n)$ over a simple algebra. Some references are: Hel Braun [2–3], Borel [1], L. Cohn [1], Feit [2], Krieg

[1], Ramanathan [1–2], Resnikoff and Tai [1], Siegel [1, Vol. III, pp. 143–153; Vol. II, pp. 390–405], Weyl [1, Vol. IV, pp. 232–264], and Weil [3].

Before we leave the subject of fundamental domains, let us give an example of a discrete subgroup of isometries of $\mathcal{H}^n$ such that $\mathcal{H}^n/\Gamma$ is compact. This example comes from notes of D. Sullivan. Take

$$G = \{g \in GL(n+1, \mathbb{R}) \mid {}^t g \varphi g = \varphi\},$$

where

$$\varphi = \begin{pmatrix} 1 & & & 0 \\ & \ddots & & \\ & & 1 & \\ 0 & & & -\sqrt{2} \end{pmatrix}.$$

Let $\Gamma = G \cap GL(n+1, O_L)$ for $L = \mathbb{Q}(\sqrt{2})$. Now Γ is a discrete group of isometries of G/K. We can identify G with $O(n, 1)$ and G/K with one sheet of the hyperboloid consisting of the set of points $x \in \mathbb{R}^{n+1}$ such that $\varphi(x) = -1$.

It can be show that the quotient $\Gamma \backslash G/K$ is compact. We sketch the proof. Note that when $L = \mathbb{Q}(\sqrt{2})$, $O_L = \mathbb{Z}[\sqrt{2}]$ which is not discrete in $\mathbb{R}$. Thus we must work harder than usual to show that Γ is actually a discrete subgroup of G. Note the conjugations map:

$$O_L \to \mathbb{R}^2,$$

$$m + n\sqrt{2} \mapsto (m + n\sqrt{2}, m - n\sqrt{2}) \qquad \text{for } m, n \in \mathbb{Z}.$$

This induces a mapping which sends Γ discretely into $GL(n+1, \mathbb{R})^2$. Now $\gamma \in \Gamma$ leaves φ invariant and thus the form γ' leaves φ' invariant where γ' denotes the matrix formed by conjugating all entries of γ; i.e., sending $m + n\sqrt{2}$ to $m - n\sqrt{2}$. Now

$$\varphi' = \begin{pmatrix} 1 & & & 0 \\ & \ddots & & \\ & & 1 & \\ 0 & & & \sqrt{2} \end{pmatrix},$$

and thus the matrices leaving φ' invariant form a compact group. It follows that Γ is a discrete subgroup of G. Why?

To see that $K \backslash G/\Gamma$ is compact, one needs to use some version of the Hermite-Mahler compactness theorem in Exercise 13, Section 4.4.2.

A similar example is $G = \{g \in GL(4, \mathbb{R}) \mid {}^t g \chi g = \chi\}$, where

$$\chi = \begin{pmatrix} 1 & 0 & 0 & 0 \\ 0 & 1 & 0 & 0 \\ 0 & 0 & 1 & 0 \\ 0 & 0 & 0 & -7 \end{pmatrix}.$$

Let $\Gamma = G \cap GL(4, \mathbb{Z})$. It is clear that Γ is discrete. To see that the fundamental domain $K\backslash G/\Gamma$ is compact, identify $K\backslash G/\Gamma$ with a subset S of $\mathscr{P}_4/GL(4, \mathbb{Z})$ by mapping Kg to $I[g]$, as usual. Now to see S has compact closure, we need only show that $Y \in S$ implies $|Y|$ bounded above and $m_Y = \min\{Y[a] | a \in \mathbb{Z}^4 - 0\}$ bounded below. The fact that m_Y is bounded from below comes from the fact that $\chi(x) = 0$ has no solution $x \in \mathbb{Z}^4 - 0$. For the neighborhood U of 0 defined by

$$U = \{x \in \mathbb{R}^4 | \chi[x] \le 1/2\}$$

contains no point of $g(\mathbb{Z}^4 - 0)$, for $g \in G$.

To see that $\chi[x] \ne 0$ for $x \in \mathbb{Z}^4 - 0$, one looks at

$$x_1^2 + x_2^2 + x_3^2 - 7t^2 \equiv 0 \text{ (modulo 8)}.$$

Since any integer can be written as a sum of 4 squares, the quadratic form

$$\varphi = x_1^2 + \cdots + x_n^2 - 7t^2$$

does vanish on $\mathbb{Z}^n - 0$ when n is larger than 3. That's why we took the form φ for general n.

There are other references for examples in which $K\backslash G/\Gamma$ is compact; e.g., Borel [1, p. 57]. Borel notes that if G is the orthogonal group of a quadratic form over $\mathbb{Q}$ in n variables and Γ is a group of units of a lattice in $\mathbb{Q}^n$, then $G_\mathbb{R}/\Gamma$ is compact iff the form F does not represent 0 over $\mathbb{Q}$. Here Borel uses another sort of compactness criterion. See also Borel [5], Mostow [2], and Mostow and Tamagawa [1]. The last reference proves that if G is a semisimple algebraic matrix group defined over the field $\mathbb{Q}$ and having no unipotent or parabolic elements other than 1, then $G/G_\mathbb{Z}$ is compact.

Having considered the analogue of Section 4.4 for the types of arithmetic groups in Examples 1-4 above, we next begin the study of automorphic forms for these arithmetic groups. The theory of holomorphic forms has received the most attention. This restricts us here to consideration of automorphic forms for the Hilbert modular group $SL(2, O_K)$, for a totally real field K, or the Siegel modular group $Sp(n, \mathbb{Z})$, or the Hilbert-Siegel modular group. It is also possible to discuss non-holomorphic automorphic forms on these symmetric spaces. Such forms satisfy some sort of differential equation involving the invariant differential operators on the symmetric space. Harish-Chandra made a very general definition of automorphic form on a Lie group (see Borel's article in Borel and Mostow [1, p. 199]). Let us begin with a brief sketch of the holomorphic theory which models itself on that from Sections 3.4 and 3.6 of Volume I.

Example 1 (Holomorphic Hilbert Modular Forms). Some references are Blumenthal [1], Gundlach [1], O. Herrmann [1], Hirzebruch [1, 3], Maass [9-10], Resnikoff [1], Shimura [2], Siegel [4], Marie-France Vignéras [3-5], Zagier and Van der Geer [1], and Zagier [1, 2].

Suppose that K is a totally real algebraic number field of degree m with ring of integers O_K. We say that a function $f: \mathcal{H}^m \to \mathbb{C}$ is an (entire) *Hilbert modular form* of weight k belonging to $\Gamma = SL(2, O_K)$ if f has the following two properties, *assuming that m is greater than one*:

(1) $f(z)$ is holomorphic on $\mathcal{H}^m$;
(2) $f(\gamma Z) = N(cz+d)^k f(z)$, for all

$$\gamma = \begin{pmatrix} a & b \\ c & d \end{pmatrix} \in SL(2, O_K), \qquad z \in \mathcal{H}^m.$$

When f satisfies (1) and (2) we say that f is in $\mathcal{M}(SL(2, O_K), k)$. The notation in (2) means that for $z = (z^{(1)}, \ldots, z^{(m)})$, we have

$$\gamma(z) = (\gamma^{(1)} z^{(1)}, \ldots, \gamma^{(m)} z^{(m)}),$$

where $\gamma^{(j)}$ denotes the matrix obtained from γ by conjugating each entry of γ by the jth conjugation of the number field K. Here $\gamma^{(j)}$ acts on $z^{(j)}$ by fractional linear transformation. And the norm Nz is defined by:

$$Nz = \prod_{j=1}^{m} z^{(j)}, \qquad \text{for } z \in \mathcal{H}^m.$$

Of course, one might expect that, as in the case that $m = [K:\mathbb{Q}] = 1$, we should also require a Hilbert modular form to satisfy certain growth conditions at the cusps of the fundamental domain. For the cusp at infinity, one would just require that $f(z)$ be holomorphic at infinity. It was proved by Götzky [1] that this growth condition is unnecessary when $K = \mathbb{Q}(\sqrt{5})$. Gundlach generalized Götzky's result to any totally real number field (see Siegel [4]).

Siegel [4] is a good reference for the basic facts about Hilbert modular forms. For example, Siegel [4, p. 215] shows that a Hilbert modular form of weight $k < 0$ is identically zero, while a Hilbert modular form of weight 0 must be a constant. The argument is analogous to that of Hecke for $SL(2, \mathbb{Z})$.

Exercise 5. Show that mk must be an even integer if $\mathcal{M}(SL(2, O_K), k)$ is nonzero.

Now to discuss Fourier expansions of Hilbert modular forms, we need to recall the concept of different $\mathfrak{d}_K$ of a number field K (see Section I.4 of Volume I). The inverse different is the dual lattice to the lattice O_K of the number field (cf. pages 73–74 of Volume I). Here the duality is with respect to the form $\text{Tr}(\alpha\beta)$, and $\mathfrak{d}^{-1} = \{\beta \in K \mid \text{Tr}(\alpha\beta) \in \mathbb{Z} \text{ for all } \alpha \in O_K\}$. Fourier expansions are sums running over this dual lattice. Suppose now that f is in $\mathcal{M}(SL(2, O_K), k)$. Then $f(z)$ is periodic under translations from elements of the lattice O_K, since matrices

$$\begin{pmatrix} 1 & a \\ 0 & 1 \end{pmatrix} \text{ are in } SL(2, O_K) \text{ when } a \in O_K.$$

5.2. Geometry and Analysis on $\Gamma \backslash G/K$

It follows that $f(z)$ has a *Fourier expansion at the infinite cusp* of the form:
$$f(z) = c(0) + \sum_{0 \ll b \in \mathfrak{d}_K^{-1}} c(b) \exp\{2\pi i \operatorname{Tr}(zb)\},$$
where $\operatorname{Tr} z = z^{(1)} + \cdots + z^{(m)}$ for $z \in \mathscr{H}^m$ and zb is the element of $\mathscr{H}^m$ with jth coordinate $z^{(j)} b^{(j)}$. The sum is over b in the inverse different and such that $0 \ll b$, which means that b is *totally positive*; i.e., all conjugates $b^{(j)}$ are positive for $j = 1, \ldots, m$.

Fourier expansions at other cusps x_j can be described using the matrices A_j such that $A_j \infty = x_j$. One must also make use of the formula for the stabilizer of a cusp in Lemma 1.

Exercise 6. Show how to obtain the Fourier expansion at infinity of a Hilbert modular form $f(z)$, making use of the Cauchy-Riemann equations to show that the coefficients have the form $c(b) \exp\{-2\pi \operatorname{Tr}(yb)\}$, if $z = x + iy \in \mathscr{H}^m$.

One example of a Hilbert modular form is the *Eisenstein series* corresponding to an integral ideal $\mathfrak{a}$ in K defined by:
$$E_k(\mathfrak{a}, z) = \sum_{\substack{c, d/U_K \\ (c,d) = \mathfrak{a}}} N(cz + d)^{-k}, \qquad k > 2.$$

The sum is over a complete system of representatives for pairs c, d which generate the ideal $\mathfrak{a}$ under the equivalence relation:
$$(c, d) \sim (uc, ud) \qquad \text{for a unit } u \in U_K \text{ with } Nu = +1.$$

In fact, the Eisenstein series E_k will vanish identically if k is odd and K has a unit of norm -1. It is possible to use an integral test to obtain the convergence of E_k for $k > 2$ (see Siegel [4]). Moreover, Siegel [loc. cit., p. 292] proves the vanishing of the lead coefficient or constant term of the Fourier expansion of $E_k(\mathfrak{a}, z)$ with respect to the cusp corresponding to an ideal $\mathfrak{b}$ if $\mathfrak{b}$ is not in the same ideal class as $\mathfrak{a}$. This is to be expected when we recall that Proposition 1 gave the cusp-ideal class correspondence. Thus one demonstrates the linear independence of the Eisenstein series corresponding to ideals $\mathfrak{a}_1, \ldots, \mathfrak{a}_h$ representing the ideal classes in the ideal class group I_K.

Define a *cusp form* to be a Hilbert modular form f such that $f(z)$ approaches zero as z approaches any cusp of the fundamental domain. Let $\mathscr{S}(SL(2, O_K), k)$ be the vector space of cusp forms of weight k for $SL(2, O_K)$. Then we have the direct sum decomposition:
$$\mathscr{M}(SL(2, O_K), k) = \sum_{i=1}^{h} {}^{\oplus} \mathbb{C} E_k(\mathfrak{a}_i, z) \oplus \mathscr{S}(SL(2, O_K), k)$$

(see Siegel [loc. cit., p. 294]).

It is also possible to define *Poincaré series*. Look at the cusp p/s corresponding to the ideal $\mathfrak{a}$ via Proposition 1 and let $A \in SL(2, K)$ have the property that $A(\infty) = p/s$. Define the Poincaré series by:

$$f_k(\mathfrak{a}, \lambda, z) = \sum_{(c,d)=\mathfrak{a}} N(cz+d)^{-k} \exp\{2\pi i \operatorname{Tr}(\lambda \gamma z)\}.$$

Here λ is a totally positive element of the ideal $\mathfrak{a}^2 \mathfrak{d}_K^{-1}$, where $\mathfrak{d}_K$ is the different of K. The sum is over pairs of generators c, d of the ideal $\mathfrak{a}$ such that

$$\gamma = \begin{pmatrix} a & b \\ c & d \end{pmatrix} = A^{-1}\tau, \quad \text{for some } \tau \in SL(2, O_K), \quad \text{with } A(\infty) = p/q.$$

See Siegel [4, p. 230]). It can be shown that Poincaré series are cusp forms of weight k. Maass has proved that the Poincaré series and Eisenstein series generate the space $\mathcal{M}(SL(2, O_K), k)$, for $k \geq 2$. The Poincaré series can vanish identically, but not for large enough weights. See Siegel [4] for more information on the subject.

When $m = [K : \mathbb{Q}]$ is larger than one, a function $f(z)$ which is meromorphic on $\mathcal{H}^m$ is called a *Hilbert modular function* if $f(\gamma z) = f(z)$ for all $\gamma \in SL(2, O_K)$ and $z \in \mathcal{H}^m$. In the case that $m = 1$ and $K = \mathbb{Q}$, we would add a further requirement that $f(z)$ have at most a pole at the infinite cusp. This need not be assumed when $m \geq 2$. A Hilbert modular function which is holomorphic in $\mathcal{H}^m$ is automatically holomorphic at the cusps and thus must be a modular form of weight zero and therefore a constant. For there are no isolated singularities in several complex variables. Thus, when $m \geq 2$ there does not exist an analogue of the elliptic modular invariant $J(z)$ from (3.58) of Section 3.4 of Volume I. This fact was first noted by Götzky [1]. There are errors in Hecke's early papers due to the lack of knowledge of Götzky's result. These early Hecke papers seek to solve Hilbert's 12th problem which asks for an explicit construction of class fields (extension fields having abelian Galois groups) over arbitrary algebraic number fields using automorphic forms (see Hecke [1, p. 942]). Siegel [4] gives a proof that the Hilbert modular functions form an algebraic function field of n variables.

O. Herrmann [1] investigated the theory of Hecke operators for Hilbert modular forms. This theory has been extended to Picard groups by Styer [1]. Shimura [2] describes a very general theory of Hecke operators. Adelic versions of Hecke theory also exist for quite general groups (see Jacquet and Langlands [1], Gelbart [3], and Weil [4]).

The correspondence between Hilbert modular forms and Dirichlet series has been much studied (see the preceding references and Stark [1]). The situation is much like that investigated by Weil [2, Vol. III, pp. 165–172] for congruence subgroups of $SL(2, \mathbb{Z})$. One must have functional equations for L-functions that have been "twisted" by Hecke grössencharacters for K, in order to know that a corresponding function $f(z), z \in \mathcal{H}^m$ is a Hilbert modular form for $SL(2, O_K)$.

Let us just consider the simplest example of Hecke theory over number fields. Define the *theta function* corresponding to an ideal $\mathfrak{a}$ of a totally real number field K by:

$$\theta(\mathfrak{a}, z) = \sum_{a \in \mathfrak{a}} \exp\{-\pi \operatorname{Tr}(za^2)\}.$$

5.2. Geometry and Analysis on $\Gamma \backslash G/K$

By slightly altering the proof of Theorem 2 in Section 1.4 of Volume I, we can view the ideal class zeta functions (which occur as partial sums of the Dedekind zeta function) of a totally real number field as Mellin transforms of this theta function (see Hecke [1, p. 227]). Similarly Hecke obtained the analytic continuation of his L-functions by showing them to be Mellin transforms of theta functions. Generalizations of this theta function are considered by Eichler [1], Kloosterman [1], and Schoeneberg [1]. These authors also look at the effect of Hecke operators on such theta functions.

If Γ is a subgroup of $SL(2, O_K)$ without elliptic fixed points, either the Selberg trace formula or the Hirzebruch-Riemann-Roch Theorem can be used to compute the dimension of the space of Hilbert modular forms (see Ash, Mumford, Rapoport, and Tai [1], Hirzebruch [1-3], Langlands [6], and Shimizu [1]). If, for example, K is real quadratic, Γ of index a in $SL(2, O_K)/\pm I$, and Γ acts freely on $\mathcal{H}^2$, then, for $k \geq 3$, we have:

$$\dim \mathcal{S}(\Gamma, k) = \frac{k(k-2)}{2} \zeta_K(-1)a + \chi, \qquad \chi = 1 + \dim \mathcal{S}(\Gamma, 2).$$

Example 2 (Holomorphic Siegel Modular Forms). Some references for this section are Andrianov [1-5], Baily [1], Böcherer [2], Hel Braun [1], Christian [1-2], Eichler [1-3], Feit [1-2], Freitag [1], Garrett [2], Hoobler and Resnikoff [1], Igusa [1-2], Kaori Imai [1], Kalinin [1], Karel [1], Klingen [2, 4, 5], Maass [2, 4, 5], Morita [1], Resnikoff [2-3], Shimura [3], Siegel [1, 5], Tsao [1], Weissauer [1-2], and Yamazaki [1].

We will say that a function f on Siegel's upper half plane $\mathcal{H}_n$, $n > 1$, is a holomorphic *Siegel modular form* of weight k if it satisfies the following two conditions:

(1) f is holomorphic on $\mathcal{H}_n$;
(2) $f(\gamma Z) = |CZ + D|^k f(Z)$ for all

$$\gamma = \begin{pmatrix} A & B \\ C & D \end{pmatrix} \text{ in } Sp(n, \mathbb{Z}), \qquad Z \in \mathcal{H}_n.$$

One might expect to add a third condition that f must be bounded in the region $\text{Im } Z = Y \geq Y_0 > 0$, where $Y \geq Y_0$ means that $Y - Y_0$ lies in the closure of $\mathcal{P}_n$. Koecher shows that this third condition is unnecessary when n is bigger than one (see Proposition 2 and Maass [2, Section 13]).

We will write $f \in \mathcal{M}(Sp(n, \mathbb{Z}), k)$ when f satisfies (1) and (2). Thus we have another analogue of the space of ordinary modular forms which was studied in Section 3.4 of Volume I. It can be shown that when k is larger than one, $\mathcal{M}(Sp(n, \mathbb{Z}), k) \neq 0$ implies $k \in \mathbb{Z}$ and $nk \in 2\mathbb{Z}$.

Exercise 7. Prove the last statement.

It can be proved that Siegel modular forms of negative weight must vanish while those of weight zero must be constant.

Next we want to consider Fourier expansions of Siegel modular forms. First note that if X lies in the lattice of integral $n \times n$ symmetric matrices, then the dual lattice with respect to the form $\mathrm{Tr}(TX)$ consists of the $n \times n$ semi-integral symmetric matrices $T = (t_{ij})$. Here "semi-integral" means that $t_{jj} \in \mathbb{Z}$ and $t_{ij} \in \frac{1}{2}\mathbb{Z}$, when $i \neq j$.

If f is a Siegel modular form of weight k, then $f(X + iY)$ has period one in each entry of the symmetric matrix X. This implies that f has a *Fourier expansion*:

$$f(Z) = \sum_{\substack{0 \leq T = {}^tT \\ T \text{ semi-integral}}} a(T) \exp\{2\pi i \, \mathrm{Tr}(TZ)\}. \tag{2.2}$$

Here $T \geq 0$ means that $T[x] \geq 0$ for all $x \in \mathbb{R}^n$.

To see that the Fourier coefficient has the form $a(T)\exp\{-2\pi \mathrm{Tr}(TY)\}$, one must use the fact $f(Z)$ satisfies the Cauchy-Riemann equations in each variable. The sum is over semi-integral matrices because that is the dual lattice to $\mathbb{Z}^{n \times n}$ with respect to the form $\mathrm{Tr}(TX)$.

Exercise 8. Prove what we just said about (2.2).

The sum is over non-negative matrices T by Lemma 2 below.

Lemma 2 *Let $f \in \mathcal{M}(Sp(n, \mathbb{Z}), k)$ have Fourier coefficients $a(T)$ as in (2.2) above. Then $f(Z)$ is bounded in every domain $Y \geq Y_0 > 0$ (for fixed Y_0) if and only if the Fourier coefficient $a(T) \neq 0$ implies that $T \geq 0$.*

PROOF. See Maass [2, pp. 183–184].
$\Rightarrow$ Suppose that $|f(Z)| \leq C(Y_0)$ in $Y \geq Y_0 > 0$. Then

$$a(T)\exp\{-2\pi \mathrm{Tr}(TY)\} = \int_{{}^tX = X \in [0,1]^{n \times n}} f(X + iY)\exp\{-2\pi i \mathrm{Tr}(TX)\} dX.$$

In order to show that $a(T) \neq 0$ implies that $T \geq 0$, use the above equality to obtain the bound:

$$|a(T)| \leq C(Y_0)\exp\{2\pi \mathrm{Tr}(TY)\}.$$

If T is not ≥ 0, then one can show that the right hand side of this inequality can be made arbitrarily small. Thus $a(T)$ must vanish when T is not ≥ 0.
$\Leftarrow$ Suppose the Fourier expansion of $f(Z)$ has the form

$$f(Z) = \sum_{\substack{{}^tT = T \geq 0 \\ \text{semi-integral}}} a(T)\exp\{2\pi i \mathrm{Tr}(TZ)\}.$$

The convergence of the Fourier series implies that if $Y \geq Y_0 \geq aI > 0$

$$|f(Z)| \leq C\left(\frac{1}{2}Y_0\right) \sum_{T \geq 0} \exp\{-\pi a \mathrm{Tr}(T)\}.$$

The series is easily seen to converge. $\square$

5.2. Geometry and Analysis on $\Gamma \backslash G / K$

Exercise 9. Fill in the details in the proof of Lemma 2 above.

Proposition 2 (Koecher). *If $f \in \mathcal{M}(Sp(n, \mathbb{Z}), k)$, then f is bounded in any domain $Y \geq Y_0 > 0$. In particular, f is bounded in the fundamental domain $\mathscr{F}_n$ for $Sp(n, \mathbb{Z})$ which was described earlier.*

PROOF. See Maass [2, pp. 185–187]. For $n = 1$, the result is part of the definition of modular form. When n is larger than one, consider the Fourier expansion (2.2) of f. Note that $a(T[U]) = a(T)$ for all $U \in GL(n, \mathbb{Z})$. The main step in the proof of Proposition 2 is the proof of the claim that when T is not ≥ 0, then the number

$$c(T, -m) = \#\{U \in SL(n, \mathbb{Z}) | T[U] \text{ distinct}, \text{Tr}(T[U]) = -m\}$$

is greater than or equal to one for infinitely many $m \geq 1$. Therefore $a(T)$ must vanish in this case.

To prove this claim, observe that, upon setting $U = I + b\,{}^t d$, with $b, d \in \mathbb{Z}^n$, such that ${}^t bd = 0$, we have

$$|U| = 1 \quad \text{and} \quad \text{Tr}(T[U]) = \text{Tr}\, T + 2\,\text{Tr}(Tb\,{}^t d) + T[b]({}^t dd).$$

If T were not ≥ 0, then we could choose b in $\mathbb{Z}^n$ so that $T[b] < 0$ and ${}^t bd = 0$, for any d. And then $c(T, -m)$ must indeed be ≥ 1 for infinitely many m. The proof of Proposition 2 is completed using Lemma 2. □

An example of a Siegel modular form of even weight k in $\mathcal{M}(Sp(n, \mathbb{Z}), k)$ is given by the *Eisenstein series*:

$$E_k(Z) = \sum_{C,D} |CZ + D|^{-k}, \quad \text{for } k > n + 1,$$

where the sum is over coprime symmetric pairs C, D of matrices in $\mathbb{Z}^{n \times n}$ modulo the equivalence relation

$$(C \quad D) \sim (UC \quad UD), \quad \text{for } U \in GL(n, \mathbb{Z}).$$

Hel Braun [1] proved the convergence of the Eisenstein series in the stated region. See also Freitag [1, pp. 67–77] for a convergence proof using a sort of integral test. Freitag [1, p. 67] and Maass [2, Section 14] consider more general Eisenstein series involving modular forms for $Sp(r, \mathbb{Z})$, $r < n$, which were introduced by Klingen [4].

Siegel [1, Vol. II, p. 133] gives the Fourier expansion of the Eisenstein series E_k (k even and $> n + 1$) for $Sp(n, \mathbb{Z})$ (see also Baily [1]). Let

$$T[U] = \begin{pmatrix} T_1 & 0 \\ 0 & 0 \end{pmatrix} \quad \text{for } U \in GL(n, \mathbb{Z}), T_1 \in \mathbb{Z}^{r \times r},$$

and set $D(T) = |T_1|$. The term corresponding to T in the Fourier expansion of $E_k(Z)$ is:

$$(-1)^{rk/2} 2^{r(k-(r-1)/2)} \prod_{j=0}^{r-1} \frac{\pi^{k-j/2}}{\Gamma(k - j/2)} D(T)^{k-(r+1)/2} \sum_{{}^t R = R \in (\mathbb{Q}/\mathbb{Z})^{r \times r}} e^{2\pi i \, \text{Tr}(T_1 R)} v(R)^{-k}.$$

Here $v(R)$ is the product of the reduced denominators of the elementary divisors of R. An analogous result for Eisenstein series for $GL(n,\mathbb{Z})$ was discussed in Exercise 35 of Section 4.5.3 (see also Terras [3]). Siegel used his main theorem on quadratic forms to deduce the rationality of the coefficients of the Eisenstein series E_k for $Sp(n,\mathbb{Z})$. Baily [1, Ch. 12] gives another derivation. Similar and more general results of this type are obtained by Karel [1] and Tsao [1]. Kaufhold [1] finds even more explicit results for Fourier coefficients of Eisenstein series for $Sp(2,\mathbb{Z})$. When the Eisenstein series have rational coefficients and the Eisenstein series generate the full field of automorphic functions (i.e., meromorphic functions satisfying condition (2) in the definition of Siegel modular form and having weight $k = 0$) for $Sp(n,\mathbb{Z})$, then the algebraic variety which is the Satake compactification of the fundamental domain $Sp(n,\mathbb{Z})\backslash \mathscr{H}_n$ is a variety defined over $\mathbb{Q}$ (see Baily [1, p. 238]). More information on Eisenstein series can be found in the references mentioned at the beginning of this discussion of holomorphic Siegel modular forms. Some of the references consider non-holomorphic Eisenstein series as well.

An example of a modular form for a congruence subgroup of $Sp(n,\mathbb{Z})$ is the *theta function* defined for $Q \in \mathscr{P}_k \cap \mathbb{Z}^{n \times n}$, $Z \in \mathscr{H}_n$ by:

$$\theta(Z) = \sum_{A \in \mathbb{Z}^{k \times n}} \exp\{\pi i \operatorname{Tr}(ZQ[A])\}.$$

The theta function in formula (4.1) of Section 4.4.1 is simply a restriction of this symplectic theta function to $Z = iY$, $Y \in \mathscr{P}_n$. Eichler [4] shows that in the special case $n = 1$ the theta function is a modular form for a congruence subgroup of $SL(2,\mathbb{Z})$ by considering it as a special value of a theta function on $\mathscr{H}_k$. Andrianov and Maloletkin [1] generalize this result to any n— evaluating the 8th root of unity involved when k is even. Stark [4] extends these results further by evaluating the 8th root of unity in a case that can be reached by theorems on matrix primes in progressions (see also Styer [1]). Another reference for theta functions is Freitag [1].

Theta functions and zeta functions for indefinite quadratic forms have been studied by many authors (see Koecher [2], Maass [11–13], Siegel [1, Vol. I, pp. 410–443; Vol. II, pp. 41–96, 421–466; Vol. III, pp. 85–91, 105–142, 154–177], Andrianov and Maloletkin [2], and Friedberg [1]).

It is possible to obtain examples of modular forms of various types by integrating theta functions of indefinite quadratic forms. For example, one can obtain holomorphic Hilbert modular forms for a real quadratic field by integrating ordinary holomorphic modular forms for $SL(2,\mathbb{Z})$ multiplied by an appropriate theta function. One can similarly lift Maass wave forms for $SL(2,\mathbb{Z})$. And one can replace real by imaginary quadratic fields, etc. References for such constructions, often referred to as "base change", include Friedberg [1], Kudla [1], Stark [5], Tsuyumine [1], Vignéras [1], and Waldspurger [1].

Theta functions have many other applications. There are, in fact, entire books devoted to them (e.g., Igusa [1]). Siegel's main theorem on quadratic forms can be viewed as an equality between linear combinations of theta functions and generalized Eisenstein series (see Siegel [1, Vol. I, pp. 326–405, 410–443, 469–548]). Theta functions on the Siegel upper half plane can be used to obtain an expression for the generalization of elliptic integrals known as Abelian integrals (see Siegel [5]). This leads to many applications in physics. We have already mentioned the work of Sonya Kovalevsky [1] and Dubrovin, Matveev, and Novikov [1]. Other references for applications to the Korteweg-deVries equation are McKean and Trubowitz [1] and Novikov [1]. Siegel's work on quadratic forms has been connected with quantum mechanics and representation theory via the Segal-Shale-Weil representation (see Cartier's talk in Borel and Mostow [1, pp. 361–368], Lion and Vergne [1], Shale [1], Weil [2, Vol. III, pp. 1–157], and Wallach [3]). The role of theta functions in algebraic geometry and purely algebraic constructions of these functions are discussed in Mumford [1] and Shafarevitch [1]. Connections with Jordan algebras are pursued by Resnikoff [3].

In order to discuss cusp forms, Siegel defined the Φ-*operator* taking $f \in \mathcal{M}(Sp(n, \mathbb{Z}), k)$ to $f|\Phi$ in $\mathcal{M}(Sp(n-1, \mathbb{Z}), k)$ by:

$$f|\Phi(W) = \lim_{t \to \infty} f\begin{pmatrix} W & 0 \\ 0 & it \end{pmatrix}, \quad \text{for } W \in \mathcal{H}_{n-1}.$$

A Siegel modular form $f \in \mathcal{M}(Sp(n, \mathbb{Z}), k)$ is said to be a *cusp form* if $f|\Phi$ vanishes identically. Let $\mathcal{S}(Sp(n, \mathbb{Z}), k)$ denote the space of Siegel cusp forms of weight k. The Fourier expansion (2.2) of $f \in \mathcal{S}(Sp(n, \mathbb{Z}), k)$ can have no terms corresponding to singular symmetric semi-integral matrices T. Here a singular T is one with determinant equal to zero.

Exercise 10. Prove the last statement about cusp forms.

There is an opposite concept to that of cusp form—the concept of "singular form" which is a form in $\mathcal{M}(Sp(n, \mathbb{Z}), k)$ whose Fourier coefficients $a(T)$ in (2.2) vanish unless the T are singular matrices. Certain theta functions give examples. The structure of spaces of singular forms has been studied by Resnikoff [2].

It can be proved that there is a positive constant c_n depending only on n such that

$$\dim \mathcal{M}(Sp(n, \mathbb{Z}), k) \leq c_n k^{n(n+1)/2}.$$

See Eicher [3, 5], Freitag [1, p. 52], Maass [2, p. 194], and Siegel [1, Vol. II, pp. 97–137]. The main principle needed to prove such an inequality is that which gave us Theorem 4 of Section 3.5 in Volume I. This principle says: The vanishing of sufficiently many terms in the Fourier series of $f \in \mathcal{M}(Sp(n, \mathbb{Z}), k)$ implies the vanishing of f itself. Freitag [1, p. 50] shows that $\mathcal{S}(Sp(n, \mathbb{Z}), k)$ vanishes in the following situations:

$$n = 1, \quad k < 12;$$
$$n = 2, \quad k < 9;$$
$$n = 3, \quad k < 8;$$
$$n = 4, \quad k < 5.$$

It is possible to use the Selberg trace formula or the Hirzebruch-Riemann-Roch theorem to give formulas for dimensions of spaces of Siegel cusp forms for congruence subgroups of $Sp(n, \mathbb{Z})$ acting without elliptic fixed points. See Arakawa [2], Christian [4], Hirzebruch [2, Appendix], Langlands [6], Morita [1], Petra Ploch [1], and Yamazaki [1]. See also Eie [1] and Hashimoto [2] for the case of $Sp(n, \mathbb{Z})$ for small n. Let us examine one such computation—that of Arakawa [2] using the Selberg trace formula. One begins with the dimension formula of Godement which writes the dimension of the space of cusp forms as an integral of a sum over Γ. The identity and parabolic elements of Γ produce the only non-zero contributions to the trace formula. Work of Shintani [3] is needed to compute special values of zeta functions arising in the calculations.

The spaces $\mathcal{M}(Sp(2, \mathbb{Z}), k)$ were completely determined by Igusa [2]. See also Freitag [2].

Hecke operators for $Sp(n, \mathbb{Z})$ were first systematically investigated by Maass [5]. Let m be a positive integer and define

$$M_n = \{g \in \mathbb{Z}^{2n \times 2n} | {}^t g J_n g = m J_n\}, \quad \text{for } J_n \text{ as in the definition of } Sp(n, \mathbb{Z}).$$

Then we will characterize the mth *Hecke operator* by what it does to a form $f \in \mathcal{M}(Sp(n, \mathbb{Z}), k)$, namely:

$$T(m)f(Z) = m^{nk - n(n+1)/2} \sum_{\gamma = \left(\begin{smallmatrix} * & * \\ C & D \end{smallmatrix}\right) \in Sp(n, \mathbb{Z}) \backslash M_n} |CZ + D|^{-k} f(\gamma Z).$$

It is possible to show that symplectic Hecke operators have similar properties to those of Hecke operators for $GL(n, \mathbb{Z})$ which were obtained in Theorem 2 of Section 4.5.2. The Euler products involved are more complicated though. Some other references for these Hecke operators are Andrianov [1–5], Freitag [1], and Shimura [2].

There are many sorts of *Dirichlet series* associated to Siegel modular forms $f \in \mathcal{M}(Sp(n, \mathbb{Z}), k)$. For simplicity, let us assume that f is a cusp form. Then given an automorphic form v for $GL(n, \mathbb{Z})$ in $\mathcal{A}(GL(n, \mathbb{Z}), \lambda)$ as in Section 4.5.2, we can consider the following Mellin transform:

$$M(f, v) = \int_{Y \in \mathcal{P}_n / GL(n, \mathbb{Z})} f(iY) v(Y) \, d\mu_n(Y).$$

Suppose that $f(Z)$ has the Fourier expansion (2.2) above and that

$$v(Y) = |Y|^s u(Y^0), \quad \text{for } Y^0 = |Y|^{-1/n} Y \in \mathcal{P}_n, \quad s \in \mathbb{C}.$$

Set

$$u^*(W) = u(W^{-1}).$$

As in the proof of part (5) of Theorem 2 in Section 4.5.2, we have:

5.2. Geometry and Analysis on $\Gamma \backslash G/K$

$$M(f,v) = \sum_{T>0} \int_{Y \in \mathscr{P}_n/GL(n,\mathbb{Z})} a(T)\exp\{-2\pi \operatorname{Tr}(TY)\}v(Y)\,d\mu_n(Y)$$

$$= (2\pi)^{-s}\Gamma_n(r(u,s)) \sum_{0 < T/GL(n,\mathbb{Z})} a(T)|T|^{-s}u^*(T),$$

for some $r = r(u,s) \in \mathbb{C}^n$. Thus it is natural to associate to $f \in \mathscr{S}(Sp(n,\mathbb{Z}),k)$, with Fourier expansion (2.2) having Fourier coefficients $a(T)$, the *Dirichlet series*:

$$L(f,v) = \sum_{0 < T \text{ symmetric, semi-integral}/GL(n,\mathbb{Z})} a(T)v(T^{-1}).$$

Maass [2, Section 15] obtains the analytic continuation and functional equation of $L(f,v)$, even when f is not a cusp form. See the proof of part (5) of Theorem 2 of Section 4.5.2 for a discussion of similar analytic continuation.

Harmonic analysis on $\mathscr{P}_n/GL(n,\mathbb{Z})$ gives a converse to this Hecke correspondence between f and $L(f,v)$, as Imai [1] shows when $n = 2$. Weissauer [1] obtains a converse result for congruence subgroups of $Sp(n,\mathbb{Z})$, for general n. It would be nice not to have to know that there are Dirichlet series with functional equations for all the v in $\mathscr{A}(GL(n,\mathbb{Z}),\lambda)$. Just how many such v are necessary is a very interesting question. Similar questions exist for Weil's theory of the Hecke correspondence for congruence subgroups of $SL(2,\mathbb{Z})$ (see Section 3.6 of Volume I).

Andrianov [1, 2, 4] investigates various sorts of Dirichlet series associated to eigenfunctions of Hecke operators. And the language of adelic representation theory leads to the same sort of results. (See Piatetski-Shapiro's talk in Borel and Casselman [1, Vol. 1, pp. 185–188]. See also Piatetski-Shapiro [3].)

Example 3 (Eisenstein Series for $GL(2, O_K)$). References for this subject include: Asai [1], Efrat and Sarnak [1], Elstrodt, Grunewald, and Mennicke [4], Fueter [1], Gelbart [3], Grosswald [2], Hecke [1], Hoffstein [1], Hunter [1], Jacquet and Langlands [1], Mennicke [1], Mordell [1], Ramanathan [1–2], Sarnak [1], Siegel [1, Vol. I, pp. 173–179], Stark [5], Tamagawa [3], Terras [12–15], and Weil [4].

Let K be an algebraic number field of degree m over $\mathbb{Q}$. We use the notation set up earlier during the discussion of the fundamental domain for $GL(n, O_K)$. The *Epstein zeta function* for $GL(2, O_K)$ is defined for $Y \in \mathscr{P}_2^K$, $\mathfrak{a}$ an ideal of K, and s a complex number with $\operatorname{Re} s > 1$, by:

$$Z(\mathfrak{a}, Y, s) = \sum_{0 \neq b \in \mathfrak{a}^2/U_K} N(Y\{b\})^{-s}. \tag{2.3}$$

Here

$$N(Y\{b\}) = \prod_{j=1}^{r_1+r_2} (\overline{{}^t b^{(j)}} Y^{(j)} b^{(j)})^{e_j},$$

where

$$e_j = \begin{cases} 1, & j = 1, \ldots, r_1, \\ 2, & j = r_1 + 1, \ldots, r_1 + r_2. \end{cases}$$

And the sum in (2.3) is over a complete system of non-zero column vectors in $\mathfrak{a}^2$ inequivalent under the equivalence relation

$${}^t b = (b_1, b_2) \sim (b_1 u, b_2 u) \quad \text{for } u \in U_K = \text{the group of units of } O_K.$$

In order to prove the convergence of Epstein's zeta function (2.3), one could devise an integral test similar to that used in the case that K is the field of rational numbers (see Corollary 2 to Proposition 2 in Section 4.4.4). Related methods are used by Siegel [4, p. 290] and Godement in Borel and Mostow [1, p. 207]. It is also possible to deduce the convergence from bounds on theta functions as in Ramanathan [2, p. 54].

Exercise 11. Obtain the analytic continuation and functional equation of Epstein's zeta function for $GL(2, O_K)$ in (2.3) by imitating Riemann's proof of the analytic continuation of $\zeta(s)$ given in Section 1.4 of Volume I. See also Hecke's proof of the analytic continuation of the Dedekind zeta function in Lang [4, pp. 255–258]. You will find that $Z(\mathfrak{a}, Y, s)$ has a simple pole at $s = 1$ and a functional equation relating it to $Z(\mathfrak{a}', Y^{-1}, 1 - s)$, where $\mathfrak{a}' = (\mathfrak{a}\mathfrak{d}_K)^{-1}$, if $\mathfrak{d}_K$ is the different of K.

We have a *simultaneous Iwasawa decomposition* of the vector $Y \in \mathscr{P}_2^K$

$$Y = \begin{pmatrix} v & 0 \\ 0 & w \end{pmatrix} \begin{bmatrix} 1 & q \\ 0 & 1 \end{bmatrix}, \quad \text{with } v, w \in \mathscr{P}_1^K, q \in \mathbb{R} \otimes_{\mathbb{Q}} K. \tag{2.4}$$

This means that

$$Y^{(j)} = \begin{pmatrix} v^{(j)} & 0 \\ 0 & w^{(j)} \end{pmatrix} \begin{bmatrix} 1 & q^{(j)} \\ 0 & 1 \end{bmatrix}, \quad \text{for } v^{(j)}, w^{(j)} > 0, q^{(j)} \in E_j,$$

$j = 1, \ldots, r_1 + r_2$. Here $E_j = \mathbb{R}$, for $j = 1, \ldots, r_1$ and $E_j = \mathbb{C}$, for $j = r_1 + 1, \ldots, r_1 + r_2$.

Clearly the Epstein zeta function $Z(\mathfrak{a}, Y, s)$ has the *invariance property*:

$$Z(\mathfrak{a}, Y, s) = Z(\mathfrak{a}, Y\{\gamma\}, s) \quad \text{for all } \gamma \in GL(2, O_K), Y \in \mathscr{P}_2^K.$$

It follows that if we view $Z(\mathfrak{a}, Y, s)$ as a function of the q-variable in the Iwasawa decomposition (2.4) of Y, we are looking at a function that is periodic modulo O_K. Thus we can obtain a Fourier expansion of $Z(\mathfrak{a}, Y, s)$ in the q-variable.

Let us eliminate the dependence on the ideal $\mathfrak{a}$ by defining for $\mathfrak{b} \in C$:

$$Z^*(Y, s) = \sum_{C \in I_K} N\mathfrak{b}^{2s} Z(\mathfrak{b}, Y, s). \tag{2.5}$$

Because the ideal class group I_K is finite, so is the sum in (2.5) provided that $s \neq 1$. Note that the sum is also independent of the choice of the ideal $\mathfrak{b}$ in the ideal class C.

5.2. Geometry and Analysis on $\Gamma \backslash G/K$

Set

$$A = 2^{-r_2}\pi^{-m/2}D_K^{1/2}, \qquad D_K = \text{the absolute value of the discriminant of } K,$$

and define *Dedekind's zeta function* by

$$\zeta_K(s) = \sum_{\text{ideals } \mathfrak{b} \text{ of } O_K} N\mathfrak{b}^{-s}, \qquad \operatorname{Re} s > 1.$$

The functional equation of Dedekind's zeta function is:

$$\Lambda_K(s) = A^s \Gamma(s/2)^{r_1} \Gamma(s)^{r_2} \zeta_K(s) = \Lambda_K(1-s).$$

See the corollary to Theorem 2 of Section 1.4 in Volume I. Motivated by this functional equation, we define:

$$\Lambda^*(Y, s) = A^{2s} \Gamma(s)^{r_1} \Gamma(2s)^{r_2} Z^*(Y, s). \tag{2.6}$$

Theorem 1 (Fourier Expansion of Epstein's Zeta Function for K). *Using the notation of formulas (2.3)–(2.6), we have the Fourier expansion:*

$$\Lambda^*(Y, s) = Nv^{-s}\Lambda_K(2s) + Nv^{-1/2}Nw^{1/2-s}\Lambda_K(2s-1)$$

$$+ \frac{2^{r_1+r_2}D_K^{s-1/2}}{Nv^{1/2}N|Y|^{-1/4+s/2}} \sum_{0 \neq u \in \mathfrak{d}_K^{-1}} |Nu|^{s-1/2}\sigma_{1-2s}(u\mathfrak{d}_K)e^{2\pi i \operatorname{Tr}(qu)}$$

$$\times \prod_{j=1}^{r_1+r_2} K_{e_j(s-1/2)}\left(2\pi e_j \sqrt{\left(\frac{w}{v}u^2\right)^{(j)}}\right),$$

where $K_s(y)$ is the ordinary K-Bessel function and for any ideal $\mathfrak{b}$ in O_K, $\sigma_s(\mathfrak{b})$ is the divisor function:

$$\sigma_s(\mathfrak{b}) = \sum_{\mathfrak{c}|\mathfrak{b}} N\mathfrak{c}^s.$$

PROOF. See Terras [13–14]. The idea is to generalize Exercise 4 of Section 3.5 in Volume I. Set

$$\Lambda(\mathfrak{a}, Y, s) = A^{2s}\Gamma(s)^{r_1}\Gamma(2s)^{r_2}Z(\mathfrak{a}, Y, s).$$

Then, if Y has the Iwasawa decomposition (2.4), it follows that:

$$Y\begin{Bmatrix}a\\b\end{Bmatrix} = v\{a+qb\} + w\{b\}.$$

Thus

$$\Lambda(\mathfrak{a}, Y, s) = Nv^{-s}\Lambda_K(\mathfrak{a}, 2s) + A^{2s}\Gamma(s)^{r_1}\Gamma(2s)^{r_2}$$

$$\times \sum_{\substack{0 \neq b \in \mathfrak{a}/U_K \\ a \in \mathfrak{a}}} N(v\{a+qb\} + w\{b\})^{-s},$$

if $\operatorname{Re} s > 1$ and we define

$$\Lambda_K(\mathfrak{a}, 2s) = A^{2s}\Gamma(s)^{r_1}\Gamma(2s)^{r_2} \sum_{0 \neq b \in \mathfrak{a}/U_K} |N\mathfrak{b}|^{-2s}.$$

The Poisson summation formula from Section 1.3 of Volume I or Weil [5, p. 106] shows that the sum over $a \in \mathfrak{a}$ equals the sum of Fourier transforms over $c \in \mathfrak{a}' = (\mathfrak{a}\mathfrak{d}_K)^{-1}$ which is the dual ideal to $\mathfrak{a}$. The Fourier transforms here are:

$$\hat{f}(b,c) = \int_{x \in K \otimes_{\mathbb{Q}} \mathbb{R}} N(v\{x+qb\} + w\{b\})^{-s} \exp\{-2\pi i \operatorname{Tr}^t cx\} \, d\mu(x).$$

And the measure $d\mu(x)$ is chosen so that:

$$\int_{K \otimes_{\mathbb{Q}} \mathbb{R}/\mathfrak{a}} d\mu(x) = 1.$$

Now the ideal $\mathfrak{a}$ has an integral basis; i.e.,

$$\mathfrak{a} = \sum_{j=1}^{m} {}^{\oplus} \mathbb{Z} w_j \quad \text{and} \quad K \otimes_{\mathbb{Q}} \mathbb{R} = \sum_{j=1}^{m} {}^{\oplus} \mathbb{R} w_j.$$

So if

$$x = \sum_{j=1}^{m} x_j w_j, \quad x_j \in \mathbb{R},$$

we can take

$$d\mu(x) = \prod_{j=1}^{m} dx_j, \quad \text{with } dx_j = \text{Lebesgue measure on } \mathbb{R}.$$

We can also write

$$K \otimes_{\mathbb{Q}} \mathbb{R} \cong \sum_{j=1}^{r_1+r_2} {}^{\oplus} E_j, \quad \text{where } E_j = \begin{cases} \mathbb{R}, & 1 \le j \le r_1, \\ \mathbb{C}, & r_1+1 \le j \le r_1+r_2. \end{cases}$$

Therefore we can define the mapping:

$$T : K \otimes_{\mathbb{Q}} \mathbb{R} \to \sum_{j=1}^{r_1+r_2} {}^{\oplus} E_j,$$

$$T : \sum_{j=1}^{m} x_j w_j \mapsto y = (y^{(1)}, \ldots, y^{(r_1+r_2)}),$$

where

$$y^{(i)} = \sum_{j=1}^{m} x_j w_j^{(i)}.$$

Exercise 12. Show that the Jacobian of the mapping T above is:

$$\left| \frac{\partial y}{\partial x} \right| = \left| \frac{\partial(y_1, \ldots, y_{r_1}, \operatorname{Re} y_{r_1+1}, \operatorname{Im} y_{r_1+1}, \ldots, \operatorname{Re} y_{r_1+r_2}, \operatorname{Im} y_{r_1+r_2})}{\partial(x_1, \ldots, x_m)} \right|$$

$$= 2^{-r_2} D_K^{1/2} N\mathfrak{a}.$$

Using Exercise 12, we find that

5.2. Geometry and Analysis on $\Gamma\backslash G/K$

$$\hat{f}(b,c) = D_K^{-1/2} N\mathfrak{a}^{-1} 2^{r_2} \prod_{j=1}^{r_1+r_2} \int_{E_j} N_{E_j/\mathbb{R}}((v\{y+qb\} + w\{b\})^{(j)})^{-s}$$

$$\times \exp\{-2\pi i \operatorname{Tr}_{E_j/\mathbb{R}}((cy)^{(j)})\} \, dy^{(j)}.$$

Make the change of variables $x_j = ((w\{b\})^{-1/2} t(y+qb))^{(j)}$, where $v = t^2$, $t^{(j)} > 0$, to see that:

$$\hat{f}(b,c) = D_K^{-1/2} N\mathfrak{a}^{-1} 2^{r_2} Nv^{-1/2} N(w\{b\})^{1/2-s} \exp\{2\pi i \operatorname{Tr}(cqb)\}$$

$$\times \prod_{j=1}^{r_1+r_2} \int_{E_j} (1 + \bar{x}_j x_j)^{-se_j} \exp\{-2\pi i \operatorname{Tr}_{E_j/\mathbb{R}}((ct^{-1}(w\{b\})^{1/2})^{(j)} x_j)\} \, dx_j.$$

Define

$$I_j(a,s) = \int_{E_j} (1+\bar{x}x)^{-se_j} \exp\{-2\pi i \operatorname{Tr}_{E_j/\mathbb{R}}(ax)\} \, dx.$$

By part (a) of Exercise 1 of Section 3.2 in Volume I, we find that for $j = 1, \ldots, r_1$:

$$I_j(a,s) = \begin{cases} 2\pi^{1/2} \Gamma(s)^{-1} |\pi a|^{s-1/2} K_{s-1/2}(2\pi|a|), & a \neq 0, \\ \Gamma(\tfrac{1}{2})\Gamma(s-\tfrac{1}{2})\Gamma(s)^{-1}, & a = 0. \end{cases}$$

For $j = r_1 + 1, \ldots, r_1 + r_2$, we must compute:

$$I_\mathbb{C}(a_1 + ia_2, s) = \int_{x_1 + ix_2 \in \mathbb{C}} (1 + x_1^2 + x_2^2)^{-2s} \exp(-4\pi i(a_1 x_1 - a_2 x_2)) \, dx$$

$$= k_{1,2}(2s | I, 2\pi(a_1, a_2)),$$

where $k_{1,2}$ is the k-Bessel function defined in formula (2.20) in Section 4.2.2. We can use part (2) of Theorem 2 in Section 4.2.2 to see that in terms of the K-Bessel function defined by (2.21) in Section 4.2.2, we have:

$$I_\mathbb{C}(a_1 + ia_2, s) = \Gamma(2s)^{-1} \pi K_1(1 - 2s | 4\pi^2(a_1^2 + a_2^2), 1).$$

It follows then that for $j = r_1 + 1, \ldots, r_1 + r_2$, we have the following formula when $a = a_1 + ia_2 \in \mathbb{C} = E_j$:

$$I_{E_j}(a,s)$$

$$= \begin{cases} 2^{2s}\Gamma(2s)^{-1}(\pi^2(a_1^2+a_2^2))^{s-1/2} K_{2s-1}(4\pi\sqrt{a_1^2+a_2^2}), & a = a_1 + ia_2 \neq 0, \\ \pi\Gamma(2s-1)\Gamma(2s)^{-1}, & a = 0. \end{cases}$$

Substituting these results into the original Poisson sum leads to:

$$\Lambda(\mathfrak{a}, Y, s) = Nv^{-s}\Lambda_K(\mathfrak{a}, 2s) + Nv^{-1/2} Nw^{1/2-s} N\mathfrak{a}^{-1} \Lambda_K(\mathfrak{a}, 2s-1)$$

$$+ N|Y|^{1/4-s/2} Nv^{-1/2} A^{2s-1} N\mathfrak{a}^{-1} 2^{r_1 + 2sr_2}$$

$$\times \sum_{\substack{0 \neq b \in \mathfrak{a}/U_K \\ 0 \neq c \in (\mathfrak{ad}_K)^{-1}}} \left|\frac{Nc}{Nb}\right|^{s-1/2} e^{2\pi i \operatorname{Tr}(qab)} \prod_{j=1}^{r_1+r_2} K_{e_j(s-1/2)}\left(2\pi e_j \sqrt{\frac{w^{(j)}}{v^{(j)}}} |b^{(j)}c^{(j)}|\right).$$

To complete the proof of Theorem 1, suppose that C is an ideal class of K and $\mathfrak{b} \in C$ is as in formula (2.5). Note that the equation

$$\mathfrak{a}\mathfrak{b} = bO_K$$

defines a one-to-one mapping from ideals $\mathfrak{a} \in C^{-1}$ onto elements $b \in \mathfrak{b} \bmod U_K$. Set $u = bc$, for $c \in (\mathfrak{b}\mathfrak{d}_K)^{-1}$. Then define the map

$$(\mathfrak{b}/U_K) \times (\mathfrak{b}\mathfrak{d}_K)^{-1} \xrightarrow{L} (C^{-1}) \times (\mathfrak{d}_K^{-1})$$

$$(b, c) \mapsto (\mathfrak{a}, u = bc) = L(b, c).$$

The map L is easily seen to be one-to-one. It is not onto, since the image consists of $(\mathfrak{a}, u)$ such that $\mathfrak{a}$ divides $\mathfrak{d}_K u$.

Finally observe that

$$N\mathfrak{d}^{2s-1} \left| \frac{Nc}{Nb} \right|^{s-1/2} = N\mathfrak{a}^{1-2s} |Nu|^{s-1/2}.$$

This completes the proof of Theorem 1. $\square$

Corollary 1 (Relations Between $\zeta_K(s)$ and $\zeta_K(s-1)$). *Set*

$$M_s(z) = K_s(z) + 2z \frac{d}{dz} K_s(z)$$

and

$$T(s, u) = M_{e_1 s}(2\pi e_1 |u^{(1)}|) \prod_{j=2}^{r_1+r_2} K_{e_j s}(2\pi e_j |u^{(j)}|).$$

Then

$$(1-s)\Lambda_K(2s-1) + s\Lambda_K(2s)$$
$$= -2^{r_1+r_2-1} D_K^{s-1/2} \sum_{0 \neq u \in \mathfrak{d}_K^{-1}} |Nu|^{s-1/2} \sigma_{1-2s}(u\mathfrak{d}_K) T(s - \tfrac{1}{2}, u).$$

PROOF. This is the analogue of a generalization of part (b) of Exercise 7 in Section 3.5 of Volume I.

Substitute

$$Y = \begin{pmatrix} v & 0 \\ 0 & 1 \end{pmatrix} \text{ in } Z^*(Y, s)$$

and use the following functional equation:

$$Z^*\left(\begin{pmatrix} v & 0 \\ 0 & 1 \end{pmatrix}, s \right) = Z^*\left(\begin{pmatrix} 1 & 0 \\ 0 & v \end{pmatrix}, s \right)$$

plus Theorem 1 to deduce that

5.2. Geometry and Analysis on $\Gamma \backslash G/K$

$$Nv^{-s}\Lambda_K(2s) + Nv^{-1/2}\Lambda_K(2s-1) + \frac{2^{r_1+r_2}D_K^{s-1/2}}{Nv^{1/4+s/2}}$$

$$\times \sum |Nu|^{s-1/2}\sigma_{1-2s}(u\mathfrak{d}_K)\prod K_{e_j(s-1/2)}\left(2\pi e_j \frac{|u^{(j)}|}{\sqrt{v^{(j)}}}\right)$$

$$= \Lambda_K(2s) + Nv^{1/2-s}\Lambda_K(2s-1) + \frac{2^{r_1+r_2}D_K^{s-1/2}}{Nv^{-1/4+s/2}}$$

$$\times \sum |Nu|^{s-1/2}\sigma_{1-2s}(u\mathfrak{d}_K)\prod K_{e_j(s-1/2)}(2\pi e_j\sqrt{v^{(j)}}|u^{(j)}|).$$

Differentiate this equation with respect to v_1 and set all $v_j = 1$, $j = 1, \ldots, r_1 + r_2$ to finish the proof of Corollary 1. □

Corollary 2 (A Formula for the Product of the Class Number and the Regulator). *Let K be any algebraic number field of degree m, with*

$w_K = $ *the number of roots of unity in K,* $\qquad R_K = $ *the regulator of K,*

$h_K = $ *the class number of K,* $\qquad D_K = $ *the absolute value of the discriminant,*

$e_1 = 1$ *if the field K has any real conjugate fields and 2 otherwise,*

$\mathfrak{d}_K = $ *the different,* $\qquad r_2 = $ *the number of complex conjugate fields,*

$\zeta_K = $ *the Dedekind zeta function,* $\qquad T(s, u)$ *as defined in Corollary* 1.

Then

$$h_K R_K/w_K = 2(2\pi)^{-m}D_K\zeta_K(2) + 2^{r_2}D_K^{1/2}\sum_{0\neq u\in\mathfrak{d}_K^{-1}}|Nu|^{1/2}\sigma_{-1}(u\mathfrak{d}_K)T(\tfrac{1}{2},u).$$

PROOF. Let s approach 1 in Corollary 1. □

Exercise 13. Complete the proof of Corollary 2.

Corollary 3. *If K is a totally real algebraic number field, then, using the notation of Corollary 2, we have:*

$$h_K R_K = 4(2\pi)^{-m}D_K\zeta_K(2) - D_K^{1/2}2^{3-m}\pi$$

$$\times \sum_{0\neq u\in\mathfrak{d}_K^{-1}}|u^{(1)}|\sigma_{-1}(u\mathfrak{d}_K)\exp\{-2\pi(|u^{(1)}| + \cdots + |u^{(m)}|)\}.$$

PROOF. Since $K_{1/2}(z) = (2z/\pi)^{-1/2}e^{-z}$, we see that $M_{1/2}(z) = -(2\pi z)^{1/2}e^{-z}$. The result follows easily then from Corollary 2. □

Note that Corollary 3 gives an easy upper bound for the product of the class number and the regulator, which should be compared with that obtained by Lang [4, p. 261]. A lower bound is more difficult to obtain.

When $K = \mathbb{Q}$, Corollary 1 gives formulas relating $\zeta(2n)$ and $\zeta(2n+1)$ (see

Exercise 7 of Section 3.5, Vol. I). Formulas of this sort have been studied by many authors, without, however, leading to information on the rationality, irrationality, algebraicity or transcendence of $\zeta(2n+1)$, $n = 1, 2, \ldots$. See Hunter [1].

Siegel [1, Vol. I, pp. 173–179] used the Fourier expansion of Eisenstein series for $GL(2, O_K)$ to obtain the analytic continuation and functional equation of the Dedekind zeta function. Mordell [1, pp. 518ff.] also derives the Fourier expansion of the Eisenstein series for $GL(2, O_K)$.

Hoffstein [1] has used Fourier expansions of Eisenstein series for $GL(2, O_K)$, K a real quadratic field, to study the real zeros of these series. Asai [1] uses such Fourier expansions to generalize Kronecker's limit formula (see Exercise 6 of Section 3.5 in Vol. I). See also Zagier [1].

Grosswald [2] obtains results related to that in Corollary 1—formulas for the Dedekind zeta function involving the Meijer's G-function. Elstrodt, Grunewald, and Mennicke [4] consider these Fourier expansions in the case that K is an imaginary quadratic field.

Our final goal in this section is to describe the relation between the Epstein zeta function (2.3) and the Eisenstein series for $SL(2, O_K)$. First recall that Proposition 1 gave a correspondence between the cusps of $\mathscr{SP}_2^K/SL(2, O_K)$ and ideal classes in the ideal class group I_K:

$$\hat{K}/SL(2, O_K) \qquad \leftrightarrow \qquad I_K.$$

$$\text{represented by cusps} \qquad \text{represented by ideals}$$

$$x_1, \ldots, x_h \qquad\qquad \mathfrak{a}_1, \ldots, \mathfrak{a}_h$$

The map was obtained by setting $x_i = p_i/s_i$ with $p_i, s_i \in O_K$,

$$A_i = \begin{pmatrix} p_i & u_i \\ s_i & v_i \end{pmatrix} \in SL(2, K), \quad (p_i, s_i) = \begin{pmatrix} \text{the ideal generated} \\ \text{by } p_i \text{ and } s_i \end{pmatrix} = \mathfrak{a}_i,$$

$$x_i = A_i \infty, \quad u_i, v_i \in \mathfrak{a}_i^{-1}.$$

We showed in Lemma 1 that

$$\Gamma_{x_i} = \{\gamma \in SL(2, O_K) | \gamma x_i = x_i\}$$

$$= \left\{ A_i \begin{pmatrix} w & z \\ 0 & w^{-1} \end{pmatrix} A_i^{-1} \;\middle|\; z \in \mathfrak{a}_i^{-2}, w \in U_K = \text{units of } O_K \right\}.$$

We can now define an *Eisenstein series corresponding to the cusp* x_i (cf. Kubota [1]):

$$E_i(Y, s) = N\mathfrak{a}_i^{2s} \sum_{\gamma \in SL(2, O_K)/\Gamma_{x_i}} N(v(Y\{\gamma A_i\}))^{-s}, \qquad \text{for Re } s > 1. \quad (2.7)$$

Here we use the notation that if $Y \in \mathscr{P}_2^K$ has Iwasawa decomposition (2.4), we write $v(Y)$ for the v-coordinate of Y. It is also the upper left entry of Y. Now taking the v-part of $Y\{\gamma A_i\}$ amounts to taking $Y\{g\}$, where g is the first column

5.2. Geometry and Analysis on $\Gamma\backslash G/K$

of γA_i. Since such a g must generate $\mathfrak{a}_i$, we find that

$$E_i(Y,s) = N\mathfrak{a}_i^{2s} \sum_{\substack{g \in \mathfrak{a}_i^2/U_K \\ \text{entries of } g \text{ generate } \mathfrak{a}_i}} N(Y\{g\})^{-s}, \quad \text{for Re } s > 1. \quad (2.8)$$

Exercise 14. Prove formula (2.8) for all $Y \in \mathscr{P}_2^K$. Then show that the Eisenstein series E_i is dependent only on the ideal class containing the ideal $\mathfrak{a}_i$, and not on the choice of $\mathfrak{a}_i$ in that ideal class. Finally, prove that, if we define for an ideal class $C \in I_K$ the *ideal class zeta function*:

$$\zeta(C,s) = \sum_{\mathfrak{c} \in C} N\mathfrak{c}^{-s}, \quad \text{for Re } s > 1,$$

then we have a relation between Epstein's zeta function (2.3) and the Eisenstein series:

$$Z(O_K, Y, S) = \sum_{i=1}^h \zeta(C_i, 2s) E_i(Y, s),$$

using the notation C_i for the ideal class containing $\mathfrak{a}_i$, $i = 1, \ldots, h$.

In order to generalize the relation obtained in Exercise 14 between $Z(O_K, Y, s)$ and the Eisenstein series E_i to $Z(\mathfrak{a}_i, Y, s)$, we need a *matrix of ideal class zeta functions*:

$$M_K(s) = (\zeta(C(i,j), s))_{1 \le i,j \le h}, \quad (2.9)$$

where $C(i,j)$ is the ideal class containing the ideal $\mathfrak{a}_j \mathfrak{a}_i^{-1}$. Define also the *column vector of Epstein zeta functions*:

$$\vec{Z}(Y,s) = {}^t(Z(\mathfrak{a}_1, Y, s), \ldots, Z(\mathfrak{a}_h, Y, s)), \quad (2.10)$$

and the *column vector of Eisenstein series*:

$$\vec{E}(Y,s) = {}^t(E_1(Y,s), \ldots, E_h(Y,s)). \quad (2.11)$$

Proposition 3 (The Relation Between Epstein's Zeta Function and the Eisenstein Series for $SL(2, O_K)$). *Using the notation (2.3), (2.7), (2.9)–(2.11), we have the following equality for Re $s > 1$:*

$$\vec{Z}(Y,s) = M_K(s) \vec{E}(Y,s).$$

PROOF. We have the following chain of equalities:

$$Z(\mathfrak{a}_i, Y, s) = N\mathfrak{a}_i^{2s} \sum_{0 \ne g \in \mathfrak{a}_i^2/U_K} N(Y\{g\})^{-s} = N\mathfrak{a}_i^{2s} \sum_{\mathfrak{a}_i | \mathfrak{b}} \sum_{\substack{g \in \mathfrak{b}^2/U_K \\ (g_1, g_2) = \mathfrak{b}}} N(Y\{g\})^{-s}$$

$$= \sum_{\mathfrak{a}_i | \mathfrak{b}} \left(\frac{N\mathfrak{b}}{N\mathfrak{a}_i}\right)^{-2s} N\mathfrak{b}^{2s} \sum_{\substack{g \in \mathfrak{b}^2/U_K \\ (g_1, g_2) = \mathfrak{b}}} N(Y\{g\})^{-s}.$$

Then set $\mathfrak{b}/\mathfrak{a}_i = \mathfrak{c}$ and observe that $\mathfrak{c}$ runs through all integral ideals in the class $C(i,j)$, to complete the proof. □

The formula in Proposition 3 raises certain questions:

1. Is it possible to diagonalize the matrix $M_K(s)$ using characters of the ideal class group?
2. What does this have to do with Hecke operators for $GL(2, O_K)$?
3. What does this have to do with the analogue of Siegel's integral formula for $GL(n, O_K)$? See Proposition 2 of Section 4.4.4 for the integral formula when $K = \mathbb{Q}$.

Here we shall discuss only question 1. Hecke operators for these groups are treated by Hermann [1], Shimura [2], and Styer [1]. Siegel's integral formula for $SL(n, O_K)$, K imaginary quadratic, is considered by Hunter [1].

See also Elstrodt, Grunewald, and Mennicke [4] as well as Efrat and Sarnak [1].

The ideal class group I_K is a finite abelian group of order h. Let $\hat{I}_K$ denote the *dual group* of characters $\chi \colon I_K \to \mathbb{T}$, where $\mathbb{T}$ is the circle group of complex numbers of norm 1; i.e.,

$$\mathbb{T} = \{z \in \mathbb{C} \,|\, |z| = 1\}.$$

That is, χ is a homomorphism of multiplicative groups. Then the dual group is:

$$\hat{I}_K = \{\chi_1, \ldots, \chi_h\}.$$

We can diagonalize the matrix M_K using Fourier transforms on I_K. Suppose that C_i is the ideal class containing the ideal $\mathfrak{a}_i$, for $i = 1, \ldots, h$. Define

$$Z(C_i) = Z(\mathfrak{a}_i, Y, s), \qquad E(C_i) = E(\mathfrak{a}_i, Y, s), \qquad \zeta(C) = \zeta(C, 2s). \quad (2.12)$$

Our proof of Proposition 3 rested on the equation:

$$Z(C_i) = \sum_{j=1}^{h} \zeta(C_j/C_i) E(C_j). \quad (2.13)$$

Now define convolution of functions $f \colon I_K \to \mathbb{C}$ by:

$$(f * g)(C_i) = h^{-1} \sum_{j=1}^{h} f(C_i/C_j) g(C_j). \quad (2.14)$$

Thus formula (2.13) says that

$$Z(C) = h \zeta(C^{-1}) * E(C). \quad (2.15)$$

As for the group of real numbers (see part (4) of Theorem 1 of Section 1.2 in Volume I), the Fourier transform can be used to simplify this convolution equation. We define the *Fourier transform* of a function

$$f \colon I_K \to \mathbb{C}$$

at the character $\chi \in \hat{I}_K$ by:

$$\hat{f}(\chi) = h^{-1} \sum_{y \in I_K} f(y) \overline{\chi(y)}. \quad (2.16)$$

5.2. Geometry and Analysis on $\Gamma\backslash G/K$

Since I_K is a finite abelian group, there are no convergence problems. In fact, the theory of Fourier transforms on finite abelian groups has many applications, since it is just what is needed for the fast Fourier transform, an idea which has speeded computation of such transforms immensely. See Brigham [1] and Terras [16] for more information on finite and fast Fourier transforms.

Proposition 4 (Some Properties of the Fourier Transform on I_K).

(1) Convolution.
$$\widehat{f*g}(\chi) = \hat{f}(\chi) \cdot \hat{g}(\chi).$$

(2) Inversion.
$$f(x) = \sum_{\chi \in \hat{I}_K} \hat{f}(\chi)\chi(x), \quad \text{for all } x \in I_K.$$

PROOF.

(1) Note that
$$\widehat{f*g}(\chi) = h^{-2} \sum_{z \in I_K} \sum_{y \in I_K} f(zy^{-1})g(y)\overline{\chi(z)}$$
$$= h^{-2} \sum_{y \in I_K} g(y) \sum_{w = zy^{-1} \in I_K} f(w)\overline{\chi(wy)} = \hat{f}(\chi) \cdot \hat{g}(\chi).$$

(2) Observe that
$$h^{-1} \sum_{\chi \in \hat{I}_K} \chi(x) \sum_{y \in I_K} f(y)\overline{\chi(y)} = \sum_{y \in I_K} f(y) h^{-1} \sum_{\chi \in \hat{I}_K} \chi(xy^{-1}) = f(x),$$

since we have:
$$h^{-1} \sum_{\chi \in \hat{I}_K} \chi(xy^{-1}) = \begin{cases} 0, & x \neq y, \\ 1, & x = y; \end{cases} \quad \text{and} \quad \overline{\chi(y)} = \chi(y)^{-1}. \tag{2.17}$$

□

Exercise 15. Prove formula (2.17) above.

For
$$\chi \in \hat{I}_K, \quad Y \in \mathscr{P}_2^K, \quad s \in \mathbb{C} \quad \text{with} \quad \operatorname{Re} s > 1,$$
define the *zeta function*:
$$Z(\chi, Y, s) = \sum_{\mathfrak{a}} N\mathfrak{a}^{2s} \chi(\mathfrak{a}) \sum_{0 \neq g \in \mathfrak{a}^2/U_K} N(Y\{g\})^{-s}, \tag{2.18}$$

where the outer sum is over all ideals $\mathfrak{a}$ of O_K and the character χ of the ideal class group is regarded in the obvious way as a function of ideals. Then $Z(\chi, Y, s)$ is the Fourier transform of $Z(C)$ (times h), where $Z(C)$ is defined in formula (2.12).

Similarly, define the *Eisenstein series* associated to $\chi \in \hat{I}_K$, $Y \in \mathcal{P}_2^K$, and $s \in \mathbb{C}$ with $\operatorname{Re} s > 1$ by:

$$E(\chi, Y, s) = \sum_{\mathfrak{a}} N\mathfrak{a}^{2s}\chi(\mathfrak{a}) \sum_{\substack{g \in \mathfrak{a}^2/U_K \\ (g_1, g_2) = \mathfrak{a}}} N(Y\{g\})^{-s}, \tag{2.19}$$

where the outer sum is over all ideals $\mathfrak{a}$ of O_K and the inner sum is over column vectors $g = {}^t(g_1, g_2)$ such that the ideal $\mathfrak{a}$ is generated by g_1 and g_2 and the vectors g form a complete set of representatives for the equivalence relation obtained from multiplication by units. Then $E(\chi, Y, s)$ is h times the Fourier transform of $E(C)$ defined by (2.12).

Proposition 5 (The Diagonalization of the Relation Between Epstein's Zeta Function and the Eisenstein Series for $SL(2, O_K)$). *Using the definitions* (2.18) *and* (2.19) *and setting*

$$L(\chi, s) = \sum_{\mathfrak{a}} \chi(\mathfrak{a})N\mathfrak{a}^{-s}, \quad \text{for } \operatorname{Re} s > 1,$$

with the sum running over all ideals $\mathfrak{a}$ of O_K, we have

$$Z(\bar{\chi}, Y, s) = L(\chi, s)E(\bar{\chi}, Y, s).$$

PROOF. This is just the convolution property in Proposition 4 for the special case of the functions from Proposition 3. □

Our discussion of nonholomorphic automorphic forms for $GL(2, O_K)$ is now at an end, although there still remains much to do if we wish to extend all of Chapters 3 and 4 to $GL(n, O_K)$. For we have not even begun the theory of Hecke operators, the Hecke correspondence, the Selberg trace formula. Sarnak [1] applies the Selberg trace formula for $SL(2, O_K)$, K imaginary quadratic of class number one, to extend his results on the asymptotics of units in number fields. Elstrodt, Grunewald, and Mennicke [1–4] give applications of the noneuclidean Poisson sum formula and the trace formula for $SL(2, O_K)$, K imaginary quadratic. See Efrat [2], Müller [1], Zograf [1], and Shimizu [1] for the totally real case. See Gelbart [3], Jacquet-Langlands [1], and Weil [4] for an adelic version of the subject. Many papers on automorphic forms are reviewed in LeVeque [1] and Guy [1].

APPENDIX
Corrections to Volume I

p. 5, line 4. Replace U by H.
p. 5, line 16. The x's should all be lower case.
p. 11, line 9. texts, not tests.
p. 16, line 1 in Table 1.1. Replace dy_1 by dx.
p. 16, line 11.

$$\widehat{S * T}.$$

p. 21, line -11. Replace $L^1(\mathscr{R})$ with $L^1(\mathbb{R})$.
p. 24, line 6. We should have defined a "Borel set" to be a member of the smallest σ-algebra containing the compact sets in a topological space.
p. 27, line 14. There is an s missing inside the parentheses of exp:

$$\exp(2\pi i x s/\sqrt{n}).$$

p. 29, line 10. Shriftstück.
p. 29, line 23. willkürliche.
p. 51, line 3. Though.
p. 57, Table 1.3, third entry under $f(y)$. Replace a by $-a$.
p. 62, line 1. For *some* s with $\operatorname{Re} s > 1$.
p. 65, line 8. like *if* $n > 1$. Moreover, there are forms Y in $\mathscr{P}_n$ with $|Y| = 1$ and m_Y.
p. 67, line -12. *A* generator, not *the* generator.
p. 69, line -4. Insert "for $s > 1$" at the end of the line.
p. 71, line 11. The integrand should be $y^{s-1} \exp(-ay)$.
p. 75, line -13. B_n.
p. 134, line 14. The ${}^t g$ on the left needs a larger g.
p. 137, line 4. $-u[(x - ir/u)^2 + q^2 + (r/u)^2]$.

p. 146, line −10. The result on eigenfunctions was proved by Helgason.
p. 149, line 14. Replace *da* by *db*.
p. 158, line −6. The.
p. 160. The argument should be changed as described in 4.2.3, pp. 110–111.
p. 174, line −5. Add "except ±1" at the end of the sentence.
p. 176, line −11. Replace $\mathbb{R}$ by $\mathbb{Z}$.
p. 178, line −5. Replace "unless p" by "unless p is Γ equivalent to".
p. 184, line −7. Replace "$z \to \gamma z$" by "$z \to \gamma z, \gamma \varepsilon \Gamma$".
p. 186, line −11. not solvable in *nonnegative* integers".
p. 187, line 16. The 6 multiplier systems result was found by Rademacher and Zuckerman in 1938.
p. 190, line −1. called.
p. 193, line 8. $S_P = S_{P-1}$.
p. 195, line −2. The factor in front of the summation should be $(z/i)^{-1/2}$.
p. 203, line 18. Do.
p. 204, footnote. The word "automorphic" should be replaced by "holomorphic."
p. 207, line 12. Replace sentence with: You also need to know that a row vector in $\mathbb{Z}^2$ can be completed to a matrix in $SL(2, \mathbb{Z})$ if and only if its greatest common divisor is 1.
p. 209, line 15. Replace §3.1 by §3.2.
p. 216. In Exercise 12(a), replace $o(A)$ by $o(1)$.
p. 227, line −13. $|f(x + iy_0/2)|$.
p. 227, line −8. estimated.
p. 237, line 4. Replace one v by u under the summation symbol.
p. 238, formula (3.87). Replace c by 0 in the matrix.
p. 240, lines −11 to −6. The Mertens conjecture says that the absolute value of the summatory function of the Möbius function from elementary number theory is $\leq \sqrt{x}$. The reference for a disproof is: A.M. Odlyzko and H.J.J. Te Riele, Disproof of the Mertens conjecture, *J. Reine Angew. Math.*, 357 (1985), 138–160.
p. 241, line 7. LeVeque. See also Guy's Reviews.
p. 242, line 8. Another) is needed.
p. 250, line 4. We should have said:
Ultimately the points move on *vertical* line segments (one for each divisor of N) as the points go up to the cusp at infinity.
p. 250, line 6. 15 should be 51.
p. 255, line −4. Exercise 3 not 4.
p. 289, Exercise 26. $\varphi(\tfrac{1}{2}) = -1$.
p. 297, eigenvalues $\Delta e_y = -4\pi^2 \|y\|^2 e_y$.
p. 301, line 3. McGraw.
p. 307, Elstrodt, Grunewald, and Mennicke [1], $PSL_2(\mathbb{Z}[i])$.
p. 308, Freitag [1], 1983.

APPENDIX. Corrections to Volume I

p. 308, R. Gangolli.
p. 312, O. Herrmann.
p. 317, Lehmer [4] = [1].
p. 317, LeVeque.
p. 323, line 3. P. Sarnak, *Ann. Math.*, *151* (1983), 253–295.

Bibliography for Volume II

L. Ahlfors, *Möbius Transformations in Several Dimensions*. Lecture Notes. University of Minnesota, 1981.

T.W. Anderson, *An Introduction to Multivariate Statistical Analysis*. Wiley, N.Y., 1958.

D. Andreoli. See D. Wallace.

A.N. Andrianov, On zeta functions of Rankin type associated with Siegel modular forms. *Lecture Notes in Math. 627*. Springer-Verlag, N.Y., 1977, 325–338.

———, Dirichlet series with Euler product in the theory of Siegel modular forms of genus 2. *Proc. Steklov Inst. Math.*, *112* (1971), 70–93.

———, Spherical functions for $GL(n)$ over local fields and summation of Hecke series. *Math. U.S.S.R. Sbornik*, *12* (1970), 429–452.

———, Euler products corresponding to Siegel modular forms of genus two. *Russian Math. Surveys*, *29* (1974), 43–110.

———, *Quadratic Forms and Hecke Operators*. Springer-Verlag, N.Y., 1987.

A.N. Andrianov & G.N. Maloletkin, Behavior of theta series of degree N under modular substitutions. *Izv. Akad. Nauk. S.S.S.R.*, *39* (1975), 243–258.

———, Behavior of theta-series of genus n of indeterminate quadratic forms under modular substitution (Russian). *Trudy Mat. Inst. Steklov*, *148* (1978), 5–15, 271.

T. Apostol. *Calculus, Vols. I, II*. Blaisdell, Waltham, Mass., 1967.

T. Arakawa, Dirichlet series corresponding to Siegel's modular forms. *Math. Ann.*, *238* (1978), 157–174.

———, The dimension of the space of cusp forms on the Siegel upper half plane of degree 2 related to a quaternion unitary group. *J. Math. Soc. Japan*, *33* (1981), 125–145.

G. Arfken, *Mathematical Methods for Physicists*. Academic, N.Y., 1970.

J. Arthur, The trace formula for noncompact quotient. *Proc. Internatl. Cong. Math.*, Warsaw, 1983.

———, The trace formula in invariant form. *Ann. of Math. 114* (1981), 1–74.

———, Automorphic representations and number theory, *Canad. Math. Soc. Conf. Proc. 1*. Amer. Math. Soc., Providence, 1981, 3–54.

E. Artin, *Collected Papers*. Addison-Wesley, Reading, Mass., 1965.

T. Asai, On a certain function analogous to $\log|\eta(z)|$. *Nagoya Math. J.*, *40* (1970), 193–211.

A. Ash, Cohomology of congruence subgroups of $SL_n(\mathbb{Z})$. *Math. Ann.*, 249 (1980), 55–73.

A. Ash, D. Grayson, & P. Green, Computations of cuspidal cohomology of congruence subgroups of $SL(3, \mathbb{Z})$. *J. of Number Theory*, 19 (1984), 412–436.

A. Ash, D. Mumford, M. Rapoport, & Y. Tai, *Smooth Compactifications of Locally Symmetric Spaces*. Math. Sci. Press, Brookline, Mass., 1975.

W.L. Baily, *Introductory Lectures on Automorphic Forms*. Princeton University Press, Princeton, 1973.

W.L. Baily & A. Borel, Compactification of arithmetic quotients of bounded symmetric domains. *Ann. of Math.*, 84 (1966), 442–528.

E.S. Barnes, The complete enumeration of extreme senary forms. *Phil. Trans. Royal Soc., London*, 249 (1957), 461–506.

P. Barrucand, H. Williams, & L. Baniuk, A computational technique for determining the class number of a pure cubic field. *Math. Comp.*, 30 (1976), 312–323.

H.J. Bartels, Nichteuklidische Gitterpunktprobleme und Gleichverteilung in linear algebraischen Gruppen. *Comment. Math. Helvetici*, 57 (1982), 158–172.

A.O. Barut & R. Rączka, *Theory of Group Representations and Applications*. Polish Scientific Publishers, Warsaw, 1977.

H. Bass, Algebraic K-theory. *Proc. Internatl. Cong. Math., Vancouver*, 1974, Vol. I, 277–283.

J.G.F. Belinfante & B. Kolman, *A Survey of Lie Groups and Lie Algebras and Computational Methods*. S.I.A.M., Philadelphia, Pa., 1972.

T. Bengtson, Bessel functions on $\mathscr{P}_n$. *Pacific. J. Math.*, 108 (1983), 19–30.

F.A. Berezin, Laplace operators on a semisimple Lie group. *A.M.S. Transl. (2)*, 21 (1962), 239–339.

F.A. Berezin & I.M. Gelfand, Some remarks on the theory of spherical functions on symmetric Riemannian manifolds. *A.M.S. Transl. (2)*, 21 (1962), 193–238.

A. Berman & R.J. Plemmons, *Nonnegative Matrices in the Mathematical Sciences*. Academic, N.Y., 1979.

T.S. Bhanu-Murthy, Plancherel's measure for the factor space $SL(n, \mathbb{R})/SO(n)$. *Dokl. Akad. Nauk., S.S.S.R.*, 133 (1960), 503–506.

L. Bianchi, Geometrische Darstellung der Gruppen linearer Substitutionen mit ganzen complexen Coefficienten nebst Anwendungen auf die Zahlentheorie. *Math. Ann.*, 38 (1891), 313–333.

G.L. Blankenship, Perturbation theory for stochastic ordinary differential equations with applications to optical waveguide analysis, in *Applications of Lie Group Theory to Nonlinear Network Problems*. Western Periodicals Co., N. Hollywood, Calif., 1974, 51–77.

O. Blumenthal, Über Modulfunktionen von mehreren Veränderlichen. *Math. Ann.*, 56 (1903), 509–548; 58 (1904), 497–527.

S. Böcherer, Über die Funktionalgleichung automorpher L-Funktionen zur Siegelschen Modulgruppe, preprint, *J. Reine Angew. Math.*, 362 (1985), 46–168.

―――――, Über die Fourierkoeffizienten der Siegelschen Eisensteinreihen. *Manuscripta Math.*, 45 (1984), 273–288.

S. Bochner, Bessel functions and modular relations of higher type and hyperbolic differential equations. *Comm. Sém. Math. Univ. Lund, Tome suppl.* (1952), 12–20.

S. Bochner & W.T. Martin, *Several Complex Variables*. Princeton University Press, Princeton, N.J., 1948.

A. Borel, *Introduction aux Groupes Arithmétiques*. Hermann, Paris, 1969.

―――――, Arithmetic properties of algebraic groups. *Proc. Internatl. Cong. Math.*, Stockholm, 1962.

―――――, Les fonctions automorphes de plusieurs variables complexes. *Bull. Soc. Math. France*, 80 (1952), 167–182.

———, Les espaces hermitiens symétriques. *Séminaire Bourbaki*, Paris, 1952.

———, Compact Clifford-Klein forms of symmetric spaces. *Topology*, 2 (1963), 111–121.

A. Borel & W. Casselman, *Automorphic Forms, Representations, and L-Functions*, Proc. Symp. Pure Math. 33. A.M.S., Providence, R.I., 1979.

A. Borel & G. Mostow, *Algebraic Groups and Discontinuous Subgroups*, Proc. Symp. Pure Math. Amer. Math. Soc., Providence, 1966.

A. Borel & J.-P. Serre, Corners and arithmetic groups. *Comm. Math. Helv.*, 48 (1973), 436–491.

A. Borel & N. Wallach, *Continuous Cohomology, Discrete Subgroups and Representations of Reductive Groups*. Princeton University Press, Princeton, 1980.

P. Bougerol, Comportement asymptotique des puissances de convolution d'une probabilité sur un espace symétrique. *Astérisque*, 74 (1980), 29–45.

———, *Un Mini-cours sur les Couples de Gelfand*. Publications du Laboratoire de Statistiques et Probabilités de l'Université Paul Sabatier, N° 01-83, 1983.

H. Braun, Konvergenz verallgemeinerter Eisensteinscher Reihen. *Math. Zeitschr.*, 44 (1939), 387–397.

———, Der Basissatz für Hermitesche Modulformen. *Abh. aus dem Math. Sem. d. Univ. Hamburg*, 19 (1955), 134–148.

———, Hermitian modular functions, I, II. *Ann. of Math.*, 50, 51 (1949, 1950), 827–855, 92–104.

A.J. Brentjes, Multi-dimensional continued fraction algorithms, in H.W. Lenstra and R. Tijdeman (Eds.), *Computational Methods in Number Theory*. Math. Centrum, Amsterdam, 1982, 287–320.

E.O. Brigham, *The Fast Fourier Transform*. Prentice-Hall, Englewood Cliffs, N.J., 1974.

T. Broecker & T. tom Dieck, *Representations of Compact Lie Groups*. Springer-Verlag, N.Y., 1985.

D. Bump, *Automorphic Forms on $GL(3)$*, Lecture Notes in Math. 1083. Springer-Verlag, N.Y., 1984.

D. Bump & S. Friedberg, The exterior square automorphic L-functions on $GL(n)$, preprint.

———, On Mellin transforms of unramified Whittaker functions on $GL(3, \mathbb{C})$, preprint.

D. Bump, S. Friedberg, & D. Goldfeld, Poincaré series and Kloosterman sums, in *The Selberg Trace Formula and Related Topics*, Contemp. Math., Vol. 53. A.M.S., Providence, 1986, 39–49.

D. Bump & D. Goldfeld, A Kronecker limit formula for cubic fields, in R.A. Rankin (Ed.), *Modular Forms*. Horwood, Chichester (distrib. Wiley), 1984, 43–49.

D. Bump & J. Hoffstein, Cubic metaplectic forms on $GL(3)$, *Invent. Math.* 84 (1986), 481–505.

R. Burridge & G. Papanicolaou, The geometry of coupled mode propagation in one-dimensional random media. *Comm. Pure Appl. Math.*, 25 (1972), 715–757.

C.J. Bushnell & I. Reiner, L-Functions of arithmetic orders and asymptotic distribution of ideals. *J. für die reine und angew. Math.*, 327 (1981), 156–183.

E. Cartan, Sur une classe remarquable d'espaces de Riemann. *Bull. Soc. Math. France*, 54 (1926), 214–264.

———, Sur une classe remarquable d'espaces de Riemann. *Bull. Soc. Math. France*, 55 (1927), 114–134.

———, Sur la détermination d'un système orthogonal complet dans un espace de Riemann symétrique clos. *Rend. Circ. Mat. Palermo*, 53 (1929), 217–252.

R.W. Carter, *Simple Groups of Lie Type*. Wiley, N.Y., 1972.

W. Casselman, $GL(n)$, in A. Fröhlich (Ed.), *Algebraic Number Fields*. Academic, London, 1977, 663–704.

J.W.S. Cassels, *An Introduction to the Geometry of Numbers*. Springer, Berlin, 1959.

———, *Rational Quadratic Forms.* Academic, N.Y., 1978.
J.W.S. Cassels & A. Fröhlich, *Algebraic Number Theory.* Thompson, Washington, D.C., 1967.
C.-L. Chai, Siegel moduli schemes and their compactifications over $\mathbb{C}$, in G. Cornell and J. Silverman (Eds.), *Arithmetic Geometry.* Springer-Verlag, N.Y., 1986.
C. Chevalley, *Theory of Lie Groups.* Princeton University Press, Princeton, 1946.
Y. Choquet-Bruhat, C. DeWitt-Morette, & M. Dillard-Bleick, *Analysis, Manifolds, and Physics.* North-Holland, N.Y., 1977.
S. Chowla & A. Selberg, On Epstein's zeta function. *J. Reine Angew. Math.*, 227 (1967), 86–110.
U. Christian, Siegelsche Modulfunktionen. *Lectures U. Göttingen,* 1974–1975.
———, *Selberg's Zeta-, L-, and Eisensteinseries, Lecture Notes in Math. 1030.* Springer-Verlag, N.Y., 1983.
———, Maaßsche L-Reihen und eine Identität für Gaußsche Summen. *Abh. Math. Sem. Univ. Hamburg,* 54 (1984), 29–32.
———, Berechnung des Ranges der Schar der Spitzenformen zur Modulgrouppe zweiten Grades und Stufe $q > 2$. *J. Reine Angew. Math.,* 277 (1975), 130–154; 296 (1977), 108–118.
J.E. Cohen, Ergodic theorems in demography. *Bull. Amer. Math. Soc.,* 1 (1979), 275–295.
J.E. Cohen, H. Kesten, & C.M. Newman (Eds.), *Random Matrices and Their Applications, Contemporary Math., Vol. 50.* Amer. Math. Soc., Providence, 1986.
H. Cohn, On the shape of the fundamental domain of the Hilbert modular group. *Proc. Symp. Pure Math., VIII.* Amer. Math. Soc., Providence, 1965, 190–202.
L. Cohn, *The Dimension of Spaces of Automorphic Forms on a Certain Two-Dimensional Complex Domain, Memoirs Amer. Math. Soc.,* 158 (1975).
R. Cooke, *The Mathematics of Sonya Kovalevskaya.* New York, Springer-Verlag, 1984.
R. Courant & D. Hilbert, *Methods of Mathematical Physics, Vol. I.* Wiley-Interscience, N.Y., 1961.
H. Cramér, *Mathematical Methods of Statistics,* Princeton University Press, Princeton, N.J., 1946.
C.W. Curtis, Representations of finite groups of Lie type. *Bull. Amer. Math. Soc.,* 1 (1979), 721–757.
T.W. Cusick & L. Schoenfeld, A table of fundamental pairs of units in totally real cubic fields. *Math. Comp.,* 48 (1987), 147–158.
H. Davenport, *Selected Topics in the Geometry of Numbers.* Lectures, Stanford University, 1950.
B.N. Delone & D.K. Faddeev, *The Theory of Irrationalities of the Third Degree, Transl. Math. Monographs, 10.* Amer. Math. Soc., Providence, 1964.
B.N. Delone & S.S. Ryskov, Extremal problems in the theory of positive quadratic forms. *Proc. Steklov Inst. Math.,* 112 (1971), 211–231.
M. Deuring, *Algebren.* Springer-Verlag, N.Y., 1968.
B. Diehl, Die analytische Fortsetzung der Eisensteinreihe zur Siegelschen Modulgruppe. *J. für die Reine und Angew. Math.,* 317 (1980), 40–73.
J. Dieudonné, *Treatise on Analysis, Vols. I–VI,* Academic, N.Y., 1969–78.
H. Donnelly, On the cuspidal spectrum for finite volume symmetric spaces. *J. Differential Geometry,* 17 (1982), 239–253.
B.A. Dubrovin, V.B. Matveev, & S.P. Novikov, Non-linear equations of Korteweg-deVries type, finite-zone linear operators, and abelian varieties. *Russ. Math. Surveys,* 31 (1976), 59–146.
W. Duke, Hecke's representation for L-functions for GL_n, preprint.
H. Dym & H.P. McKean, *Fourier Series and Integrals.* Academic, N.Y., 1972.
F.J. Dyson, Unfashionable pursuits. *Math. Intelligencer,* 5 (1983), 47–54.
I. Efrat, On a $GL(3)$-analog of $|\eta(z)|$, preprint.

———, *The Selberg Trace Formula for* $PSL_2(\mathbb{R}^n)$, *Memoirs of the Amer. Math. Soc.*, Vol. 65, Number 359. Amer. Math. Soc., Providence, 1987.

I. Efrat & P. Sarnak, The determinant of the Eisenstein matrix and Hilbert class fields. *Trans. Amer. Math. Soc.*, 290 (1985), 815–824.

L. Ehrenpreis & F. Mautner, Some properties of the Fourier transform on semisimple Lie groups, I–III. *Ann. Math.*, 61 (1955), 406–439; *Trans. Amer. Math. Soc.*, 84, 90 (1957, 1959), 1–55, 431–483.

M. Eichler, On theta functions of real algebraic number fields. *Acta Arith.*, 33 (1977), 269–292.

———, Über die Anzahl der linear unabhängigen Siegelschen Modulformen von gegebenem Gewicht. *Math. Ann.*, 213, (1975), 281–291.

———, Zur Begründung der Theorie der automorphen Funktionen in mehreren Variablen. *Aeq. Math.*, 3 (1969), 93–111.

———, *Introduction to the Theory of Algebraic Numbers and Functions*. Academic, N.Y., 1966.

M. Eie, *Dimensions of Spaces of Siegel Cusp Forms of Degree 2 and 3, Memoirs Amer. Math. Soc.*, 50, 1984.

P.D.T.A. Elliott, C. Moreno, & F. Shahidi, On the absolute value of Ramanujan's τ-function. *Math. Ann.*, 266 (1984), 507–511.

J. Elstrodt, F. Grunewald, & J. Mennicke, On the group $PSL_2(\mathbb{Z}[i])$, in J.V. Armitage (Ed.), *Journées Arithmétiques 1980 Exeter*, LMS Lecture Notes, Cambridge University Press, Cambridge, 1982.

———, Discontinuous groups on 3-dimensional hyperbolic space: Analytical theory and arithmetic applications. *Russian Math. Surveys*, 38 (1983), 137–168.

———, $PSL(2)$ over imaginary quadratic integers. *Astérisque*, 94 (1983), 43–60.

———, Eisenstein series on three-dimensional hyperbolic space and imaginary quadratic number fields. *J. für die reine und angew. Math.*, 360 (1985), 160–213.

P. Epstein, Zur Theorie allgemeiner Zetafunktionen, I, II. *Math. Ann.*, 56, 63 (1903, 1907), 614–644, 205–216.

R.D. Farrell, *Techniques of Multivariate Calculus, Lecture Notes in Math.*, Vol. 520. Springer-Verlag, N.Y., 1976.

———, *Multivariate Calculation, Use of the Continuous Groups*. Springer-Verlag, N.Y., 1985.

P. Feit, Locating the poles of Eisenstein series of level 1 on SL_n, Sp_n and $SU(n,n)$, preprint.

———, *Poles and Residues of Eisenstein Series for Symplectic and Unitary Groups, Memoirs of the Amer. Math. Soc.*, 346. A.M.S., Providence, 1986.

L. Fejes Tóth, *Regular Figures*. MacMillan, N.Y., 1964.

W. Feller, *An Introduction to Probability Theory and Its Applications*, Vols. I, II, Wiley, N.Y., 1950, 1966.

H.R.P. Ferguson & R.W. Forcade, Generalization of the Euclidean algorithm for real numbers to all dimensions higher than 2. *J. für die reine und angew Math.*, 334 (1984), 171–181.

R.A. Fisher, The sampling distribution of some statistics obtained from nonlinear equations. *Ann. Eugenics*, 9 (1939), 238–249.

M. Flensted-Jensen, Spherical functions on a real semisimple Lie group. A method of reduction to the complex case. *J. Functional Anal.*, 30 (1978), 106–146.

Y.Z. Flicker, *The Trace Formula and Base Change for GL(3), Lecture Notes in Math.* 927. Springer-Verlag, N.Y., 1982.

E. Freitag, *Siegelsche Modulfunktionen*. Springer-Verlag, N.Y., 1983.

———, Zur Theorie der Modulformen zweiten Grades. *Nachr. Akad. Wiss. Göttingen, II, Math.-Phys. Kl.* (1965), 151–157.

S. Friedberg, A global approach to the Rankin-Selberg convolution for $GL(3, \mathbb{Z})$. *Trans. Amer. Math. Soc.*, 300 (1987), 159–174.

———, *Lectures on Modular Forms and Theta Series Correspondences.* Middle East Technical University Foundation, Ankara, Turkey, 1985.
R. Fueter, Uber automorphe Funktionen der Picard'schen Gruppe I. *Comm. Math. Helve., 3* (1931), 42–68.
H. Funk, Beiträge zur Theorie der Kugelfunktionen. *Math. Ann.,* 77 (1916), 136–152.
H. Furstenberg, Noncommuting random products. *Trans. Amer. Math. Soc., 108* (1963), 377–428.
———, A Poisson formula for semi-simple Lie groups. *Ann. of Math.,* 77 (1963), 335–386.
R. Gangolli, Spectra of discrete uniform subgroups, in W. Boothby and G. Weiss (Eds.), *Geometry and Analysis on Symmetric Spaces.* Dekker, N.Y., 1972, 93–117.
———, Spherical functions on semisimple Lie groups, in W. Boothby and G. Weiss (Eds.), *Geometry and Analysis on Symmetric Spaces.* Dekker, N.Y., 1972, 41–92.
———, On the Plancherel formula and the Paley-Wiener theorem for spherical functions on semisimple Lie groups. *Ann. of Math.,* 93 (1971), 150–165.
———, Isotropic infinitely divisible measures on symmetric spaces. *Acta Math., 111* (1964), 213–246.
R. Gangolli & and G. Warner, On Selberg's trace formula. *J. Math. Soc. Japan,* 27 (1975), 328–343.
———, Zeta functions of Selberg's type for some noncompact quotients of symmetric spaces of rank one. *Nagoya Math. J., 78* (1980), 1–44.
P.R. Garabedian, *Partial Differential Equations.* Wiley, N.Y., 1964.
P. Garrett, Decomposition of Eisenstein series: Rankin triple products, *Ann. Math. 125* (1987), 209–237.
———, Arithmetic properties of Fourier-Jacobi expansions of automorphic forms in several variables. *Amer. J. Math., 103* (1981), 1103–1134.
C.F. Gauss, *Werke.* Königlichen Gesellshaft der Wissenshaften, Göttingen, 1870–1927.
S. Gelbart, Bessel functions, representation theory, and automorphic functions, in *Proc. Symp. Pure Math.,* Vol. 26. A.M.S., Providence, 1973, 343–345.
———, An elementary introduction to the Langlands program. *Bull. Amer. Math. Soc., 10* (1984), 177–220.
———, *Automorphic Forms on Adele Groups,* Princeton University Press, Princeton, 1975.
S. Gelbart & H. Jacquet, A relation between automorphic representations of $GL(2)$ and $GL(3)$. *Ann. Sci. Ecole Norm. Sup., 11* (1978), 471–552.
I.M. Gelfand, Spherical functions on symmetric spaces. *Dokl. Akad. Nauk. S.S.S.R., 70* (1950), 5–8.
I.M. Gelfand, M.I. Graev, & I.I. Piatetski-Shapiro, *Representation Theory and Automorphic Functions.* Saunders, Philadelphia, 1966.
I.M. Gelfand & M.A. Naimark, *Unitary Representations of the Classical Groups* (German translation). Akademie-Verlag, Berlin, 1957.
P. Gérardin, On harmonic functions on symmetric spaces and buildings. *Canad. Math. Soc. Conf. Proc., 1* A.M.S., Providence, 1981, 79–92.
S.G. Gindikin, Analysis in homogeneous domains. *Russ. Math. Surv., 19* (1964), 1–90.
S. Gindikin & F. Karpelevic, Plancherel measures of Riemannian symmetric spaces of non-positive curvature. *Sov. Math. Dokl., 3* (1962), 962–965.
J. Giraud, Surfaces d'Hilbert-Blumenthal, I, II, III, *Springer Lecture Notes in Math.,* Vol. 868. Springer, N.Y., 1981, 1–18, 19–34, 35–37.
R. Godement, A theory of spherical functions. *Trans. Amer. Math. Soc.,* 73 (1952), 496–556.
———, Introduction aux travaux de A. Selberg. *Séminaire Bourbaki, exp. 144,* Paris, 1957.
———, Introduction à la théorie de Langlands. *Séminaire Bourbaki, exp. 244,* Paris, 1962.

———, *Notes on Jacquet-Langlands Theory*. Institute for Advanced Study, Princeton, N.J., 1970.

———, Une généralization du théorème de la moyenne pour les fonctions harmoniques. *C.R. Acad. Sci. Paris, 234* (1952), 2137–2139.

———, La formule des traces de Selberg considerée comme source de problèmes mathématiques. *Séminaire Bourbaki, exp. 244*, Paris, 1962.

R. Godement & H. Jacquet, *Zeta Functions of Simple Algebras, Lecture Notes in Math.*, 260. Springer-Verlag, N.Y., 1972.

D. Goldfeld, J. Hoffstein, & S.J. Patterson, On automorphic functions of half-integral weight with applications to elliptic curves, in N. Koblitz (Ed.), *Number Theory Related to Fermat's Last Theorem*. Birkhäuser, Boston, 1982, 153–193.

R. Goodman, Horospherical functions on symmetric spaces. *Canad. Math. Soc. Conf. Proc., 1.* A.M.S., Providence, 1981, 125–133.

R. Goodman & N. Wallach, Conical vectors and Whittaker vectors. *J. Functional Anal., 39* (1980), 199–279.

E. Gottschling, Explizite Bestimmung der Randflächen des Fundamentalbereiches der Modulgruppe zweiten Grades. *Math. Ann., 138* (1959), 103–124.

F. Götzky, Über eine zahlentheoretische Anwendung von Modulfunktionen zweier Veränderlicher. *Math. Ann., 100* (1928), 411–437.

D. Grenier, *Fundamental Domains for $\mathscr{P}_n/GL(n, \mathbb{Z})$ and Applications in Number Theory*, Ph.D. thesis, U.C.S.D., 1986.

———, Fundamental domains for the general linear group, preprint.

D. Grenier, D. Gordon, & A. Terras, Hecke operators and the fundamental domain for $SL(3, \mathbb{Z})$. *Math. Comp., 48* (1987), 159–178.

W. Greub, S. Halperin, & R. Vanstone, *Connections, Curvature, and Cohomology, Vol. II*. Academic, N.Y., 1973.

K.I. Gross, W.J. Holman, & R.A. Kunze, A new class of Bessel functions and applications in harmonic analysis. *Proc. Symp. Pure Math., Vol. 33*. A.M.S., Providence, 1979, 407–415.

K.I. Gross & D. St. P. Richards, Special functions of matrix argument, I: algebraic induction, zonal polynomials and hypergeometric functions. *Trans. Amer. Math. Soc., 301* (1987), 781–811.

E. Grosswald, *Topics from the Theory of Numbers*. MacMillan, N.Y., 1966.

———, Relations between values at integer arguments of Dirichlet series satisfying functional equations. *Proc. Symp. Pure Math., 24.* Amer. Math. Soc., Providence, 1973.

K.-B. Gundlach, Die Bestimmung der Funktionen zu einigen Hilbertschen Modulgruppen, *J. für die Reine und Angew. Math., 220* (1965), 109–153.

R.K. Guy, *Reviews in Number Theory, 1973–1983*, A.M.S., Providence, 1984.

R.W. Hamming, *Coding and Information Theory*. Prentice-Hall, Englewood Cliffs, N.J., 1980.

W. Hammond, The modular groups of Hilbert and Siegel. *Amer. J. Math., 88* (1966), 497–516.

H. Hancock, *Development of the Minkowski Geometry of Numbers, Vols. I, II*, Dover, N.Y., 1939.

G. Harder, A Gauss-Bonnet formula for discrete arithmetically defined groups. *Ann. Sci. Éc. Norm. Sup., 4,* (1971), 409–455.

Harish-Chandra, *Automorphic Forms on Semi-Simple Lie Groups, Lecture Notes in Math., Vol. 62*. Springer-Verlag, N.Y., 1968.

———, *Collected Papers, I–IV*, Springer-Verlag, N.Y., 1984.

K. Hashimoto, The dimension of the spaces of cusp forms on Siegel upper half plane of degree 2, I. *J. Fac. Sci. Univ. Tokyo, 30* (1983), 403–488; II. *Math. Ann., 266* (1984), 539–559.

M. Hashizume, Whittaker models for real reductive groups. *Japan J. Math., 5* (1979), 349–401.

D. Healy, *A Relationship Between Harmonic Analysis on $SU(2)$ and on $SL(2, \mathbb{C})/SU(2)$*, Ph.D. thesis, U.C.S.D., 1986.

D.R. Heath-Brown & S.J. Patterson, The distribution of Kummer sums at prime arguments. *J. Reine und Angew. Math., 310* (1979), 111–130.

E. Hecke, *Mathematische Werke*, Vandenhoeck und Ruprecht, Göttingen, 1970.

D. Hejhal, *The Selberg Trace Formula for $PSL(2, \mathbb{R})$, Vols. I, II, Lecture Notes in Math., Vols. 548, 1001.* Springer-Verlag, N.Y., 1976, 1983.

———, The Selberg trace formula and the Riemann zeta function. *Duke Math. J., 43* (1976), 441–482.

———, Some observations concerning eigenvalues of the Laplacian and Dirichlet L-Series, in H. Halberstam and C. Hooley (Eds.), *Recent Progress in Analytic Number Theory.* Academic, London, 1981, 95–110.

———, Roots of quadratic congruences and eigenvalues of the non-Euclidean Laplacian, in *The Selberg Trace Formula and Related Topics, Contemporary Math., Vol. 53.* A.M.S., Providence, 1986, 277–339.

D. Hejhal, P. Sarnak, & A. Terras (Eds.), *The Selberg Trace Formula and Related Topics, Contemporary Math., Vol. 53.* A.M.S., Providence, 1986.

S. Helgason, Lie groups and symmetric spaces, in C.M. DeWitt and J.A. Wheeler (Eds.), *Battelle Rencontres.* Benjamin, N.Y., 1968, 1–71.

———, *Differential Geometry and Symmetric Spaces.* Academic, N.Y., 1962.

———, *Differential Geometry, Lie Groups and Symmetric Spaces.* Academic, N.Y., 1978.

———, *Groups and Geometric Analysis.* Academic, N.Y., 1984.

———, *Topics in Harmonic Analysis on Homogeneous Spaces.* Birkhäuser, Boston, 1981.

———, Functions on symmetric spaces. *Proc. Symp. Pure Math., 26.* Amer. Math. Soc., Providence, 1973, 101–146.

———, Analysis on Lie groups and homogeneous spaces. *Amer. Math. Soc. Regional Conf., 14.* A.M.S., Providence, 1971, corrected, 1977.

———, A duality for symmetric spaces with applications to group representations. III. tangent space analysis. *Adv. in Math., 36* (1980), 297–323.

———, *The Radon Transform.* Birkhäuser, Boston, 1980.

———, An analogue of the Paley-Wiener theorem for the Fourier transform on certain symmetric spaces. *Math. Ann., 165* (1966), 297–308.

S. Helgason & K. Johnson, The bounded spherical functions on symmetric spaces. *Adv. in Math., 3* (1969), 586–593.

J.W. Helton, Non-Euclidean functional analysis and electronics. *Bull. Amer. Math. Soc., 7* (1982), 1–64.

———, The distance of a function to H^∞ in the Poincaré metric. Electrical power transfer. *Journal of Functional Anal., 38* (1980), 273–314.

———, A simple test to determine gain bandwidth limitations. *Proc. I.E.E.E. Internatl. Conf. on Circuits and Systems*, 1977.

R.A. Herb & J.A. Wolf, The Plancherel theorem for general semisimple Lie groups, MSRI 015-84-7, Math. Sciences Research Inst., Berkeley, California, 1984.

R. Hermann, *Lie Groups for Physicists.* Benjamin, N.Y., 1966.

———, *Fourier Analysis on Groups and Partial Wave Analysis.* Benjamin, N.Y., 1969.

C. Hermite, *Oeuvres, Vols. I–IV.* Gauthiers-Villars, Paris, 1905–1917.

O. Herrmann, Über Hilbertsche Modulfunktionen und die Dirichletschen Reihen mit Eulerscher Produktentwicklung. *Math. Ann., 127* (1954), 357–400.

C. Herz, Bessel functions of matrix argument. *Ann. of Math., 61* (1955), 474–523.

K. Hey, *Analytische Zahlentheorie in Systemen Hyperkomplexer Zahlen.* Inaug.-Diss., Hamburg, 1929.

F. Hirzebruch, Hilbert modular surfaces, *L'Enseignement Math., 21* (1973), Université Genève.

———, *Topological Methods in Algebraic Geometry.* Springer-Verlag, N.Y., 1966.

———, The ring of Hilbert modular forms for real quadratic fields of small discriminant. *Lecture Notes in Math.*, *627*. Springer-Verlag, N.Y., 1977, 288–323.

F. Hirzebruch & G. Van der Geer, *Lectures on Hilbert Modular Surfaces*. Presses Université de Montréal, Montréal, Quebec, 1981.

F. Hirzebruch & D. Zagier, Intersection numbers of curves on Hilbert modular surfaces and modular forms of Nebentypus. *Inv. Math.*, *36* (1976), 57–113.

———, Classification of Hilbert modular surfaces, in *Complex Analysis and Algebraic Geometry*. Iwanami Shoten, Tokyo, 1977, 43–77.

J. Hoffstein, Real zeros of Eisenstein series. *Math. Z.*, *181* (1982), 179–190.

R.T. Hoobler & H.L. Resnikoff, Normal connections for automorphic embeddings, preprint.

C. Hooley, On the distribution of the roots of polynomial congruences. *Mathematika*, *11* (1964), 39–49.

L. Hörmander, *An Introduction to Complex Analysis in Several Variables*. Van Nostrand, Princeton, N.J., 1966.

P.L. Hsu, On the distribution of the roots of certain determinantal equations. *Ann. Eugenics*, *9* (1939), 250–258.

L.K. Hua, *Harmonic Analysis of Functions of Several Complex Variables in the Classical Domains*, Transl. of Math. Monographs, 6. A.M.S., Providence, R.I., 1963.

G. Humbert, Théorie de la réduction des formes quadratiques définis positives dans un corps algébrique K fini. *Comm. Math. Helv.*, *12* (1939/40), 263–306.

———, Sur la mesure des classes d'Hermite de discriminant donné dans un corps quadratique imaginaire, et sur certains volumes non euclidiens. *Comptes Rendus, Paris*, *169* (1919), 448–454.

J. Humphreys, *Arithmetic Groups, Lecture Notes in Math.*, *789*, Springer-Verlag, N.Y., 1980.

J. Hunter, *Harmonic Analysis over Imaginary Quadratic Number Fields*, Ph.D. thesis, U.C.S.D., 1982.

N. Hurt, *Geometric Quantization in Action*. D. Reidel, Amsterdam, 1983.

———, Propagators in quantum mechanics on multiply connected spaces, in *Lecture Notes in Physics*, *Vol. 50*. Springer-Verlag, N.Y., 1976, 182–192.

J.-I. Igusa, *Theta Functions*, Springer-Verlag, N.Y., 1964.

———, On Siegel modular forms of genus 2, I, II. *Amer. J. Math.*, *84* (1962), 175–200; *86* (1964), 392–412.

K. Imai, Generalization of Hecke's correspondence to Siegel modular forms. *Amer. J. Math.*, *102* (1980), 903–936.

K. Imai & A. Terras, Fourier expansions of Eisenstein series for $GL(3, \mathbb{Z})$. *Trans. Amer. Math. Soc.*, *273* (1982), 679–694.

A.E. Ingham, An integral which occurs in statistics. *Proc. Cambridge Phil. Soc.*, *29* (1933), 271–276.

K. Iwasawa, On some types of topological groups. *Ann. of Math.*, *50* (1949), 507–558.

H. Jacquet, Les fonctions de Whittaker associées aux groupes de Chevalley. *Bull. Soc. Math. France*, *95* (1967), 243–309.

H. Jacquet & R.P. Langlands, *Automorphic Forms on $GL(2)$, Lecture Notes in Math.* 114. Springer-Verlag, N.Y., 1970.

H. Jacquet, I.I. Piatetski-Shapiro, & J. Shalika, Automorphic forms on $GL(3)$, I, II. *Ann. of Math.*, *109* (1979), 169–212, 213–258.

———, Rankin-Selberg convolutions. *Amer. J. Math.* *105* (1983), 367–464.

H. Jacquet & J. Shalika, On Euler products and the classification of automorphic representations, I, II. *Amer. J. Math.*, *103* (1981), 499–558, 777–815.

———, A non-vanishing theorem for zeta functions of $GL(n)$. *Inv. Math.*, *38* (1976), 1–16.

A.J. James, Special functions of matrix and single argument in statistics, in R. Askey (Ed.), *Theory and Applications of Special Functions*. Academic, N.Y., 1975, 497–520.

———, Distributions of matrix variates and latent roots derived from normal samples. *Ann. Math. Statist.*, *35* (1964), 475–501.

———, Zonal polynomials of the real positive definite symmetric matrices. *Ann. of Math.*, *75* (1961), 456–469.

J.M. Jauch, Projective representations of the Poincaré group, in E.M. Loebl (Ed.), *Group Theory and Its Applications.* Academic, N.Y., 1968, 131–182.

F. John, *Plane Waves and Spherical Means Applied to Partial Differential Equations.* Wiley-Interscience, N.Y., 1955.

G.A. Kabatiansky & V.I. Levenshtein, Bounds for packings on the sphere and in space. *Problems of Information Transmission*, *14* (1978), 1–17.

V.I. Kalinin, Eisenstein series on the symplectic group. *Math. U.S.S.R. Sbornik*, *32* (1977), 449–476.

M. Karel, Eisenstein series and fields of definition. *Compositio Math.*, *32* (1976), 225–291.

F.I. Karpelevich, V.N. Tutubalin, & M.G. Shur, Limit theorems for the compositions of distributions in the Lobachevsky plane and space. *Theory of Probability and its Applications*, *4* (1959), 399–402.

G. Kaufhold, Dirichletsche Reihen mit Funktionalgleichung in der Theorie der Modulfunktion 2. Grades. *Math. Ann.*, *137* (1959), 454–476.

D.A. Kazhdan & S.J. Patterson, *Metaplectic Forms, Inst. des Hautes Etudes Scientifiques, Publ. Math.* 59, Paris, 1984.

L. Keen (Ed.), *The Legacy of Sonya Kovalevskaya, Contemporary Math.*, *64*. Amer. Math. Soc., Providence, 1987.

O-H. Keller, Geometrie der Zahlen. *Enzyklop. der Math. Wissenschaften, I.2.2. Aufl. Heft. 11, III.*

A.B. Kirillov, Unitary representations of nilpotent Lie groups. *Russ. Math. Surv.*, *17* (1962), 53–104.

———, *Elements of the Theory of Representations.* Springer-Verlag, N.Y., 1976.

H. Klingen, Volumbestimmung des Fundamentalbereichs der Hilbertschen Modulgruppe *n*-ten Grades. *J. für die Reine und Angew. Math.*, *206* (1961), 9–19.

———, Eisensteinreihen zur Hilbertschen Modulgruppe *n*-ten Grades. *Nachr. Akad. Wiss. Göttingen*, (1960), 87–104.

———, Über die Werte der Dedekindschen Zetafunktion. *Math. Ann.*, *145* (1962), 265–272.

———, Zum Darstellungssatz für Siegelsche Modulformen. *Math. Z.*, *102* (1967), 30–43; *105* (1968), 399–400.

———, On Eisenstein series and some applications, in *Automorphic Forms of Several Variables, Katata Conference, 1983.* Birkhäuser, 1984.

H.D. Kloosterman, Thetareihen in total reellen algebraischen Zahlkörpern. *Math. Ann.*, *103* (1930), 279–299.

A. Hibner Koblitz, *A Convergence of Lives. Sofia Kovalevskaia: Scientist, Writer, Revolutionary.* Birkhäuser, Boston, 1983.

P. Kochina, *Love and Mathematics: Sofya Kovalevskaya.* Mir, Moscow, 1985.

M. Koecher, Über Dirichlet-Reihen mit Funktionalgleichung. *J. Reine und Angew. Math.*, *192* (1953), 1–23.

———, Über Thetareihen indefiniter quadratischer Formen. *Math. Nachr.*, *9*, (1953), 51–85.

———, Gruppen und Lie Algebren rationaler Funktionen, *Math. Z.*, *109* (1969), 349–392.

A.N. Kolmogorov & S.V. Fomin, *Introductory Real Analysis.* Dover, N.Y., 1975.

T. Koornwinder, Jacobi functions and analysis on noncompact semisimple Lie groups, in R. Askey (Ed.), *Special Functions, Group Theoretical Aspects and Applications.* D. Reidel, Boston, 1984, pp. 1–85.

A. Koranyi, A survey of harmonic functions on symmetric spaces. *Proc. Symp. Pure*

Math., Vol. 35. Amer. Math. Soc., Providence, R.I., 1979, 323–344.
A. Korkine & G. Zolotareff, Sur les formes quadratiques positives. *Math. Ann. 11* (1877), 242–292.
B. Kostant, On Whittaker vectors and representation theory. *Inv. Math.*, 48 (1978), 101–184.
S. Kovalevsky, Sur le problème de la rotation d'un corps solide autour d'un point fixe. *Acta Math. 12* (1889), 177–232.
A. Krieg, *Modular Forms on Half-Spaces of Quaternions, Lecture Notes in Math., Vol. 1143.* Springer-Verlag, N.Y., 1985.
———, *Lectures on Hecke Algebras*, U.C.S.D., 1988.
T. Kubota, *Elementary Theory of Eisenstein Series.* Wiley, N.Y., 1973.
———, Some results concerning reciprocity law and real analytic automorphic functions. *Proc. Symp. Pure Math., Vol. 20.* American Math. Soc., Providence, R.I., 1971, 382–395.
———, *On Automorphic Functions and the Reciprocity Law in a Number Field, Lectures in Math., Kyoto University.* Kinokuniya Book Store Co., Ltd., Tokyo, Japan, 1969.
———, Über diskontinuierliche Gruppen Picardschen Typus und zugehörige Eisensteinsche Reihen. *Nagoya Math. J.*, 32 (1968), 259–271.
———, On a special kind of Dirichlet series. *J. Math. Soc. Japan*, 20 (1968), 193–207.
S. Kudla, Relations between automorphic forms produced by theta functions. *Lecture Notes in Math., 627.* Springer-Verlag, N.Y., 1977, 277–285.
———, Theta functions and Hilbert modular forms. *Nagoya Math. J.*, 69 (1978), 97–106.
J.C. Lagarias & A. Odlyzko, Solving low-density subset sum problems. *Proc. 24th Annual IEEE Symp. on Found. Comp. Science*, (1983), 1–10.
J.-L. Lagrange, *Oeuvres, Vols. I–XIV.* Gauthier-Villars, Paris, MDCCCXCII–MDCCCLVII.
S. Lang, *Real Analysis.* Addison-Wesley, Reading, Mass., 1983.
———, *Analysis I.* Addison-Wesley, Reading, Mass., 1968.
———, $SL_2(\mathbb{R})$. Addison-Wesley, Reading, Mass., 1975.
———, *Algebraic Number Theory.* Addison-Wesley, Reading, Mass., 1968.
R. Langlands, *Eisenstein Series, Lecture Notes in Math., Vol. 544.* Springer-Verlag, N.Y., 1976.
———, Review of Osborne and Warner, Eisenstein Systems. *Bull. Amer. Math. Soc.* 9 (1983), 351–361.
———, *Base Change for GL(2).* Princeton University Press, Princeton, 1980.
———, Problems in the theory of automorphic forms. *Lecture Notes in Math. 170.* Springer-Verlag, N.Y., 1970, 18–61.
———, *Euler Products.* Yale University Press, New Haven, 1967.
———, The dimension of spaces of holomorphic forms. *Amer. J. Math.* 85 (1963), 99–125.
———, L-functions and automorphic representations. *Proc. Internatl. Cong. Math.*, Helsinki, 1978.
N.N. Lebedev, *Special Functions and Their Applications.* Dover, N.Y., 1972.
R. Lee & J. Schwermer, Cohomology of arithmetic subgroups of SL_3 at infinity. *J. für die Reine und Angew. Math.*, 330 (1982), 100–131.
R. Lee & R.H. Szczarba, Homology and cohomology of congruence subgroups. *Proc. Natl. Acad. Sci. U.S.A.*, 72 (1975), 651–653.
W.J. LeVeque, *Reviews in Number Theory*, A.M.S., Providence, 1974.
G. Lion & M. Vergne, *The Weil Representation, Maslov Index and Theta Series.* Birkhäuser, Boston, 1980.
K. Lonngren & A. Scott (Eds.), *Solitons in Action.* Academic, N.Y., 1978.
O. Loos, *Symmetric Spaces, I, II.* Benjamin, N.Y., 1969.

A. Lubotzky, R. Phillips, & P. Sarnak, Hecke operators and distributing points on the sphere, I, preprint.

H. Maass, *Lectures on Modular Forms of One Complex Variable.* Tata Institute of Fundamental Research, Bombay, India, 1964.

———, *Siegel's Modular Forms and Dirichlet Series*, Lecture Notes in Math., Vol. 216. Springer-Verlag, N.Y., 1971.

———, Über eine neue Art von nichtanalytischen automorphen Funktionen und die Bestimmung Dirichletscher Reihen durch Funktionalgleichung. *Math. Ann., 121* (1949), 141–183.

———, Modulformen zweiten Grades und Dirichletreihen. *Math. Ann., 122* (1950), 90–108.

———, Die Primzahlen in der theorie der Siegelschen Modulfunktionen. *Math. Ann., 117* (1940), 538–578.

———, Spherical functions and quadratic forms. *J. Indian Math. Soc., 20* (1956), 117–162.

———, Zetafunktionen mit Grössencharakteren und Kugelfunktionen. *Math. Ann., 134* (1957), 1–32.

———, Zur Theorie der Kugelfunktionen einer Matrix-variablen. *Math. Ann., 135* (1958), 391–416.

———, Zur Theorie der automorphen Funktionen von n Veränderlichen. *Math. Ann., 117,* (1940), 538–578.

———, Über Gruppen von hyperabelschen Transformationen. *Sitz.-Ber. der Heidelberg Akad. Wiss., Math.-Nat., Kl. 2 Abh.* (1940).

———, Modulformen zu indefiniten quadratischen Formen, *Math. Scand., 17,* (1965), 41–55.

———, Automorphe Funktionen und indefinite quadratische Formen. *Sitz.-Ber. der Heidelberg Akad. Wiss., Math.-Nat. Kl., 1 Abh.* (1949).

———, Über die räumliche Verteilung der Punkte in Gittern mit indefiniter Metrik. *Math. Ann., 138* (1959), 287–315.

I. Macdonald, Some conjectures for root systems. *S.I.A.M. J. Math. Anal., 13* (1982), 988–1007.

———, *Symmetric Functions and Hall Polynomials.* Clarenden Press, Oxford, 1979.

G. Mackey, *Unitary Group Representations in Physics, Probability and Number Theory.* Benjamin/Cummings, Reading, Mass., 1978.

———, *Harmonic Analysis as the Exploitation of Symmetry—A Historical Survey*, Rice Univ. Studies, Vol. 64. Houston, Texas, 1978.

———, *The Theory of Group Representations.* University of Chicago Press, Chicago, Ill., 1976.

K. Maurin, *General Eigenfunction Expansions and Unitary Representations of Topological Groups.* Polish Scientific Publ., Warsaw, 1968.

F. Mautner, Spherical functions and Hecke operators, in *Lie Groups and Their Representations, Proc. Summer School Bolya Janos Math. Soc., Budapest, 1971.* Halsted, N.Y., 1975, 555–576.

———, Geodesic flows on symmetric Riemannian spaces. *Ann. of Math., 65* (1957), 416–431.

H.P. McKean & E. Trubowitz, Hill's operator and hyperelliptic function theory in the presence of infinitely many branch points. *Comm. Pure Appl. Math., 29* (1976), 143–226.

M.L. Mehta, *Random Matrices and the Statistical Theory of Energy Levels.* Academic, N.Y., 1967.

J. Mennicke, *Vorträge über Selbergs Spurformel I.* University of Bielefeld, W. Germany.

———, Lectures on discontinuous groups on 3-dimensional hyperbolic space. Modular Forms Conf., Durham, 1983.

P. Menotti & E. Onofri, The action of $SU(N)$ lattice gauge theory in terms of the heat kernel on the group manifold, in Claudio Rebbi (Ed.), *Lattice Gauge Theories and Monte Carlo Simulation*. World Scientific Publ., Singapore, 1983, 447–459.

J. Milnor, Hilbert's problem 18. *Proc. Symp. Pure Math.*, 28. Amer. Math. Soc., Providence, 1976, 491–506.

———, Hyperbolic geometry: the first 150 years. *Bull. Amer. Math. Soc.*, 6 (1982), 9–24.

J. Milnor & D. Husemoller, *Symmetric Bilinear Forms*. Springer-Verlag, N.Y., 1973.

H. Minkowski, *Gesammelte Abhandlungen*. Chelsea, N.Y., 1911 (reprinted 1967).

C.W. Misner, K.S. Thorne, & J.A. Wheeler, *Gravitation*. Freeman, San Francisco, 1973.

M.I. Monastyrsky & A.M. Perelomov, Coherent states and bounded homogeneous domains. *Reports on Math. Phys.*, 6 (1974), 1–14.

H. Montgomery, The pair correlation of zeros of the zeta function. *Proc. Symp. Pure Math.*, Vol. 24. A.M.S., Providence, 1973, 181–193.

C. Moore, Compactifications of symmetric spaces. *Amer. J. Math.*, 86 (1964), 201–218.

———, Representations of solvable and nilpotent groups and harmonic analysis on nil and solvimanifolds. *Proc. Symp. Pure Math.*, Vol. 26. A.M.S., Providence, 1973, 3–44.

L.J. Mordell, On Hecke's modular functions, zeta functions, and some other analytic functions in the theory of numbers. *Proc. London Math. Soc.*, 32 (1931), 501–556.

C. Moreno & F. Shahidi, The L-functions $L(s, Sym^m(r), \pi)$. *Canad. Math. Bull.*, 28 (1985), 405–410.

———, The fourth moment of the Ramanujan tau function. *Math. Ann.*, 266 (1983), 233–239.

Y. Morita, An explicit formula for the dimension of spaces of Siegel modular forms of degree 2. *J. Fac. Sci. U. Tokyo*, 21 (1974), 167–248.

D.F. Morrison, *Multivariate Statistical Methods*. McGraw-Hill, N.Y., 1976.

G. Mostow, Some new decomposition theorems for semisimple Lie groups. *Memoirs, A.M.S.*, 14 (1955), 31–54.

———, Discrete subgroups of Lie groups. *Advances in Math.*, 15 (1975), 112–123.

G. Mostow & T. Tamagawa, On the compactness of arithmetically defined homogeneous spaces. *Ann. of Math.*, 76 (1962), 440–463.

R.J. Muirhead, *Aspects of Multivariate Statistical Theory*. Wiley, N.Y., 1978.

W. Müller, Signature defects of cusps of Hilbert modular varieties and values of L-series at $s = 1$. Report Math., Akad. der Wiss. der D.D.R., Inst. für Math., Berlin, 1983.

D. Mumford, *Tata Lectures on Theta, I, II*. Birkhäuser, Boston, 1982, 1984.

Y. Namikawa, *Toroidal Compactification of Siegel Spaces, Lecture Notes in Math.*, 812. Springer-Verlag, N.Y., 1980.

S.P. Novikov, A method for solving the periodic problem for the KdV equation and its generalizations. *Rocky Mt. J. Math.*, 8 (1978), 83–93.

N.V. Novikova, Korkin-Zolotarev reduction domains of positive quadratic forms in $n \leq 8$ variables and a reduction algorithm for these domains. *Soviet Math. Dokl.*, 27 (1983), 557–560.

M.E. Novodvorsky & I.I. Piatetski-Shapiro, Rankin-Selberg method in the theory of automorphic forms. *Proc. Symp. Pure Math.* 30. Amer. Math. Soc., Providence, 1976, 297–301.

M.S. Osborne & G. Warner, *The Theory of Eisenstein Systems*. Academic, N.Y., 1981.

A. Perelomov, *Generalized Coherent States and Their Applications*. Springer-Verlag, N.Y., 1986.

I.I. Piatetski-Shapiro, Cuspidal automorphic representations associated to parabolic subgroups and Ramanujan conjecture, in N. Koblitz (Ed.), *Number Theory Related to Fermat's Last Theorem*. Birkhäuser, Boston, 1982.

———, *Automorphic Functions and the Geometry of the Classical Domains*. Gordon and Breach, N.Y., 1969.

———, Euler subgroups, in I.M. Gelfand (Ed.), *Lie Groups and Their Representations, Summer School Bolyai Janos Math. Soc.* Halsted, N.Y., 1975, 597–620.

E. Picard, Sur un groupe de transformations des points de l'espace situés du même coté d'un plan. *Bull. Soc. Math. de France, 12* (1844), 43–47.

P. Ploch, Bestimmung von Konjugiertenklassen und Beweis von Verschwindungssätzen, die bei der Berechnung des Ranges der Schar der Spitzenformen zur Siegelschen Modulgruppe vierten Grades und Stufe $q \geq 3$ Auftreten, Dissertation, Göttingen, 1985.

H. Pollard, *The Theory of Algebraic Numbers.* Math. Assoc. of America, Washington, D.C., 1961.

A.M. Polyakov, Quantum geometry of bosonic strings. *Phys. Lett., 103B* (1981), 207; Quantum geometry of fermionic strings. *Phys. Lett., 103B* (1981), 211.

S.J. Press, *Applied Multivariate Analysis.* Holt, Rinehart and Winston, Inc. N.Y., 1972.

N.V. Proskurin, Expansions of automorphic functions. *J. Sov. Math., 26* (1984), 1908–1921.

N.J. Pullman, *Matrix Theory and Its Applications.* Dekker, N.Y., 1976.

M.S. Raghunathan, *Discrete Subgroups of Lie Groups.* Springer-Verlag, N.Y., 1972.

K.G. Ramanathan, Quadratic forms over involutorial division algebras II. *Math. Ann., 143* (1961), 293–332.

———, Zeta functions of quadratic forms. *Acta Arith.* 7 (1961), 39–69.

M. Reed & B. Simon, *Methods of Modern Mathematical Physics, I, Functional Analysis.* Academic, N.Y., 1972.

A. Regev, letter of 8/30/79, Math. Dept., Weizmann Inst. Science, Rehovot, Israel.

H. Resnikoff, On the graded ring of Hilbert modular forms associated with $\mathbb{Q}(\sqrt{5})$. *Math. Ann., 203* (1974), 161–170.

———, Automorphic forms of singular weight are singular forms. *Math. Ann., 215* (1975), 175–193.

———, Theta functions for Jordan algebras. *Inv. Math., 31,* (1975), 87–104.

———, Differential geometry and color perception. *J. Math. Biology, 1* (1974), 97–131.

H. Resnikoff & Y.-S. Tai, On the structure of a graded ring of automorphic forms on the 2-dimensional complex ball, I, II. *Math. Ann., 238* (1978), 97–117; *258* (1982), 367–382.

J. Rhodes, *Modular Forms on p-adic Planes*, Ph.D. thesis, M.I.T., 1986.

W. Roelcke, Über die Wellengleichung bei Grenzkreisgruppen erster Art. *Sitzber. Akad. Heidelberg, Math.-naturwiss. Kl. 1953/55.*

C.A. Rogers, *Packing and Covering.* Cambridge University Press, Cambridge, 1964.

J. Rosenberg, A quick proof of Harish-Chandra's Plancherel theorem for spherical functions on a semisimple Lie groups. *Proc. Amer. Math. Soc., 63* (1977), 143–149.

S.N. Roy, P-statistics or some generalizations in analysis of variance appropriate to multivariate problems. *Sankyha, 4* (1939), 381–396.

J.A. Rush, On Hilbert's 18*th* problem: packing, preprint.

J.A. Rush & N. Sloane, An improvement to the Minkowski-Hlawka bound for packing superballs, *Mathematika, 34* (1987), 8–18.

S.S. Ryskov, The geometry of positive quadratic forms. *A.M.S. Transl. (2), 109.* Amer. Math. Soc., Providence, 1977, 27–32.

———, The theory of Hermite-Minkowski reduction of positive quadratic forms. *J. Sov. Math., 6* (1976), 651–676.

S.S. Ryskov & E.P. Baranovskii, Classical methods in the theory of lattice packings. *Russian Math. Surveys, 34* (1979), 1–68.

A.A. Sagle & R.E. Walde, *Introduction to Lie Groups and Lie Algebras.* Academic, N.Y., 1973.

H. Saito, *Automorphic Forms and Algebraic Extensions of Number Fields, Lecture Notes in Math., 8.* Kinokuniya Book Store, Tokyo, Japan, 1975.

P. Sarnak, The arithmetic and geometry of some hyperbolic three-manifolds. *Acta Math.*, 151 (1983), 253–295.
I. Satake, Theory of spherical functions on reductive algebraic groups over p-adic fields. *Publ. Math.*, 18. Inst. des Hautes Etudes, Paris, 1963.
———, Review of Ash, Mumford, Rapoport and Tai in *Math. Reviews*, 56, #15642.
———, On the compactification of the Siegel space. *J. Indian Math. Soc.*, 20 (1956), 259–281.
G. Schiffman, Intégrales d'entrelacement et fonctions de Whittaker. *Bull. Soc. Math. France*, 99 (1971), 3–72.
B. Schoeneberg, Das Verhalten von mehrfachen Thetareihen bei Modulsubstitutionen. *Math. Ann.*, 116 (1939), 511–523.
R.L.E. Schwarzenberger, *N-Dimensional Crystallography*. Pitman, San Francisco, 1980.
J. Schwermer, *Eisensteinreihen und die Kohomologie von Kongruenzuntergruppen von $SL(n, \mathbb{Z})$*, Bonner Math. Schriften 99, 1977.
———, *Kohomologie arithmetisch definierter Gruppen und Eisensteinreihen*, Lecture Notes in Math. 988. Springer-Verlag, N.Y., 1983.
L.A. Seeber, *Untersuchungen über die Eigenschaften der Positive Ternaren Quadratischen Formen*. Freiburg, 1831.
A. Selberg, Harmonic analysis and discontinuous groups in weakly symmetric Riemannian spaces with applications to Dirichlet series. *J. Indian Math. Soc.*, 20 (1956), 47–87.
———, Remarks on a multiple integral. *Norsk Mat. Tidsskr.*, 26 (1944), 71–78 (in Norwegian).
———, Discontinuous groups and harmonic analysis. *Proc. Internatl. Cong. of Math.*, Stockholm, 1962, 177–189.
———, A new type of zeta function connected with quadratic forms. Report of the Institute in the Theory of Numbers, University of Colorado, Boulder, 1959, 207–210.
Séminaire H. Cartan, 1957/58, *Fonctions Automorphes*. Benjamin, N.Y., 1967.
J.-P. Serre, Cohomologie des groupes discrets, in *Prospects in Math*. Princeton University Press, Princeton, N.J., 1971, 77–169.
I.R. Shafarevitch, *Basic Algebraic Geometry*. Springer-Verlag, N.Y., 1974.
F. Shahidi, On the Ramanujan conjecture and finiteness of poles for certain L-functions, *Ann. Math.*, to appear.
D. Shale, Linear symmetries of free boson fields. *Trans. Amer. Math. Soc.*, 103 (1962), 149–167.
J.A. Shalika, The multiplicity one theorem for $GL(n)$. *Ann. of Math.*, 100 (1974), 171–193.
H. Shimizu, On discontinuous groups operating on the product of the upper half planes. *Ann. of Math.*, 77 (1963), 33–71.
G. Shimura, Confluent hypergeometric functions on tube domains. *Math. Ann.*, 260 (1982), 269–302.
———, *Introduction to the Arithmetic Theory of Automorphic Functions*. Princeton University Press, Princeton, 1971.
———, On Eisenstein series. *Duke Math. J.*, 50 (1983), 417–476.
T. Shintani, On an explicit formula for class 1 "Whittaker functions" over $\mathscr{P}$-adic fields. *Proc. Japan Acad.*, 52 (1976), 180–182.
———, On "liftings" of holomorphic automorphic forms (a representation-theoretic interpretation of the recent work of Saito), *U.S.-Japan Sem.*, Ann Arbor, Michigan, 1975.
———, On zeta-functions associated with the vector space of quadratic forms. *J. Fac. Sci. U. Tokyo*, 22 (1975), 25–65.
C.L. Siegel, *Gesammelte Abhandlungen*, Vols. I–IV. Springer-Verlag, N.Y., 1966, 1979.
———, *Lectures on Quadratic Forms*. Tata Institute of Fundamental Research, Bombay, 1957.

———, *Geometry of Numbers*. Lectures, New York University, 1945–1946.
———, *Lectures on Advanced Analytic Number Theory*. Tata Institute of Fundamental Research, Bombay, India, 1957.
———, *Topics in Complex Function Theory*. Wiley-Interscience, N.Y., 1969–1973.
F. Sigrist, Sphere packing. *Math. Intelligencer*, 5 (1983), 34–38.
N.J.A. Sloane, The packing of spheres. *Scientific American*, 250 (1984), 116–125.
———, Binary codes, lattices, and sphere-packings. *Combinatorial Surveys, Proc. 6th British Combinat. Conf.* Academic, London, 1977, 117–164.
L. Solomon, Partially ordered sets with colors. *Proc. Symp. Pure Math., Vol. 34.* A.M.S., Providence, 1979, 309–329.
C. Soulé, The cohomology of $SL(3, \mathbb{Z})$. *Topology*, 17 (1978), 1–22.
———, Cohomology of $SL_3(\mathbb{Z})$. *Comptes Rendus Acad. Sci. Paris*, 280 (1975), 251–254.
H. Stark, M.I.T. and U.C.S.D number theory course lecture notes.
———, Fourier coefficients of Maass wave forms, in R.A. Rankin (Ed.), *Modular Forms* Horwood, Chichester (distributed by Wiley), 1984, 263–269.
———, Modular forms and related objects. *Canadian Math. Soc. Conf. Proc.*, 7 (1987), 421–455.
———, On the transformation formula for the symplectic theta function and applications. *J. Fac. Sci. U. Tokyo*, 29 (1982), 1–12.
———, On modular forms from L-functions in number theory, I, II, to appear.
G. Strang, *Linear Algebra and Its Applications*. Academic, N.Y., 1976.
G. Strang & G.J. Fix, *An Analysis of the Finite Element Method*. Prentice-Hall, Englewood Cliffs, N.J., 1973.
H. Strassberg, L functions for $GL(n)$, *Math. Ann.*, 245 (1979), 23–36.
R. Styer, *Hecke Theory over Complex Quadratic Fields*, Ph.D. thesis, M.I.T., 1981.
A. Takemura, *Zonal Polynomials, Inst. of Math. Stat. Lecture Notes*, Vol. 4. Inst. of Math. Stat., Hayward, California, 1984.
L.A. Takhtadzhyan & A.I. Vinogradov, Theory of Eisenstein series for the group $SL(3, \mathbb{R})$ and its application to a binary problem. *J. Soviet Math.* 18 (1982), 293–324.
T. Tamagawa, On Selberg's trace formula. *J. Fac. Sci. U. Tokyo, Sec. I*, 8 (1960), 363–386.
———, On the zeta-functions of a division algebra. *Ann. of Math.*, 77 (1963), 387–405.
———, On some extensions of Epstein's Z-series. *Proc. Internatl. Symp. on Alg. No. Theory*, Tokyo-Nikko, 1955, 259–261.
P.O. Tammela, The Minkowski reduction domain for positive quadratic forms of seven variables. *Proc. Steklov Inst. Math.*, 67 (1977), 108–143 (translation, *J. Soviet Math.*, 16 (1981), 836–857).
Y. Tanigawa, Selberg trace formula for Picard groups. *Proc. Internatl. Symp. on Alg. Number Theory*, Tokyo, 1977, 229–242.
A. Terras, Integral formulas and integral tests for series of positive matrices. *Pac J. Math.*, 89 (1980), 471–490.
———, Special functions for the symmetric space of positive matrices. *S.I.A.M. J. Math. Anal.*, 16 (1985), 620–640.
———, Fourier coefficients of Eisenstein series of one complex variable for the special linear group. *Trans. A.M.S.*, 205 (1975), 97–114.
———, The Chowla-Selberg method for Fourier expansion of higher rank Eisenstein series. *Canad. Math. Bull.*, 28 (1985), 280–294.
———, Some simple aspects of the theory of automorphic forms for $GL(n, \mathbb{Z})$. *J. Contemp. Math.*, 53 (1986), 409–447.
———, A generalization of Epstein's zeta function. *Nagoya J. Math.*, 42 (1971), 173–188.
———, Functional equations of generalized Epstein zeta functions in several variables. *Nagoya J. Math.*, 44 (1971), 89–95.

———, On automorphic forms for the general linear group. *Rocky Mt. J. of Math.*, *12* (1982), 123–143.

———, Real zeroes of Epstein's zeta function for ternary positive quadratic forms. *Illinois J. Math.*, *23* (1979), 1–14.

———, The minima of quadratic forms and the behavior of Epstein and Dedekind zeta functions. *J. Number Theory*, *12* (1980), 258–272.

———, Bessel series expansions of the Epstein zeta function and the functional equation. *Trans. Amer. Math. Soc.*, *183* (1973), 477–486.

———, Applications of special functions for the general linear group to number theory. *Sém. Delange-Pisot-Poitou, 1976–1977,* exp. 23.

———, The Fourier expansion of Epstein's zeta function for totally real algebraic number fields and some consequences for Dedekind's zeta function. *Acta Arith.*, *30* (1976), 187–197.

———, The Fourier expansion of Epstein's zeta function over an algebraic number field and its consequences for algebraic number theory. *Acta Arith.*, *32* (1977), 37–53.

———, A relation between $\zeta_K(s)$ and $\zeta_K(s-1)$ for any algebraic number field K, in A. Fröhlich (Ed.), *Algebraic Number Fields*. Academic, N.Y., 1977, 475–483.

———, *An Introduction to Number Theory with the Aid of a Computer*, U.C.S.D. Lecture Notes.

E. Thomas & A.T. Vasquez, On the resolution of cusp singularities and the Shintani decomposition in totally real cubic number fields. *Math. Ann.*, *247* (1980), 1–20.

T.M. Thompson, *From Error-Correcting Codes Through Sphere Packings to Simple Groups*. Math. Assoc. of America, Washington, D.C., 1983.

L.-C. Tsao, The rationality of the Fourier coefficients of certain Eisenstein series on tube domains. *Comp. Math.*, *32* (1976), 225–291.

S. Tsumumine, Construction of modular forms by means of transformation formulas for theta series. *Tsukuba J. Math.* 3 (1979), 59–80.

J. Tunnell, Artin's conjecture for representations of octahedral type. *Bull. Amer. Math. Soc.*, *5* (1981), 173–175.

———, On the local Langlands conjecture for $GL(2)$. *Inv. Math.*, *46* (1978), 179–200.

G. Van der Geer & D. Zagier, The Hilbert modular group for the field $\mathbb{Q}(\sqrt{13})$. *Inv. Math.*, *42* (1977), 93–133.

B.L. Van der Waerden, *Studien zur Theorie der Quadratischen Formen*. Birkhäuser, Basel, 1968.

———, Die Reduktionstheorie der positiven quadratischen Formen. *Acta Math.*, *96* (1956), 265–309.

———, Punktverteilungen auf der Kugel und Informationstheorie. *Die Naturwissenschaften*, *48* (1961), 189–192.

V.S. Varadarajan, *Lie Groups, Lie Algebras, and Their Representations*. Prentice-Hall, Englewood Cliffs, N.J., 1974.

———, *Harmonic Analysis on Real Reductive Groups, Lecture Notes in Math.*, Vol. 576. Springer-Verlag, N.Y., 1977.

———, Eigenfunction expansions on semisimple Lie groups, in A. Figà-Talamanca (Ed.), *Harmonic Analysis and Group Representations*, C.I.M.E. Intl. Math. Summer Center. Liguori editoro, Napoli, Italy, 1982, 351–422.

A.B. Venkov, On the trace formula for $SL(3, \mathbb{Z})$. *J. Soviet Math.*, *12* (1979), 384–424.

B.A. Venkov, Über die Reduction positiver quadratischer Formen. *Izv. Akad. Nauk. S.S.S.R.*, *4* (1940), 37–52.

M. Vergne, Representations of Lie groups and the orbit method, in B. Srinivasan and J. Sally (Eds.), *Emmy Noether in Bryn Mawr*. Springer-Verlag, N.Y., 1983, 59–101.

M.-F. Vignéras, *Arithmétique des Algèbres de Quaternions, Lecture Notes in Math.*, 800. Springer-Verlag, N.Y., 1980.

———, Séries théta des formes quadratiques indéfinies. *Lecture Notes in Math.*, 627. Springer-Verlag, N.Y., 1977, 227–240.

———, Invariants numériques des groups de Hilbert. *Math. Ann.*, *224* (1976), 189–215.

———, Genre arithmétique des groupes modulaires de Hilbert et nombres de classes de quaternions. *Sém. de Théorie de Nombres, 1975–1976, Univ. Bordeaux*, Exp. 28, Talence, 1976.

———, Invariants des groupes modulaires de Hilbert sur un corps quadratique. *Sém. Delange-Pisot-Poitou, 1975–1976*, Exp. 69, Paris, 1977.

N.J. Vilenkin, *Special Functions and the Theory of Group Representations*, Transl. Math. Monographs, 22. A.M.S., Providence, 1968.

G.F. Voronoi, Propriétés des formes quadratiques positives parfaites. *J. Reine Angew. Math.*, *133* (1908), 97–178.

———, Nouvelles applications des paramètres continus à la thèorie des formes quadratiques. *J. Reine Angew. Math.*, *134* (1908), 198–287; *136* (1909), 67–178.

J.-L. Waldspurger, Formes quadratiques à 4 variables et relèvement. *Acta Arith. 36* (1980), 377–405.

D. Wallace (Andreoli), *Selberg's Trace Formula and Units in Higher Degree Number Fields*, Ph.D. thesis, U.C.S.D., 1982.

———, Conjugacy classes of hyperbolic matrices in $SL(n, \mathbb{Z})$ and ideal classes in an order. *Trans. A.M.S.*, *283* (1984), 177–184.

———, Explicit form of the hyperbolic term in the Selberg trace formula for $SL(3, \mathbb{R})$ and Pell's equation for hyperbolics in $SL(3, \mathbb{R})$, *J. Number Theory*, *29* (1986), 127–133.

———, A preliminary version of the Selberg trace formula for $PSL(3, \mathbb{Z}) \backslash PSL(3, \mathbb{R})/SO(3, \mathbb{R})$, in *The Selberg Trace Formula and Related Topics*, Contemporary Math., Vol. 53. A.M.S., Providence, 1986, 11–15.

———, Maximal parabolic terms in the Selberg trace formula for $SL(3, \mathbb{Z}) \backslash SL(3, \mathbb{R})/SO(3, \mathbb{R})$, preprint.

———, Minimal parabolic terms in the Selberg trace formula for $SL(3, \mathbb{Z}) \backslash SL(3, \mathbb{R})/SO(3, \mathbb{R})$, preprint.

N. Wallach, Lecture Notes, 1981 NSF-CBMS Regional Conf. on Representations of Semisimple Lie Groups and Applications to Analysis, Geometry and Number Theory, unpublished.

———, *Harmonic Analysis on Homogeneous Spaces*. Dekker, N.Y., 1973.

———, *Symplectic Geometry and Fourier Analysis*. Math. Sci. Press, Brookline, Mass, 1977.

G. Warner, *Harmonic Analysis on Semi-Simple Lie Groups, I, II*. Springer-Verlag, N.Y., 1972.

———, Selberg's trace formula for non-uniform lattices: the rank one case. *Studies in Algebra and Number Theory*, Advances in Math., Suppl. Studies, 6. Academic, N.Y., 1979, 1–142.

J.C. Watkins, Functional central limit theorems and their associated large deviation principles: Products of random matrices, preprint.

A. Wawrzyńczyk, *Group Representations and Special Functions*. Reidel, Boston, 1984.

A. Weil, *L'intégration dans les Groupes Topologiques*. Hermann, Paris, 1965.

———, *Oeuvres Scientifiques, Collected Papers (1926–1978)*, Vols. *I–III*. Springer-Verlag, N.Y., 1979.

———, *Discontinuous Subgroups of Classical Groups*. University of Chicago, 1958.

———, *Dirichlet Series and Automorphic Forms*, Lecture Notes in Math. 189. Springer-Verlag, N.Y., 1971.

———, *Basic Number Theory*. Springer-Verlag, N.Y., 1974.

R. Weissauer, Siegel modular forms and Dirichlet series, preprint.

———, Eisensteinreihen vom Gewicht $n + 1$ sur Siegelschen Modulgruppe n-ten Grades. *Math. Ann.*, *268* (1984), 357–377.

D. Weisser, The arithmetic genus of the Hilbert modular variety and the elliptic fixed points of the Hilbert modular group. *Math. Ann., 257* (1981), 9–22.

H. Weyl, *Gesammelte Abhandlungen.* Springer-Verlag, N.Y., 1968.

E.P. Wigner, On a generalization of Euler's angles, in E.M. Loebl (Ed.), *Group Theory and Its Applications.* Academic, N.Y., 1968, 119–129.

H. Williams & J. Broere, A computational technique for evaluating $L(1,\chi)$ and the class number of a real quadratic field. *Math. Comp., 30* (1976), 887–893.

J. Wishart, The generalized product moment distribution in samples from a normal multivariate population. *Biometrika, 20A* (1928), 32–43.

S.-T. Wong, *Analytic Continuation of Eisenstein Series*, Ph.D. thesis, University of Illinois, 1987.

B.G. Wybourne, *Classical Groups for Physicists.* Wiley, N.Y., 1974.

T. Yamazaki, On Siegel modular forms of degree 2. *Amer. J. Math. 98* (1973), 39–53.

D. Zagier, A Kronecker limit formula for real quadratic fields. *Math. Ann., 213* (1975), 153–184.

——, Modular forms associated to real quadratic fields. *Inv. Math., 30* (1975), 1–46.

R.J. Zimmer, *Ergodic Theory and Semisimple Groups.* Birkhäuser, Boston, 1984.

P.G. Zograf, The Selberg trace formula for the Hilbert modular group of a real quadratic algebraic number field (Russian). *Zap. Nauchn. Sem. Leningrad, Otdel Mat. Inst. Steklov (LOMI), 100* (1980), 26–47, 173.

Mathematics tries to replace reality with a dream of order. It is perhaps for this reason that mathematicians are often such strange and socially inept people. To devote oneself to mathematics is to turn away from the physical world and meditate about an ideal world of thoughts. The striking thing is that these pure mathematical meditations can in fact make fairly good predictions about messy matter. Eclipses are predicted; bridges are built; computers function,...

Index

A
A, $\mathfrak{a}$, abelian Lie group and its Lie algebra, 6, 20, 33–35, 74, 279, 284, 295
$\mathfrak{a}_n$, $\mathfrak{sl}(n+1, \mathbb{C})$, Lie algebra of the special linear group, 266, 271–272
$\mathscr{A}(\Gamma, \lambda)$, space of automorphic forms, 182–245
Abelian integral, 119, 315, 337
Adelic theory, 5–6, 62, 115, 182, 216, 241, 257, 322, 332, 339, 350
Adjoint of a differential operator, 44, 88, 208
Adjoint representations, 261, 265–266, 279, 284, 295, 297
Algebraic integer, 314, 316, 320–325
Algebraic number field, 5, 116, 127, 182, 211, 213, 215, 259, 314–326, 340–350
Analytic continuation, 188–197, 200, 206–211, 213, 246, 340, 346
Arc length, 7, 14, 24, 33–34, 36, 287, 292, 294, 305, 313
Artin conjecture, L-functions, and reciprocity, 5, 115–116, 182
Arthur polynomials, 256
Arzelà-Ascoli theorem, 249
Associated parabolic subgroups, 246
Asymptotics of
  Eisenstein series, 216–245, 253
  K-Bessel functions, 53
  number-theoretical quantities, 5, 248, 315, 350
  spherical functions, 72–82, 100–105, 309
Asymptotics/functional equations principle, 87, 89, 99–101, 216, 253
Automorphic form, 4, 115, 181–245, 329–350
Automorphic function, 332, 336

B
$\mathfrak{b}_n$, $\mathfrak{so}(2n+1, \mathbb{C})$, the Lie algebra of an orthogonal group, 266
Babylonian reduction, 14
Base change, 116, 336
Bessel functions
  I-Bessel, 222
  J-Bessel, 58–59, 72, 84
  K-Bessel, 49–64, 105, 222–226, 232, 234–238, 240, 248, 341, 343–345
Beta function, 48, 52, 88, 91, 104–105, 237–239, 309
Borel or minimal parabolic subgroup, 22, 101, 189, 197, 218, 228–230, 304, 307
Boson field, 259–315
Boundary of a fundamental domain, 216–221, 253

Boundary of a symmetric space, 22, 88, 101, 111–112, 304, 307–308
Bounded symmetric domain, 274–275
Bruhat decomposition, 101, 217–218, 227–229, 239, 304

C

$\mathfrak{c}_n$, $\mathfrak{sp}(n, \mathbb{C})$, the Lie algebra of the symplectic group, 266, 272
Calculus of variations, 8
Campbell (-Baker) -Hausdorff formula, 74, 263
Cartan classification of symmetric spaces, 266, 276–277
Cartan decomposition, 268–278, 283
Cartan involution, 74, 268–273, 276, 280, 295
Cartan's fixed point theorem, 276
Cartan subalgebra, $\mathfrak{h}$, maximal abelian subalgebra of a complex semisimple Lie algebra, 298
Cauchy-Riemann equations, 331, 334
Cayley transform, 153, 275, 311
Centralizer, 218, 228, 254, 279, 295
Central limit theorem, 72–73, 106–110
Change of variables summary, 33–36
Character, 50, 61, 348
Character group or dual group of a finite abelian group, 348
Characteristic function of a random variable, 108
Chi-square distribution, 84
Chowla-Selberg method of obtaining Fourier expansions of Eisenstein series, 221, 223–227
Circle problem, 248
Class field, 332
Class group or ideal class group, 323–326, 331, 340, 346, 348–350
Classification of complex simple Lie algebras, 266
Classification of symmetric spaces, 266, 276–277
Class number, 4, 116, 127, 319, 345
Class one representation, 70
Class one spherical function, 65
Closed geodesic in fundamental domain, 116

Coding theory, 124
Coherent states, 3, 259
Cohomology of arithmetic groups, 184
Color perception, 314
Compact fundamental domain, 254, 328–329
Compactification, 22, 119, 145, 216, 221, 327, 336
Compact operator, 249
Compact real form of a complex semisimple Lie algebra, 271, 274
Compact semisimple Lie group, algebra, K, $\mathfrak{k}$, 268, 271, 296, 299
Compact symmetric space, 268, 270
Complete Riemannian manifold, 267
Complexification of a Lie algebra, 270–271, 274, 298
Composition of random variables, 107
Conditional distribution, 36
Cone, 11
Confluent hypergeometric function, 49, 59–60, 105, 216, 222
Congruence subgroup, 184, 201–202, 216, 248, 332, 336, 338–339
Conjugacy class, 254
Conjugacy of maximal compact subgroups, 276
Conjugation of
  an algebraic number field, 320–323
  a complex simple Lie algebra, 267, 270–271
  quaternions, 311
Constant term in Fourier expansion, 184, 215–245, 249, 331
Continued fractions, 4, 117, 161–162
Continuous spectrum, 221, 247, 254, 256
Convex hull, 194, 197
Convex set, 125, 307
Convolution, 25–26, 66–70, 88, 92, 106–108, 249, 348–349
Coprime symmetric pair, 326, 335
Correlation, 36
Covariance, 36, 108
Crystallography, 125
Cubic number field, 5, 115, 211, 256
Curvature, 275
Cusp of fundamental domain, 215, 319, 323–326

Cusp form, 115, 182–184, 245, 252–253, 331, 337
Cuspidal spectrum, 246–247

D

$\mathfrak{d}_n$, $\mathfrak{so}(2n, \mathbb{C})$, Lie algebra of an orthogonal group, 266
$D(G/K)$, $D(\mathcal{P}_n)$, invariant differential operators on G/K, $\mathcal{P}_n$, 18, 26–33, 39–40, 44–50, 65–69, 88, 181–186, 207–210, 246, 248, 304–306, 308
Dedekind eta function, 178, 215
Dedekind zeta function, 4, 259, 315, 319, 322, 333, 340–341, 344–345
Densely wound line in a torus, 163, 264
Density of a random variable, 107
Determinant one surface in
  $\mathcal{P}_n$, $\mathcal{SP}_n$, 2, 17, 23, 89, 164, 184–186
  $\mathcal{M}_n$, $\mathcal{SM}_n$, 164, 184
  $\mathcal{F}_n$, $\mathcal{SF}_n$, 150–161, 163–164
  $\mathcal{P}_n^K$, $\mathcal{SP}_n^K$, 323
Dictionary of Lie groups, Lie algebras, 264–267, 269, 296
Different, 330–331, 340, 345
Differential, 11, 15, 24, 259
Differential of
  action of G on G/K, 11, 270, 286
  exp, 263–264
  Int(g), 265–266, 299
  multiplication, 294
Differential operators on a symmetric space, 18, 26–33, 39–40, 44–50, 65–69, 88, 181–186, 207–210, 246, 248, 304–306, 308
Dimensions of spaces of automorphic forms, 184, 257, 274, 315, 333, 337–338
Dirac delta function, 69, 107
Dirac sequence or family, 69, 89, 98, 107, 250
Dirichlet L-function, 350
Dirichlet unit theorem, 3, 127, 255
Discrete, discontinuous subgroup, 5, 314–329
Discrete spectrum, 183–184, 246–248, 252
Discriminant of a number field, 127, 320, 322, 341

Divisor function, 215–216, 222, 227, 238, 341
Dual group or group of characters of a finite abelian group, 348
Dual lattice or ideal, 330, 334, 342
Duality between compact and non-compact sysmmetric spaces, 274–275
Dynkin diagram, 124, 266

E

Eisenstein series, $E(v, s|Y)$, $E_P(s|Y)$, $E_{(n)}(s|Y)$, $E(f|Y)$, $E_k(\mathfrak{a}, z)$, $E_k(Z)$, $E_i(Y, s)$, $\vec{E}(Y, s)$, $E(\chi, Y, s)$, 51, 61, 114, 116, 159, 172, 178, 181–245, 247, 331, 335, 337, 339–350
Edge form, 143–144
Eichler-Selberg trace formula, 256–257
Eigenfunctions and eigenvalues, 2, 44–47, 49–51, 65, 67, 69, 73, 76, 86–87, 99, 183, 185, 200, 210, 213, 226–227, 248, 308–311, 335
Eigenvalues of Hecke operators, 213, 226–227, 238
Eigenvalues of the Laplacian, 39–40, 44–49, 109, 210, 248
Eigenvalues of symmetric or Hermitian matrices, 3–4, 292
Electrical engineering, 259
Elementary divisor theory, elementary row and column operations on a matrix, 127–128, 174, 186, 222, 225, 232, 240, 336
Elliptic conjugacy class, 256, 338
Elliptic curve, 216
Epstein zeta function, $Z(Y, S)$, $Z_{m,n-m}(X, s)$, $Z(f, s|Y)$, $Z(f|Y)$, $Z(\mathfrak{a}, Y, s)$, $\vec{Z}(f|Y)$, 114, 116, 171–172, 191–192, 206, 211, 223–224, 237, 246, 251, 339–350
Ergodic theory, 161–163
Eta function, 178, 215
Euler angles, 23, 79–80, 289–291
Euler characteristic, 327
Euler's differential operator, 76, 78
Euler-Lagrange equation, 8
Euler product, 115–116, 200–201, 338
Expectation or mean, 36, 108

Exponential, 12, 74–75, 261–266, 284, 287, 302
Extreme form, 144

F
Face-centered cubic lattice, 123–124
Factor analysis, 85
Finite element method, 10
Finite Fourier transform, 348–349
Fixed point property, 276, 307
Flag manifold, 112, 304
Fourier analysis on
 finite abelian group, 348–349
 fundamental domain, 115, 181–183, 216, 245–257
 symmetric space, 58, 66–67, 86–106, 183, 274, 309–310, 314
Fourier expansions of automorphic forms, 49, 51, 59–62, 118–119, 211, 215–245, 330–331, 334–338, 340–341, 346
Functional equations of
 Dedekind zeta function, 341
 Eisenstein series, 61, 191, 194–195, 200, 209–210, 214, 217, 224, 243–244, 340
 K-Bessel function, 58, 63
 spherical function, 70, 93, 98, 309
 Whittaker functions, 61
Fundamental cone, 145
Fundamental domain, 3–5, 113–181, 201, 215, 258, 316–329
Fundamental set, 137
Fundamental solution of the heat equation, 106–110, 246, 311, 314
Funk-Hecke theorem, 70

G
G, $\mathfrak{g}$, semisimple real (usually noncompact) Lie group and its Lie algebra, 3, 8–9, 18, 182, 261, 265, 267–271, 296, 299
G/K or K/G, noncompact symmetric space, 8–9, 267, 269–271, 313
$GL(n, \mathbb{R})$, $\mathfrak{gl}(n, \mathbb{R})$, general linear group and its Lie algebra, 2, 7–9, 261–262, 289, 297, 301

$GL(n, \mathbb{Z})$, modular group, 3, 113–125, 127, 130–131, 145
$GL(n, O_K)$, modular group over a number field K, 314–315, 322–323, 339
Γ, discrete or discontinuous group of isometries in G, 1, 314
Γ, gamma function, 41, 52, 87, 105, 114, 121, 182, 207, 210–211, 310, 339, 341
$\Gamma(N)$, congruence subgroup, 184, 201–202, 216, 248, 332, 336, 338–339
Gauss-Bonnet formula, 118, 327
Gaussian elimination, 14
Gaussian integers, $\mathbb{Z}[i]$, 316
Gauss sum, 216, 315
Gelfand's characterizations of spherical functions, 69–70
General linear group, $GL(n, \mathbb{R})$, 2, 7–9, 261–262, 289, 297, 301
Generators of discrete groups, 145, 161, 317, 327
Geodesic, curve minimizing distance, 3, 11–17, 116, 162, 267, 287, 291–294
Geodesic polar coordinates, polar coordinates, 23–24, 32–34, 90–92, 288–291, 293, 313
Geodesic-reversing isometry, 2, 18, 28, 44, 260, 267, 269
Gindikin-Karpelevic formula, 309
Godement mean value theorem, 305
Gram-Schmidt orthonormalization, 14, 42
Grassmann variety, 112
Greatest common divisors of matrices, 173
Great Green Arkleseizure, 1
Green's functions, 25
Green's identity, 256
Grenier's fundamental domain, $\mathscr{F}_n$, 146–164
Grössencharacter, 182, 211, 215, 332
Group action, 8–9, 22, 285–286, 313, 316, 321, 323
Group representation, 39, 51, 61, 64, 70, 82, 106, 115–116, 120, 182, 216, 259, 337, 348–349

Index

H

H, Poincaré or hyperbolic upper half plane, 11, 22, 65, 72, 99, 106, 108, 246, 275, 321

$\mathcal{H}^c$, quaternionic upper half plane or hyperbolic 3-space, 256–258, 274, 312–313, 316

$\mathcal{H}_n$, Siegel upper half plane or space, 215, 256, 273, 275, 285, 288, 292, 301, 305, 326

$\mathcal{H}$, Helgason-Fourier transform on a symmetric space, 66–67, 86–106, 183, 246–248, 256, 309, 314

$\mathcal{H}^n$, hyperbolic n-space, 321, 328, 330

Haar measure, 6, 25, 34, 43, 165, 298–299

Hankel transform, 58

Harish-Chandra c-function, 88, 90–91, 99, 101–105, 238, 309

Harish-Chandra integral formula for spherical functions, 70, 309

Harish-Chandra-Selberg theorem on invariant differential operators, 29, 305

Harish transform, 93, 256

Harmonic analysis, Fourier analysis on finite abelian group, 348–349
  fundamental domain, 115, 181–183, 216, 245–257
  symmetric space, 58, 66–67, 86–106, 183, 274, 309–310, 314

Harmonic function, 22–23, 305–308

Heat equation, 106–110, 246, 310, 314

Hecke correspondence between modular forms and Dirichlet series, 5, 42, 114–115, 184, 200–201, 215, 243, 248, 332, 338–339, 350

Hecke L-functions with grössencharacter, 5, 184, 211, 215, 333

Hecke operators, 65, 116, 153–162, 176, 182, 184, 187, 198–214, 226, 238, 332–333, 338, 348, 350

Helgason transform, H, 66–67, 86–106, 183, 246–248, 256, 309, 314

Hermitian symmetric space, 275

Hermite-Mahler compactness theorem, 145, 328

Highest point method, 147, 152–153, 326

Hilbert cusp form, 331
Hilbert modular form, 329–333, 336
Hilbert modular function, 332
Hilbert modular group, 320–321
Hilbert problems 18 and 12, 120, 332
Hilbert-Siegel modular group, 327
Hirzebruch-Riemann-Roch theorem, 274, 333, 338
Homogeneous space, G/H, 2, 6–7, 9, 267, 269–270, 272–274
Homomorphism of Lie algebras, 260, 264
Hopf-Rinow theorem, 267
Horocycle, 161, 163
Huyghen's principle, 107, 311
Hyperbolic element of Γ, 116, 255
Hypergeometric functions, 38, 49, 58–60, 72–73, 76–78, 85–86, 105, 216, 222

I

I_K, ideal class group of an algebraic number field K, 323–326, 331, 340, 346, 348–350

Ideal class group, 323–326, 331, 340, 346, 348–350
Ideal class zeta function, 347
Ideal in a Lie algebra, 260
Ideal in a number field, 323–326
Incomplete gamma function, 48, 172, 192
Incomplete theta series, 253
Independent random variables, 107
Integral basis of an ideal or Z-basis, 342
Integral formulas on G, G/K, $\mathcal{P}_n$, etc., 7, 18–26, 33–36, 164–181, 299–304
Integral test, 3, 7, 171, 178, 184–185, 187–188, 212, 331, 335, 340
Int(g), 265–266
Invariant differential operator, 18, 26–33, 39–40, 44–50, 65–69, 88, 181–186, 207–210, 246, 248, 260, 304–305, 308
Invariant integral operator, 25–26, 66–70, 88, 92, 249
Invariant random variable under K, 107
Invariant vector field, 260
Inversion of Fourier transform, 88, 309, 314, 349

Involutive automorphism or conjugation of a Lie group or algebra, 74, 267–273, 276, 280, 295
Irreducible homogeneous bounded symmetric domain, 275
Irreducible symmetric space, 275–276
Iwasawa decomposition, 13–14, 19–22, 30–33, 35–37, 39, 44–51, 54–58, 63, 73–74, 101, 133, 135–139, 144, 146–151, 153, 156, 163, 169, 180, 185, 189, 191, 197, 216–221, 223, 225, 232–233, 235, 240, 247, 250, 259, 279–288, 292, 299, 304, 307, 312, 340–341, 346

J

J, modular invariant, 332
J, Bessel function, 58–59, 72, 84
Jacobi identity for Lie algebras, 260
Jacobi transformation, 14
Jacobian of
 the action of G on G/K, 286, 301
 the action of G on the boundary of G/K, 22, 35, 304
 exp, 263–264, 302
 Iwasawa decomposition, 19–22, 33–35, 299, 307
 polar coordinates, 23–24, 33–35, 302–303
Jordan algebra, 275, 337
Jordan canonical form, 265

K

K, maximal compact subgroup of G with Lie algebra $\mathfrak{k}$, 8, 20, 33–35, 74, 267–270, 277, 279, 284, 295–296
K, algebraic number field, 314–326, 340–350
K, Bessel function, $K(s|A, B)$, $k(s|Y, A)$, 49–64, 105, 222–226, 232, 234–238, 240, 248, 341, 343–345
Killing form, 11, 261–262, 265, 268–269, 271, 279, 287, 294–295, 299
Kirillov theory, 50, 61
Koecher zeta function, 4, 114, 178, 186–187, 199, 205–206, 210, 244–245
Kontorovich-Lebedev inversion formula, 58, 89

Korteweg-deVries equation, 119, 337
Kronecker limit formula, 215, 224, 346

L

Langlands program, 5, 115
Laplace transform, 40–41, 58, 86
Laplacian, Δ, 7, 26–27, 30–36, 49, 61, 106, 184–186, 216, 305, 310, 313
Lattice, 120
Lattice packing of spheres, 4, 7, 120–125, 144, 171
LDU factorization, 14
Leech lattice, 124
Legendre function, 64–65, 91
L-function, 4–5, 63, 115–116, 155, 182, 184, 200–201, 211, 215, 235, 238–239, 245, 315, 332, 350
Lie algebra, 73, 260–266
Lie bracket, 73, 260
Lie group, 3, 260–266
Lifting, 116, 336
Likelihood function, 37
Lorentz-type group, $O(p, q)$, $SO(p, q)$, 276, 284, 290

M

M, M', centralizer and normalizer of $\mathfrak{a}$ in K and their Lie algebra $\mathfrak{m}$, 22, 228, 295–298
$\mathscr{M}(\Gamma, k)$, space of modular forms, 330, 333, 338
$\mathscr{M}_n$, Minkowski's fundamental domain, 125–147, 164–170, 178–180, 207–210, 326
Maass-Selberg relations, 256
Maass wave form, 187, 215, 259, 336
Maass zeta function, 213
Macdonald-Dyson conjecture, 42
Mahler's inequality in the geometry of numbers, 135
Matrix decomposition, matrix coset representatives, 174, 180, 186, 190, 197, 199, 212–213, 229, 231, 239
Matrix of regression coefficients, 37
Maximal boundary, 112, 304
Maximum likelihood estimate, 37
Mean, expectation, 36, 108

Index   381

Mean value theorem for harmonic functions, 305
Mehler-Fock inversion formula, 89, 91
Mehta conjecture, 42
Mellin transform of K-Bessel function, 63, 243
Mellin transform of θ, 114, 199, 207, 209, 223, 225, 241, 333
Mellin transform on a symmetric space or fundamental domain, Fourier analysis on a symmetric space, 58, 66–67, 86–106, 114–116, 181–183, 201, 216, 245–257, 309–310, 314, 338
Method of inserting larger parabolic subgroups, 188–197
Method of theta functions, 189, 206–211
Minimal parabolic subgroup or Borel subgroup, 22, 101, 189, 197, 218, 228–230, 304, 307
Mini-max principle, 10
Minimum m_Y of a positive symmetric matrix over $\mathbb{Z}^n - 0$, 120–121, 126, 145, 170
Minkowski-Hlawka theorem in the geometry of numbers, 170, 177
Minkowski-reduced matrices, 128–129
Minkowski's fundamental domain, $\mathscr{M}_n$, 125–147, 164–170, 178–180, 207–210
Minkowski's fundamental lemma in the geometry of numbers, 126–127
Minkowski's inequality for successive minima of $Y \in \mathscr{P}_n$, 135
Modular function associated with Haar measure, 7, 165, 299
Modular group, $GL(n, O_K)$, $SL(n, O_K)$, $Sp(n, \mathbb{Z})$, 113–257, 314–350
Modular or automorphic form or function, 181–257, 329–350
Monomial symmetric function, 77
Multivariate statistics, 2–3, 36–38, 83–86, 106

N

N, $\mathfrak{n}$, nilpotent Lie group and its Lie algebra, 6, 20, 33–35, 60–61, 74, 218, 225, 241, 249–252, 256, 279–280, 284, 295–296, 299

$\bar{N}$, $\bar{\mathfrak{n}}$, opposite group, algebra, to N, 89, 101, 229, 279, 304
Nilpotent Lie group, algebra, N, $\mathfrak{n}$, 6, 20, 33–35, 60–61, 74, 218, 225, 241, 249–252, 256, 279–280, 284, 295–296, 299
Noncompact symmetric space, 270–276
Normal density, 36, 109
Normalization of a power function, 44
Normalizer M' of $\mathfrak{a}$ in K, 228, 295, 297–298
Normally distributed, 36
Normal real form of a semisimple complex Lie algebra, 271–273, 279
Norm of an element of $\mathscr{H}^m$, 330
Norm of an ideal, 341–347
Norm of a quaternion, 311
Numerical integration, 124–155

O

O_K, ring of integers of algebraic number field K, 314, 322
$O(n)$, $\mathfrak{o}(n)$, orthogonal or rotation group and its Lie algebra, 8, 42, 266
$O(p, q)$, Lorentz-type group, 265
One parameter subgroup, 262
Opposite nilpotent group, $\bar{N}$, 89, 101, 229, 279, 304
Orbital integral, 110, 254–256
Orbit method, 111

P

P, $\mathfrak{p}$, from the Cartan decomposition, 74, 268, 271–273, 279, 295
$\mathscr{P}_n$, positive real $n \times n$ symmetric matrices, 1–2, 8–9, 301, 305, 322
$\mathscr{P}_n^*$, positive $2n \times 2n$ symplectic matrices, 273, 276, 285, 291–292, 305, 326
$\mathscr{P}_n^c$, positive $n \times n$ Hermitian matrices, 322–323
$\mathscr{P}_n^K$, positive $n \times n$ matrices over a number field K, 322–323
p_s, power function, 38–40, 96, 103, 207, 308
p-adic numbers, 25, 199
Paley-Wiener theorem, 87, 89, 93, 101
Parabolic elements, 256, 338

Parabolic subgroup, 51, 101, 178, 184, 186–191, 197–198, 211, 214, 217–221, 227–230, 232, 235, 245–247, 252–253
Paradox associated with balls in $\mathbb{R}^n$ for large n, 170–171
Partial correlation, 36–37
Partial covariance, 36–37
Partial Iwasawa decomposition, 13, 19, 30–33, 35–37, 44–51, 54–58, 63, 133, 136, 139, 147–151, 163, 169, 180, 191, 197, 216–221, 223, 225, 232–233, 235, 240, 247, 286–287, 292, 340
Partitions, 178
Pauli matrices, 312
Perpendicular bisectors method, 162, 326
Picard group, 258, 316, 332
Plancherel formula, 51, 88–89, 111, 309, 314
Plancherel or spectral measure, 88–89, 309, 314
Poincaré series, 331–332
Poincaré upper half plane, H, 11, 22, 65, 72, 99, 106, 108, 246, 275, 321
Point-pair invariant, 26
Point spectrum or discrete spectrum, 183–184, 246–248, 252
Poisson integral formula, 23, 306–307
Poisson Kernel, 23, 307
Poisson sum formula, 5, 166–167, 176, 207, 225, 233, 235, 245–249, 253–254, 315, 342, 350
Polar coordinates, geodesic polar coordinates, Euler angles, 23–24, 32–34, 90–92, 288–291, 293, 298, 301–303, 313
Poles of Koecher's zeta function, 244–245
Positive matrix, 1–2, 8–9, 312, 322–323
Positive restricted root, 280
Positive spherical function, 70
Potential theory, 305–308
Power function, 38–40, 96, 103, 207, 308
Probability density, 36, 107
Pseudo algebraic group, 277

Q
Quadratic form, symmetric matrix, 1, 4, 42, 118–119, 259, 328–329
Quadratic number field, 5, 116–117, 248, 315–321
Quantum mechanics, 3, 120, 165, 254, 259, 312, 315, 336–337
Quantum statistical mechanics, 42, 259
Quaternion, 311
Quaternionic upper half plane, $\mathscr{H}^c$, 256–258, 274, 311–314, 316
Quotient or homogeneous space, G/H, 2, 6–7, 9, 267–274

R
Radon transform, 7, 311
Ramanujan sum, 232
Random variable, 36, 83, 107
Rank of a symmetric space, $\dim_\mathbb{R} \mathfrak{a}$, 279, 305
Rankin-Selberg method, 181
Rayleigh-Ritz method, 10
Real form of a Lie algebra over $\mathbb{C}$, 270–271
Reduction algorithm, 128, 147, 152–153, 161
Reductive Lie group, 18
Regression coefficient, 37
Regulator of a number field, 255, 345
Representation of a group, 39, 51, 61, 64, 70, 82, 106, 115–116, 182, 216, 259, 337, 348–349
Representation of an integer by a quadratic form, 4, 118–119
Residual spectrum, 247, 254
Residues of Eisenstein series, 191–193, 198, 206, 211, 246–247, 320
Resolvent kernel, 25
Restricted root, 279, 295, 298
Riemannian manifold, differentiable manifold with a notion of arc length, 11, 260
Riemann-Lebesgue lemma, 92
Riemann method of theta functions, 206–211
Riemann metric tensor, arc length, 7, 11, 260, 269–270, 287, 294
Riemann-Roch theorem, Hirzebruch-

Index 383

Riemann-Roch theorem, 119, 274, 333, 338
Riemann zeta function, 4, 114, 118, 166, 168, 175, 187, 191, 193, 195, 199, 205–206, 211–212, 238–239, 327, 340
Riesz type integrals, 111
Right spherical function, 38
Roelcke-Selberg-Mellin inversion formula, 115, 246
Root, 124, 217–218, 279–281, 295
Root space decomposition, 279
Rotating top, 119, 259
Rotation group, $O(n)$ or $SO(n)$, 8, 42, 266

S

$SL(n, \mathbb{R})$, $\mathfrak{sl}(n, \mathbb{R})$, special linear group, algebra, 17, 164–165, 173, 255, 265–266, 269–272, 276, 279, 281, 304
$SL(n, \mathbb{Z})$, modular group, 116, 151, 165–167, 173, 245–257
$SL(n, O_K)$, modular group over a number field, 316–326, 329–333
$SO(n)$, $\mathfrak{so}(n)$, special orthogonal group, algebra, 266, 271–272, 276, 289–290, 312
$SO^*(2n)$, 276
$SO(p,q)$, $\mathfrak{so}(p,q)$, Lorentz-type groups, algebra, 276, 284, 290
$SO_o(p,q)$, connected component of I in $SO(p,q)$, 276
$Sp(n)$, $\mathfrak{sp}(n)$, compact symplectic group, algebra, 272–274, 276
$Sp(n, \mathbb{R})$, $\mathfrak{sp}(n, \mathbb{R})$, symplectic group, algebra, 17, 256, 265–266, 272–273, 276, 279, 281, 285–287, 291–292, 297, 301, 310
$Sp(n, \mathbb{Z})$, 4, 115, 155, 189, 199, 216, 221–222, 314, 326–328, 333–339
$Sp(p, q)$, $\mathfrak{sp}(p, q)$, 273, 276
$SU(n)$, $\mathfrak{su}(n)$, special unitary group, algebra, 271–272, 276, 291, 312
$SU^*(n)$, $\mathfrak{su}^*(n)$, 271–272, 276
$SU(p, q)$, $\mathfrak{su}(p,q)$, 271–272, 276, 282–284, 298
$\mathscr{SF}_n$, determinant one surface in Grenier fundamental domain, 150–161, 163–164
$\mathscr{SM}_n$, determinant one surface in Minkowski's fundamental domain, 129, 143–144, 164, 167–168, 210
$\mathscr{SP}_n$, determinant one surface in positive matrix space, 2, 17, 23, 89, 113, 246, 269–270
$\mathscr{SP}_n^c$, determinant one surface in positive Hermitian matrix space, 312
$\mathscr{SP}_n^K$, determinant one surface over a number field, 323
$\mathscr{S}(\Gamma, k)$, cusp forms, 331, 337
Satake compactification, 216, 221, 336
Schwartz space, 89, 107, 245
Sectional curvature, 275
Segal-Shale-Weil representation, 119, 337
Selberg's analytic continuation of Eisenstein series, 188–197, 200, 206–211, 213
Selberg's basic lemma on eigenfunctions of invariant differential and integral operators, 67, 248, 253
Selberg's differential operators, 208–209
Selberg Eisenstein series, 188–197, 195, 237–238
Selberg integral, 42–44
Selberg trace formula, 4, 116, 147, 221, 254–259, 315, 333, 338, 350
Selberg transform, 66
Selberg's zeta function, 188
Semisimple Lie group, algebra, 3, 8–9, 18, 182, 261, 265, 268–271, 296, 299
Serre conjecture, 128
Siegel cusp form, 337
Siegel integral formula, 165–181, 319, 322, 348
Siegel modular form, 5, 51, 114–115, 119, 182, 222, 248, 333–339
Siegel modular function, 336
Siegel modular group, $Sp(n, \mathbb{Z})$, 4, 115, 155, 189, 199, 216, 221–222, 240, 245, 256, 258, 314–315, 326–328
Siegel Φ-operator, 245, 337
Siegel set, 137
Siegel singular form, 337
Siegel theorem on quadratic forms, 118–120, 165, 337

Siegel upper half plane or space, $\mathscr{H}_n$, 215, 256, 258, 273, 275, 285, 288, 326
Simple associative algebra, 4, 114, 175, 206
Simple Lie algebra, group, 18, 261, 266, 271–276
Simply connected Lie group, 264, 270
Singular series, 216, 222, 240
Singular Siegel modular form, 337
Singular value decomposition, 23
Solitons, 119, 259, 315
Special linear group, algebra, $SL(n, \mathbb{R})$, $\mathfrak{sl}(n, \mathbb{R})$, 17, 265–266, 269–272, 276, 279, 281, 304
Special orthogonal group, algebra, $SO(n)$, $\mathfrak{so}(n)$, 266, 271–272, 276, 289–290, 312
Special unitary group, algebra, $SU(n)$, $\mathfrak{su}(n)$, 271–272, 276, 291, 312
Spectral resolution of differential operators, Fourier analysis on symmetric spaces and fundamental domains, 58, 66–67, 86–106, 115, 172, 181–183, 216, 245–257, 274, 309–310, 314
Spectral theorem for a symmetric matrix, 9–10, 289
Sphere, 275
Sphere packing, 4, 7, 120–125, 144, 171
Spherical function, 64–83, 92–93, 212, 309, 313
Spherical tranform, 66, 89, 92
Split or normal real form of a complex Lie algebra, 281
Stiefel manifold, 58
Strings, 259
Subalgebra of a Lie algebra, 260, 265
Successive minima of a quadratic form, 130, 133–135
Surface area of unit sphere in $\mathbb{R}^n$, 122
Symmetric polynomials, 29, 44, 49, 72
Symmetric space, 2, 18, 260
   Cartan's classification, 276
   compact, Euclidean or noncompact type, 268–270, 274–276
   decomposition, 267–268
   of types I–IV, 276
Symplectic group, algebra, $Sp(n, \mathbb{R})$, $\mathfrak{sp}(n, \mathbb{R})$, 17

T
T_n, triangular group, 34–35, 39
Taylor's expansion, 67–68, 78–82, 259, 262–264
Tessellation, 161, 317–321
θ, theta function, 114, 119, 189, 198, 206–211, 223, 225, 246, 315, 332, 336, 340
θ function for indefinite quadratic form, 336
Total differential, 27
Totally geodesic submanifold, 17
Totally positive element of a number field, 331
Totally real number field, 321
Trace formula, Selberg trace formula, 4, 116, 147, 221, 254–259, 315, 333, 338, 350
Triangular group, T_n, 34–35, 39
Truncation of fundamental domain, 147, 256
Tube domain, 194
Twisted trace formula, 116

U
$U(n)$, $\mathfrak{u}(n)$, unitary group and its Lie algebra, 273, 277–278
$U(p,q)$, $\mathfrak{u}(p,q)$, 290
Unimodular group, 7, 165, 299
Unitary group, $U(n)$, $SU(n)$, 273, 277–278
Unit disc, generalized, 275, 311
Unit group in an algebraic number field, U_K, 4, 116, 127, 256, 315, 321–322, 324, 350
Unit sphere in $\mathbb{R}^n$, volume, 121–122
Universal covering group, 270
Universal enveloping algebra, 305

V
Vector field, 260
Vibrating string, 10
Volume
   of fundamental domain for $GL(n, \mathbb{Z})$ in $\mathscr{SP}_n$, 138, 163–168, 206
   Hilbert modular group, 322
   Picard group, 319

Siegel modular group, 327
of fundamental domain for Γ, 4, 118, 164–168, 255
on G/H, 7, 165
on G/K, 300–301
on $\mathcal{H}_2^c$, 313
on $\mathcal{H}_n$, 301
of $O(n)$, 42
on $\mathcal{P}_n$, 18–24, 33–36
on a Rienmannian manifold, 7, 18–24, 33–36, 301, 313
of the unit sphere in $\mathbb{R}^n$, 121–122
Voronoi map, points, polyhedron, 122, 124, 144–145

W

W, Weyl group, 24, 194, 218, 221, 227–228, 289, 291, 295–298
Wave equation, 107, 311
Weakly symmetric space, 18, 28
Weight of a modular form, 330, 333
Weyl chamber, 295, 304, 308
Weyl character formula, 82
Weyl group, 24, 194, 218, 221, 227–228, 289, 291, 295–298

Whittaker function, 60–62, 104–105, 222, 240–241
Wishart distribution, 50, 83–86
Wishart's integral formula, 83, 177, 179

Z

$\mathbb{Z}$-basis, integral basis of an ideal, 342
Zeta function
of algebra $\mathbb{Q}^{n \times n}$, 114, 175, 206
Dedekind, 4, 259, 315, 319, 322, 333, 340–341, 344–345
for Eisenstein series, 211–212
Epstein, 114, 116, 171–172, 191–192, 206, 211, 223–224, 237, 246, 251, 339–350
Koecher, 4, 114, 178, 186–187, 199, 205–206, 210, 244–245
Riemann, 4, 114, 118, 166, 168, 175, 187, 191, 193, 195, 199, 205–206, 211–212, 238–239, 327, 340
Zeta function (Riemann) at odd integer argument, 4, 118, 168, 193
Zonal polynomial, 76, 86
Zonal spherical function, 65, 309, 313